DYNAMICAL TUNNELING

Theory and Experiment

DYNAMICAL TUNNELING

Theory and Experiment

Edited by
Srihari Keshavamurthy
Peter Schlagheck

CRC Press
Taylor & Francis Group
Boca Raton London New York

CRC Press is an imprint of the
Taylor & Francis Group, an **informa** business

CRC Press
Taylor & Francis Group
6000 Broken Sound Parkway NW, Suite 300
Boca Raton, FL 33487-2742

First issued in paperback 2017

© 2011 by Taylor and Francis Group, LLC
CRC Press is an imprint of Taylor & Francis Group, an Informa business

No claim to original U.S. Government works

ISBN 13: 978-1-138-11350-3 (pbk)
ISBN 13: 978-1-4398-1665-3 (hbk)

Visit the Taylor & Francis Web site at
http://www.taylorandfrancis.com

and the CRC Press Web site at
http://www.crcpress.com

Contents

Foreword

From my personal perspective, it is fundamentally interesting to consider physical phenomena which occur quantum mechanically, but not classically, or the reverse, occur classically, but not quantum mechanically. Tunneling falls squarely into this scheme and as such is quite a fascinating topic. Although tunneling was recognized a long, long time ago and was worked on extensively in many physical contexts, it has nevertheless gone through a revolution of sorts beginning in the very early 1980s during which it went from merely being fascinating to being way more fascinating, as I might be tempted to put it crudely. I would identify two advances at the heart of this revolution. The first is the conceptual introduction of dynamical tunneling by Davis and Heller. This generalization of barrier tunneling to tunneling through a separatrix allows for a unified picture of, say, what is happening in a rapidly rotating quantum pendulum tunneling between clockwise and counterclockwise motions, a problem without a clearly identifiable potential barrier, and a particle tunneling back and forth in a double well, a problem with a clearly identifiable potential barrier. The dynamics of either possesses a separatrix from a classical phase-space perspective. Generalizing a bit further from there, Ozorio de Almeida's treatment of tunneling induced by a classical resonance followed not long thereafter.

Davis and Heller were concerned with tunneling problems that necessarily require a multi-degree-of-freedom perspective. This is in fact what I call the second advance, that is, not shying away from such problems. Later in the 1980s, Wilkinson initiated analytical work on such systems and we began our studies on chaos-assisted tunneling (Lin and Ballentine were also seeing the same effects). Prior to this point, a number of multi-degree-of-freedom problems were properly solved via mapping back to a barrier problem in an effective single collective degree of freedom. Think of fission. However, once one begins to consider tunneling problems of an inescapable, multi-degree-of-freedom nature, then all the complex features of classical dynamics, regular and chaotic motion, homoclinic tangles, periodic orbits, diffusion, and so on, may rear their heads in a variety of ways, and as many of us have found over the years, they all seem to in some context or another. It is this realization which, in my view, greatly enriches the subject, and I have no sympathy for the notion that these "complexities" represent unfortunate complications. It also leads to an openness to consider new experimental problems, which may be found with ultracold atoms and optical lattices. Instead of avoiding multidimensional arrangements, there is a motivation for creating them or seeking them out.

Back in 1997, the National Institute for Nuclear Theory held an international program organized by Bohigas, Leggett, and Tomsovic entitled "Tunneling in Complex Systems" where a significant community interested in tunneling problems came together and discussed such topics as dynamical tunneling, chaos-assisted tunneling and ionization, chaotic tunneling, multidimensional complex trajectories, and possible or even speculative connections to various experiments involving magnetic systems or the decay of superdeformed nuclei. My feeling in hindsight is that it was clear that there was still an enormous amount we did not understand about dynamical tunneling problems at that time. Perhaps going a bit further, it was not at all obvious whether some of the known unsolved problems would be solved anytime soon either. Many appeared on their face to be rather intractable. In terms of an analytical theory, understanding chaos-assisted tunneling is yet to begin. My recollection is that this was also before most, if not all, of the beautiful work done with ultracold atoms and optical lattices.

Nevertheless, I guess it should not come as an enormous surprise that a great deal of progress has been made since then, but a pleasant surprise it has been. New ideas, new contexts, new theories, new experiments, they are all there. Personally, and among other things not being mentioned in a short foreword, I have been impressed with the work on resonance-assisted tunneling beginning with Ullmo's group and tunneling between regular and chaotic states from Ketzmerick's group, all of which are required for a complete theory of chaos-assisted tunneling. The experiments from Phillips' and Raizen's groups impress as well, though the effective size of $\hbar$ in those works muddies the interpretations. A book is most certainly called for and this one fits the bill; one where there is an overview and a consolidation of presentation of the current state of dynamical tunneling. Anyone generally interested in tunneling really should be familiar with the works presented here. I look forward to seeing this book in print.

Steven Tomsovic
Indian Institute of Technology Madras

Preface

Tunneling has remained a special phenomenon, a quintessential quantum effect, starting with the early days of quantum theory. Nearly a century's worth of theoretical and experimental studies have highlighted the crucial role of tunneling in various physical phenomena. The far-reaching implications of tunneling are evident in diverse fields including nuclear, atomic, molecular physics, and, more recently, in the area of mesoscopic science. Despite the obvious relevance of this topic to a wide range of disciplines, an interdisciplinary scientific community devoted to tunneling has not yet developed to a satisfactory degree. One may attribute this, at first glance, to the apparent simplicity of a generic tunneling process, which basically involves "only" a quantum particle that crosses a classical barrier due to the evanescent components of its wave function. The quantitative description of this seemingly simple process, however, can become rather intricate and rich if more than one particle and/or more than one spatial dimension are effectively involved. This is especially the case for "dynamical tunneling," which essentially denotes classically forbidden transitions through dynamical rather than energetic barriers, that is barriers that are formed by constraints of the underlying classical dynamics related to exact or approximate constants of motion. This book is devoted to the study of dynamical tunneling—mechanisms and consequences.

Apart from providing a resource for the state of the art along various research lines related to dynamical tunneling, the present book is also an attempt to establish and connect the "tunneling community" across the borders of the various fields and disciplines in physics. Previous steps in this direction include, among others, the book *Tunneling in Complex Systems* edited by Steven Tomsovic (World Scientific, 1998), as well as several international workshops, such as the "International Symposium of Complexified Dynamics, Tunneling and Chaos" (Kusatsu, Japan, 2005) and, most recently, the conference and summer school "Tunneling and Scattering in Complex Systems—From Single to Many Particle Physics" (Dresden, Germany, 2009). Incidentally, the selection of specific topics and contributors for this book coincides to a large extent with the spectrum of talks and lectures at this latter event in Dresden—although this book was not intended to serve as a proceeding of the Dresden conference. We were certainly not able to cover the entire breadth of the subject of dynamical tunneling in this book, but had to make a selection for the sake of coherence: semiclassical aspects of dynamical tunneling as well as tunneling with cold atoms and molecular manifestations are therefore more strongly represented than, for instance, nonclassical processes in electronic mesoscopic physics.

As always, this book would not have been possible without the support of a number of colleagues whom we would like to thank. Specifically, we are indebted to our contact persons at Taylor & Francis CRC, Lance Wobus and David Fausel, for their initiatives, encouragement, patience, and assistance in details concerning the formalities of the publication process. Thanks are also due to Shashi Kumar from Glyph International for his valuable technical help concerning the assembly of the individual chapters. P.S. would, further, like to take this opportunity and thank his co-organizers of the Dresden workshop, Arnd Bäcker and Markus Oberthaler, as well as the Max-Planck Institute for the Physics of Complex Systems for hosting and financing the workshop. He would also like to acknowledge support from the DFG Forschergruppe 760 "Scattering Systems with Complex Dynamics" which provided a unique framework for interdisciplinary research on complex dynamics in open systems in general and on dynamical tunneling in particular during the last three years. The success of the Forschergruppe is evident from at least two of the contributions (Chapters 6 and 8) that are a part of this volume.

Last but not least, we have to mention our contributors, the authors of the chapters of this book, who agreed to provide review articles on aspects of their specific research on dynamical tunneling.

Their willingness to participate in the cross-refereeing process of the book as well as constant words of encouragement has made our task easy. It has been a unique and wonderful experience for us. We thank them for their contributions and hope that this book will serve as a useful resource on various aspects of dynamical tunneling.

Srihari Keshavamurthy
Peter Schlagheck

Editors

Srihari Keshavamurthy is a theoretical chemist in the department of chemistry at the Indian Institute of Technology (IIT) Kanpur, India. He got his BSc degree from the University of Madras, India; MS from the Villanova University, Pennsylvania; and PhD from University of California, Berkeley. After a postdoc at Cornell University, he joined the IIT Kanpur in December 1996. His primary interest is to understand the mechanisms of chemical reaction dynamics and control from the classical-quantum correspondence perspective.

Peter Schlagheck is a theoretical physicist in the department of physics at the University of Liège, Belgium. He got his PhD in 1999 at the Technical University of Munich, Germany. After a postdoc at the Université Paris Sud, France from 1999 to 2001, he became an assistant at the University of Regensburg, Germany in 2002. In 2009, he obtained a faculty position at the University of Liège. His research interests include the transport of ultracold atoms and tunneling in the presence of chaos.

Contributors

Joachim Ankerhold
Institut für Theoretische Physik
Universität Ulm
Ulm, Germany

Ennio Arimondo
Dipartimento di
Fisica Enrico Fermi
Università di Pisa
Pisa, Italy

Stephan Arlinghaus
Institut für Physik
Carl von Ossietzky Universität
Oldenburg, Germany

Arnd Bäcker
Institut für Theoretische Physik
Technische Universität Dresden
Dresden, Germany

Stephen C. Creagh
School of Mathematical Sciences
University of Nottingham
Nottingham, United Kingdom

Sergej Flach
Max-Planck Institute for the Physics of
Complex Systems
Dresden, Germany

Eric J. Heller
Department of Physics and Department of
Chemistry and Chemical Biology
Harvard University
Cambridge, Massachusetts

Winfried K. Hensinger
Department of Physics and Astronomy
University of Sussex
Sussex, United Kingdom

Martin Holthaus
Institut für Physik
Carl von Ossietzky Universität
Oldenburg, Germany

Kensuke S. Ikeda
Department of Physics
Ritsumeikan University
Kusatsu, Japan

Srihari Keshavamurthy
Department of Chemistry
Indian Institute of Technology
Kanpur
Kanpur, India

Roland Ketzmerick
Institut für Theoretische Physik
Technische Universität Dresden
Dresden, Germany

Matthias Langemeyer
Institut für Physik
Carl von Ossietzky Universität
Oldenburg, Germany

David M. Leitner
Department of Chemistry
University of Nevada
Reno, Nevada

Steffen Löck
Institut für Theoretische Physik
Technische Universität Dresden
Dresden, Germany

Stefano Longhi
Dipartimento di Fisica
Politecnico di Milano
Milano, Italy

Amaury Mouchet
Laboratoire de Mathématiques
    et de Physique Théorique
Université François Rabelais de Tours
Tours, France

Ricardo A. Pinto
Department of Electrical Engineering
University of California Riverside
Riverside, California

Mark G. Raizen
Department of Physics and Center
    for Nonlinear Dynamics
University of Texas
Austin, Texas

Peter Schlagheck
Département de Physique
Université de Liège
Liège, Belgium

Akira Shudo
Department of Physics
Tokyo Metropolitan University
Tokyo, Japan

Daniel A. Steck
Oregon Center for Optics and
    Department of Physics
University of Oregon
Eugene, Oregon

Kin'ya Takahashi
The Physics Laboratories
Kyushu Institute of Technology
Iizuka, Japan

Denis Ullmo
Laboratoire de Physique Théorique
    et Modèles Statistiques
Université Paris-Sud
Orsay, France

Alvise Verso
Institut für Theoretische Physik
Universität Ulm
Ulm, Germany

Sandro Wimberger
Institut für Theoretische Physik
Universität Heidelberg
Heidelberg, Germany

1 An Overview of Dynamical Tunneling

Eric J. Heller

CONTENTS

This chapter represents a personal perspective on dynamical tunneling. It does not attempt to review all the significant recent developments in the subject, many of which are already represented in this volume. Rather, it weaves a path through some case histories, some simple but emblematic, others more complex and involving experimental counterparts.

The whole concept of tunneling is by its very nature semiclassical. Quantum amplitudes do what they need to do according to quantum mechanics, and it is up to us to label something as tunneling or not. We do so by comparison with classical mechanics, making tunneling an intrinsically semi-classical idea. Potential barrier tunneling, a process wherein quantum eigenstates penetrate regions of coordinate space which are classically forbidden, is the simplest kind of tunneling, found in every elementary textbook on quantum mechanics.

Dynamical tunneling got its name by its contrast with barrier tunneling. Classical mechanical trajectories sometimes refuse to go where they are allowed to by constraints other than potential barriers, constraints such as action or angular momentum. In other words, even in the absence of potential barriers, the dynamics may refuse to carry the trajectories from region A to region B, and if the quantum amplitudes respect no such constraint in the quantized version of the same system, we say the quantum system is "dynamically tunneling." It is well known, for example, that when a plane wave meets a potential barrier, some of it is backscattered even if the total energy lies above the highest point of the potential barrier. Even better known is the wave amplitude found in the shadow region behind a barrier. These are examples of dynamical tunneling, although in both cases historically these phenomena have been called "diffraction." The trouble with the term diffraction is that it is highly abused: one can find it describing all sorts of phenomena including scattering which is perfectly classically allowed.

We prefer the umbrella term suggested long ago by Bill Miller, namely "classically forbidden processes." Such processes are usefully divided into two types: barrier tunneling and dynamical tunneling.

One might easily get the impression in a book like this that quantum mechanics happily explores everywhere that classical trajectories do, and then some, due to tunneling. But of course this is not the case: certain types of Anderson localization represent quantum reluctance to go where classical trajectories do. A very simple example of quantum localization (which we take to be an umbrella term of which Anderson localization is one example) is a particle in a "box and corridor" shaped as in Figure 1.1. All but a measure zero of the classical orbits in this system manage to escape; however, it is intuitively obvious that the very lowest stationary states of the system are quantum mechanically localized to the vicinity of the box. The reason is that the zero point energy cost of being in the corridor is greater than the total energy of states which are confined to the larger box. Therefore, the first few quantum mechanical eigenstates will be confined to the box and a short distance into the corridor.

1.1 MIRRORS AND SHADOWS

An impenetrable segment of a barrier wall, illuminated by trajectories from in front, possesses a hard shadow, a refuge from the hail of projectiles. As is well known in all sorts of wave theories including quantum mechanics, waves do not respect such hard shadows. Since the dynamics (in this case ballistic straight-line trajectories) creates a hard shadow, penetration of the waves into the shadow region is an example of dynamical tunneling. An equally good but less commonly used example is the reflection from the front surface of the mirror, from one point to another, along a path that cannot be reached by a specular bounce.

Consider the source point A and the receiving point B as shown in Figure 1.2. According to the Kirchhoff approximation, we may consider the total illumination at B to be a sum of all the amplitudes which reach a point x on the mirror, and from there scatter to the point B. The amplitude

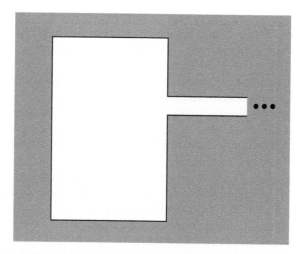

FIGURE 1.1 Box and corridor (the lines represent hard walls, the potential is flat inside, and the corridor extends forever to the right) which supports many bound states below the energy of the first state which propagates down the corridor. All but a set of measure zero trajectories escape, however, independent of their energy.

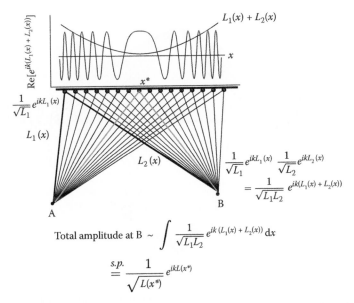

FIGURE 1.2 Schematic showing the construction of the amplitudes scattered from the point A to the point B by a mirror. The Kirchhoff construction supposes this amplitude to be decomposed in terms of waves arriving at various points along the mirror, and then from these points to a final position here the point B. The total amplitude at B is then obtained by adding up all the amplitudes along the mirror. Alternately, approximating the integral by stationary phase results in a compact formula involving only the specular path shown in gray.

of illumination of each point x is the expected form $\sim 1/\sqrt{L_1(x)}\, e^{ikL_1(x)}$ where L_1 is the distance from the source to the point at x, and k is the wavenumber. Illuminated by this amplitude, each point x radiates a spherical wave which reaches the point B, further attenuated by traveling a distance $L_2(x)$; the overall contribution at B for that impact point on the mirror is therefore $\sim 1/\sqrt{L_1}\, e^{ikL_1(x)} 1/\sqrt{L_2}\, e^{ikL_2(x)} = 1/\sqrt{L_1 L_2}\, e^{ik(L_1(x)+L_2(x))}$. Adding up all such impact points, we have the amplitude for scattering off the mirror from A to B as

$$a = \int \frac{1}{\sqrt{L_1 L_2}} e^{ik(L_1(x)+L_2(x))}\, dx = \frac{1}{\sqrt{L(x^*)}} e^{ikL(x^*)}, \tag{1.1}$$

where the integral is done by stationary phase over points x along the mirror. It is seen that the total path length from A to a point on the mirror at x and then on to B is a minimum at the specular scattering point. This should ring all sorts of bells (minimum action, etc.).

A straightforward stationary phase integration gives the right-hand side of Equation 1.1, where $L(x^*)$ is the total length of the classical, specular bounce off the mirror, and x^* is the specular, minimum action point. Thus, the Kirchhoff approximation gives exactly what we would have written down to begin with.

However, suppose now that the mirror is too short, or the points A and B are moved over, such that the would-be specular point is no longer on the mirror. According to what Miller would call the "primitive semiclassical approximation," a term pregnant with the possibility of doing better, meaning stationary phase for everything in sight, the amplitude for the indirect path from A to B is now zero. Stationary phase and the primitive semiclassical approximation always arrives at a classical path; since there is no classical path which bounces, the correct answer at this level is indeed zero.

The Kirchhoff approximation, the integral in Equation 1.1, is not necessarily intended to be evaluated by stationary phase. It is a perfectly good integral in its own right, and its exact evaluation

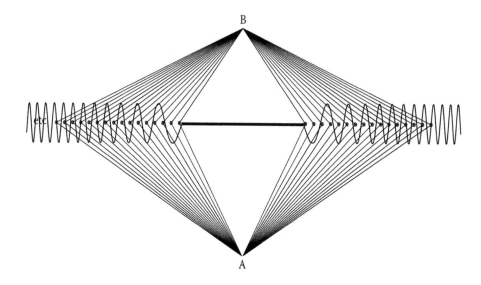

FIGURE 1.3 A construction similar to Figure 1.2, carried out for the shadow region.

would give a finite amplitude for scattering off the mirror from A to B. The resulting approximation is somehow semiclassical, but not primitive any longer; some would say this is a "uniform"
approximation. The Kirchhoff integral in this case has a strong flavor of the more general initial
value representations, a term and technique also introduced in the 1970s by Miller.

The Kirchhoff approximation to illumination in the shadow, the dark side of the barrier, can be
handled in much the same way, using points x to either side of the mirror extending to infinity, and
writing the amplitude in the shadow region as an integral over all the illuminated points not on the
mirror but lying on a line which includes it (see Figure 1.3).

Books on the theory of the stationary phase approximation point out that the subleading contributions after the stationary phase points, if any, come from any sharp cutoff in the integrals. Our
finite mirror gives precisely this type of cutoff. An end of the mirror may be thought of as a spherical source contributing a scattered amplitude of a magnitude which can be estimated by figuring the
flux through a region about one wavelength long.

This example illustrates one of the paradigms for dynamical tunneling: stationary phase gives no
contribution, since by definition there is no classical path connecting the regions in question. The
amplitude is then written as a uniform integral, which is evaluated by direct means, not stationary
phase. The result is a uniform semiclassical approximation, in some sense partly semiclassical and
partly quantum mechanical. In any case, such uniform or initial value methods often give an accurate
description of dynamical tunneling events.

1.2   ABOVE-BARRIER REFLECTION

Probably the most important paradigm for dynamical tunneling is above-barrier reflection, alluded
to at the beginning of this chapter. The reason for the prominence of this paradigm is that many
dynamical tunneling processes which appear to have nothing to do with the above barrier reflection
can be recast as such by canonical transformations. Once the above-barrier reflection is understood,
its characteristic signature can be seen in the phase space of quite different dynamical systems. The
common denominator is a phase space separatrix, crossing it in a way which is classically forbidden,
and quite analogous to barrier tunneling (Figure 1.4). In that figure, the dark gray path corresponds
to "ordinary" barrier tunneling, that is, jumping across the barrier to an equivalent energy contour on

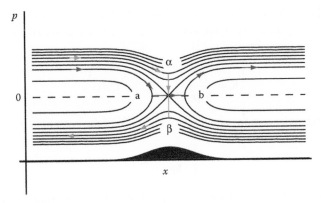

FIGURE 1.4 Plot of equal energy contours for the Hamiltonian $p^2/2 + exp[-x^2]$ in p,x phase space. The barrier is shown filled black at the bottom.

the other side by tunneling in position. Above-barrier reflection corresponds to the gray path, or in a trajectory which would classically have slowed down but not turned around at the barrier, and is in fact turned around by jumping to an equivalent contour or negative momentum. The trajectory can be thought of as tunneling in momentum, as opposed to the position tunneling required in ordinary barrier tunneling. Indeed, for suitable barriers and energies, the probability amplitude for such a barrier reflection can be written in terms of a tunneling integral involving the exponential of the classical imaginary action

$$-\int_{\alpha}^{\beta} |q|\,dp$$

from α to β, which is a direct analog to the usual barrier tunneling obtained from the exponential of

$$-\int_{a}^{b} |p|\,dq$$

from a to b. As shown earlier [1], the uniformized semiclassical reflection coefficient derived in this way is

$$\mathcal{R} = \frac{e^{-\frac{2}{\hbar}\Im\int_{-\alpha}^{\beta}x(p')dp'}}{1 + e^{-\frac{2}{\hbar}\Im\int_{-\alpha}^{\beta}x(p')dp'}}, \tag{1.2}$$

where $\alpha = -\beta$ is the classical value for momentum at the center of the potential, $x(p)$ is the classical solution for the position in terms of momentum, and $\Im$ stands for the imaginary part.

However, the usual tunneling formula for the potential barrier breaks down if the energy is too near the top of the barrier, which also implies that it fails at any energy if the barrier is not very high, since in that case the energy is necessarily either close to the top of the barrier or already well above it, in which case it is not barrier tunneling. The dynamical tunneling integral involving a path along momentum also breaks down for energies too near the top of the barrier, but if the energy is high enough it does not break down for weak barriers, since the action integral is still quite large. Semiclassics is expected to break down when tunneling action integrals become as small as the order of Planck's constant.

One of the first treatments of above barrier tunneling was Landau's in his famous quantum text. He used perturbation theory, arguing that, in one dimension, a plane wave $\psi_1 = \exp[ikx]$ couples to the backscattered wave $\psi_2 = \exp[-ikx]$ weakly, allowing one to use first-order perturbation theory—that is, the Born approximation of scattering theory—with an amplitude proportional to

$\int \psi_1^* \psi_2 V(x)\,dx = \int \exp[2ikx] V(x)\,dx$; this is seen to be the Fourier transform of the potential evaluated at the backscattering momentum transfer. Choosing states normalized to unit flux, the reflection coefficient in the Born approximation is [2]

$$\mathcal{R} = \frac{m^2}{\hbar^2 p^2} \left| \int_{-\infty}^{\infty} V(x) e^{\frac{2i}{\hbar} px}\,dx \right|^2. \tag{1.3}$$

This formula is valid under certain "smallness" conditions of the potential: we need

$$\left| \frac{V(x)a}{\hbar v} \right| \ll 1,$$

where $v = p\hbar/m$ is the classical velocity of the particle and a is the width of the potential, which essentially requires the scattering potential to be weak compared to the kinetic energy of the incoming particle [2].

This led Maitra and myself [1] to the idea of using a distorted wave Born approximation based on the above-barrier, WKB primitive semiclassical wave function instead of a plane wave. The WKB wave function has no backscattering in it since it is based on the above-barrier classical trajectory, which passes over the barrier. Nonetheless, it is a much better approximation to the true wave function in the vicinity of the barrier, and a better place to start for perturbation theory. The idea can be implemented by asking the question: for what potential is the WKB wave function exact? The answer is

$$V_{eff}(x,E) = V(x) - \hbar^2 \left[\frac{5}{32m} \left(\frac{V'(x)}{E - V(x)} \right)^2 + \frac{V''(x)}{8m(E - V(x))} \right]. \tag{1.4}$$

Thus, we have

$$\left(-\frac{\hbar^2}{2m} \frac{d^2}{dx^2} + V_{eff}(x,E) - E \right) \psi_{WKB}^{\pm}(x) \equiv (H_{WKB} - E)\psi_{WKB}^{\pm} = 0. \tag{1.5}$$

This allows us to use $W(x,E) = V_{eff}(x,E) - V(x)$ as a perturbation which "turns off" quantum reflection if added to the original Hamiltonian—a wave sent in from the left has no reflected component at energy E; it is perfectly "impedance matched." We note that $V_{eff}(x,E)$ is energy dependent and blows up at classical turning points, as does the WKB wave function. At energies above the potential maximum both the WKB wave function and effective potential are well behaved.

The idea of turning off dynamical tunneling with a perturbation is very appealing, especially since as we shall see the above-barrier reflection paradigm is so ubiquitous. However, this program has not been carried out in a wider context, and whether it can be done or not is not clear.

In any case, the approach works very well, as shown, for example, in Figure 1.5.

1.3 ROTATIONAL TUNNELING

One of the most beautiful examples of semiclassical separatrix dynamical tunneling was developed by Harter and coworkers [3] to treat rotational tunneling in polyatomic molecules in the 1980s. This work is far less well known than it should be. Given what has already been said, it is possible to describe the idea almost entirely in pictures (see Figure 1.7). However, the semiclassical perturbation idea, if it is applicable at all, has not yet been applied to this case.

Before we turn to Harter's work, let us consider a simpler situation, very close to the above-barrier reflection but translated to a case of rotation (Figure 1.6). We merely fold the real line back on itself to get a circle, and introduce the same kind of bump in the potential. Before the bump is

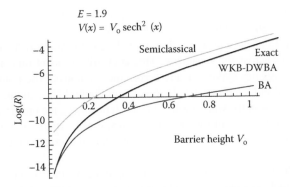

FIGURE 1.5 Comparison of the exact, Born approximation (BA), semiclassical action integral (semiclassical, Equation 1.2), and the WKB distorted wave Born approximation for a sech^2 barrier potential.

introduced, the rotational energy levels come in degenerate pairs, except for the ground state. The above-barrier reflection problem is extremely similar to the case of a plane wave, and we need not go into it deeply, except to point out that the rotor flips its angular momentum (which is no longer conserved because of the bump) sinusoidally in time, exactly as one would expect for two-state tunnel splitting.

Harter began his developments by introducing what he called the rotational energy surface, which is obtained (in the case of a rigid asymmetric top) by spinning up the rotor along all possible axes, until a fixed angular momentum J is reached. Then, the rotational energy for that axis and for that J is read off so to speak, and plotted as a radial distance from the origin. The angles of the rotation axis relative to the principal axes of the rotor are used for the angle of the radial vector. The tip of the vector traces out a surface, Harter's rotational energy surface. Figure 1.7 shows the case for a prolate rigid asymmetric rotor like a water molecule.

For a fixed magnitude of angular momentum, the rotor attains its highest rotational energy if spun about its low moment of inertia axis. This is the pink region of the rotational energy surface, corresponding to the diagram at the lower right. The rotor attains its lowest rotational energy if spun about its high moment of inertia axis, and of course the intermediate access gives an intermediate energy.

The shape and contours of the rotational energy surface make obvious many things about rigid rotors. The elliptical shape of the contours (which are strictly followed in the classical dynamics—that is, a trajectory must adhere to one and only one of them) near the high and low moment of

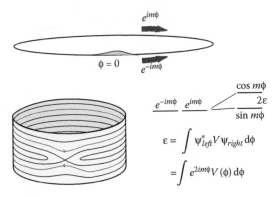

FIGURE 1.6 Separatrix formation and tunnel splitting of a simple rotor. The above-barrier splitting of the levels is an example of dynamical tunneling.

inertia axes show their stability; the "barrier top" separatrix motion near the intermediate access immediately reveals its well-known dynamical instability.

Most importantly for our present discussion, tunneling paths are drawn along the two topologically distinct directions across the separatrix. This is clearly resonant tunneling, since an exactly equivalent contour always lies symmetrically across the separatrix. The tunneling corresponds to interesting classically forbidden processes. In the case of the pink, low moment of inertia axis tunneling, the hydrogen atoms in the water molecule switch positions, the one above becoming the one below and vice versa. In the case of the gray, low moment of inertia region, they can switch positions, with the one that was leading in the direction of rotation becoming the one that is following, and vice versa.

Tunnel paths are shown in Figure 1.7. The tunneling integrals involved depend of course on the pair of level curves which are connected, and naturally the parameters of the asymmetric rotor. The tunneling splits otherwise degenerate pairs of levels; these splittings are well known of course in the spectrum of the asymmetric top as a function of the asymmetry parameter. For many cases, the action integrals involved are large and the splitting is fantastically small, meaning the system is for all intents and purposes symmetry broken. For other values of the parameters, the action integrals are smaller and the splittings are reasonably large, implying that the kind of classically forbidden oscillations just discussed are really happening for molecules rotating in the gas phase. This is dynamical tunneling par excellence!

The tunnel splitting is largest and the action integrals smallest when the shortest distances need to be traveled across the separatrix, meaning that motion starting near the intermediate axis rotation tunnels most readily, as seen in Figure 1.8.

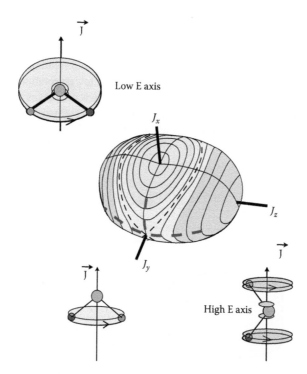

FIGURE 1.7 Harter's rotational energy surface for the case of a prolate asymmetric rotor like the water molecule shown here.

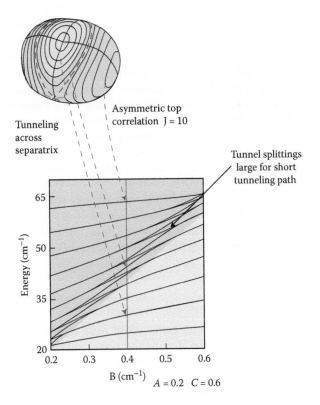

FIGURE 1.8 Harter's rotational energy surface and the spectrum of quantum levels as a function of this asymmetry parameter B. Note that the tunneling is easiest, the splittings largest, starting near the separatrix.

1.4 VIBRATIONAL TUNNELING

Quantum systems obey conservation laws demanded by symmetry just as classical systems do. Thus, molecules, for example, cannot tunnel to a different angular momentum in a system with no torques or to a different energy in an isolated system. But classical systems can have good integrals of the motion not demanded by symmetry. The quantum system can break these by tunneling, mixing the states which would otherwise coincide one for one with classical motion. Just as in the case of quantum rotating systems doing classically forbidden things like switching which end is pointing up, so too can one type of vibration turn into another. In the simplest cases, there may be symmetry creating a "double well" pair (or higher multiplicity) of states which could be symmetry broken classically. In more general circumstances, given a high density of primitive semiclassical states of many different zeroth-order quantum numbers, the quantum system will be able to tunnel among them, provided the interactions are sufficiently large compared to the splittings. In fact, the quantum system may achieve a kind of ergodicity by tunneling which the classical system lacks entirely.

   These questions are complicated by Arnold diffusion, a slow classical process of diffusion in phase space through successive narrow resonance zones. Interestingly, the process can be so slow that quantum tunneling may ignore it and find its own way around phase space!

   A classical example of vibrational tunneling is given by a two-degree-of-freedom system with Hamiltonian [4]:

$$H = \frac{p_x^2}{2} + \frac{p_y^2}{2} + \frac{1}{2}\omega_x^2 x^2 + \frac{1}{2}\omega_y^2 y^2 + \gamma x^2 y. \tag{1.6}$$

In the figures shown here, we took $\omega_x = 1.1$, $\omega_y = 1.0$, and $\gamma = -0.11$.

A periodic orbit exists by symmetry for $x = 0, p_x = 0$ (i.e., trajectories along the y-axis). Because the zeroth-order frequencies in the normal coordinates along x and y at low energies are not low-order resonant, this periodic orbit is stable at low energy, but by $E = 7.9$, the orbit has become unstable (hyperbolic). As the Poincaré surface of section in Figure 1.9 shows, two new, stable islands are born when the original periodic orbit goes unstable. These are the "eyes" of the figure-eight pattern—our classic separatrix, looking like a double well phase space except for rotated. This analogy suggests the presence of tunneling pairs of states, narrowly split in energy [5,6]. Lawton and Child [5] succeeded in an intuitive semiclassical tunneling estimate, and Davis and Heller coined the term "dynamical tunneling" in Davis and Heller (1981) [6]; both papers concern themselves with this type of potential with the "local mode" islands, so called because in the molecular version—water, for example—the symmetry broken state corresponds to one O–H stretching while the other remains nearly at rest. This can happen because the anharmonicity takes the frequency of the O–H stretch with considerable energy in it out of resonance with the O–H stretch with little energy.

The classical motion near the stable islands and two quantum eigenstates, and narrowly split pair of even and odd parity about the symmetry line $x = 0$ are shown in Figure 1.10 on the far left and far right, respectively.

1.5 AVOIDED CROSSINGS

Under certain rather general circumstances, collisions of pairs of levels resulting in an avoided crossing as a function of a parameter in the Hamiltonian signal interaction of those two levels, and if they come from very different parts of phase space, signal the dynamical tunneling from one to the other. In a Hamiltonian with two or more degrees of freedom, the density of states at moderate energies gets very high, and almost all the near coincidences of energy levels indeed involve states residing in different parts of phase space, assuming the Hamiltonian is not giving rise to chaotic motion, but rather mostly integrable motion.

Indeed, all the symmetric tunnel splittings we have investigated so far, coming from rotational tunneling and the vibrational Hamiltonian above, which is symmetric under the transformation $x \rightarrow -x$, have involved tunnel splitting of degenerate levels. In some sense, these are pairs of levels caught in the act of being in the middle of an avoided crossing.

We can begin this discussion with the previous example, by introducing a symmetry breaking potential term in the Hamiltonian with a parameter β: $\beta y^2 x$. The situation for β taking on the small value 5×10^{-5} is shown in Figure 1.11.

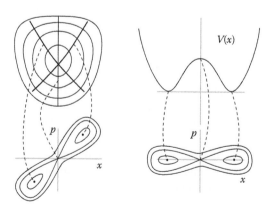

FIGURE 1.9   (Left) The phase space of the Hamiltonian given in Equation 1.6, showing an unstable, hyperbolic central orbit and two stable islands with a separatrix between them. The analogy with the double well (right) is clear.

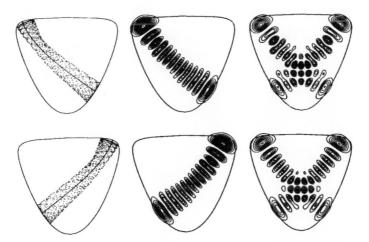

FIGURE 1.10 Classical (left) and quantum (right) fixed energy states for the Hamiltonian given in Equation 1.6. If the two eigenstates are added, we get the top middle result; if they are subtracted from one another, we get the bottom middle result, in exact analogy with the addition and subtraction of tunneling pairs of eigenstates of a double well.

By making the perturbation to be on the order of the tunnel splitting, the fully mixed levels for $\beta = 0$ can be partially localized. The energies of the two equivalent local mode states start to diverge for larger β. Clearly, as a function of β, there is an avoided crossing. At the closest approach of the two energy levels, for $\beta = 0$, the two classical like local mode states are fully mixed. This is a simple example of what could become a ubiquitous process in few body systems: dynamical tunneling to access most or all of phase space. However, it is a subtle business to decide what is "most" or "all" depending upon what prior constraints one wants to assign to the possible flow in phase space. For example, some zeroth-order states will fall at energies too far to be reached by the small tunneling interactions; some method must be assigned for penalizing or not penalizing the system for failing to access this state [7].

Figure 1.12 shows a succession of three values of a parameter μ in a Hamiltonian which need not be specified here. The states are degenerate in zero order for $\mu = 1$. This is a clear case of an avoided crossing due to dynamical tunneling: the classical analog states are evident on either side of the figure, and the mixing of them by tunneling interactions is obvious in the middle pair. The middle pair of eigenstates is not degenerate but split by a small amount, and thus this region of the parameter corresponds to an avoid crossing.

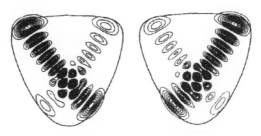

FIGURE 1.11 By breaking the right–left symmetry of the potential slightly, the tunneling pair begins to localize, one for each side. This is again an exact analogy to what would happen for a slight symmetry broken double well. The two local mode states acquire a different energy due to the splitting term $\beta y^2 x$; as a function of β, there is thus an avoided crossing.

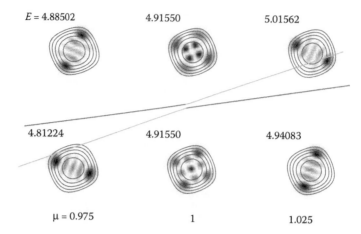

FIGURE 1.12   Avoided crossing induced by dynamical tunneling; and the tunneling interaction becomes comparable to the zeroth-order splitting, the zero-order states mix and the level curves avoid each other by the tunnel splitting.

A more general tunneling scenario, at least in two dimensions, is shown in Figure 1.13. This is a classic Poincaré surface of section, revealing resonance zones lying between more or less unperturbed regions. The resonance zones are themselves mostly integrable, however, a canonical transformation is required to find the constant actions inside them. This problem has received considerable attention from a semiclassical perspective from several groups [8–10]. Similar problems involving tunneling across nonintegrable regimes in phase space were considered by Maitra and Heller [11] and later Brodier et al. [12]. A classic study of chaos induced tunneling, which is partly a dynamical tunneling issue in the sense that trajectories are classically stuck in integrable zones, was initiated by Tomsovic and Ullmo [13].

1.6   IMPLICATIONS FOR SPECTRA

Dynamical tunneling interactions and their consequent avoided crossings cause a mixing of states. This has important experimental implications, since not all states are equal when it comes to coupling to the outside world. For example, optical dipole moments can vary tremendously from one zeroth-order state to the next. If tunneling interactions mix a state carrying a strong dipole from an

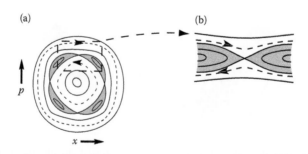

FIGURE 1.13   A Poincaré surface of section for a typical system with narrow resonance islands (a); locally, the surface of section is clearly analogous to the potential barrier we began with (b), tunneling between states of different actions across the separatrix corresponding to a reflection above the barrier.

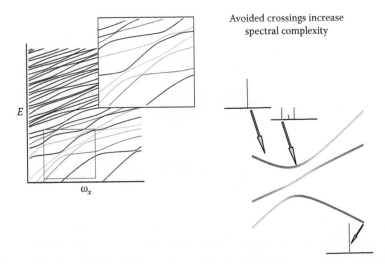

Avoided crossings increase spectral complexity

FIGURE 1.14 Avoided crossing as a function of a parameter, in this case causing dilution of dipole oscillator strength into two neighboring states involved in the avoided crossing.

initial state into many others nearby in energy, then a strong spectral line may be shared by many nearby eigenstates, all of which have some of the character of the original zero to order state [14].

An example is given by the Hamiltonian $\frac{p_x^2}{2} + \frac{p_y^2}{2} + \frac{1}{2}\omega_x^2 x^2 + \frac{1}{2}\omega_y^2 y^2 + \gamma x y^2$, where ω_x is used as the parameter. As ω_x is increased, all the energy levels rise, but obvious collisions are taking place (see Figure 1.14). We suppose that one unmixed state carries most of the oscillator strength. Labeling the potential curve belonging to this state with the color dark gray in the right-hand part of the figure, we see an admixture of colors develop in the curves as a function of ω_x. There is a corresponding dilution of a single spectral line into three separate lines. If ω_x continues to increase however, the dark gray character of one of the curves returns and a single line is seen again in the spectrum. This example heralds a much wider and more general class of problems, where many more than two states may be involved simultaneously, and also where the avoided crossing is much more accidental—not due to some obvious symmetries such as in the case of rotational tunneling or the double well, and so on.

Figure 1.15 shows schematically what might happen in a system with more degrees of freedom. An "extremal" state which by its very nature might carry most of the oscillator strength is coupled by tunneling interactions into other states, which themselves are coupled to more states, and so on. The bottom line is a fractionated spectrum of essentially unassignable lines, not respecting any but the obvious symmetries.

1.7 SEMICLASSICAL RESONANCE ANALYSIS [15,16]

Consider an integrable single resonance Hamiltonian for two degrees of freedom, following Ramachandran and Kay (1993) [17]:

$$H(\mathbf{I},\theta) = \omega_1 I_1 + \beta_1 I_1^2 + \omega_2 I_2 + \beta_2 I_2^2 + \lambda I_1 I_2 \, v(\theta_1,\theta_2) \tag{1.7}$$

$$\equiv H_0 + V(I_1,I_2,\theta_1,\theta_2), \tag{1.8}$$

where $v(\theta_1,\theta_2)$ is periodic in the angle variables θ_i. The base frequency is given by the parameter ω_i and the diagonal anharmonicity is controlled by β_i. The potential terms $v(\theta_1,\theta_2)$ can usefully be expanded as

$$v(\theta_1,\theta_2) = \sum_{n_1,n_2}' v_{n_1,n_2}(n_1\theta_1 - n_2\theta_2), \tag{1.9}$$

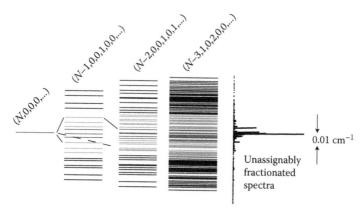

FIGURE 1.15 A hierarchy of tunneling interactions starting from a state with large oscillator strength leads finally to a fractionated spectrum of narrow bandwidth, the narrowness reflecting the weakness of the tunneling interactions. This kind of spectral fractionation is not common in two-degree-of-freedom systems, since their density of states is typically insufficient. It should however be commonplace in systems with a few degrees of freedom, at moderate energies were classical chaos is not strong or dominant, but the densities of states are high.

where (n_1, n_2) are relatively prime numbers possessing no common factors (a prime over the summation sign reminds us of this). Then the functions $v_{n_1,n_2}(n_1\theta_1 - n_2\theta_2)$ contain all the $(n_1 : n_2)$ resonance interaction and are expanded as

$$v_{n_1,n_2}(n_1\theta_1 - n_2\theta_2) = \sum_p v_{n_1,n_2}^{(p)} \exp[ip(n_1\theta_1 - n_2\theta_2)] \qquad (1.10)$$

so that the potential has the expansion

$$v(\theta_1,\theta_2) = \sum_p \sum_{n_1,n_2}' v_{n_1,n_2}^{(p)} \exp[ip(n_1\theta_1 - n_2\theta_2)]. \qquad (1.11)$$

For example, a 2:2 resonance zone would be induced by $v_{1:1}^{(\pm 2)}$.

Canonical transformation to new action-angle coordinates $(J_1, J_2, \phi_1, \phi_2)$ is accomplished via the generator

$$F(\mathbf{J},\theta) = \frac{1}{2}J_1(\theta_1 - \theta_2) + \frac{1}{2}J_2(\theta_1 + \theta_2), \qquad (1.12)$$

which gives

$$\phi_1 = \frac{1}{2}(\theta_1 - \theta_2); \quad \phi_2 = \frac{1}{2}(\theta_1 + \theta_2)$$
$$J_1 = (I_1 - I_2); \quad J_2 = (I_1 + I_2). \qquad (1.13)$$

The Hamiltonian in the new coordinates is

$$H = \omega^- J_1 + \omega^+ J_2 + \beta^+ J_1^2 + \beta^+ J_2^2 + 2\beta^- J_1 J_2 + \frac{\lambda}{2}(J_2^2 - J_1^2) v_{1:1}^{(2)} \cos(4\phi_1)$$
$$\equiv H_0 + \frac{\lambda}{2}(J_2^2 - J_1^2) v_{1:1}^{(2)} \cos(4\phi_1), \qquad (1.14)$$

where $\omega^\pm = (\omega_1 \pm \omega_2)/2$ and $\beta^\pm = (\beta_1 \pm \beta_2)$. It is clear that J_2 is a constant of the motion because ϕ_2 is absent in H. The phase portrait in the J_1, ϕ_1 plane for fixed energy E is shown in Figure 1.16,

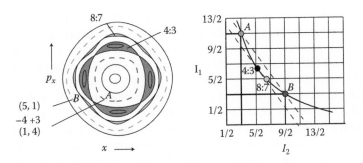

FIGURE 1.16 A 4:3 resonance mediates the tunneling between the $A = (3/2, 11/2)$ torus and the $B = (9/2, 3/2)$ torus. The solid heavy line shows the level curve of constant energy in action space; note that $E_A = E_B$. Somewhere in between points A and B will lie the actions where $\omega_1/\omega_2 = -(\partial H/\partial I_1)/(\partial H/\partial I_2) = -\partial I_2/\partial I_1|_E$; this is denoted by 4:3 and a black dot. At this point on the level curve, the tangent is parallel to the line drawn between A and B.

top. A single 2:2 resonance zone is seen. Now suppose the tori labeled A and B have exactly the same energy at the primitive (EBK) level that is, $H(I_1^A, I_2^A) = H(I_1^B, I_2^B)$ where $I_1^A = (n_1^A + \alpha_1^A)h$, and so on, where n_1^A is an integer and α_1^A arises from the well-known Maslov phase corrections. The heavy lines to either side of the resonance zone are degenerate tori which are EBK quantized. Dynamical tunneling is expected between them. Even if the EBK energies differ, the states will be mixed if the tunneling interaction is larger or on the order of the EBK splitting.

We now demonstrate that the dynamical tunneling causing the avoided crossing is once again a case of an above-barrier reflection amenable to a perturbative treatment.

Suppose two states with quantum numbers (1,5) and (4,1) are involved in an avoided crossing. The resonance zone appears between the tori involved in the tunneling for the reason seen in Figure 1.16. A 4:3 resonance mediates the tunneling between the A torus with actions $(1 + 1/2, 5 + 1/2) = (3/2, 11/2)$ and the B torus with actions $(9/2, 3/2)$. The heavy line delineates the level curve of constant energy in action space; note that $E_A = E_B$. By the mean value theorem, somewhere in between points A and B must lie actions where the slope S of the constant energy curve is equal to the slope of the line connecting the two sets of actions.

$$-S = \omega_1/\omega_2 = -\frac{\partial H/\partial I_1}{\partial H/\partial I_2} = -\frac{\partial I_2}{\partial I_1}\bigg|_E; \qquad (1.15)$$

this is denoted by 4:3 and a black dot. At this point on the level curve, the tangent is parallel to the line drawn between A and B.

$S = -4/3$ corresponds to the zone in classical phase space where the resonance (4:3 in this example) exists. In this way, we can always find an appropriate resonance zone which will connect two tori by over-the-barrier dynamical tunneling.

Even if the quantum numbers of the states involved in the avoided crossing are very different, say differing by n quanta in action I_1 and m quanta in I_2, there is in general an n:m classical resonance island chain lying between the tori corresponding to the states which is the agent of the quantum tunneling. In a generic coupled but nearly integrable phase space, an island chain will exist for every rational winding number. In this sense, classical phase-space structures are the cause even of very narrow avoided crossings between states of very different character.

1.7.1 CALCULATION OF THE TUNNELING INTERACTION

There are several ways to calculate the tunneling between the tori caused by the intervening resonance zone. They all begin with a resonance analysis as just described. In two closely related approaches, one gets the tunneling by stopping short of the fully semiclassical analysis, substituting a little quantum mechanics into the one-dimensional effective Hamiltonian produced by the resonance analysis. This produces a "uniform" approximation; in essence, one is doing full quantum mechanics on a classically preprocessed Hamiltonian. One approach quantizes the (J_1, ϕ_1) Hamiltonian for fixed J_2 with the Ansatz $J_1 \rightarrow -i\hbar d/d\phi_1 + \hbar$ [8,9]. The resulting Schrödinger equation is then solved numerically, or analytically if possible. The pendulum-like Hamiltonian 1.14 contains the nonclassical below barrier tunneling (and above barrier reflection). A closely related scheme uses the Heisenberg (matrix) formulation of quantum mechanics by finding matrix elements of the resonant interaction term and diagonalizing. Bohr correspondence in invoked in the following way: the resonant term depends on both actions and angles of the nonresonant part of the Hamiltonian, as in Equation 1.7. The actions appearing in the resonant term are set to their mean values $(\bar{J}_1, \bar{J}_2)$ and the off diagonal tunneling matrix element becomes

$$\langle J_1, J_2 | H | J_1', J_2' \rangle = \frac{1}{(2\pi)^2} \int\int e^{ip(n_1\theta_1 - n_2\theta_2)} V(\bar{J}_1, \bar{J}_2, \theta_1, \theta_2)\, d\theta_1\, d\theta_2 \equiv V(\bar{J}_1, \bar{J}_2)_{n1,n2}^{(p)}, \qquad (1.16)$$

with $\bar{J}_1 = (J_1 + J_1')/2$, and so on. The method is uniform since the matrix of EBK diagonal energies and off-diagonal couplings is numerically diagonalized. An example of the use of a procedure much like that just described is found in Roberts and Jaffe (1993) [10].

1.8 RADIATIVE TRANSITIONS

Suppose we apply weak monochromatic radiation to an anharmonic oscillator, resonant (in the quantum sense) with a high-lying eigenstate with $h\nu \gg \hbar\omega$, where ω is the local frequency of the oscillator near its minimum and ν is the frequency of the radiation. This process is classically forbidden: the frequency of the perturbation is much too high to pump much energy in or out of the oscillator. Essentially nothing happens classically, while quantum mechanically we know a certain fraction of the amplitude will be promoted to the excited state. The classically nonresonant absorption is a classically forbidden process, and a fine example of dynamical tunneling. This discussion applies to any nonresonant single-photon process, such as photoionization of Helium or the photoelectric effect. Indeed, the usual textbook discussion of the photoelectric effect, touted as an example of the need for quantum mechanics, is a statement that the photoemission process fails classically under the circumstances of the experiment (weak, classically nonresonant light), yet is allowed quantum mechanically.

It is interesting to assign a harmonic oscillator of angular frequency $2\pi\nu$ to the field, as in quantum electrodynamics, and treat it as a degree of freedom coupled to the anharmonic potential. Suppose there are four quanta in the field mode, and zero in the oscillator initially. This state is by design almost degenerate with three quanta in the field and four in the oscillator. Although this implies the existence of a 4:1 resonance in the classical phase space (see below), the resonance will generally not directly affect a trajectory representing the initial oscillator and in any case will be so narrow (for weak coupling) that it barely excites the oscillator. The usual energy diagram, as in Figure 1.17a, can be augmented in two ways, in coordinate space Figure 1.17b, and in phase space, Figure 1.17c.

First, we sketch a schematic two-dimensional oscillator potential illustrating the low-frequency mostly x-direction (horizontal) motion of the anharmonic potential and the much tighter high-frequency y-motion of the field oscillator. Eigenstates corresponding to the initial and final state as described above are sketched as well. Classically, the initial motion with all four quanta in the field mode and ground-state energy in the oscillator is indicated by the up–down arrow, and the final state motion with three quanta in the field and four in the anharmonic oscillator is indicated by the right–left arrow. These two motions do not communicate classically, but quantum mechanically there is a nonvanishing matrix element connecting them.

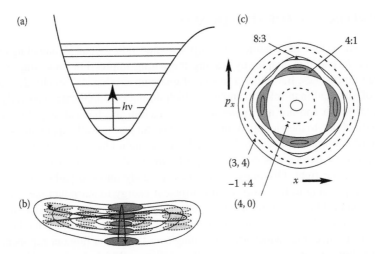

FIGURE 1.17 (a) The usual energy representation of a one-photon, multiquantum transition. (b) A coordinate space picture, illustrating the potential, the wavefunctions of the field oscillator, the anharmonic oscillator, and arrows indicating the qualitative classical motion corresponding to the initial and final states. (c) Poincaré surface of section for the field oscillator–anharmonic oscillator problem, showing resonance zones in phase space corresponding to the 4:1 resonance. However, even in the presence of the resonance and the coupling, the initial torus representing energy mostly in the field oscillator is almost undistorted from what it would be without the coupling to the field. Motion is confined to the invariant surfaces seen here, corresponding to the classical "no-go" for the transition.

Figure 1.17c, displays the phase-space view of this, with the convention that a $y = 0$ (field oscillator) surface of section is constructed, in which x and p_x are plotted every time the trajectory penetrates $y = 0$ with $p_y > 0$. Resonance islands appear corresponding to the 4:1 resonance and higher order resonances. Now we can understand the relevance of the above-barrier reflection problem to dynamical tunneling: the local phase-space structure near the islands is the same as the above-barrier problem. This means that we can use perturbation theory or distorted wave perturbation theory to determine the tunneling interaction between nearly degenerate states with different actions, in this case (4,0) and (3,4), just as we did for the barrier reflection problem.

1.9 COLLISIONAL DYNAMICAL TUNNELING

Many years ago, Miller [18] noted that classically forbidden events appear in semiclassical scattering theory ("classical S-matrix theory") as action changes of the target which fail to reach a full quantum jump (action change of h or greater), for all initial conditions. Then, the "primitive" semiclassical cross-section will be zero. However, Miller also gave uniform expressions which correctly bootstrap the finite quantum transition amplitude from the failed classical attempt. (This same failure would become classically allowed if $\hbar$ were allowed to become much smaller.) This is one of the first clearly worked out examples of dynamical tunneling—something that happens quantum mechanically but fails to happen classically, however with no potential barrier to blame.

A related example involves vibrational relaxation in neat O_2, as studied, for example, by Faltermeir et al. [19]. A classical study of the relaxation rate gives 0.0005 s^{-1}, many orders of magnitude slower than the experimental rate of 360 s^{-1}. For all practical purposes, the vibrational relaxation of an excited O_2 molecule in cold, liquid O_2 is classically forbidden, yet it proceeds slowly quantum mechanically. Under the conditions of the experiment, the vibrational frequency of the O_2 is $\omega_{vib.} = 1552$ cm^{-1}; however, typical environment frequencies are a very off-resonant 50 cm^{-1}. Dynamical tunneling is the agent of the relaxation of neat O_2. There are many other examples of this sort.

1.10 TUNNELING IN THE TIME DOMAIN

Our examples have so far been confined to energy domain studies—that is, tunneling of eigenstates. We do not want to end this chapter by leaving the wrong impression; tunneling is not a matter of fixed energy eigenstates, but rather a matter of quantum mechanical amplitude finding its way from one set of classical manifolds to another, which do not communicate classically. That is, the manifolds never touch and are not connected, but quantum mechanical amplitude finds its way from the vicinity of one to the vicinity of the other. This can just as well happen in the time domain. Some years ago, the problem of the off-diagonal time domain Green's function $G(x_2, x_1; t)$ in the vicinity of a potential barrier was considered [20]. One of the coordinates lies to the left of the barrier, and one to the right. Of course, there are classically allowed paths which connect these two points, corresponding to trajectories with sufficient energy to surmount the barrier; these are included in the manifold that corresponds to specific initial position, since this manifold has infinite momentum uncertainty. Because of the exponential instability near the barrier top however, at long times, the classical trajectories which are connecting over the barrier thin out exponentially fast. Of course, the tunneling amplitudes are exponentially small, too. Which dominates? The exponentially small amplitude contribution from over the barrier allowed trajectories or the exponentially small tunneling under the barrier? We termed this the "battle of the exponentials"; we found that tunneling

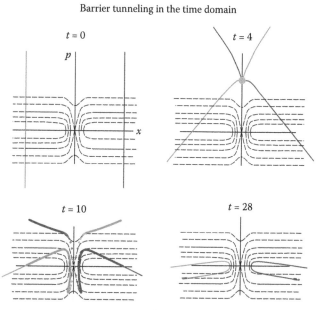

FIGURE 1.18 The battle of the exponentials. Two exponentials fight it out for dominance in the semiclassical Green's function for propagation across the top of a barrier. For early times, at $t = 4$, for example, the amplitude is dominated by trajectories which have enough energy to cross over the barrier. In this series of pictures, the amplitude for crossing the barrier in time t is divided up as forward propagation from the left for time $t/2$ and backward propagation from the right for the same time. This gives a symmetrical appearance and is equivalent to the propagation from one side for the full-time. The gray manifold from the left, and the dark gray one from the right, consist of many thousands of trajectories with slightly different initial conditions. Already by time $t = 10$ the classically allowed contributions have thinned out enormously in the barrier vicinity, and by $t = 28$ their contribution is truly infinitesimal. However, the trajectories on either side that can contribute to tunneling are quite dense, at $t = 28$, for example, the Green's function for transfer across the barrier is entirely dominated by tunneling contributions under the barrier.

paths start to dominate over the classically allowed paths after some finite time [20]. The situation is depicted in Figure 1.18. Even though it is potential barrier tunneling which eventually dominates in this case, example serves to show that tunneling can be a dominant factor in the time domain as well as the energy domain.

1.11 CONCLUSION

We have rambled through the subject of dynamical tunneling, setting the context and discussing some paradigm examples. There are many other examples which we know of; dynamical tunneling is truly ubiquitous!

Much more work remains to be done. It remains to be seen whether well-defined semiclassical methods even exist in most instances of dynamical tunneling. Certainly they are not routine yet, and the examples which are known are all rather special in one respect or another. This problem is certainly also not solved for multidimensional tunneling through potential barriers. Nonetheless, even if a successful semiclassical description based on action integrals over imaginary paths, and so on, cannot be found, interpretation of the process as dynamical tunneling remains intact. That is unambiguous: there are processes which do not happen classically and do happen quantum mechanically.

It is hoped that the examples given here will help inspire even more interest in dynamical tunneling and tunneling in general. The contributions to this volume certainly show it is a lively field.

REFERENCES

1. N. T. Maitra and E. J. Heller, Semiclassical perturbation approach to quantum reflection. *Phys. Rev. A*, 54, 4763, 1997.
2. L. D. Landau and E. M. Lifshitz, *Quantum Mechanics (Non-relativistic Theory)*, Pergamon Press, New York, 1977.
3. W. G. Harter and C. Patterson, Rotational energy surfaces and high-J eigenvalue structure of polyatomic molecules. *J. Chem. Phys.*, 80, 4241, 1984.
4. (a) E. J. Heller, E. B. Stechel, and M. J. Davis, Molecular spectra, Fermi resonances, and classical motion. *J. Chem. Phys.*, 73, 4720 (1980); (b) J. H. Frederick, E. J. Heller, J. L. Ozment, and D. W. Pratt, Ring torsional dynamics and spectroscopy of benzophenone—a new twist. *J. Chem. Phys.*, 88, 2169, 1988; (c) M. J. Davis and E. J. Heller, Quantum dynamical tunneling in bound states. *J. Chem. Phys.* 75, 246–254, 1981.
5. R. T. Lawton and M. S. Child, Local and normal stretching vibrational states of H_2O classical and semiclassical considerations. *Mol. Phys.*, 44, 709, 1981.
6. M. J. Davis and E. J. Heller, Quantum dynamical tunneling in bound states. *J. Chem. Phys.*, 75, 246, 1981.
7. E. J. Heller, Quantum intramolecular dynamics—criteria for stochastic and nonstochastic flow. *J. Chem. Phys.*, 72, 1337–1347, 1980.
8. D. Farrelly and T. Uzer, Semiclassical quantization of slightly nonresonant systems—avoided crossings, dynamic tunneling, and molecular spectra. *J. Chem. Phys.*, 85, 308, 1986.
9. A. M. Ozorio de Almeida, Tunneling and the semiclassical spectrum for an isolated classical resonance. *J. Phys. Chem.*, 88, 6139, 1984.
10. F. L. Roberts and C. Jaffe, The correspondence between classical nonlinear resonances and quantum mechanical Fermi resonances. *J. Chem. Phys.*, 99, 2495, 1993.
11. N. T. Maitra and E. J. Heller, Semiclassical amplitudes: Supercaustics and the whisker map. *Phys. Rev. A*, 61, 3620, 2000.
12. O. Brodier, P. Schlagheck, and D. Ullmo, Resonance-assisted tunneling in near-integrable systems. *Phys. Rev Lett*, 87, 064101, 2001.
13. S. Tomsovic and D. Ullmo, Chaos-assisted tunneling. *Phys. Rev. E*, 50, 145, 1994.
14. E. J. Heller, Dynamical tunneling and molecular spectra. *J. Phys. Chem.*, 99, 2625, 1995.

15. M. Carioli, E. J. Heller, and K. Møller, Intrinsic resonance representation of quantum mechanics. *J. Chem. Phys.*, 106, 8564, 1997.

16. E. J. Heller, The many faces of tunneling. *J. Phys. Chem.*, 103, 10433, 1999.

17. B. Ramachandran and K. Kay, The influence of classical resonances on quantum energy levels. *J. Chem. Phys.*, 99, 3659, 1993.

18. W. H. Miller, Semiclassical theory of atom-diatom collisions—path integrals and classical S-matrix. *J. Chem. Phys.*, 53, 1949, 1970.

19. B. Faltermeier, R. Protz, and M. Maier, Concentration and temperature dependence of electronic and vibrational energy relaxation of O_2 in liquid mixtures. *Chem. Phys.*, 62, 377, 1981.

20. N. Maitra and E. J. Heller, Barrier tunneling and reflection in the time and energy domains: The battle of the exponentials. *Phys. Rev. Lett.*, 78, 3035, 1997.

2 Dynamical Tunneling with a Bose–Einstein Condensate

Winfried K. Hensinger

CONTENTS

2.1 INTRODUCTION

Dynamical tunneling is a phenomenon that can be observed in certain dynamical systems at the borderline of quantum and classical dynamics. Rather than tunneling through an energy barrier (as in the case of "conventional" tunneling), a different constant of motion forbids this classically forbidden process. It is mostly observed in dynamical systems that exhibit a mixed classical phase space. In a mixed phase space, the dynamics of a particle can be either chaotic or regular depending on the initial position and momentum of the particle.

Jules Henri Poincaré invented a very convenient means to analyze the phase space of a driven dynamical system utilizing classical equations, known as Poincaré section [1]. This classical picture treats atoms as point objects with no inherent position and momentum width that is required in quantum mechanics by Heisenberg's uncertainty relation. Momentum and position of a particle are plotted stroboscopically with the stroboscopic period being equal to the modulation period. Figure 2.1 shows a Poincaré section [13] for the driven pendulum, corresponding to atoms in a modulated standing wave. The presence of ellipses in the Poincaré section indicates the existence of Kolmogorov, Arnold, and Moser (KAM) surfaces. The crossing of these surfaces by an atom is forbidden by classical mechanics.

In experiments carried out by collaborators and me at the University of Queensland [2–4] and the National Institute of Standards and Technology (NIST) [5], we realized the dynamical system of a driven pendulum which encompasses a mixed phase space. It is a textbook example for nonlinear Hamiltonian systems, and we accomplished a number of experiments where we were able to show various aspects of quantum and classical nonlinear dynamics [3–5]. Not all dynamical systems exhibit quantum effects on reasonable timescales (reasonable means observable within experimental constraints and smaller than the specific decoherence time). In order for that to occur, the typical action of motion for the particular system (product of typical position and momentum variation of the particle) must be approaching Planck's constant. In 1992, Graham et al. [6] proposed utilizing ultracold atoms to explore nonlinear dynamics as they realized that the system would be indeed capable of exhibiting both classical and quantum dynamics. They realized that this would allow them to investigate the borderland of "quantum" and "classical" dynamics. In

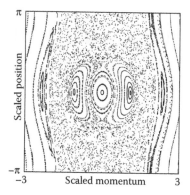

FIGURE 2.1 Poincaré section for a classical particle in an amplitude-modulated optical standing wave, exhibiting driven-pendulum dynamics. Momentum and position of the particle along one single well of a standing wave are plotted stroboscopically with the stroboscopic period being equal to the modulation period. The central region consists of small-amplitude regular motion. Chaos (dotted region) separates this region from two regions of regular motion (represented in the Poincaré section as sets of closed curves) located left and right of the center along momentum $p = 0$. Further out in momentum are two stable regions of motion known as librations. At the edges are bands of regular motion corresponding to above barrier motion. The Poincaré section is plotted for modulation parameter $\varepsilon = 0.20$ and scaled well depth $\kappa = 1.20$. (Adapted from W. K. Hensinger et al., *Phys. Rev. A*, 70, 013408, 2004.)

an attempt to give a meaningful description to this field of study, it has often been referred to as "quantum chaos," even though this expression is somewhat misleading as there is no chaos in quantum physics. A slightly better terminology is "quantum nonlinear dynamics" (QNLD). Cold atoms are a very useful system to investigate QNLD as the typical action of motion can be varied [4] and made sufficiently small, but also, because the system is well isolated from the environment, therefore exhibiting quantum effects on appropriate time scales. Dynamical tunneling was first proposed in the 1980s [7] and some years later, Dyrting et al. [8] predicted it to occur in the dynamics of cold atoms in a modulated standing wave—mimicking the dynamics of a driven pendulum [4].

In order to give a more intuitive understanding of dynamical tunneling, Eric Heller uses the example of a weightless ball bouncing between two semicircular mirrors [9] (see Figure 2.2). While the classical trajectory of the ball remains confined between the two mirrors, quantum mechanics will allow the trajectory to leave the cavity. This occurs not by tunneling through the actual mirror, however, by evolution of the trajectory in such a way that the particle can escape the cavity mirrors.

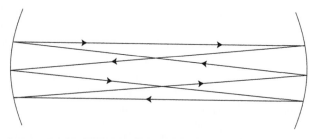

FIGURE 2.2 The trajectory of a weightless particle stays confined between two mirrors in the case of classical dynamics. Quantum physics will eventually allow the trajectory to leave the cavity.

2.2 EXPERIMENTS WITH A BOSE–EINSTEIN CONDENSATE

In our experiment [5], dynamical tunneling was observed in the dynamics of a Bose–Einstein condensate located within an amplitude-modulated standing wave. This experiment is indeed closely related to the experiment by Steck and coworkers [10], which is discussed in Chapter 3 of this book. In our experiment, an off-resonant modulated standing wave forms a potential for the condensate. It is possible to load the condensate into a particular region of regular motion by appropriate choice of initial momentum and position. A region of regular motion is a classical phase-space feature and corresponds in this case to a particle oscillating with a frequency equal to the amplitude modulation frequency [3,4]. If the system would follow purely classical dynamics, the particle would carry out this particular motion indefinitely. However, quantum physics leads to a tunneling effect in which atoms coherently tunnel into another region of regular motion, corresponding to an oscillation 180° out of phase. This is shown in Figure 2.3.

In the experiment carried out at NIST [5], a sodium Bose–Einstein condensate was utilized. It was produced using a magnetic optic trap consisting of an array of laser beams and magnetic fields. The atoms are then further cooled down using the method of evaporative cooling which eventually results in Bose–Einstein condensation [11]. The Bose–Einstein condensate was carefully loaded into a region of regular motion. The condensate was left to interact with the modulated standing wave for a number of modulation periods. Single frequency modulation of the intensity of the standing wave can be described by an effective Hamiltonian for the center-of-mass motion (a derivation is provided in Hensinger et al. (2003) [4]) given by

$$H = \frac{p_x^2}{2m} + \frac{\hbar\Omega_{\text{eff}}}{4}\left(1 - 2\varepsilon\sin\left(\omega t + \phi\right)\right)\sin^2(kx), \tag{2.1}$$

where the effective Rabi frequency is $\Omega_{\text{eff}} = \Omega^2/\delta$, $\Omega = \Gamma\sqrt{I/I_{\text{sat}}}$ is the resonant Rabi frequency, ε is the modulation parameter, ω is the modulation angular frequency, Γ is the inverse spontaneous lifetime, δ is the detuning of the standing wave, t is the time, p_x is the momentum component of the atom along the standing wave, and k is the wave number. Here I is the spatial mean of the intensity of the unmodulated standing wave (which is half of the peak intensity); so $\Omega = \Gamma\sqrt{I_{\text{peak}}/2I_{\text{sat}}}$ and $I_{\text{sat}} = hc\Gamma/\lambda^3$ is the saturation intensity. λ is the wavelength of the standing wave. ϕ determines the start phase of the amplitude modulation. Using scaled variables [12], the Hamiltonian is given by

$$\mathcal{H} = p^2/2 + 2\kappa(1 - 2\varepsilon\sin\left(\tau + \phi\right))\sin^2(q/2), \tag{2.2}$$

where $\mathcal{H} = (4k^2/m\omega^2)H$, $q = 2kx$, and $p = (2k/m\omega)p_x$. The driving amplitude is given by

$$\kappa = \omega_r\Omega_{\text{eff}}/\omega^2 = \frac{\hbar k^2\Omega_{\text{eff}}}{2\omega^2 m} = \frac{4U_0\omega_r^2}{\omega^2}, \tag{2.3}$$

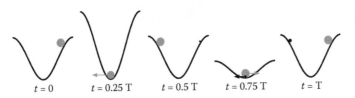

$t = 0 \qquad t = 0.25\,T \qquad t = 0.5\,T \qquad t = 0.75\,T \qquad t = T$

FIGURE 2.3 Illustration of regions of regular motion in phase space and dynamical tunneling. The ball (corresponding to the Bose–Einstein condensate) oscillates with a period equal to the modulation period T of the well. Due to quantum mechanics, atoms can tunnel into the motion 180° out of phase—this effect is referred to as the dynamical tunneling.

where $\omega_r = \hbar k^2/2m$ is the recoil frequency, $\tau = t\omega$ is the scaled time variable, and U_0 is the well depth in units of $\hbar\omega_r$.

The optical standing wave was turned off with a particular phase so that the atoms are located at the bottom of the well, therefore having maximum velocity [4]. The atomic position distribution measured after 1.5 m of free flight then allowed us to deduce the atomic momentum distribution of the atoms when the potential was turned off [4]. Figure 2.3 is a simplified depiction of the experiment, showing only one individual well. In the actual experiment a number of adjacent wells were loaded. Dynamics of atoms in adjacent wells were evolving simultaneously. As the atomic wavefunction is coherent and localized over multiple wells (the experiment is carried out with a Bose–Einstein condensate and the optical lattice is not deep enough to produce a Mott insulator), one obtains diffraction peaks located at integer multiples of $2\,\hbar k$ instead of a continuous momentum distribution. Further experimental details can be found in more detailed papers [4,5].

Figure 2.4 shows momentum distributions that are taken after 0.25 (front), 2.25 (middle), and 5.25 (back) modulation periods for one set of experimental parameters. This stroboscopic view of the experiment allows convenient visualization of the experiments. The momentum distribution after 0.25 modulation periods shows a pair of diffraction peaks at -4 and -6 $\hbar k$. Classically the atoms should remain in the region of regular motion leaving the stroboscopically measured momentum distribution unchanged. However, after 2.25 modulation periods, about half of the atoms have appeared with opposite momenta, which corresponds to the other region of regular motion. By 5.25 modulation periods, most of the atoms are in the other region of regular motion. At 9.25 modulation periods, the atoms have returned to the original region of regular motion. This transfer of atoms back and forth between the two regions of regular motion constitutes dynamical tunneling.

Figure 2.5a shows the mean atomic momentum after multiples of the modulation period for the same parameters used to obtain Figure 2.4. The circles (diamonds) correspond to turning off the standing wave at the maximum (minimum) of the amplitude modulation (see Figure 2.3 at $t = 0.25$ T and $t = 0.75$ T). Atoms, inhabiting only one region of regular motion, are at the bottom of the well at both turn-off phases. However, they move in opposite directions for the two different turn-off phases if they are part of the same region of regular motion. Classically, the mean momentum should approximately remain constant for the same turn-off phase. However, we observed an oscillation of the mean momentum indicating the occurrence of dynamical tunneling. The tunneling process is coherent as atoms tunnel back and forth between the two states of motion. By taking the Fourier transform of the data, a tunneling period of 10.3(2) modulation periods was found. Closely

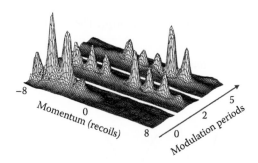

FIGURE 2.4 Three atomic momentum distributions of the Bose–Einstein condensate after oscillation within the wells of an optical lattice. The pictures are taken after 0.25, 3.25, and 5.25 modulation periods. Classically, one would expect most of the atoms to remain in a state with negative momentum (the original region of regular motion) as shown after 0.25 modulation periods. However, atoms tunnel into the state with opposite momentum (the other region of regular motion) as shown after 3.25 and 5.25 modulation periods. These experiments were taken for modulation parameter $\varepsilon = 0.29$, well depth $\kappa = 1.66$, and a modulation frequency $\omega/2 = 250$ kHz.

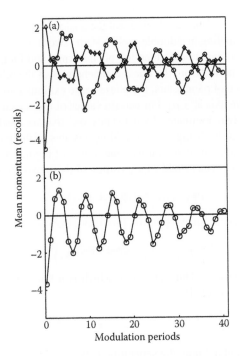

FIGURE 2.5 Mean momentum of the atomic momentum distribution as a function of the number of modulation periods. (a) The circles (diamonds) correspond to turning off the standing wave at the maximum (minimum) of the amplitude modulation (see Figure 2.3 at $t = 0.25\ T$ and $t = 0.75\ T$). (b) For a different set of parameters ($\varepsilon = 0.30$, well depth $\kappa = 1.82$, and a modulation frequency $\omega/2 = 222$ kHz), we observe a different tunneling period.

analyzing Figure 2.5, it is interesting to note that there is no significant (above background) zero momentum peak even in the case of approximately zero mean momentum (when the atoms have tunneled half-way). This indicates that at half the tunnel period, the system is in a coherent superposition of two distinguishable classical motions: one with positive momenta and one with equal but opposite momenta.

2.3 QUANTUM DESCRIPTION OF DYNAMICAL TUNNELING

Quantum theory predicts dynamical tunneling to occur for various values system parameters such as well depth, the degree of modulation, and modulation frequency and also predicts a strong sensitivity of the tunneling period and amplitude on these parameters. An example of this is shown in Figure 2.5b, where for a different set of parameters, we found a tunneling period of six modulation periods with a significantly longer decay time. We have carefully analyzed the parameter space and underlying mechanism in dynamical tunneling, showing how the tunneling frequency depends on the system parameters [13]. In order to understand such dependencies, it is best to describe the quantum dynamics of this periodically driven Hamiltonian system in terms of the eigenstates of the Floquet operator F, which evolves the system in time by one modulation period. It is indeed possible to associate the Floquet eigenstates with classical phase-space features such as regions of regular motion. However, when the typical classical action is not sufficiently large compared to $\hbar$, the phase-space representation of the Floquet eigenstates do not necessarily completely match with classical phase-space features [8,14]. However, it is possible to associate initial states localized around the corresponding classical region of regular motion with superpositions of a small number of Floquet eigenstates. One can then reveal the analogy of the dynamical tunneling experiments and conventional tunneling in a double-well system. Two states of opposite parity are usually responsible for the observed dynamical tunneling phenomenon and Floquet analysis is very useful to determine the tunneling period from the quasi-eigenenergies of the relevant Floquet states.

For an appropriate choice of parameters, the phase space exhibits two regions of regular motion, which lie on the momentum axis at $\pm p_0$. For certain values of the scaled well depth κ and modulation parameter ε, there are two dominant Floquet states $|\phi_\pm\rangle$ that are localized on regions of regular motion but are distinguished by being even or odd eigenstates of the parity operator that changes the sign of momentum. A state localized on just one region of regular motion is therefore likely to have dominant support on an even or odd superposition of these two Floquet states:

$$|\psi(\pm p_0)\rangle \approx (|\phi_+\rangle \pm |\phi_-\rangle)/\sqrt{2}. \tag{2.4}$$

Repeated application of the Floquet operator provides stroboscopic evolution of the quantum system,

$$F|\phi_\pm\rangle = e^{(-i2\pi\omega\phi_\pm/8\omega_r)}|\phi_\pm\rangle. \tag{2.5}$$

$\phi_\pm$ are the Floquet quasienergies. Thus, at a time which is n times the period of modulation, the state initially localized on $+p_0$, evolves to

$$|\psi(n)\rangle \approx \left(e^{(-i2\pi n\omega\phi_+/8\omega_r)}|\phi_+\rangle + e^{(-i2\pi n\omega\phi_-/8\omega_r)}|\phi_-\rangle\right)/\sqrt{2}. \tag{2.6}$$

Ignoring an overall phase and defining the separation between Floquet quasienergies as

$$\Delta\phi = \phi_- - \phi_+, \tag{2.7}$$

one obtains

$$|\psi(n)\rangle \approx \left(|\phi_+\rangle + e^{(-i2\pi n\omega\Delta\phi/8\omega_r)}|\phi_-\rangle\right)/\sqrt{2}. \tag{2.8}$$

At

$$n = 8\omega_r/(2\omega\Delta\phi) \tag{2.9}$$

periods, the state will form the antisymmetric superposition of Floquet states and thus is localized on the other region of regular motion at $-p_0$. In other words, the atoms have tunneled from one region of regular motion to another one with a frequency given by the energy difference between the eigenstates. By calculating Floquet states and energies for a particular set of parameters and checking which Floquet states are required to describe the initial Bose–Einstein condensate wave packet, it is possible to predict tunneling periods.

2.4 CONCLUSIONS AND OUTLOOK

The question whether classical chaos can in some sense "assist" the tunneling has been subjected to some debate. Most recently, Osovski and Moiseyev concluded that indeed our experiment [5] and the experiment by Steck et al. [10] exhibit some evidence of a "fingerprint of chaos in the quantum dynamics" [15]. In an earlier experiment, Dembowski et al. also reported first evidence of a chaos-assisted tunneling process in a microwave annular billard [16]. More recently, Chaudhury et al. [17] managed to provide experimental evidence that there is a connection between the presence of classical chaos and the emergence of entanglement. Furthermore, Chaudhury et al. [17] demonstrated that the sensitivity of the dynamics to perturbation of the Hamiltonian depends whether the initial state is located in a classically chaotic or regular region of motion.

In all these experiments it was possible to vary the effective typical action of motion for the system within reach of $\hbar$. This means that it is possible to experimentally investigate the quantum–classical transition. The observation of dynamical tunneling along with related theory and experiments demonstrate the usefulness of Bose–Einstein condensates as test-bed for quantum nonlinear dynamics and quantum chaos. There are many directions to investigate; for example, experiments could be extended to include two-dimensional potentials about which little is known in both the classical and the quantum domain.

ACKNOWLEDGMENTS

I would like to acknowledge and thank my collaborators who made very significant contributions toward making the dynamical tunneling experiments reported in Hensinger et al. (2001) [5] succeed, in particular, H. Häffner, A. Browaeys, N. R. Heckenberg, K. Helmerson, C. McKenzie, G. J. Milburn, W. D. Phillips, S. L. Rolston, H. Rubinsztein-Dunlop, and B. Upcroft. I would also like to acknowledge and thank P. Julienne, A. Mouchet, and D. Delande for collaborating with me toward gaining a better theoretical understanding of the underlying mechanisms of dynamical tunneling [13].

REFERENCES

1. A. J. Lichtenberg and M. A. Liebermann, *Regular and Stochastic Motion, Applied Mathematical Sciences*, Vol. 38, Springer-Verlag, New York, NY, 1988.

2. W. K. Hensinger, A. G. Truscott, B. Upcroft, M. Hug, H. M. Wiseman, N. R. Heckenberg, and H. Rubinsztein-Dunlop, Experimental study of the quantum driven pendulum and its classical analog in atom optics, *Phys. Rev. A*, 64, 033407, 2001.

3. W. K. Hensinger, B. Upcroft, G. J. Milburn, N. R. Heckenberg, and H. Rubinsztein-Dunlop, Multiple bifurcations in atom optics, *Phys. Rev. A*, 64, 063408, 2001.

4. W. K. Hensinger, N. R. Heckenberg, G. J. Milburn, and H. Rubinsztein-Dunlop, Experimental tests of quantum nonlinear dynamics in atom optics, *J. Opt. B: Quant. Semiclassical Opt.*, 5, R83–R120, 2003.

5. W. K. Hensinger, H. Häffner, A. Browaeys, N. R. Heckenberg, K. Helmerson, C. McKenzie, G. J. Milburn, et al., Dynamical tunnelling of ultracold atoms, *Nature*, 412 (5 July), 52–55, 2001.

6. R. Graham, M. Schlautmann, and P. Zoller, Dynamical localization of atomic-beam deflection by a modulated standing light wave, *Phys. Rev. A*, 45, R19–R22, 1992.

7. M. J. Davis and E. J. Heller, Quantum dynamical tunneling in bound states, *J. Chem. Phys.*, 75, 246–254, 1981.

8. S. Dyrting, G. Milburn, and C. A. Holmes, Nonlinear quantum dynamics at a classical second order resonance, *Phys. Rev. E*, 48, 969–978, 1993.

9. E. J. Heller, Quantum physics: Air juggling and other tricks, *Nature*, 412 (5 July), 33–34, 2001.

10. D. A. Steck, W. H. Oskay, and M. G. Raizen, Observation of chaos-assisted tunneling between islands of stability, *Science*, 293 (13 July), 274–278, 2002.

11. M. Kozuma, L. Deng, E. W. Hagley, J. Wen, R. Lutwak, K. Helmerson, S. L. Rolston, and W. D. Phillips, Coherent splitting of Bose–Einstein condensate with optically induced Bragg diffraction, *Phys. Rev. Lett.*, 82, 871–875, 1999.

12. F. L. Moore, J. C. Robinson, C. F. Bharucha, B. Sundaram, and M. G. Raizen, Atom optics realization of the quantum δ-kicked rotor, *Phys. Rev. Lett.*, 75, 4598–4601, 1995.

13. W. K. Hensinger, A. Mouchet, P. S. Julienne, D. Delande, N. R. Heckenberg, and H. Rubinsztein-Dunlop, Analysis of dynamical tunneling experiments with a Bose–Einstein condensate, *Phys. Rev. A*, 70, 013408, 2004.

14. A. Mouchet, C. Miniatura, R. Kaiser, B. Grémaud, and D. Delande, Chaos-assisted tunneling with cold atoms, *Phys. Rev. E*, 64, 016221, 2001.

15. S. Osovski and N. Moiseyev, Fingerprints of classical chaos in manipulation of cold atoms in the dynamical tunneling experiments, *Phys. Rev. A*, 72, 033603, 2005.

16. C. Dembowski, H.-D. Gräf, A. Heine, R. Hofferbert, H. Rehfeld, and A. Richter, First experimental evidence for chaos-assisted tunneling in a microwave annular billiard, *Phys. Rev. Lett.*, 84, 867, 2000.

17. S. Chaudhury, A. Smith, B. E. Anderson, S. Ghose, and P. S. Jessen, Quantum signatures of chaos in a kicked top, *Nature*, 461, 768–771, 2009.

3 Chaos-Assisted Dynamical Tunneling in Atom Optics

Daniel A. Steck and Mark G. Raizen

CONTENTS

3.1 INTRODUCTION

The study of quantum tunneling dates back to the early days of quantum mechanics, and has been a major topic of investigation in both theory and experiment. In recent years, there has been significant effort to understand tunneling phenomena for systems that exhibit chaos in their classical counterpart [13,15,18,29,42]. In particular, tunneling was predicted for mixed systems where chaotic and stable regions are intermingled [14]. The tunneling in this case is between two stable regions (referred to as *nonlinear resonances*, or *islands of stability*) in the classical phase space. The classical transport between these islands is forbidden by "dynamical barriers" in phase space. In contrast, quantum tunneling can couple the two islands so that a wave packet oscillates coherently between the two symmetry-related regions. This phenomenon is called *dynamical tunneling* because it occurs in phase space, without a fixed potential barrier. In the case of a mixed phase space, the presence of the chaotic region can enhance the rate of dynamical tunneling, a phenomenon known as *chaos-assisted tunneling* [53]. In this chapter, we review our experimental observation of chaos-assisted tunneling with ultracold cesium atoms [48–51]. We first give a brief overview of some basic concepts and of earlier work in order to provide a context for our results.

Dynamical tunneling was originally introduced in the context of a two-dimensional, time-independent potential [14]. Subsequently, it was shown that the combination of chaos and discrete symmetries could greatly increase the tunneling rate [53]. The presence of regular islands is also important for producing coherent tunneling, as they cause localization of the Floquet states [47]. This enhancement was understood in terms of a three-state process, where the tunneling doublet interacts with a third state associated with the chaotic region. The term "chaos-assisted tunneling" was introduced [53] to distinguish this process from the original notion of dynamical tunneling—tunneling as the interference of two states—though now the term "dynamical tunneling" also encompasses these more complicated tunneling processes.

Previous experimental work on dynamical and chaos-assisted tunneling was limited to wave analogies of tunneling. These include microwave billiards [15] and acoustic resonators [42]. The first direct observation of chaos-assisted tunneling was reported by our group [49], in parallel with a similar study of dynamical tunneling with Bose-condensed sodium atoms [30]. The role of chaos in these experiments has been considered in later theoretical analyses [4,37,40].

3.2 ATOMS IN OPTICAL LATTICE POTENTIALS

We begin by considering the basic workhorse of the atom-optical tunneling experiments in this chapter: the motion of an atom in a 1D lattice of far-detuned laser light. Expanding on the argument in Graham et al. (1992) [22], our basic conclusion will be that under the proper conditions, the atom behaves as a point particle. Furthermore, the "reduced" atom moves under the influence of the effective center-of-mass Hamiltonian

$$H_{\text{eff}} = \frac{p^2}{2m} + V_0 \cos(2k_{\text{L}}x), \tag{3.1}$$

where m is the atomic mass, k_{L} is the wave number of the laser light, and the potential amplitude V_0 is proportional to the laser intensity and inversely proportional to the detuning from the nearest atomic resonance (here, one of the components of the cesium D_2 spectral line). This Hamiltonian is simply that of the plane pendulum.

3.2.1 COUPLED INTERNAL AND EXTERNAL DYNAMICS

The basic model we will consider is a two-level atom moving in a 1D standing wave of light. The standing wave is given by the sum of two counterpropagating (but otherwise identical) traveling waves,

$$\begin{aligned} \mathbf{E}(x,t) &= \hat{z}E_0[\cos(k_{\text{L}}x - \omega_{\text{L}}t) + \cos(k_{\text{L}}x + \omega_{\text{L}}t)] \\ &= \hat{z}E_0 \cos(k_{\text{L}}x)\left(e^{-i\omega_{\text{L}}t} + e^{i\omega_{\text{L}}t}\right), \end{aligned} \tag{3.2}$$

where E_0 is the amplitude of either constituent traveling wave and ω_{L} is the laser frequency. The atom then evolves according to the Hamiltonian

$$H = \frac{p^2}{2m} + \hbar\omega_0\sigma^\dagger\sigma + \frac{\hbar\Omega}{2}(\sigma e^{i\omega_{\text{L}}t} + \sigma^\dagger e^{-i\omega_{\text{L}}t})\cos k_{\text{L}}x, \tag{3.3}$$

where the excited and ground internal atomic states are $|e\rangle$ and $|g\rangle$, respectively, $\sigma := |g\rangle\langle e|$ is the atomic annihilation operator, and ω_0 is the frequency of the atomic resonance. The first term is the center-of-mass kinetic energy of the atom, which is implicitly proportional to the identity operator in the internal atomic Hilbert space. The second term represents the internal free evolution of the atom, where we have taken the ground-state energy to be zero. The final term is the coupling of the atom to the laser field in the dipole and rotating-wave approximations, where we assume a nearly resonant laser, $|\omega_{\text{L}} - \omega_0| \ll \omega_{\text{L}} + \omega_0$. The maximum Rabi frequency

$$\Omega := -\frac{2\langle e|d_z|g\rangle E_0}{\hbar} \tag{3.4}$$

gives the strength of the atom–field coupling, where **d** is the dipole operator for the atom (we assume to be spherically symmetric).

We can simplify this Hamiltonian by making a transformation into the rotating frame of the laser field. This arises by expanding the internal atomic state as $|\psi\rangle = c_g|g\rangle + c_e|e\rangle$, and then rewriting the equations of motion in terms of the slowly varying amplitude $\tilde{c}_e := c_e e^{i\omega t}$. This procedure leads to the effective Hamiltonian

$$\tilde{H} = \frac{p^2}{2m} - \hbar\Delta_L\sigma^\dagger\sigma + \frac{\hbar\Omega}{2}(\sigma + \sigma^\dagger)\cos k_L x, \tag{3.5}$$

where $\Delta_L := \omega_L - \omega_0$ is the detuning of the laser from the atomic resonance. This representation removes the explicit time dependence from the problem.

Since we assume that the detuning from resonance is large (i.e., $\Delta_L \gg \Gamma$, where $1/\Gamma$ is the natural lifetime of $|e\rangle$), we will neglect spontaneous emission and use the Schrödinger equation $\tilde{H}|\psi\rangle = i\hbar\partial_t|\psi\rangle$ to describe the atomic evolution. Decomposing the state vector $|\psi\rangle$ into a product of internal and external states,

$$|\psi\rangle = |\psi_e(t)\rangle\,|e\rangle + |\psi_g(t)\rangle\,|g\rangle, \tag{3.6}$$

where the $|\psi_i(t)\rangle$ are states in the center-of-mass space of the atom. In the following, we will associate all time dependence of the atomic state with the center-of-mass components of the state vector. Defining the coefficients $\psi_i(x,t) := \langle x|\psi_i(t)\rangle$ leads to the coupled pair of equations

$$i\hbar\partial_t\psi_e = \frac{p^2}{2m}\psi_e + \left(\frac{\hbar\Omega}{2}\cos k_L x\right)\psi_g - \hbar\Delta_L\psi_e$$

$$i\hbar\partial_t\psi_g = \frac{p^2}{2m}\psi_g + \left(\frac{\hbar\Omega}{2}\cos k_L x\right)\psi_e. \tag{3.7}$$

for the wave functions $\psi_\alpha(x,t)$.

3.2.2 ADIABATIC APPROXIMATION

The equations of motion (Equation 3.7) can be simplified via the *adiabatic approximation*, which we motivate by examining the various timescales in the evolution of ψ_e and ψ_g. The kinetic energy terms in Equations 3.7 induce variations on a timescale corresponding to several recoil frequencies $\omega_r := \hbar k_L^2/2m$, where $\omega_r = 2\pi \times 2.07$ kHz for cesium. However, the pump-field terms induce motion on a timescale corresponding to the Rabi frequency (typically from zero to several hundred MHz), and the free evolution term induces motion of ψ_e on a timescale corresponding to Δ_L (typically from several to many GHz); together, these terms induce internal atomic oscillations at the *generalized Rabi frequency* $\Omega_{gen}(x) := \sqrt{\Omega^2\cos^2 k_L x + \Delta_L^2} \simeq \Delta_L$. Furthermore, in between these long and short timescales of external and internal atomic motion lies the damping timescale due to coupling with the vacuum, which corresponds to the natural decay rate Γ (for cesium, $\Gamma/2\pi = 5.2$ MHz). Because we are primarily interested in the slow center-of-mass atomic motion, and the internal atomic motion takes place for times much shorter than the damping time, it is a good approximation to assume that the internal motion is damped instantaneously to equilibrium (i.e., $\partial_t\psi_e = 0$, because ψ_e is the variable that carries the natural internal free-evolution time dependence at frequency Δ_L, whereas ψ_g has no natural internal oscillation, because the state $|g\rangle$ is at zero energy). This approximation then gives a relation between ψ_e and ψ_g:

$$\left(\hbar\Delta_L - \frac{p^2}{2m}\right)\psi_e = \left(\frac{\hbar\Omega}{2}\cos k_L x\right)\psi_g. \tag{3.8}$$

We can then use this constraint to eliminate ψ_e in the second of Equations 3.7, and with $\hbar\Delta_L \gg p^2/2m$, this equation becomes

$$i\hbar\partial_t\psi_g = \left(\frac{p^2}{2m}\right)\psi_g + V_0\cos(2k_L x)\psi_g, \tag{3.9}$$

where we have shifted the zero of the potential energy, and

$$V_0 := \frac{\hbar\Omega^2}{8\Delta_L} = \frac{|\langle e|d_z|g\rangle|^2 E_0^2}{2\hbar\Delta_L}. \tag{3.10}$$

Since the detuning is large, nearly all the population is contained in $|g\rangle$, so the excited state completely drops out of the problem. Hence, the atom obeys the Schrödinger equation with the center-of-mass Hamiltonian

$$H = \frac{p^2}{2m} + V_0\cos(2k_L x), \tag{3.11}$$

and the atom behaves like a point particle in a sinusoidal potential, where the strength of the potential is given by Equation 3.10. In the experiments, we will of course consider a slightly more complicated Hamiltonian, where the laser intensity (and thus the potential scale V_0) is modulated sinusoidally in time.

3.3 QUANTUM-STATE PREPARATION

Now we will discuss the preparation of localized initial wave packets in phase space. An overall schematic view of the procedure is illustrated in Figure 3.1, which shows the condition of the state in

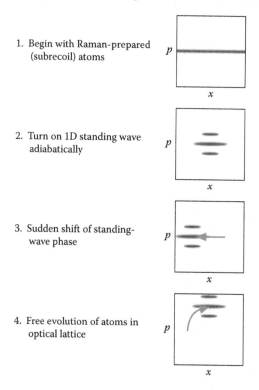

FIGURE 3.1 Schematic picture of the state-preparation sequence, beginning with the atoms prepared by subrecoil Raman velocity selection. The influence on the atoms in phase space is illustrated. The "striped" character of the distributions is a result of the discrete nature of the momentum transfer to the atoms.

phase space at various points in the process. This state-preparation procedure begins with laser cooling in a standard magneto-optic trap [38], and then further cooling in a 3D optical lattice [16,51]. A stimulated Raman velocity-selection process [31,39,51] then prepares a quantum state that is subrecoil in momentum but delocalized in space (i.e., coherent over multiple optical wavelengths). The 1D optical lattice is then turned on adiabatically, causing the atoms to become localized in the potential wells, at the expense of some broadening in momentum. Because the initial momentum distribution was narrow compared to the photon recoil momentum $\hbar k_L$, the resulting phase-space distribution had the discrete structure shown in Figure 3.1. This structure can be understood intuitively in the discrete momentum transfer (note that the potential $V_0 \cos 2k_L x$ is a linear combination of momentum-shift operators of momentum steps $\pm 2\hbar k_L$) from the lattice as it is turned on and also indicates coherence of the wave packet over multiple potential wells. For adiabatic processes, the atom stays in the same energy band, and thus the atoms are loaded completely into the lowest band of the optical lattice. For deep wells (as used in the experiment), the lowest band is approximately the harmonic oscillator ground state (repeated in each well), and thus the overall distribution envelope is approximately a minimum-uncertainty Gaussian wave packet, modulo the standing-wave period. The structure of subrecoil "slices" in the distribution out of an overall Gaussian profile is important in the tunneling experiments, and we will return to this issue in the following discussion.

After the atoms become localized in the potential wells of the lattice, the phase of the standing wave is shifted by about $1/4$ of the lattice period, displacing the atoms onto the gradients of the potential. They then evolve in the stationary optical lattice, where they returned to the potential minima, acquiring momentum in the process. In a harmonic potential, this procedure amounts to a boost of the wave packet in momentum, where the distance in momentum is set by the magnitude of the displacement. The anharmonicities in the optical lattice lead to a slight distortion of the wave packet, although it is still mostly Gaussian. More importantly, the subrecoil structure of the wave packet is preserved because the motional control is induced solely by the lattice.

To make this procedure more concrete, typical experimental parameters are as follows: the Raman π-pulse selection time was 800 μs, giving a velocity slice with a half width at half minimum (HWHM) of $0.03 \times 2\hbar k_L$; the lattice phase was shifted by 0.25 of the lattice period, and the atoms evolved in the lattice for 6 μs, which was the time after which the atomic momentum was maximized; and the resulting distribution (in momentum) was peaked at $4.1 \times 2\hbar k_L$, with a width $\sigma_p = 1.1 \times 2\hbar k_L$.

After the state-preparation sequence, the atoms were exposed to a temporally modulated optical lattice, realizing a (classically) chaotic modulated pendulum, where the dynamics of interest occurred. The atoms were then allowed to drift freely in the dark for 20 ms, and the freezing molasses and CCD camera enabled a measurement of the atomic momentum distribution by imaging the atomic fluorescence for 20 ms. Thus, the atomic momentum distribution is the basic observable in the experiment that we present below.

3.4 TUNNELING IN ATOM OPTICS

The basic experimental system that we used to study tunneling is essentially an ensemble of cold atoms in an amplitude-modulated optical lattice, along with the state-preparation procedure outlined above. To produce a relatively simple classical phase space with mixed stable and chaotic regions, the amplitude modulation of the potential was relatively smooth:

$$H = \frac{p^2}{2m} - 2V_0 \cos^2\left(\frac{\pi t}{T}\right) \cos(2k_L x). \qquad (3.12)$$

This Hamiltonian is again that of the pendulum, but with a sinusoidal variation of the potential amplitude in time from zero to $2V_0$ with period T. We can make a transformation into scaled units as follows:

$$
\begin{aligned}
x' &= 2k_{\mathrm{L}}x, \\
p'/\hbar &= p/2\hbar k_{\mathrm{L}}, \\
t' &= t/T, \\
H' &= (\hbar T/\hbar)H, \\
\alpha &:= (\hbar T/\hbar)V_0, \\
\hbar &:= 8\omega_{\mathrm{r}}T.
\end{aligned}
\tag{3.13}
$$

We have chosen the timescaling such that the scaled modulation period is unity, α is the scaled amplitude of the potential, and $\hbar$ is the effective Planck constant in the scaled units. The Hamiltonian in scaled units, after dropping the primes, is

$$
H = \frac{p^2}{2} - 2\alpha\cos^2(\pi t)\cos(x),
\tag{3.14}
$$

with the rescaled Schrödinger equation given by $H\psi = i\hbar\partial_t\psi$.

We can rewrite the potential as the sum of three pendulum-like terms with time-independent amplitude:

$$
V(x,t) = -\alpha\cos(x) - \frac{\alpha}{2}\cos(x + 2\pi t) - \frac{\alpha}{2}\cos(x - 2\pi t).
\tag{3.15}
$$

Thus, the modulated potential can be regarded as a combination of three pendulum potentials; two of these potentials are moving with momentum $\pm 2\pi$, and the third is stationary. When this system is sampled at integer times, these three terms produce primary resonances in phase space centered at $(x,p) = (0,\pm 2\pi)$ and $(0,0)$. This structure is evident in the phase space in Figure 3.2 for the small value $\alpha = 0.4$ (small secondary resonances due to nonlinear mixing of the primary resonances are also visible). For larger α, the resonances interact, producing a phase-space structure of bands of chaos surrounding the three main islands of stability. The tunneling that we consider here is between the two outer islands of stability, which are related to each other by reflection symmetry through the origin $(x,p) = (0,0)$. In configuration space, tunneling occurs between a state of coherent motion in

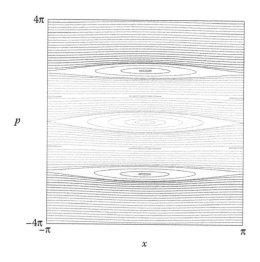

FIGURE 3.2 Phase portrait of amplitude-modulated pendulum for a small amplitude $\alpha = 0.4$, showing the three primary resonances.

only one direction and a state of the oppositely directed motion. These two states each correspond to being tightly bound to one of the two moving components of the lattice. The center island does not directly participate in tunneling.

In the experiment, we generally observed tunneling in the strongly driven regime, where large chaotic regions are present. A schematic representation of the initial condition for $\bar{k} = 2.08$ (corresponding to a 20 μs modulation period) and an 800 μs Raman-pulse duration is shown in Figure 3.3 with the classical phase space for the typical experimental value of $\alpha = 10.5$. In this regime, the center island has mostly dissolved into the chaotic sea, making this a clean configuration for studying tunneling between the remaining two islands. The two islands are located $8 \times 2\hbar k_L$ apart in momentum. The measured evolution of the momentum distribution in this case is plotted in Figure 3.4, where the distribution was sampled every two modulation periods out to 80 periods. Four of these distributions are also shown in more detail in Figure 3.5. Four coherent oscillations of the atoms between the islands are apparent before the transport is damped out. During the first oscillation, nearly half of the atoms appear in the secondary (tunneled) peak.

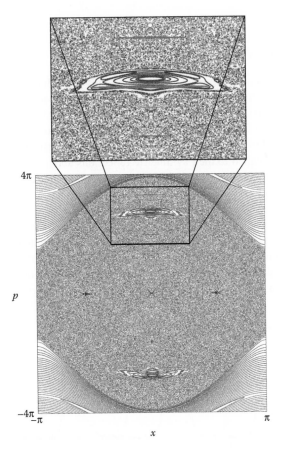

FIGURE 3.3 (See color insert.) Phase space corresponding to the experimental conditions for the data in Figure 3.4 ($\alpha = 10.5$). A schematic representation of the atomic initial state is superimposed in red on the upper island ($\bar{k} = 2.08$), showing the subrecoil structure that we expect from the state-preparation procedure. A magnified view of the upper island and initial quantum state is also shown. Color version available online.

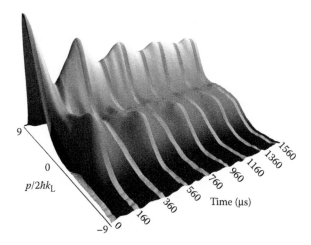

FIGURE 3.4 Observation of coherent tunneling oscillations in the momentum-distribution evolution between the two symmetry-related islands of stability, as shown in Figure 3.3. The two island centers are separated in momentum by $8 \times 2\hbar k_L$. In this plot, the distribution was sampled every 40 μs (every two modulation periods).

3.4.1 BROKEN SYMMETRY

The subrecoil Raman velocity selection is important not only to produce a nearly uncertainty-limited wave packet, but also to satisfy a quantum symmetry required to observe tunneling. This symmetry stems from the discrete translational symmetry of the potential, which causes momentum transitions to occur in discrete steps of k (or $2\hbar k_L$ in unscaled units). Thus, the momentum state $|nk + \delta\rangle$ (where n is an integer) is coupled to the $|-nk + \delta\rangle$ state via two-photon transitions. For $0 < |\delta| < k/2$, these states are therefore not coupled to their symmetric reflections about $p = 0$. In the language of barrier tunneling in the double-well potential, this situation is equivalent to an *asymmetric* double well, because the potential couples two states with a difference of $2nk\delta$ in energy. Thus, complete tunneling only fully occurs for the $|nk\rangle$ momentum states and is suppressed for states off this integer ladder. A deviation in momentum from this symmetric ladder is equivalent to a broken time-reversal symmetry [10], and the symmetric/antisymmetric character of the tunneling doublet can be sensitive to this broken symmetry [13]. This symmetry condition is automatically fulfilled for a rotor, because the periodic boundary conditions select the tunneling states, but in the case of a particle in an extended potential, as in the present experiment, careful state preparation is required to populate

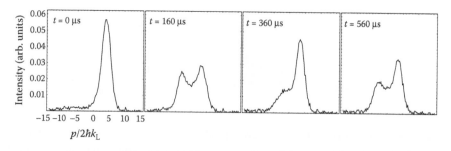

FIGURE 3.5 Detailed view of the first four highlighted distributions in Figure 3.4, where it is clear that a significant fraction of the atoms tunnel to the other island.

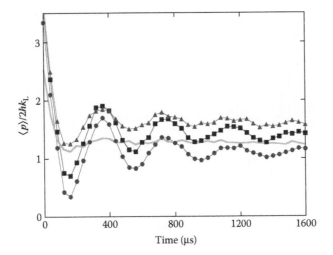

FIGURE 3.6 Comparison of tunneling oscillations for different Raman π-pulse durations, and thus selected velocity widths ($\alpha = 10.5$, $k = 2.08$). The longest (800 μs) Raman pulse has the narrowest (HWHM of $0.03 \times 2\hbar k_L$) width of momentum slices, and best tunneling (circles). Other data, in order of decreasing Raman pulse duration: 400 μs, $0.06 \times 2\hbar k_L$ (squares); 200 μs, $0.12 \times 2\hbar k_L$ (triangles). Also shown is the case of no Raman velocity selection (heavy solid line).

only the proper states. The subrecoil velocity selection, coupled with the rest of the state-preparation sequence, fulfills the simultaneous goals of producing a wave packet localized on an island of stability and populating only states with momentum nearly an integer multiple of k.

The importance of the subrecoil momentum selection is demonstrated in Figure 3.6, where the evolution of $\langle p \rangle$ corresponding to the data in Figure 3.4 (with an 800 μs Raman selection pulse) is shown, along with data for 400 and 200 μs Raman pulses. Shorter Raman pulses result in wider velocity slices, so that fewer atoms fulfill the symmetry condition, and thus the tunneling oscillations are suppressed as the Raman pulse duration decreases. Also shown is the case where the experiment was performed without any Raman velocity selection, and the state-preparation sequence in the 1D lattice was performed immediately after cooling in the 3D lattice. The momentum distribution after

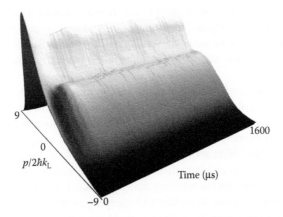

FIGURE 3.7 Evolution of the momentum distribution as in Figure 3.4, but without Raman velocity selection. The tunneling oscillations are suppressed here.

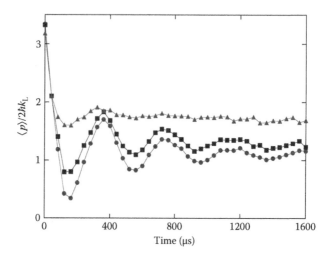

FIGURE 3.8 Tunneling oscillations for different Raman detunings ($\alpha = 10.5$, $\bar{k} = 2.08$). The strongest oscillations observed (circles) correspond to Raman velocity selection at $p = 0$. The other two cases are $p = 0.05 \times 2\hbar k_{\mathrm{L}}$ (squares) and $p = 0.12 \times 2\hbar k_{\mathrm{L}}$ (triangles).

the atoms were released from the 3D lattice was not subrecoil, so the prepared wave packet was no longer minimum uncertainty (the wave-packet area in phase space was about three times the size of a minimum-uncertainty state). More importantly, though, there was no subrecoil structure in this last case, so that the tunneling oscillations are completely absent in the figure. The evolution of the momentum distribution in this case is shown in Figure 3.7. There is perhaps half of a tunneling oscillation at the beginning of the evolution, but the oscillations are again clearly suppressed. Most of the atoms have suppressed tunneling amplitudes, because they are too far away from the proper tunneling momenta. Also, because there is a continuum of states populated near the symmetric ladder, the different momentum classes tunnel at slightly different rates. This situation provides another mechanism for damping of the coherent oscillations, similar to broadened excitation of a two-level atom.

We also studied this broken symmetry more directly by fixing the velocity-selection width at the minimum value and varying the locations of the velocity slices within the Gaussian profile. This was accomplished by slightly varying the detuning of the Raman pulse before loading the atoms into the standing wave. The experimental results are shown in Figure 3.8, where the data with the optimum Raman detuning are compared to data with two other Raman detunings. As the detuning increases, the tunneling oscillations are again suppressed, being almost fully destroyed for an offset corresponding to $0.12 \times 2\hbar k_{\mathrm{L}}$ in momentum. The tunneling is evidently quite sensitive to this broken symmetry.

Figure 3.9 shows simulations of the tunneling oscillations that model the Raman tag widths in Figure 3.6 as well as oscillations in the limit of arbitrarily narrow velocity selection (i.e., the rotor case). The simulation assumes an overall profile of a minimum-uncertainty wave packet with the same center and momentum width as in the experiment, along with ideal Raman π-pulse momentum-selection profiles. With no width, there are no signs of damping, and the tunneling is nearly complete. With wider momentum slices, a smaller fraction of the atoms successfully tunnels, and the tunneling oscillations become increasingly damped. The Raman tagging thus explains a substantial part of the incomplete tunneling and damping in the experiment. In principle, then, a Raman tagging pulse even longer than 800 μs could have provided more complete tunneling, although such a pulse was impractical, as the atoms would have fallen too far with respect to the beams over the course of the experimental sequence.

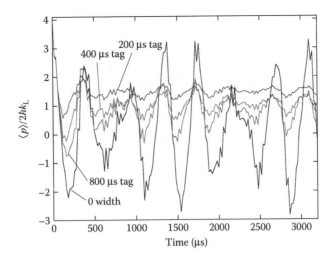

FIGURE 3.9 Effects of the Raman tag width on the tunneling signal, for $\alpha = 10.5$ and $\bar{k} = 2.077$ (simulation). The average momentum $\langle p \rangle$ is plotted at each modulation period for a single, minimum-uncertainty wave packet with an overall Gaussian envelope (out of which the Raman-selected slices are taken) centered at $(x_0, p_0) = (0, 4.1 \times 2\hbar k_L)$, with $\sigma_p = 1.1 \times 2\hbar k_L$, to model the experimental conditions in Figure 3.6.

3.4.2   TUNNELING DEPENDENCE ON WAVE-PACKET LOCATION

To argue that the observed tunneling was indeed between islands of stability, it was important to demonstrate that the tunneling is sensitive to the location of the wave packet in phase space. Just after the state-preparation sequence for the above experiments, the wave packet was moving. Thus, it was possible to displace the initial wave packet in the x-direction in phase space simply by inserting a time delay between the usual state-preparation procedure and the amplitude-modulated lattice phase of the experiment. Doing so produced a shift of the wave-packet center, where the distance was proportional to the time delay, along with a shear of the profile of the wave packet due to dispersion effects. Figure 3.10 shows the usual zero-delay case compared to data with three different time delays, corresponding to displacements of $1/4$, $1/2$, and one full period of the lattice potential. Schematic plots of the initial conditions in the classical phase space are shown in Figure 3.11 for these four cases. The tunneling oscillations are strongest for zero time delay, when the wave packet was centered on the island. For the $1/4$-period displacement, the wave packet was centered in the chaotic region next to the island, and the tunneling oscillations are significantly suppressed. For the $1/2$-period displacement, the wave packet was centered in the outer stability region, and the tunneling oscillations are almost completely suppressed. For the longest time delay, the wave packet was displaced by a full period of the potential and thus is again centered on the island. The tunneling oscillations return in this case, but with smaller amplitude due to the stretched character of the wave packet after the dispersive free evolution. Hence, it is clear that the islands of stability were important in supporting the tunneling in this experiment.

We have also displaced the center of the wave packet in the p-direction in phase space by changing the amplitude of the lattice phase shift during the state preparation (and adjusting the subsequent evolution period in the lattice accordingly). For the experimental parameters here, we varied the wave packet center in steps of $0.5 \times 2\hbar k_L$, and we observed strong tunneling when the wave packet was centered at $p/2\hbar k_L = 3$, 3.5, and 4, while tunneling was suppressed at the other values outside this range.

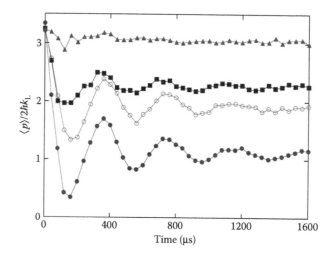

FIGURE 3.10 Comparison of tunneling dynamics ($\alpha = 10.5$, $k = 2.08$) for different horizontal displacements of the initial condition, as illustrated in Figure 3.11. The strongest oscillations occur for zero drift time (filled circles), where the initial wave packet is centered on the island of stability as in Figure 3.3. The oscillations are progressively suppressed for displacements of $\pi/2$ (3.8 μs delay, squares) and π (7.6 μs delay, triangles), but recur somewhat for a 2π displacement (15.1 μs delay, open circles).

3.5 CHAOS-ASSISTED TUNNELING

In considering the dynamical tunneling phenomenon in the experiment, we have thus far focused only on the role of the islands of stability in the tunneling. However, as we will now discuss, the chaotic region surrounding the islands is important in enhancing the tunneling process, and we will argue that the tunneling in the experiment is due to *chaos-assisted tunneling*.

The possibility of tunneling enhancement by classical chaos was first introduced in a numerical study by Lin and Ballentine [36], where it was found that the tunneling rate between islands of stability in the periodically driven, double-well potential could be orders of magnitude larger than the tunneling rate in the undriven (integrable) double well. While the presence of two islands related by a discrete symmetry is important in supporting the tunneling in this system [44], the authors attributed the increased tunneling rate to the presence of the chaotic region in the classical phase space. It was subsequently shown that the tunneling rate is correlated with the degree of overlap of the tunneling states with the chaotic region [55], which also points to the role of the chaotic sea as a catalyst for the tunneling. This enhancement of the tunneling was understood in works of Bohigas et al. [6] and Tomsovic and Ullmo [53] (where the term "chaos-assisted tunneling" was introduced) in terms of an avoided crossing of the tunneling doublet with a third level associated with the chaotic region, which can greatly increase the tunnel splitting. Because the (quasi)energies of the chaotic states exhibit strong and irregular dependence of the system parameters, the tunneling rate also exhibits irregular fluctuations over orders of magnitude [40,45,58], sometimes reaching zero for exact crossings of the tunneling doublet (the "coherent destruction of tunneling" [24]). The smooth, universal dependence of the tunneling rate on $\hbar$, as mentioned above for the double-well tunneling, is therefore lost for chaos-assisted tunneling. In addition to this three-state picture, chaos-assisted tunneling has also been understood in terms of the dominance of indirect paths, which are multistep paths that traverse the chaotic region, over direct paths, which tunnel in a single step and are responsible for regular (two-state) tunneling [19]. Thus, chaos-assisted tunneling occurs as small portions of the population from the initial wave packet break off, transport through the chaotic region, and then accumulate

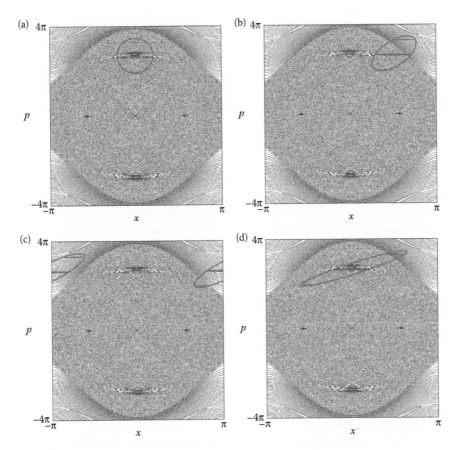

FIGURE 3.11 (See color insert.) Initial conditions in phase space for the four time delays (a–d) used in obtaining the data of Figure 3.10. The large ellipse around the three narrow population slices in each case marks the overall profile of the wave packet to guide the eye. Color version available online.

in the symmetric region, without a large population building up in the chaotic region [52,53]. By contrast, direct tunneling occurs with an always negligible population in the intermediate region.

The sense in which we mean "chaos-assisted tunneling" here is the influence of the chaotic region on tunneling transport between symmetry-related regions in phase space, but this term has also been applied in the sense of open systems, where the tunneling implies an escape from a bound state. In this vein, chaos-assisted tunneling has been invoked to explain fluctuations in the energy and rate of ionization of Rydberg atoms in microwave fields [57], and also to explain mode lifetimes in weakly deformed optical microresonators [43].

Previous experimental work in chaos-assisted tunneling has been performed in the spectroscopy of a microwave resonator in the shape of an annular billiard [15]. The authors measured the dependence of the quasidoublet splittings on the locations of the states in phase space and on the eccentricity of the cavity, demonstrating an enhancement in the splitting for states near the border between the stable and chaotic regions. Chaos-assisted tunneling has also been invoked to explain features in the decay of superdeformed nuclear states to normal deformed states [1], although the interpretation here is not entirely straightforward [52]. It is also worth noting that another atom-optics experiment studies tunneling of atoms in an optical lattice of double wells [27], where the classical description is chaotic as a result of the coupling of the center-of-mass motion to the spin state of the atom [21]. Recently, dynamical tunneling between stable islands has been observed in this system, as well as erratic behavior in the chaotic region [12]. Other experiments [20,56] consider the transport in the resonant tunneling diode, where a strong magnetic field induces chaos in the classical limit. However, the tunneling here is enhanced by energy resonances of states on either side of a barrier (corresponding to periodic orbits in the chaotic quantum-well region), and thus the tunneling is not enhanced by the chaos in the sense we mean here.

3.5.1 SINGLET–DOUBLET CROSSINGS

We will now review the simplified three-state model introduced in Bohigas et al. (1993) [6] and Tomsovic and Ullmo (1994) [53] because of its importance in understanding chaos-assisted tunneling and its utility in interpreting the experimental data. Because we are considering a periodically driven system, though, we will consider a Floquet–Hamiltonian model as in Kohler et al. (1998) [32]. Floquet states are eigenstates of the unitary evolution operator $U(t + 1, t)$ over one period of the modulation, with eigenvalue $\exp(-i\varepsilon_n/k)$, where ε_n is the *quasienergy*. The eigenstates can also be written as

$$|\psi_n(t)\rangle = e^{-i\varepsilon_n t/k}|\chi_n(t)\rangle, \tag{3.16}$$

where the state $|\chi_n(t)\rangle$ is periodic in time with the same period as the modulation. Thus, the quasienergies represent the phase evolution of the Floquet states (in a stroboscopic sense), just as the energies govern the phase evolution of the energy eigenstates for autonomous systems. The periodic states $|\chi_n(t)\rangle$ are also eigenstates of the Floquet Hamiltonian [40],

$$\mathcal{H} := H - i\hbar\partial_t, \tag{3.17}$$

with eigenvalue ε_n. We will therefore construct a Floquet Hamiltonian model that captures the essence of chaos-assisted tunneling.

Consider a doublet of tunneling states, localized on the two islands of stability (regular regions), with quasienergies ε_r and $\varepsilon_r + \delta_r$, so that δ_r parameterizes the tunneling rate in the absence of interaction with other levels. These states have opposite parity, and for the sake of concreteness, we can take the state with quasienergy ε_r to be of even parity. We also consider a third state in the chaotic region (although we note that three-level crossings can also be induced by states in other stable regions [7,8,19]), with quasienergy $\varepsilon_r + \Delta_c$. We may assume that this state has even parity. Also, note that states in the chaotic region do not generally occur in narrowly spaced doublets, so we can ignore the corresponding odd-parity state. The chaotic state does not interact with the odd member of the tunneling doublet, but we assume that there is some nonzero interaction between the two even states. We may then write the model Floquet Hamiltonian as [6,32,53]

$$\mathcal{H} = \begin{pmatrix} \varepsilon_r + \delta_r & 0 & 0 \\ 0 & \varepsilon_r & \beta/2 \\ 0 & \beta/2 & \varepsilon_r + \Delta_c \end{pmatrix}, \tag{3.18}$$

where β represents the coupling between the chaotic state and the even regular state. Thus, the two coupled states undergo an avoided crossing, with quasienergy solutions of the same form as in the two-state case. The regular (odd) state remains unchanged. This behavior is illustrated in Figure 3.12. In the case where the coupling energy $\beta/2$ is large compared to the two-level splitting δ_r (which is the case when the regular states have substantial overlap with the surrounding chaotic region), the tunnel splitting can be greatly enhanced, becoming of the order of $\beta/2$ between the odd state and either of the even states near the center of the crossing. As one might expect in an avoided crossing, the even regular state and the chaotic state exchange their character as Δ_c is swept through zero, as verified numerically in Latka et al. (1994) [35]. Thus, near the center of the crossing, the two even states each have population both in the islands and in the chaotic region, whereas away from

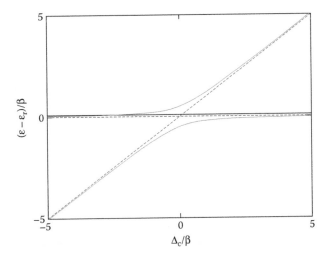

FIGURE 3.12 Illustration of a three-level avoided crossing of a tunneling doublet with a third (chaotic) state, as described by model (Equation 3.18). The behavior of the three quasienergies is shown as a function of the detuning Δ_c of the chaotic state, for an unperturbed doublet splitting $\delta_r/\beta = 0.1$. The chaotic state interacts with the regular state of the same parity (both appearing as solid, curved lines), and the other tunneling state (straight, solid line) is unaffected by the crossing in this simple model. The dashed lines show the two repelling states in the absence of any coupling.

the crossing it is possible to clearly distinguish a predominantly regular and predominantly chaotic even state. In a singlet–doublet crossing, one expects a complicated time dependence, compared to the sinusoidal two-state tunneling, because three states will be excited by a wave packet localized on a single island. In general, the three splittings will all be different, leading to complicated beating in the time domain [32].

3.5.2 COMPARISON WITH INTEGRABLE TUNNELING

The dynamical tunneling that we have studied is between two oppositely directed modes of motion. In unmodulated optical lattices, however, Bragg scattering is a well-known dynamical tunneling mechanism, producing oscillations between symmetric pairs of plane-wave states with momenta $\pm n\hbar k_L$, where n is the *order* of the scattering process. Bragg scattering produces similar results to the tunneling that we have described, including sensitivity to the same broken symmetry that we discussed above, even though there is no classical chaos without a modulation of the lattice. It was therefore important to demonstrate that the tunneling here is not simply Bragg scattering, but that the amplitude modulation has a substantial effect on the tunneling dynamics. We have done this already to a certain extent by demonstrating that the initial state must be centered on the island of stability for tunneling to occur (since Bragg scattering occurs between states not localized in position, it should not be sensitive to spatial displacements of the initial condition). However, a direct comparison between tunneling in chaotic and integrable systems is also illuminating.

A sensible integrable counterpart of the modulated system arises by using the optical lattice with constant amplitude, where the potential depth is taken to be V_0. Doing so produces a pendulum, such that the lattice intensity is the same, on average, as in the amplitude-modulated system. The phase space for the pendulum corresponding to the experimental conditions in Figure 3.3 is shown in Figure 3.13, along with the same initial condition as before. The wave packet is centered outside the separatrix, so that classical transport to the opposite momentum region is also forbidden here. However, high-order Bragg scattering, which is a manifestation of quantum above-barrier reflection

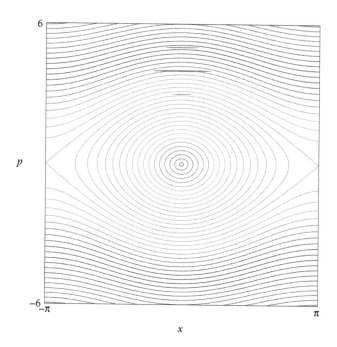

FIGURE 3.13 Phase space of the pendulum, with the same average potential amplitude as the modulated-pendulum case in Figure 3.3. The same initial condition is also shown here. The initial state is centered outside the separatrix, so that classical transport to the opposite (symmetric) momentum region is also forbidden here.

[28], allows quantum oscillatory transport between these momentum regions. in our scaled units, the Bragg oscillation frequency is

$$\Omega'_{\text{B},n} = \frac{\alpha^n}{\bar{k}^{2n-1}[(n-1)!]^2}, \tag{3.19}$$

when adapted to the scaled units of the amplitude-modulated pendulum. In this form, it is not obvious that Bragg scattering has the expected universal dependence $\exp(-S/\bar{k})$ for two-state tunneling that we mentioned above. Since tunneling occurs from initial momentum $(n/2)\bar{k}$ to $-(n/2)\bar{k}$ as an nth-order scattering process, the order n is effectively a function of $\bar{k}$. Then, in the semiclassical limit of large n, the Bragg rate becomes

$$\Omega'_{\text{B},n} \simeq \frac{1}{2\pi}\left[\frac{(2p)^2}{\alpha e^2}\right]^{-p/\bar{k}}, \tag{3.20}$$

which is consistent with the expected scaling with $\bar{k}$. Note that the factor in the square braces is greater than unity, since to be in the Bragg regime (where population in the intermediate states can be adiabatically eliminated) the wave packet must be outside the pendulum separatrix, so $|p| > 2\sqrt{\alpha}$.

The tunneling oscillations of Figure 3.4 are compared with the behavior of the corresponding pendulum in Figure 3.14. No tunneling oscillations are visible in the integrable case over the timescale studied in the experiment. Since the initial distribution is peaked near $4 \times 2\hbar k_{\text{L}}$, the dominant transport process in the pendulum is eighth-order Bragg scattering. For $n = 8$, $\alpha = 10.5$, and $\bar{k} = 2.08$, the Bragg period is about 1 s, which is much longer than the $400\,\mu$s period of the tunneling oscillations in the chaotic case (and thus the experimental Bragg measurement is in accord with our expectations).

We have also demonstrated tunneling in a parameter regime that is closer to the classical limit ($\bar{k} = 1.04$), as shown in Figure 3.15. The initial distribution here is peaked around $8 \times 2\hbar k_{\text{L}}$, and

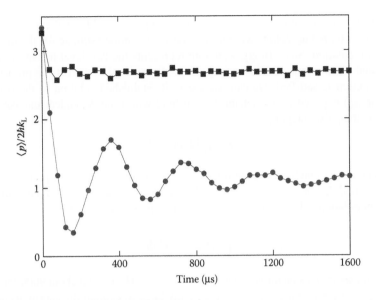

FIGURE 3.14 Comparison of chaos-assisted tunneling oscillations ($\alpha = 10.5$, $k = 2.08$, circles) to transport in the corresponding quantum pendulum (squares).

so this coherent, 32-photon process is similar to 16th-order Bragg scattering. The expected Bragg period here is 20 years, which is long compared to the 250 μs period of the tunneling in the chaotic case, and is even long compared to the coherence time of the authors. Thus, it is clear that in some sense the chaos enhances the transport, in that the tunnel splittings are much larger in the chaotic case than in the corresponding integrable case.

Of course, it could be the case that the amplitude modulation enhances the two-level tunneling rate without the influence of a third, chaotic state, especially in view of the rapid dependence of

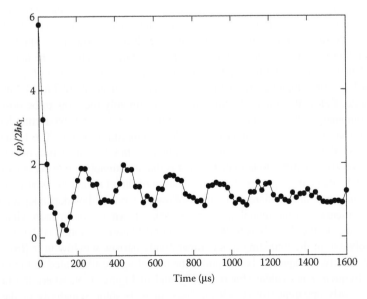

FIGURE 3.15 Tunneling oscillations for $\alpha = 11.2$, $k = 1.04$ (10 μs modulation period). The corresponding two-state (integrable) tunneling mechanism has an expected period of 20 years.

the Bragg splitting on the lattice intensity. Although we provide additional experimental evidence for chaos-assisted tunneling below, we will now derive a simple estimate for the direct tunneling rate with the modulation. Since Bragg scattering represents the two-level transport mechanism in this system, and corresponds to a resonantly coupled two-level system if the proper momentum symmetry condition is satisfied, we can use the well-established solution to the two-level atom (without damping) exposed to a resonant driving field with time-dependent intensity [2]. In this case, we define the pulse integral,

$$\phi = \int_0^t \Omega_{\mathrm{B},n}(t')\,\mathrm{d}t', \tag{3.21}$$

in terms of which the tunneled population is $\sin^2(\phi/2)$. Since the tunneling period is substantially longer than the modulation period, we can simply average the Bragg rate over a modulation period, and thus the modulation enhances the two-level tunneling rate by a factor

$$\int_0^1 [2\cos^2(\pi t)]^n\,\mathrm{d}t \tag{3.22}$$

for nth-order transport. This factor is about 50 for the $k = 2.08$ case and about 9000 for the $k = 1.04$ case. Neither of these numerical values is sufficiently large to explain the enormous differences in the tunneling rates in the integrable and chaotic cases.

3.5.3   Tunneling Dependence on Parameter Variations

To establish that the tunneling in the modulated lattice is chaos-assisted, it is also impor-tant to examine the dependence of the tunneling as the two experimental parameters (α and k) are varied. As we noted above, the dependence of the tunneling rate should be very different for direct and chaos-assisted tunneling. In this section, we examine the variation of the tunneling as a function of α for two different values of k. Operationally, α is a much more convenient parameter to vary, because it only requires a change in laser intensity, whereas k is more difficult because it requires changing both the laser intensity and the modulation period (to keep α fixed) as well as a new set of parameters for the state-preparation procedure to maintain the initial condition at the same phase-space location. While we do not necessarily expect to see rapid variations in the tunneling rate as we vary α, due to inhomogeneous broadening (different atoms see different optical intensities, depend-ing on their transverse location in the optical lattice, leading to about a 5% spread in α over the atomic sample), there are nevertheless signatures of three-state tunneling in the data.

The dependence of the tunneling oscillations in the measured evolution of $\langle p(t) \rangle$ is shown in Figure 3.16 for $k = 2.08$. Tunneling is visible in the range of α from about 7 to 14, but is suppressed outside this range. Below this range the tunneling is presumably too slow to be observed (see the Floquet-spectrum analysis below), and above this range the outer islands have completely dissolved into the chaotic sea, where we no longer expect clean tunneling to occur. The tunneling rates for these data are plotted in Figure 3.17. The tunneling rate does not fluctuate strongly as α changes, but there are two interesting features to notice. The first is that the tunneling rate *decreases* as a function of α. This dependence is the opposite of our expectation of direct tunneling, where as we have seen above the tunneling rate should increase with α, following a power-law dependence. This behavior is thus strong evidence that the tunneling is chaos-assisted, where one or more chaotic levels has a definite influence on the doublet splitting. The second feature to notice is that two frequencies are clearly resolvable in the tunneling in a comparatively narrow window in α (from about 8.5 to 10.5). The one- and two-frequency behaviors of the tunneling are illustrated in Figure 3.18, where one tunneling frequency is evident (for $\alpha = 8.0$), and in Figure 3.19, where the beating of two frequencies is clearly apparent (for $\alpha = 9.7$). Thus, there is some sensitivity of the tunneling to variations in α in this regime. This behavior is also consistent with the three-state model near the center of a singlet–doublet crossing. In this model, the initial wave packet populates a regular state

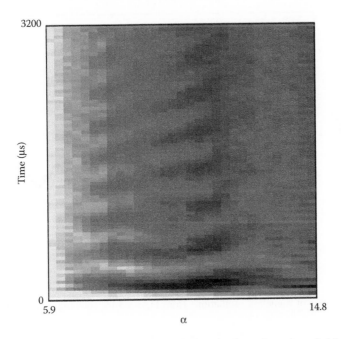

FIGURE 3.16 Dependence of the tunneling as the optical lattice intensity α is varied for $k = 2.08$ (20 μs modulation period). The shade indicates the value of $\langle p \rangle$, with dark representing the most negative values and light the most positive.

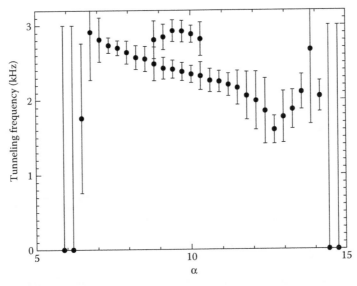

FIGURE 3.17 Dependence of the tunneling rate on the well depth α, for $k = 2.08$ (20 μs modulation period), as extracted from the data in Figure 3.16. In the range of α from 8.9 to 10.3, two distinct frequencies can be resolved in the tunneling data. The zero-frequency data points at the edges of the plot indicate that no tunneling frequency could be extracted from the data at these locations.

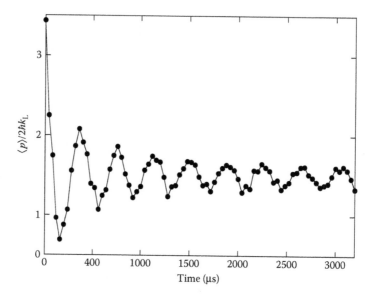

FIGURE 3.18 Example of tunneling oscillations from Figure 3.16, showing a single tunneling frequency ($\alpha = 8.0$, $k = 2.08$).

(localized on the islands) and two hybrid states, which have population in both the islands and in the chaotic sea. There should thus be two frequencies associated with the tunneling, corresponding to the two splittings between the regular state and the two hybrid states. In general, these two splittings will not be equal, but should be similar near the center of the avoided crossing, leading to two-frequency beating in the tunneling dynamics.

The variation of the tunneling behavior in the $k = 1.04$ case is plotted in Figure 3.20, with the extracted tunneling rates plotted in Figure 3.21. The observed tunneling rates appear to have

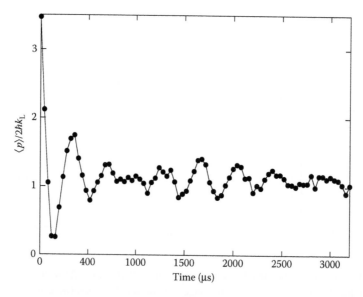

FIGURE 3.19 Example of tunneling oscillations from Figure 3.16, where two tunneling frequencies are clearly present ($\alpha = 9.7$, $k = 2.08$).

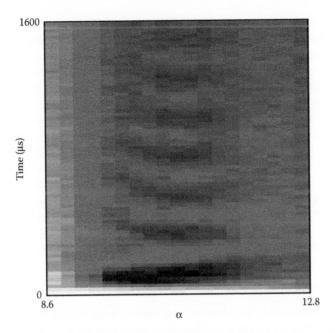

FIGURE 3.20 Dependence of the tunneling as the optical lattice intensity α is varied, as in Figure 3.16, but for $\bar{k} = 1.04$ (10 μs modulation period). The tunneling here occurs only in a narrow range of α.

weaker dependence on α than in the $\bar{k} = 2.08$ case. However, the tunneling is only visible in a much narrower interval in α, from about 9.5 to 12.5. Thus, in a sense, the tunneling here is more sensitive to variation in α than in the $\bar{k} = 2.08$ case.

One question that remains is why the tunneling rate does not go smoothly to zero at the edges of the α intervals where tunneling is observed, especially at the lower end of the interval where the

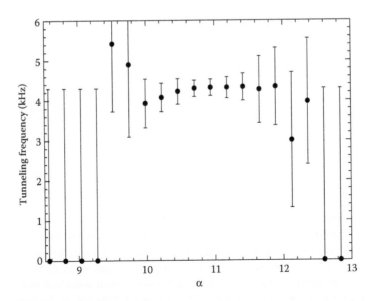

FIGURE 3.21 Dependence of the tunneling rate on the well depth α, for $\bar{k} = 1.04$ (10 μs modulation period), as extracted from the data in Figure 3.20.

tunnel splitting is expected to become very small. In the experiment, the disappearance of the tunneling as α is swept comes about as the oscillations decrease in amplitude and become damped more quickly, until the oscillations are no longer discernible. One possible explanation is the change in the location of the two islands, which move to larger momentum as α increases (see the next section for an empirical expression for the island locations). In the experiment, the initial condition was held fixed as α is swept, so that there may have been less overlap with the tunneling Floquet states if the islands moved too far. However, over the intervals where tunneling was observed, the islands moved only by $\pm 0.3 \times 2\hbar k_L$ for both the $\bar{k} = 2.08$ and the $\bar{k} = 1.04$ data sets, which is a substantially smaller amount than the respective $\sigma_p = 1.7 \times 2\hbar k_L$ and $\sigma_p = 2.1 \times 2\hbar k_L$ momentum uncertainties of the initial conditions in the two cases. Thus, misalignment of the initial conditions does not account for the disappearance of the tunneling at the extreme α values here. Another possible explanation lies in the suggestion [34] that three-level tunneling is more robust to a symmetry-breaking interaction than two-level tunneling. Since the range of populated quasimomenta (and thus the degree of broken symmetry) is fixed by the Raman velocity selection, the tunneling away from the avoided crossings may simply disappear, as opposed to being manifested as a slow tunneling oscillation.

3.5.4 FLOQUET SPECTRA

In the context of understanding the observed tunneling dependence on α, it is useful to consider the quasienergy spectrum for this system. Computed spectra for the $\bar{k} = 2.077$ and $\bar{k} = 1.039$ cases are plotted in Figures 3.22 and 3.23, respectively. These spectra only show the states with definite parity, falling on the symmetric ladder of momentum states $p = n\bar{k}$ (for integer n), corresponding to

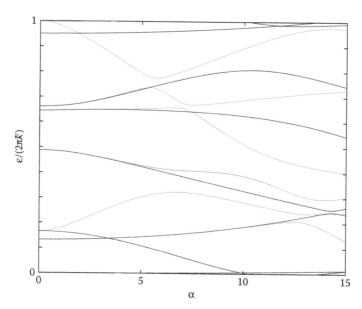

FIGURE 3.22 **(See color insert.)** Calculated quasienergy spectrum for $\bar{k} = 2.077$, corresponding to a 20 μs modulation period. Quasienergies that correspond to states with large momentum (that do not interact with the states shown in this range of α) are suppressed, and the quasienergies shown are for the symmetric momentum ladder (zero quasimomentum). The quasienergies for even-parity Floquet states are shown in green, while the odd-parity states are shown in blue. The even (orange) and odd (red) states with maximal overlap with the outer stable islands are shown, up to the point where the islands bifurcate, as described in the text. The avoided-crossing behavior of the tunneling states is apparent over a broad range of α, where two chaotic states have a clear influence on the tunneling-doublet splitting. Color version available online.

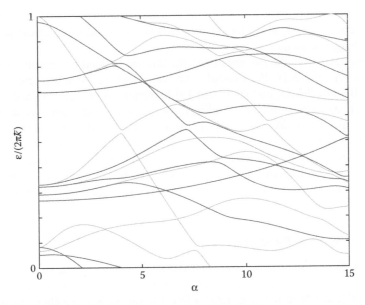

FIGURE 3.23 **(See color insert.)** Calculated quasienergy spectrum for $k = 1.039$, corresponding to a 10 μs modulation period. Quasienergies that correspond to states with large momentum (that do not interact with the states shown in this range of α) are suppressed, and the quasienergies shown are for the symmetric momentum ladder (zero quasimomentum). The quasienergies for even-parity Floquet states are shown in green, while the odd-parity states are shown in blue. The even (orange) and odd (red) states with maximal overlap with the outer stable islands are shown, up to the point where the islands bifurcate, as described in the text. Several avoided crossings of the tunneling doublet with chaotic states are apparent, although the splitting only becomes very large around $\alpha = 10$. Color version available online.

zero quasimomentum. The quasienergies were calculated by numerically constructing the unitary evolution operator for one period of the modulation and then diagonalizing the resulting operator. The even and odd tunneling states are also highlighted in these spectra. These states were identified by finding the states with maximum overlap with a minimum-uncertainty Gaussian wave packet that was centered on the fixed point of the island and had the same aspect ratio as the elliptical trajectories near the fixed point (i.e., where the linearized equations of motion are valid). The tunneling states are not identified for α above the point where the outer islands become unstable, as it is difficult to assign states to the island remnants.

In the spectrum for $k = 2.077$, the first avoided crossing (with an even-parity state of smaller quasienergy) does not occur until about $\alpha = 7$, where the splitting also first becomes significant. This behavior is consistent with the experimental data in Figure 3.16, where tunneling oscillations are also first observed around $\alpha = 7$. Beyond this point, the two even-parity states maintain a similar distance from the odd tunneling state, and this holds true in the regime where two tunneling frequencies are visible in the data. These two even states then move back toward each other (and the odd tunneling state) as they interact with two other even states, and this behavior may explain the decreasing tunneling rate as a function of α, although it is again difficult to pinpoint the tunneling states in this regime of large α.

In the $k = 1.039$ spectrum, the singlet–doublet crossings are more readily apparent. There are several clear avoided crossings involving the tunneling doublet in the range shown, but it is not until the final avoided crossing before the islands become unstable that the splitting becomes large enough to observe experimentally. The experimental observation of tunneling beginning with $\alpha = 9.5$ is thus consistent with the spectrum, although another significant avoided crossing in the spectrum suggests that tunneling might also be visible in a very narrow region around $\alpha = 8$. The experimental tunneling stops around $\alpha = 12.5$, where the spectrum has become quite complicated and the tunneling doublet can no longer be identified.

The tunneling rates from the calculated spectra here are in good agreement with the observed rates in Figures 3.17 and 3.21. For example, the two calculated tunneling rates for $\alpha = 10$ and $\bar{k} = 2.08$ are 3.0 and 2.3 kHz, and the calculated tunneling rate for $\alpha = 11$ and $\bar{k} = 1.04$ is 4.0 kHz, all of which match the observed tunneling rates reasonably well. However, it should be noted that while these spectra provide a useful basis for understanding the data, an interpretation based solely on these spectra would most likely be too simplistic to be very useful. An accurate model would at minimum need to take into account the excitation of multiple Floquet states by the initial condition, the range of quasimomenta populated after the Raman velocity selection (as we have done in Figure 3.9), and the averaging over a range of α due to the transverse profile of the optical lattice beam.

3.5.5 FLOQUET STATES IN PHASE SPACE

It is also illuminating to visualize the relevant quasienergy states in phase space. (For an expanded treatment, see Steck (2002) [48]; for other visualizations see Averbukh et al. (2002) [4], Luter and Reichl (2002) [37], and Mouchet et al. (2001) [40].) In this system we are restricted to doing so via calculations, as the experiments only permit easy access to the momentum distributions, not the atomic positions; however, very recently the reconstruction of dynamically tunneling quantum states in a classically chaotic phase space has been demonstrated [12].

Here we will consider as an example the case of $\bar{k} = 2.077$, for which we already discussed the Floquet spectrum in Figure 3.22. In this spectrum we can clearly see a doublet of states localized on the outer two islands of stability, where an avoided crossing with a third state with slightly lower quasienergy occurs around $\alpha = 7$. Already we can anticipate that this third state is associated with the chaotic region, because it has diverged from its partner of opposite parity already at around $\alpha = 1$ (regular states occur in quasidegenerate doublets).

We can visualize the states by plotting Husimi distributions of these three states. In assessing the relative importance of the states in the tunneling dynamics—beyond the avoided-crossing structure in the spectrum—it is interesting to know how each overlaps with the island of stability. A natural choice in the context of the experiment is to use the overlap of each state with the near-minimum-uncertainty initial condition from the experiment. However, generally speaking, it is most useful to compute overlaps in a way that is independent of any particular realization. However, a natural "island-adapted" state is a Gaussian state, centered on the fixed point of an outer island of stability, with the same aspect ratio as the KAM tori in the immediate neighborhood of the fixed point. Thus, we will consider the overlaps of the quasienergy states with these "natural Gaussians."

Another issue in setting up the visualization with Husimi distributions comes from their very definition. Since the Husimi distribution is a convolution of the Wigner distribution with a coherent state, the calculation of a Husimi distribution involves the rather arbitrary choice of the aspect ratio of the coherent state. The system here gives no preferred aspect ratio, except perhaps for the natural Gaussian mentioned above or the ground state of the unmodulated lattice; both of these choices lead to highly stretched states in phase space. Here we make the somewhat arbitrary choice of a convolution kernel with uncertainties $\Delta p = \sqrt{\pi \bar{k}}$ and $\Delta x = \sqrt{\bar{k}/4\pi}$, which give good visual representations of the states in phase space.

A typical set of tunneling states, in this example for $\alpha = 9.7, \bar{k} = 2.077$ is shown in Figure 3.24 with the corresponding classical phase space. The odd-parity state ($\varepsilon/2\pi\bar{k} = 0.33$, 37% overlap, Figure 3.24 upper right) has the biggest calculated overlap with the outer islands of stability, and visually reflects the two-island structure quite directly. The two even-parity states involved in the avoided crossing are somewhat more subtle. The state with the greater overlap with the islands ($\varepsilon/2\pi\bar{k} = 0.39$, 22% overlap, Figure 3.24 lower left) is spread throughout the chaotic region, and

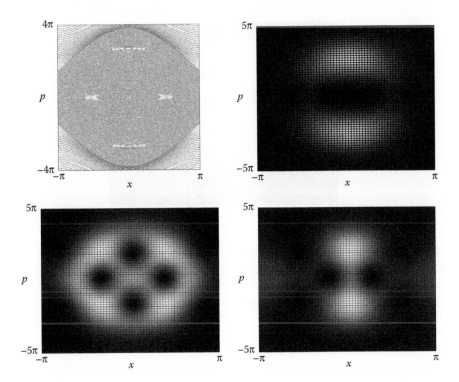

FIGURE 3.24 Classical phase space (upper left), and three Husimi distributions for $\alpha = 9.7$, $k = 2.077$. The quasienergies and populations for the three states are $\varepsilon/2\pi k = 0.33\%$ and 37% (upper right); $\varepsilon/2\pi k = 0.39\%$ and 22% (lower left); and $\varepsilon/2\pi k = 0.29\%$ and 16% (lower right).

in fact appears to avoid the stable islands. The other even-parity state ($\varepsilon/2\pi k = 0.29$, 16% overlap, Figure 3.24 lower right) again appears to reflect the structure of the islands, but the smaller overlap with the islands is due to the shift of the two lobes of the state toward the origin. The mixed character of the even states—the characteristic feature of the avoided crossing in chaos-assisted tunneling—is most apparent in the third state, where substantial probability in the chaotic region is visible in the plot.

3.5.6 High Temporal Resolution Measurements

All of the data so far in this chapter have been sampled only at a particular phase of the periodic driving, corresponding to integer times in the Hamiltonian (Equation 3.14). We will now study the dynamics on a much finer timescale, which will reveal additional interesting aspects of the tunneling dynamics. Figures 3.25 and 3.26 show the tunneling dynamics for $k = 2.08$ (for two different values of α), and Figure 3.27 shows the tunneling dynamics for $k = 1.04$; in all three figures, the momentum distribution was sampled 10 times per modulation period, and the duration of the measurement covers approximately one full period of the amplitude modulation. Besides the island-tunneling process, which is visible as the slowest oscillation, there are two other oscillatory motions that are common to the three plots. The more obvious of these features appears as a fast oscillation of the initial peak, with the same period as the modulation of the potential. As the atoms tunnel to the other island, the tunneled peak oscillates in a complementary fashion. This motion can be understood in terms of the classical phase-space dynamics. A particular phase space for this system is "stroboscopic" in that it assumes a particular sampling phase for the dynamics. To understand the present phenomenon, though, it is necessary to examine the phase space as the sampling phase

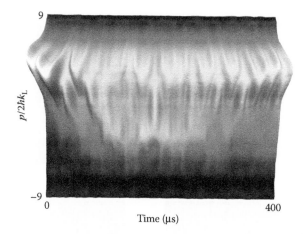

FIGURE 3.25 Experimental momentum-distribution evolution of chaos-assisted tunneling for $k = 2.08$ ($T = 20\ \mu s$) and $\alpha = 7.7$. The distribution was sampled every 2 μs out to 400 μs, covering the first full tunneling oscillation. The classical oscillations (with the same period as the modulation period) are evident, as well as more complicated oscillations into the intermediate chaotic/stable region near $p = 0$. The phase space is characterized by the two (symmetry-related) tunneling islands as well as a doublet of stable islands near $p = 0$. Color version available online.

varies, as illustrated in Figure 3.28. Because of the periodic time dependence of the potential, the time parameter acts as a third dimension in phase space. Thus the islands of stability are "flux tubes" that confine classical trajectories in the higher-dimensional phase space [3], and the islands that appear in the phase plots (Poincaré sections) are cross sections of the flux tubes. As time varies continuously, then, the islands move in opposite directions in phase space according to their mean momenta. Additionally, the islands move in the momentum direction, becoming furthest apart in

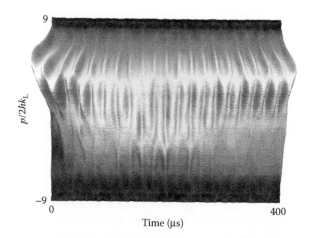

FIGURE 3.26 Experimental momentum-distribution evolution of chaos-assisted tunneling for $k = 2.08$ ($T = 20\ \mu s$) and $\alpha = 11.2$. The distribution was sampled every 2 μs out to 400 μs, covering the first full tunneling oscillation. The conditions are otherwise similar to those in Figure 3.25; the oscillations in the chaotic region occur in different locations, compared to the previous case. The phase space is characterized by the two (symmetry-related) tunneling islands with only small remnants of the island near $p = 0$.

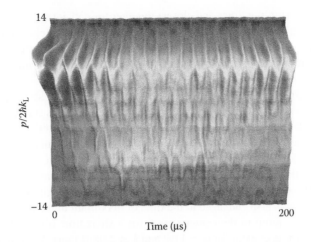

FIGURE 3.27 Experimental momentum-distribution evolution of chaos-assisted tunneling for $\bar{k} = 1.04$ ($T = 10\ \mu s$) and $\alpha = 10.5$. The distribution was sampled every 1 μs out to 200 μs, covering the first full tunneling oscillation. The oscillations in the chaotic region here are more difficult to see than in Figures 3.25 and 3.26, because of the smaller signal-to-noise ratio for these experimental conditions (the horizontal stripes are artifacts of the CCD camera). The phase space is characterized by the two (symmetry-related) tunneling islands with only small remnants of the island near $p = 0$.

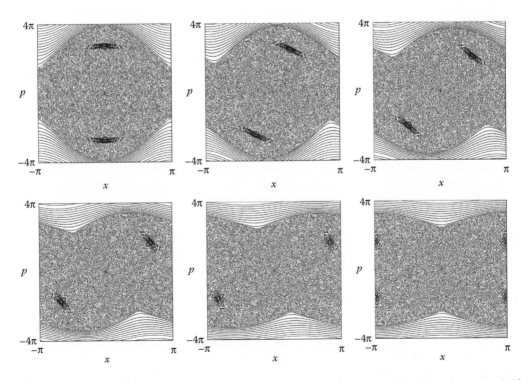

FIGURE 3.28 Classical phase space (for $\alpha = 10.5$), shown with different sampling phases in the first half of a modulation period. At the start of the modulation period, the islands of stability are separated maximally in momentum, but move inward as they drift away from $x = 0$, and return to their initial momenta by the end of the modulation period. The two islands always remain separated in momentum, and do not cross the $p = 0$ axis. The complementary motions of the two islands produce the fast oscillatory motions in Figures 3.25 through 3.27.

momentum for integer sampling times and closest together for half-integer times. This oscillation is only significant for relatively large values of α (away from the near-integrable regime), because of the mutual repulsion of the three primary resonances in phase space. Thus, the fast oscillations of the experimental momentum distributions can be attributed to the motion of the classical phase-space islands.

The second oscillatory feature is the more relevant effect for demonstrating chaos-assisted tunneling. This oscillation is slower than the classical oscillation but also is substantially faster than the tunneling oscillation. It appears as an occasional enhancement of probability in the (predominantly) chaotic region between the two islands. This effect is particularly dramatic in the case of $\alpha = 7.7$ and $k = 2.08$ (Figure 3.25). Here, the first part of the tunneling transport takes place in (at least) two steps through the chaotic sea, with the first chunk of probability crossing during the third period of the potential and the second crossing during the fifth and sixth periods of the potential. The population in the chaotic region is also enhanced at the time of maximum tunneling, where the population in the islands appears to jump in the center region for a short time (during the tenth modulation period). Similar behavior is evident for $\alpha = 11.2$ and $k = 2.08$ (Figure 3.26); in this case, this third oscillation is not as pronounced, but is still present. The details of this oscillation in the chaotic region are also slightly different than in the previous case. This is especially true at the moment of maximum tunneling, where the atoms are mostly in the two islands (unlike the case before, where the atoms were mostly in the chaotic region), but the chaotic region is populated during the modulation periods just before and after this time. In the case of $\alpha = 10.5$ and $k = 1.04$ (Figure 3.27), this oscillation is less visible because of the poorer signal-to-noise ratio (note that the atoms are spread over a much larger region in momentum for this value of k, resulting in an effectively smaller signal). Nonetheless, the tunneling again proceeds in chunks, with the transport visible as faint ridges crossing the chaotic region, especially near the ends of the first, second, fourth, and fifth modulation periods. The tunneling here in some sense resembles a Landau–Zener crossing [59], because the population crosses between the islands at the times of closest approach.

This appearance of probability in the chaotic region during the tunneling is precisely the behavior expected from the picture of chaos-assisted tunneling of Tomsovic (2001) [52] and Tomsovic and Ullmo (1994) [53] that we mentioned above, where tunneling occurs as parts of the wave packet break away from the initially populated island, transport through the chaotic sea, and then reassemble in the symmetric destination island. We also recall from the analysis of the three-level model (Equation 3.18) of chaos-assisted tunneling that near the center of the avoided crossing, the tunneling rate is given by the splitting(s) between the odd-parity regular state taken pairwise with each of the two even-parity (regular/chaotic hybrid) states, which is of the order $\beta/2$. On the other hand, it is the beating between the two hybrid states that determines the appearance of population in the chaotic region, and this beating occurs at a rate of order β. Thus, we expect the oscillation of population to be substantially faster than the tunneling oscillation. The oscillations observed in the experiment do not appear to occur with a single frequency, so it may be necessary to include couplings to other chaotic states in order to account more accurately for this phenomenon.

3.5.7 TRANSPORT IN THE STRONGLY COUPLED REGIME

For even larger α than we have considered so far, the two symmetry-related islands of stability disappear, and the quantum transport undergoes a transition to qualitatively different behavior than the above tunneling. This strongly coupled behavior is illustrated in Figures 3.29 and 3.30, where the momentum-distribution evolution is shown (sampled on a fine timescale) for two large values of α. For $\alpha = 17.0$ (Figure 3.29), the three primary resonances have disappeared, leaving a chaotic region with only very small stable structures, while for $\alpha = 18.9$ (Figure 3.30), there is a small island at the center of the phase space. The experimental measurement shows erratic oscillations of the momentum distributions on a faster timescale than the tunneling observed above.

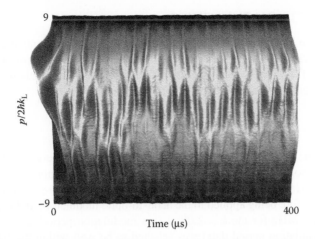

FIGURE 3.29 Experimental momentum-distribution evolution of chaos-assisted tunneling for $k = 2.08$ ($T = 20 \ \mu s$) and $\alpha = 17.0$. The distribution was sampled every 2 μs out to 400 μs. The three primary islands of stability have dissolved into the chaotic region in the classical phase space for this value of α. The experimental momentum distributions show erratic oscillations in time.

We can also understand this behavior qualitatively in terms of the Floquet states of the system. For very small α, the Floquet spectrum consists of nearly degenerate doublets associated with KAM tori, and as α increases the doublets break apart as their associated invariant structures become unstable [32,55]. In the regime that we consider here, where the stable structures have disappeared, the splittings are on the order of the mean level spacing [32] due to level repulsion of the states in the chaotic region [26]. This behavior of the splittings is apparent in the spectra in Figures 3.22 and 3.23. The Floquet states are no longer well localized in this regime, and thus the initial condition excites several states. The observed behavior is the result of complicated beating between the various populated states, and we expect a time dependence that is faster than the tunneling due to the relatively large splittings involved.

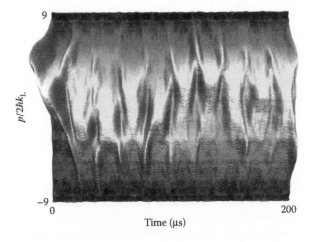

FIGURE 3.30 Experimental momentum-distribution evolution of chaos-assisted tunneling for $k = 2.08$ ($T = 20 \ \mu s$) and $\alpha = 18.9$. The distribution was sampled every 1 μs out to 200 μs. The two outer islands of stability are not present in the chaotic region in the classical phase space for this value of α. The experimental momentum distributions show erratic oscillations in time.

3.6   NOISE EFFECTS ON TUNNELING

The tunneling that we have described here is obviously an effect of quantum coherence, and tunneling in classically chaotic systems is expected to be suppressed by dissipation [23,32], measurement [46], and noise [25]. Here we consider the effects of a noisy perturbation of the optical lattice intensity, so that the atomic center-of-mass Hamiltonian becomes

$$H = \frac{p^2}{2} - 2\alpha[1 + \varsigma(t)]\cos^2(\pi t)\cos(x), \tag{3.23}$$

where $\varsigma(t)$ is a randomly fluctuating quantity with a probability distribution peaked at and symmetric about zero. This noise signal was generated digitally by picking normally distributed random deviates with a 10 MHz sampling rate. The noise was then bandwidth-limited by a digital Chebyshev low-pass filter before being applied to the acousto-optic modulator (AOM) control signal. The cutoff frequency (0.5 MHz for the $k = 2.08$, 20 μs modulation period data, and 1 MHz for the $k = 1.04$, 10 μs modulation period data) was selected to be well within the 10 MHz modulation response of the AOM driver and to make the noise spectrum the same in scaled units for different modulation periods. The rms noise levels $\langle \varsigma^2(t) \rangle^{1/2}$ that we quote correspond to the noise levels after the low-pass filter. Because the instantaneous noise level is proportional to the mean intensity, truncation effects due to noise deviations falling outside the dynamic range of the laser were rare except in the largest noise case that we consider here (62% rms). An example of the optical lattice intensity for one particular realization of the noise is illustrated in Figure 3.31.

The response of the tunneling oscillations to the noise is illustrated in Figure 3.32 for $k = 2.08$ and Figure 3.33 for $k = 1.04$ ($\alpha = 11.2$ in both cases). As one might expect, the oscillations are

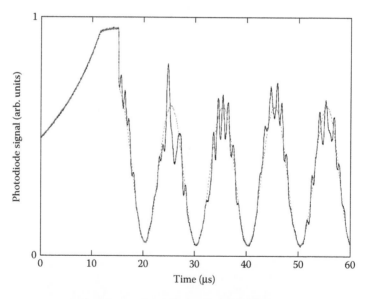

FIGURE 3.31   Illustration of amplitude noise applied to the optical lattice intensity, as measured by a fast photodiode. The end of the state-preparation sequence is visible at the beginning of the traces, where the lattice is ramped up and then remains at a high level for several μs after the lattice phase is shifted. The sinusoidal modulations begin immediately after the state preparation, and both the zero (dashed line) and 15.7% (solid line) rms deviation cases are shown here. The noise effects are most pronounced when the lattice is at the highest average intensity because the noise deviation is always proportional to the local average intensity. These traces correspond to the experimental settings for $k = 1.04$, where the modulation period is 10 μs, and the noise is filtered with a 1 MHz cutoff frequency.

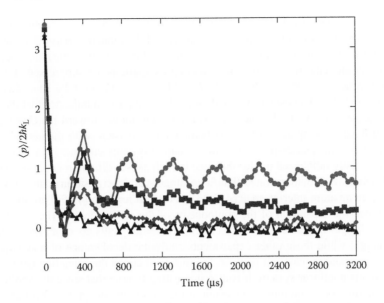

FIGURE 3.32 Effects of applied amplitude noise on the tunneling oscillations for $\alpha = 11.2$ and $k = 2.08$. The rms noise levels are 0% (circles), 15.7% (squares), 31% (diamonds), and 62% (triangles). The tunneling is only completely suppressed at the 62% level, and thus is substantially less sensitive than in the $k = 1.04$ case in Figure 3.33.

destroyed as the noise level increases, causing damping of the oscillations on progressively shorter timescales. At the largest levels of noise, classical-like behavior (with noise) is recovered, in that the tunneling oscillations are suppressed. The noise also has the "direct" effect of causing relaxation to $p = 0$, because the noise permits transitions, both quantum and classical, out of the initial island of

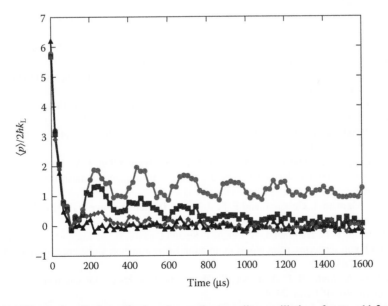

FIGURE 3.33 Effects of applied amplitude noise on the tunneling oscillations for $\alpha = 11.2$ and $k = 1.04$. The rms noise levels are 0% (circles), 7.9% (squares), 15.7% (diamonds), and 31% (triangles). The tunneling is completely suppressed at the 31% level, and thus is more sensitive than in the $k = 2.08$ case in Figure 3.32.

stability and into the chaotic sea. The more interesting feature of these data, though, is that because the value of α is fixed between the two measurements and the tunneling periods are approximately the same (in scaled units), we can compare the sensitivity of the system to the noise for two different values of $\bar{k}$. From the data, we see that the tunneling oscillations are suppressed at a much lower level of noise for the $\bar{k} = 1.04$ case than in the $\bar{k} = 2.08$ case (31% vs. 62% rms). Recalling that $\bar{k}$ is the dimensionless Planck constant in scaled units, this comparison indicates that the tunneling in this system is more sensitive to decoherence as the system moves toward the classical limit (i.e., to a larger action scale compared to $\hbar$). This behavior is consistent with theoretical expectations, because for smaller $\bar{k}$, the phase-space structure in chaotic systems saturates on a smaller scale [60], thus being more easily influenced by decoherence (which causes diffusion in phase space). Related experimental results have demonstrated that Schrödinger-cat superposition states in the phase of a cavity field [9], in an atom interferometer [11,33], and in an ion trap [41,54] are more sensitive to decoherence when the separation of the components of the state increases (i.e., as the spacing of the interference fringes decreases). The present experimental results are of a fundamentally different nature, though: while these other experiments study the decoherence of a superposition state produced by some state-preparation method, the interferences in the tunneling here are generated dynamically in this nonlinear system. It is also interesting to note that since the applied noise here leads to a fluctuating force and thus to diffusion of the atomic momenta, this form of noise mimics a continuous measurement of the atomic positions [5,17]. Thus, we might expect that the system may be more sensitive to noise that mimics a measurement of the atomic momentum, which would cause diffusion of the atomic position, rather than the momentum.

REFERENCES

1. S. Aberg. Chaos-assisted tunneling from superdeformed states. *Phys. Rev. Lett.*, 82:299, 1999.
2. L. C. Allen and J. H. Eberly. *Optical Resonance and Two-Level Atoms*. Dover, New York, NY, 1987.
3. V. Averbukh, N. Moiseyev, B. Mirbach, and H. J. Korsch. Dynamical tunneling through a chaotic region. *Z. Phys. D*, 35:247, 1995.
4. V. Averbukh, S. Osovski, and N. Moiseyev. Controlled tunneling of cold atoms: From full suppression to strong enhancement. *Phys. Rev. Lett.*, 89:253201, 2002.
5. T. Bhattacharya, S. Habib, K. Jacobs, and K. Shizume. The delta-kicked rotor: Momentum diffusion and the quantum-classical boundary. *Phys. Rev. A*, 65:032115, 2002.
6. O. Bohigas, S. Tomsovic, and D. Ullmo. Manifestations of classical phase space structures in quantum mechanics. *Phys. Rep.*, 223:43, 1993.
7. L. Bonci, A. Farusi, P. Grigolini, and R. Roncaglia. Tunneling rate fluctuations induced by nonlinear resonances: A quantitative treatment based on semiclassical arguments. *Phys. Rev. E*, 58:5689, 1998.
8. O. Brodier, P. Schlagheck, and D. Ullmo. Resonance-assisted tunneling in near-integrable systems. *Phys. Rev. Lett.*, 87:064101, 2001.
9. M. Brune, E. Hagley, J. Dreyer, X. Maître, A. Maali, C. Wunderlich, J. M. Raimond, and S. Haroche. Observing the progressive decoherence of the "meter" in a quantum measurement. *Phys. Rev. Lett.*, 77:4887, 1996.
10. G. Casati, R. Graham, I. Guarneri, and F. M. Izrailev. Tunneling between localized states in classically chaotic systems. *Phys. Lett. A*, 190:159, 1994.
11. M. S. Chapman, T. D. Hammond, A. Lenef, J. Schmiedmayer, R. A. Rubenstein, E. Smith, and D. E. Pritchard. Photon scattering from atoms in an atom interferometer: Coherence lost and regained. *Phys. Rev. Lett.*, 75:3783, 1995.
12. S. Chaudhury, A. Smith, B. E. Anderson, S. Ghose, and P. S. Jessen. Quantum signatures of chaos in a kicked top. *Nature*, 461:768, 2009.
13. B. V. Chirikov and D. L. Shepelyansky. Shnirelman peak in level spacing statistics. *Phys. Rev. Lett.*, 74:518, 1995.
14. M. J. Davis and E. J. Heller. Quantum dynamical tunneling in bound states. *J. Chem. Phys.*, 75:246, 1981.

15. C. Dembowski, H.-D. Gräf, A. Heine, R. Hofferbert, H. Rehfeld, and A. Richter. First experimental evidence for chaos-assisted tunneling in a microwave annular billiard. *Phys. Rev. Lett.*, 84:867, 2000.

16. M. T. DePue, C. McCormick, S. L. Winoto, S. Oliver, and D. S. Weiss. Unity occupation of sites in a 3D optical lattice. *Phys. Rev. Lett.*, 82:2262, 1999.

17. S. Dyrting and G. J. Milburn. Quantum chaos in atom optics: Using phase noise to model continuous momentum and position measurement. *Quant. Semiclass. Opt.*, 8:541, 1996.

18. J. H. Frederick and E. J. Heller. Ring torsional dynamics and spectroscopy of benzophenone: A new twist. *J. Chem. Phys.*, 88:2169, 1988.

19. S. D. Frischat and E. Doron. Dynamical tunneling in mixed systems. *Phys. Rev. E*, 57:1421, 1998.

20. T. M. Fromhold, L. Eaves, F. W. Sheard, M. L. Leadbeater, T. J. Foster, and P. C. Main. Magnetotunneling spectroscopy of a quantum well in the regime of classical chaos. *Phys. Rev. Lett.*, 72:2608, 1994.

21. S. Ghose, P. M. Alsing, and I. H. Deutsch. Atomic motion in magneto-optical double-well potentials: A testing ground for quantum chaos. *Phys. Rev. E*, 64:056119, 2001.

22. R. Graham, M. Schlautmann, and P. Zoller. Dynamical localization of atomic-beam deflection by a modulated standing light wave. *Phys. Rev. A*, 45:R19, 1992.

23. R. Grobe and F. Haake. Dissipative death of quantum coherences in a spin system. *Z. Phys. B*, 68:503, 1987.

24. F. Grossman, T. Dittrich, P. Jung, and P. Hänggi. Coherent destruction of tunneling. *Phys. Rev. Lett.*, 67:516, 1991.

25. F. Grossmann, T. Dittrich, P. Jung, and P. Hänggi. Coherent transport in a periodically driven bistable system. *J. Stat. Phys.*, 70:229, 1993.

26. F. Haake. *Quantum Signatures of Chaos*. Springer-Verlag, Berlin, 2 ed, 2001.

27. D. L. Haycock, P. M. Alsing, I. H. Deutsch J. Grondalski, and P. S. Jessen. Mesoscopic quantum coherence in an optical lattice. *Phys. Rev. Lett.*, 85:3365, 2000.

28. E. J. Heller. The many faces of tunneling. *J. Phys. Chem.*, 103:10433, 1999.

29. Eric J. Heller. Wavepacket dynamics and quantum chaology. In M.-J. Giannoni, A. Voros, and J. Zinn-Justin (Eds), *Chaos and Quantum Physics: Proceedings of the Les Houches Summer School, Session LII*, 1–31 August 1989, Northy-Holland, Amsterdam, 1991.

30. W. K. Hensinger, H. Häffner, A. Browaeys, N. R. Heckenberg, K. Helmerson, C. McKenzie, G. J. Milburn, W. D. Phillips, S. L. Rolston, H. Rubinsztein-Dunlop, and B. Upcroft. Dynamical tunneling of ultracold atoms. *Nature*, 412:52, 2001.

31. M. Kasevich, D. S. Weiss, E. Riis, K. Moler, S. Kasapi, and S. Chu. Atomic velocity selection using stimulated Raman transitions. *Phys. Rev. Lett.*, 66:2297, 1991.

32. S. Kohler, R. Utermann, P. Hänggi, and T. Dittrich. Coherent and incoherent chaotic tunneling near singlet–doublet crossings. *Phys. Rev. E*, 58:7219, 1998.

33. D. A. Kokorowski, A. D. Cronin, T. D. Roberts, and D. E. Pritchard. From single- to multiple-photon decoherence in an atom interferometer. *Phys. Rev. Lett.*, 86:2191, 2001.

34. M. Latka, P. Grigolini, and B. J. West. Chaos and avoided level crossings. *Phys. Rev. E*, 50:596, 1994.

35. M. Latka, P. Grigolini, and B. J. West. Control of dynamical tunneling in a bichromatically driven pendulum. *Phys. Rev. E*, 50:R3299, 1994.

36. W. A. Lin and L. E. Ballentine. Quantum tunneling and chaos in a driven anharmonic oscillator. *Phys. Rev. Lett.*, 65:2927, 1990.

37. R. Luter and L. E. Reichl. Floquet analysis of atom-optics tunneling experiments. *Phys. Rev. A*, 66:053615, 2002.

38. H. J. Metcalf and P. van der Straten. *Laser Cooling and Trapping*. Springer, New York, NY, 1999.

39. K. Moler, D. S. Weiss, M. Kasevich, and S. Chu. Theoretical analysis of velocity-selective Raman transitions. *Phys. Rev. A*, 45:342, 1992.

40. A. Mouchet, C. Miniatura, R. Kaiser, B. Grémaud, and D. Delande. Chaos assisted tunnelling with cold atoms. *Phys. Rev. E*, 64:016221, 2001.

41. C. J. Myatt, B. E. King, Q. A. Turchette, C. A. Sackett, D. Kielpinski, W. M. Itano, C. Monroe, and D. J. Wineland. Decoherence of quantum superpositions through coupling to engineered reservoirs. *Nature*, 403:269, 2000.

42. T. Neicu, K. Schaadt, and A. Kudrolli. Spectral properties of a mixed system using an acoustical resonator. *Phys. Rev. E*, 63:026206, 2001.

43. J. U. Nöckel and A. D. Stone. Ray and wave chaos in asymmetric resonant optical cavities. *Nature*, 385:45, 1997.

44. A. Peres. Dynamical quasidegeneracies and quantum tunneling. *Phys. Rev. Lett.*, 67:158, 1991. This paper is a comment on Ref. [36]; see also the response by Lin and Ballentine, *Phys. Rev. Lett.*, 67, 159, 1991.

45. R. Roncaglia, L. Bonci, F. M. Izrailev, B. J. West, and P. Grigolini. Tunneling versus chaos in the kicked Harper model. *Phys. Rev. Lett.*, 73:802, 1994.

46. B. C. Sanders and G. J. Milburn. The effect of measurement on the quantum features of a chaotic system. *Z. Phys. B*, 77:497, 1989.

47. R. Scharf and B. Sundaram. Periodic orbits in quantum standard maps. *Phys. Rev. A*, 46:3164, 1992.

48. D. A. Steck. Floquet states in chaos-assisted tunneling experiments. Unpublished. Available at http://steck.us/calculations/cat_husimi.pdf, 2002.

49. D. A. Steck, W. H. Oskay, and M. G. Raizen. Observation of chaos-assisted tunneling between islands of stability. *Science*, 293:274, 2001.

50. D. A. Steck, W. H. Oskay, and M. G. Raizen. Fluctuations and decoherence in chaos-assisted tunneling. *Phys. Rev. Lett.*, 88:120406, 2002.

51. D. A. Steck. *Quantum Chaos, Transport, and Decoherence in Atom Optics*. PhD dissertation, The University of Texas at Austin, 2001.

52. S. Tomsovic. Tunneling and chaos. *Physica Scripta*, T90:162, 2001.

53. S. Tomsovic and D. Ullmo. Chaos-assisted tunneling. *Phys. Rev. E*, 50:145, 1994.

54. Q. A. Turchette, C. J. Myatt, B. E. King, C. A. Sackett, D. Kielpinski, W. M. Itano, C. Monroe, and D. J. Wineland. Decoherence and decay of motional quantum states of a trapped atom coupled to engineered reservoirs. *Phys. Rev. A*, 62:053807, 2000.

55. R. Utermann, T. Dittrich, and P. Hänggi. Tunneling and the onset of chaos in a driven bistable system. *Phys. Rev. E*, 49:273, 1994.

56. P. B. Wilkinson, T. M. Fromhold, L. Eaves, F. W. Sheard, N. Miura, and T. Takamasu. Observation of "scarred" wavefunctions in a quantum well with chaotic electrons dynamics. *Nature*, 380:608, 1996.

57. J. Zakrzewski, D. Delande, and A. Buchleitner. Ionization via chaos assisted tunneling. *Phys. Rev. E*, 57:1458, 1998.

58. E. M. Zanardi, J. Gutiérrez, and J. M. Gomez Llorente. Mixed dynamics and tunneling. *Phys. Rev. E*, 52:4736, 1995.

59. C. Zener. Non-adiabatic crossing of energy levels. *Proc. R. Soc. London A*, 137:696, 1932.

60. W. H. Zurek. Sub-Planck structure in phase space and its relevance for quantum decoherence. *Nature*, 412:712, 2001.

4 Tractable Problems in Multidimensional Tunneling

Stephen C. Creagh

CONTENTS

4.1 INTRODUCTION

One of the appealing features of tunneling as a theoretical problem is that it poses interesting questions regarding the complexified dynamics of the classical limit. Often, these questions are of a different nature to the ones that arise in a purely dynamical systems context and can therefore push the dynamics side of the problem in novel directions, without necessarily being technically hard. In this article we explore a number of scenarios where these underlying problems of complex dynamics can be well enough understood to allow explicit analytical approximations to be given for tunneling rates. The treatment here is not intended to be comprehensive and short shrift is given to very interesting and important tunneling regimes, such as those of resonance- and chaos-assisted tunneling [3,14,16,17,40,48,78], for example. Instead, by concentrating on relatively tractable problems where answers can be given in terms of simple geometrical characterizations of the complex dynamics, we aim to achieve an understanding of problems which are important in their own right but which may also later serve as building blocks in the solution of more difficult problems.

We begin with an overview of the most fundamental tunneling problem of all: transmission across potential barriers. A one-dimensional treatment of barrier penetration based on an extension of Wentzel–Kramers–Brillouin (WKB) methods to complex coordinates provides the most familiar starting point for understanding any tunneling problem. Even this textbook problem, however, contains the beginnings of more interesting dynamical calculations in higher dimensions. We therefore provide a brief overview of the standard results available for the one-dimensional problem and highlight the parts of the calculation which have multidimensional counterparts. In particular, we center our discussion of one-dimensional tunneling on a barrier-crossing complex periodic orbit. This orbit has a natural extension to multidimensional problems, where it is commonly known as the "instanton" or "bounce orbit," and is now one of the main tools in our armoury to treat multidimensional tunneling. (Although "instanton" is often taken to mean a particular zero-energy solution, the terminology is used more loosely here to include barrier-crossing orbits at any energy.) The earliest uses of the instanton in multidimensional problems stem from independent work by Miller and George [55,57–59] and by Banks et al. [4,5] and the picture we use is related in particular to subsequent development by Auerbach and Kivelson of the so-called path decomposition expansion [2].

We characterize the multidimensional instanton orbit, and dynamics transverse to it in its immediate neighborhood in phase space, using a mapping on a Poincaré section. It is the existence of this nontrivial transverse dynamics that offers the greatest qualitative difference between one-dimensional and multidimensional barrier penetration. Although the instanton-induced Poincaré map is complex, generically mapping real initial conditions to complex images, it has within semiclassical approximation a natural quantum analog in the form of a quantum map [28]—which can be understood in terms of the formalism of Bogomolny [13] in which energy eigenstates are characterized in terms of reduced representations on Poincaré sections (and which is also related to the path-decomposition formalism of Auerbach and Kivelson (1985) [2]). The Poincaré-mapping approach allows the problem of characterizing the incident wave to be neatly separated from the problem of characterizing transmission across the barrier, while also removing the calculation from the barrier region itself, where caustics and turning points can complicate semiclassical approximation [83]. It means that we can, for example, use phase-space representations such as the Weyl symbol to visualize transmission probabilities defined from the scattering matrix, independently of the details of the incident wave itself. Decoupling the problem of transmission across the barrier from the details of the incident wave also allows us to provide statistical and other modeling of tunneling rates from chaotic quantum wells as described in Section 4.3.

The instanton picture can be extended to provide a description of the initial opening up, as energy rises above a threshold, of regions of phase space in which transmission across the barrier is classically allowed. The one-dimensional counterpart of this calculation consists of a uniform approximation for the probability of transmission across a barrier which allows smooth interpolation to be achieved between the exponentially small tunneling probabilities below threshold to the case of nearly total classically allowed transmission above threshold. A robust description of this transition can be obtained directly from the Schrödinger equation using the method of comparison equations [9,54], in which the position coordinate is stretched so as to transform the problem into a solvable one such as that of an inverted harmonic potential. Potentially more powerful, however, is an analysis of the same result based on more general transformations which act on the entire phase space. A summary of this latter approach to one-dimensional tunneling, which leads to easier analysis in the long run, is outlined in Section 4.4.

A particular advantage of the phase-space approach is that it allows generalization to multidimensional barriers. The classical picture underlying this calculation is very important to the theory of chemical reaction rates and has in recent years been the subject of renewed interest. Classical transition state theory aims to relate the classical flux through phase-space bottlenecks, such as one finds associated with transmission across multidimensional barriers, to the geometry of invariant manifolds which can be used to divide reactants from products [44,45,53,61,65,77,79,81]. Recent work has highlighted in particular the efficacy of normal forms in realizing such flux calculations in higher-dimensional problems (see, e.g., Uzer et al. (2002) [79], Waalkens et al. (2008) [81], and references therein). It turns out that the formalism underlying these calculations is also ideally suited to characterizing quantum transmission probabilities near threshold [20,21,38]. In Section 4.5 we outline how the phase-space approach to tunneling, previously described in Section 4.4, can be generalized to the multidimensional case so as to make the connection between the quantum problem and the underlying classical geometry especially explicit. We highlight in particular how an established uniform approximation of one-dimensional transmission probability (derived in Section 4.4) can be generalized to several dimensions in a formally very simple way by exploiting quantizations of the Poincaré map around the instanton (as described in Section 4.3). We can further exploit this formalism to build a phase-space picture of transmission probabilities using, as we did with below-barrier transmission, representations such as the Weyl symbol. In fact this approach allows us to see, in the quantum scattering problem, signatures of the classically reacting region and in particular how it shrinks to be replaced by a tunneling remnant as energy is decreased below threshold.

We conclude by outlining some limitations of the instanton mechanism in describing tunneling. There are, of course, a range of phenomena related to dynamical tunneling where one would expect no relationship to an instanton at all. This includes chaos-assisted tunneling [14,78] between regular states, for example. Even where barrier tunneling is concerned, however, the instanton picture may not be enough. In above-threshold tunneling, for example, the mechanisms controlling tunneling from incident waves localized away from the instanton itself may involve different classifications of the underlying complex orbits [10,11,49–51,72–76]. Even for below-threshold tunneling, when the instanton connects to a region of regular dynamics in phase space, the nature of the incident wave may itself prevent a simple description on the basis of the instanton alone. We emphasize in particular the problems posed by tunneling from Kolmogorov–Arnold–Moser (KAM)-like states, in which natural boundaries prevent a straightforward calculation of the classical data characterising the wavefunction deep enough into complex phase space [19,30]. A way out of the problems posed by natural boundaries is offered which works near enough to exactly integrable systems, although it is pointed out that there remain significant open questions in understanding this regime.

4.2 PRIMITIVE APPROXIMATIONS OF BARRIER PENETRATION

Barrier penetration provides the archetypal scenario for tunneling phenomena in physics. In the multidimensional case it provides a link between tunneling and nontrivial aspects of higher-dimensional dynamics, such as chaos or transport across dynamical bottlenecks in phase space. In this section we describe how multidimensional barrier penetration may be formulated so that it bears a strong formal resemblance to standard one-dimensional treatments, but at the same time so that the geometry of the dynamics transverse to the primary tunneling degree of freedom are also very naturally encoded. We will see in later sections that this calculation provides, in its own right, a natural connection between quantum-mechanical scattering and recent developments in the classical treatment of transition states. It also proves a means of treating tunneling from chaotic potential wells so that the nontrivial chaotic dynamics is effectively decoupled from the tunneling problem.

4.2.1 A SUMMARY OF BARRIER PENETRATION IN ONE DIMENSION

We motivate the coming treatment of multidimensional barrier penetration by recalling the standard approximations for tunneling probabilities in one dimension. Let a particle of mass m be incident on a potential barrier $V(x)$ with an energy E that is less than the maximum of $V(x)$. Then the transmission probability is approximated semiclassically by

$$T \approx e^{-\theta}, \tag{4.1}$$

where

$$\theta = \frac{2}{\hbar} \int_{x_1}^{x_2} \sqrt{2m(V(x) - E)}\, dx,$$

and where (x_1, x_2) denote the turning points. Alternatively, $S_0 \equiv \oint p\, dx = i\hbar\theta$ is the action of a complex periodic orbit obtained by using imaginary time evolution to let the particle bounce under the barrier between the turning points x_1 and x_2 (see Figure 4.1). This latter interpretation as an orbit action allows us later to generalize the tunneling action to problems where the Hamiltonian is not of kinetic-plus-potential type or to barriers in higher dimensions, where generalizations of the periodic orbit can still be defined.

The setup in Figure 4.1 illustrates that we are free to choose an initial condition p_0 for the complex periodic orbit which is removed from the barrier itself—although we must choose an appropriate time contour to make this work. The initial condition p_0 leads to periodic evolution under repeated application of the time contour in Figure 4.1b. We are free to deform this contour following Cauchy's theorem as long as branch points of the solution are not crossed. We might, for example, cut short the real-time evolution and begin the imaginary time segment before the turning point p_c is reached (this would lead to an orbit in complex position coordinates of the type illustrated later in Figure 4.8a). Depending on the potential, and for initial conditions p_0 which are not too far from the barrier, it may even be possible to generate periodic evolution using a contour that decends directly along the imaginary axis. For interesting problems (particularly in the scattering context where we would like to take p_0 arbitrarily far from the barrier), however, it is likely that a direct descent along the imaginary axis would pass on the wrong side of a branch point and lead to nonperiodic motion. The time contour shown in Figure 4.1b is guaranteed to work, however, and even for such nontrivial contours it is important to note that the net period, which is the difference between the end and starting points of the contour in the time plane, is an imaginary number. Finally, we remark that these features of the time contour also apply to multidimensional generalizations of the orbit described in the next section.

The approximation (Equation 4.1) is "primitive" because it applies when the energy E is lower than, and not close to, the maximum of $V(x)$. In this case, the breakdown of the WKB form of

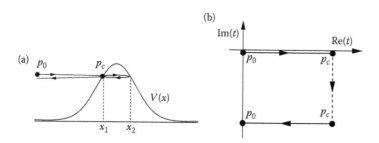

FIGURE 4.1   The complex periodic orbit whose action determines tunneling probabilities is illustrated schematically. The dashed line in (a) represents the path of the orbit during the imaginary time evolution sketched in (b). Starting at an initial condition p_c, with position coordinate x_1, evolution of time parallel to the imaginary axis results in a periodic bounce motion under the potential barrier and between turning points. (Letting the time move *downward* in the complex plane ensures that the action has positive imaginary part and contributes exponentially small terms to semiclassical approximations, otherwise the corresponding orbits would be removed by the Stokes phenomenon.) An initial condition for the periodic orbit can also be chosen away from the barrier, represented by point p_0 in (a). In this case, conjugating imaginary time evolution with real time evolution in the positive and negative directions, as illustrated in (b), also leads to periodic dynamics. This construction has a natural extension to multidimensional problems (as illustrated in Figure 4.2) and to problems which are not necessarily of kinetic-plus-potential type.

the wavefunction that occurs around each of the turning points can be treated in isolation and the transmitted wave can be obtained by extending the wavefunction into the complex x-plane and around the isolated turning points. In a detailed analysis (see, e.g., Berry and Mount (1972) [9]) it is found that, although the Stokes phenomenon leads to changes in the form of the wavefunction across certain curves emanating from the turning points in the complex x-plane, the transmitted wave itself can be obtained by a simple continuation of the incident wave, whose WKB form we write as

$$\psi_{\text{inc}}(x) \approx A(x)e^{iS(x,x_1)/\hbar}$$

and where, relative to a given reference point x_0, we denote the action

$$S(x,x_0) \equiv \int_{x_0}^{x} p\,dx.$$

We choose the reference point to be $x_0 = x_1$ when the wave is incident on the left (in Figure 4.1a). With the corresponding definition of the transmitted wave to be such that action is relative to the right turning point x_2,

$$\psi_{\text{trans}}(x) \approx A(x)e^{iS(x,x_2)/\hbar},$$

we find that a direct continuation of the incident wave to the other side of the barrier, using

$$S(x,x_1) = S(x,x_2) + S(x_2,x_1) = S(x,x_2) + i\hbar\theta/2,$$

yields

$$\psi_{\text{inc}}(x) \rightarrow t\psi_{\text{trans}}(x), \tag{4.2}$$

where the transmission amplitude t is

$$t \approx e^{-\theta/2}. \tag{4.3}$$

The probability $T = |t|^2$ is then of the form given in Equation 4.1.

4.2.2 BARRIER-CROSSING ORBITS IN MORE DIMENSIONS

How might this simple one-dimensional probability be generalized to multidimensional barriers? To answer this question we must recognize that, in the multidimensional case, the incident wave is not unique, as it effectively is in one dimension. To describe multidimensional transmission probabilities we must also provide information about the incident wave in transverse degrees of freedom. A convenient means of achieving this is by calculating Green functions or scattering matrices, for which there is a natural means of specifying an incident wave semiclassically in terms of classical trajectories with well-defined boundary conditions, ripe for complexification as demanded by a treatment of tunneling. Even in problems such as tunneling from quasibound states in chaotic potential wells, where the incident wave may be inaccessible to direct semiclassical approximation, this approach provides us with a means of decoupling the tunneling problem from that of constructing the incident wave, as described later in Section 4.3.

We find that the tunneling rate is greatest when semiclassical representations of the incident wave overlap in phase space with a small region surrounding a particular orbit which crosses the barrier with least imaginary action. This optimal orbit is in fact a natural generalization of the periodic orbit illustrated in Figure 4.1, as we now describe.

To describe multidimensional barrier penetration more concretely, let us consider a waveguide problem with configuration space coordinates $\mathbf{x} = (x,y)$ in which x represents longitudinal position along the waveguide and y represents transverse vibrations. If necessary we can let $y = (y_1,\ldots,y_d)$

be multidimensional. In chemical applications, x might be a reaction coordinate with x large and negative corresponding to decoupled reactant molecules and x large and positive corresponding to decoupled product molecules, while y describes internal vibrations of the reacting molecules.

Before describing technically how Green functions or scattering matrices are constructed, we illustrate schematically in Figure 4.2 the classical orbits that drive tunneling in this multidimensional case. The basic component is a complex periodic orbit bouncing between turning points under imaginary time evolution, illustrated as a dashed curve in Figure 4.2a, with p_c labeling a turning point of the projection of the orbit onto real configuration space. For Hamiltonians of kinetic-plus-potential type, this may be achieved by substituting $(\mathbf{x}, \mathbf{p}, t) = (\mathbf{x}, i\mathbf{u}, -i\tau)$, where $\mathbf{x}$, $\mathbf{p}$, and t, respectively represent position, momentum, and time. It is a straightforward exercise to see that real equations of motion are then obtained for $(\mathbf{x}, \mathbf{u}, \tau)$ which are equivalent to solving dynamics of the upside-down potential $-V(\mathbf{x})$. In this case, the complex periodic orbit can be identified with a real periodic orbit of the transformed problem. For a wider class of problems, such as those with time-reversal-symmetry-breaking magnetic field terms, this simple upside-down-potential picture does not work. However, an imaginary time complex periodic orbit may still be found connecting disjoint components of the real energy-level surface [28]. In this more general case, the position coordinates $\mathbf{x}$ may become complex over the course of the imaginary time evolution but one still finds turning points of the real projection of the orbit at which the phase-space coordinates of the trajectory become instantaneously real.

It is useful to be able to choose initial conditions which are removed from the barrier region, among other reasons because standard semiclassical approximation is simpler away from the caustic structures associated with turning points such as p_c. This can be achieved using the fact that, following real-time evolution, with the real point p_c serving as an initial condition, a real orbit is generated which falls into the real energy-level surface on one side of the barrier. This is illustrated by the curve connecting p_c to p_0 in Figure 4.2a, for example. Choosing initial conditions on this real extension of the barrier-crossing orbit, and following contours in the complex t-plane that are of the form illustrated in Figure 4.1b, allows us to describe a periodic evolution whose initial condition p_0

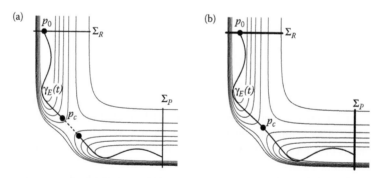

FIGURE 4.2 An illustration of the complex periodic orbits which dominate multidimensional barrier penetration, generalizing the one-dimensional construction in Figure 4.1. In (a), p_c is a real initial condition for a complex orbit which is periodic under imaginary time evolution, whose projection onto real configuration space is shown as a dashed curve. Alternatively, using p_c as an initial condition for real-time evolution generates a real orbit, shown in (a) as a solid curve intersecting the section Σ_R at p_0. Then p_0 (assuming the curve traced from p_c to p_0 has been generated by letting time evolve backward) may serve as a real initial condition for the complex periodic orbit if a time contour of the form illustrated in Figure 4.1b is followed. This whole construction generalizes to above-barrier energies, as illustrated in (b). The difference here is that evolution in imaginary time from the real initial condition p_c moves position predominantly in the imaginary direction, so that its projection onto real configuration space is not seen in this figure, but the evolution is periodic nonetheless.

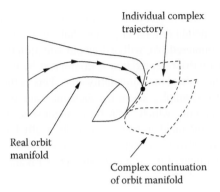

Individual complex
trajectory

Real orbit
manifold

Complex continuation
of orbit manifold

FIGURE 4.3 A schematic illustration of an individual complex trajectory in a complexified orbit family. Solid curves represent real solutions. Dashed curves represent real-position slices through the complex solutions and are used to approximate the evanescent wave on the dark side of a caustic. The surfaces shown form part of a single complex manifold in which position may become generically complex and pass from the continuous to the dashed solutions while skirting around the caustic.

can be chosen on a general surface of section Σ_R which may be far from the barrier. The orbit in full phase space corresponding to this time contour will be denoted $\gamma_E(t)$.

No single orbit suffices to achieve semiclassical approximation in the multidimensional case—one generally needs families forming a Lagrangian manifold (Figure 4.3) [15,41]. Therefore the single periodic orbit above must be generalized if it is to be used as a basis for describing multidimensional barrier penetration. A very convenient means of codifying these more general barrier-crossing orbits is to describe, in a neighborhood of the initial condition p_0 of the periodic orbit $\gamma_E(t)$, a complex Poincaré mapping [28]

$$F : \Sigma_R \to \Sigma_R.$$

Choosing a point p near p_0, evolution of time along the same contour as used for $\gamma_E(t)$ returns us to a final point once again near p, but not necessarily on Σ_R. If the section Σ_R is specified by fixing the value of a given coordinate, for example, then by tuning a final evolution of arbitrary direction in complex time we can return to a point $F(p)$ which is on (the complexified extension of) the section Σ_R. The usual arguments which show that Poincaré mappings are symplectic, modified only to allow for complex coordinates and time evolution, can be applied here to argue that $F : \Sigma_R \to \Sigma_R$ is itself a symplectic map. Note that p_0 itself is a fixed point of F and away from p_0 real initial conditions are in general mapped to complex images. As a consequence of the fact that the net time of evolution underlying p_0 is imaginary, complex conjugation of the contour amounts to reversal and one finds that the inverse map is $F^{-1} = F^*$, where F^* denotes the conjugate map $F^*(p) = (F(p^*))^*$. It is useful to denote by

$$W = DF(p_0),$$

the (complex) symplectic matrix linearizing the map F about p_0. As a result of the conjugate-inverse symmetry enjoyed by the full map F, this matrix has the property that $W^* = W^{-1}$ [28]. In practical applications, we may then take advantage of the fact that the matrix W can be generated by the dynamics of a quadratic Hamiltonian-like function $h(y, p_y)$, written as a function of transverse coordinates (y, p_y) on the section Σ_R. Note, however, that because of the property $W^* = W^{-1}$, the function $h(y, p_y)$ generates W after an *imaginary* time, which we may take to be (in a convention that simplifies notation for later uniform results) $-2\pi i$.

It should finally be stressed that this construction of a single-valued complex map around p_0 is guaranteed to work only in a neighborhood of p_0 and that further away from p_0 bifurcations may occur, or singularities be encountered. We will restrict our attention to tunneling problems in which an incident wave is such that barrier penetration is dominated by the immediate neighborhood of $\gamma_E(t)$, where the map is single-valued and singularity-free. This is the situation in the case of tunneling from chaotic wells, for example, where the incident wave is ergodic, or for the dominant reaction rates in chemical problems, which occur for incident states localized near $\gamma_E(t)$. It should be noted, however, that where the incident wave is semiclassically localized away from this neighborhood, a more complicated set of tunneling trajectories with interesting complex dynamics can ensue, as described in the treatment of the scattering matrix in Takahashi and Ikeda (2001) [72], for example.

4.2.3 BARRIER-CROSSING ORBITS AND THE SCATTERING MATRIX

Being symplectic, the complex map F allows semiclassical quantization into an evolution operator $\hat{T} : \mathcal{H}_R \to \mathcal{H}_R$ acting on a Hilbert space $\mathcal{H}_R$ which is the quantum analog of Σ_R. Concretely, this can be achieved using the Van Vleck form [15,28,41]

$$\langle y|\hat{T}|y'\rangle \approx \frac{1}{(2\pi\hbar)^{d/2}} \sqrt{\frac{\partial^2 K}{\partial y \partial y'}} e^{-K(y,y',E)/\hbar}, \tag{4.4}$$

for example, where $S(y,y',E) \equiv iK(y,y',E)$ is a generating function for the map F defined by the conditions $\partial S/\partial y = p_y$ and $\partial S/\partial y' = -p_y'$. Here, (y',p_y') denote initial position and momentum coordinates of a point on Σ_R and $(y,p_y) = F(y',p_y')$ are the coordinates of its image under F. This semiclassically defined operator turns out to be useful for a range of calculations related to multidimensional barrier penetration. It was formulated in Creagh and Whelan (1999) [28] as a basis for calculating tunneling rates from chaotic wells. As will be explained later, it can be used in particular to relate the tunneling rates of individual states to the morphology of the wavefunction [25,28,29] in a neighborhood of the trajectory $\gamma_E(t)$, or as a means of organizing the complex periodic orbits which control tunneling rates in the trace formula [27]. It has subsequently also proved useful as a basis for describing uniform approximations of quantum scattering at energies close to the barrier top [20,21,23,38], as we describe in Section 4.5. We motivate this construction in the first instance by describing how it arises naturally in semiclassical treatments of the scattering matrix.

The full quantum mechanics of barrier penetration are described by the scattering matrix, which we write in the form

$$S(E) = \begin{pmatrix} r_{RR} & t_{RP} \\ t_{PR} & r_{PP} \end{pmatrix},$$

where, for example, in the language of chemical reaction, the block t_{PR} maps asymptotic incoming states on the reacting side to asymptotic outgoing states on the product side. Explicit semiclassical approximations for the scattering matrix as sums over classical trajectories have been given by Miller [55,57] and it is precisely in this context that Miller and George [58,59] provided one of the early complex-trajectory-based treatments of multidimensional tunneling. A more complete description of how complex orbits are used to approximate the scattering matrix, and of the nontrivial complex dynamics that rule them in the deep tunneling regime, can be found in Takahashi and Ikeda (2001) [72]. Here we concentrate on an easier calculation that arises if we eschew the phase information implicit in the complete scattering matrix and ask only for probabilities of transmission across an isolated barrier. It turns out that for uniform treatments of transmission at the edge of classical reaction in particular, this reduction of the problem to probabilities rather than to amplitudes leads to simpler dynamics (and the treatment here is further simplified by restricting our attention to incoming states supported near the dominant tunneling route $\gamma_E(t)$, as described earlier).

It is natural to generalize the single transmission probability of the one-dimensional case to the matrix

$$\hat{\mathcal{R}} = t_{PR}^{\dagger} t_{PR},$$

which gives the probabilities of transmission for states incoming on the reactant side. Interpreting $\hat{\mathcal{R}}$ as an operator mapping $\mathcal{H}_R$ back to itself, and for energies below the barrier, it can be shown [20,21] that $\hat{\mathcal{R}}$ coincides with the tunneling operator $\hat{T}$, up to a change of basis:

$$\hat{\mathcal{R}} \approx \hat{T}_0 \equiv \hat{\mathcal{M}}^{\dagger} \hat{T} \hat{\mathcal{M}}. \tag{4.5}$$

The change of basis corresponding to $\hat{\mathcal{M}}$ has two aspects. First, whereas the scattering matrix is written in a basis of asymptotic modes of the transverse degree of freedom, it is useful here to write the $\hat{T}$ operator in a position basis as in Equation 4.4. Second, the operator $\hat{T}$ has been defined as a mapping on a section Σ_R which is a finite distance from the barrier whereas to define the scattering matrix properly, we need to remove the section to asymptopia and then conjugate by a mapping back to the barrier region in which coupling between the x and y degrees of freedom is switched off. None of this has a profound influence on the discussion to follow and is described in detail in Creagh (2005) [21]—here it suffices to note that the basis change is simply a matter of conjugating the classical map F with real Poincaré maps connecting Σ_R to sections far removed from the barrier region (with and without coupling between the x and y degrees of freedom).

This approximation can be justified by relating the block t_{PR} to a Green function connecting points on Σ_R to points on a section Σ_P on the product side [20]. Standard semiclassical approximations give this Green function in terms of trajectories going from Σ_R to Σ_P at energy E following a path corresponding to the the upper contour in Figure 4.4. This approximation provides a natural multidimensional generalization of Equation 4.2. The difference is that, whereas in the one-dimensional case (Equation 4.2) it suffices to complexify a single orbit in order to continue the wavefunction across the barrier, in this multidimensional case we must complexify a family of orbits, which collectively form a Lagrangian manifold (Figure 4.3). Although the orbit family has increased dimension, the qualitative structure of the underlying orbits and the nature of the time contour used to construct them are similar (assuming that we once again confine our attention to a primitive approximation of the dominant contribution). The corresponding contour for $t_{PR}^{\dagger}$ reverses the real components of this evolution while keeping the imaginary segment in the downwards direction and corresponds to the lower contour in Figure 4.4. A semiclassical treatment of the product $t_{PR}^{\dagger} t_{PR}$, which is arrived at using integration over the corresponding Green functions by the steepest-descents approximation, uses a time contour which is a concatenation of the contour for t_{PR} and

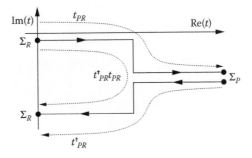

FIGURE 4.4 A schematic illustration of the time contours underlying complex orbits contributing to individually to t_{PR} and $t_{PR}^{\dagger}$ and to the combined operator $\hat{\mathcal{R}} = t_{PR}^{\dagger} t_{PR}$. The combined contour for $\hat{\mathcal{R}} = t_{PR}^{\dagger} t_{PR}$, in which the postpenetration real segments of this concatenation undo one another, is equivalent to the one shown in Figure 4.1b.

the contour for $t_{PR}^{\dagger}$ (Figure 4.4) leaving effectively a contour of the form Figure 4.1b used in the construction of $\hat{T}$.

The operator $\hat{\mathcal{R}}$ provides us with a neat means of visualizing the scattering matrix in phase space. From Equation 4.4 one can construct phase-space representations of $\hat{T}$ and therefore of $\hat{\mathcal{R}}$, such as Weyl symbols or Husimi functions, specified as functions of (y, p_y) on the Poincaré section Σ_R. A specific example of a Weyl symbol,

$$\mathcal{W}_{\hat{\mathcal{R}}}(y, p_y) = \int e^{-ip_y s/\hbar} \langle y + \frac{s}{2} | \hat{\mathcal{R}} | y - \frac{s}{2} \rangle \, ds, \tag{4.6}$$

is illustrated in Figure 4.5. This example illustrates the general feature of $\hat{\mathcal{R}}$ for energies below threshold that $\mathcal{W}_{\hat{\mathcal{R}}}(y, p_y)$ is strongly localized around the initial condition p_0 for $\gamma_E(t)$.

This localization can be further emphasized by relating $\hat{T}$ to a generating operator $\hat{h}$ by writing

$$\hat{T} = e^{-\theta} e^{-2\pi \hat{h}/\hbar}, \tag{4.7}$$

where $S_0 = i\hbar\theta$ is the action of $\gamma_E(t)$. The generating operator $\hat{h}$ is constructed as the quantization of the Lie generator $h(y, p_y)$ that generates the map F after a time $-2\pi i$. The separation of the exponent of $\hat{T}$ into the constant term $-\theta$ and the generating term $-2\pi\hat{h}/\hbar$ is such that an expansion of the classical generator $h(y, p_y)$ in displacement from p_0 begins with quadratic terms. A truncation of the generator at this order is equivalent to approximating $\hat{T}$ as a quantization of the matrix W that linearizes dynamics about $\gamma_E(t)$. Such a quadratic truncation of $h(y, p_y)$ suffices to capture the leading behavior of $\hat{T}$ around p_0 and the corresponding approximation of $\mathcal{W}_{\hat{\mathcal{R}}}(y, p_y)$ is a Gaussian centered on p_0 (note, however, that $\mathcal{W}_{\hat{\mathcal{R}}}(y, p_y) \neq e^{-\theta} e^{-2\pi h(y, p_y)/\hbar}$) [8,28,38]. The generator $h(y, p_y)$ can also be developed to higher order in displacement from p_0 [23,37] if more accurate representations are required which are valid further away from p_0. This latter approach is especially useful for the uniform approximation of transmission above a threshold energy that is described in Section 4.5.

Equation 4.7 provides a multidimensional analog of Equation 4.1, allowing us in a sense to assign transmission probability as a function of phase space (which peaks on p_0). The dominant scale of the operator $\hat{T}$ is set by the factor $e^{-\theta}$ which is a direct analog of the one-dimensional probability expressed in Equation 4.1. The operator part $e^{-2\pi\hat{h}/\hbar}$ describes how transmission probability

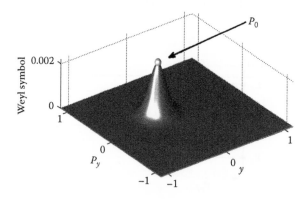

FIGURE 4.5 The Weyl symbol $\mathcal{W}_{\hat{\mathcal{R}}}(y, p_y)$ of $\hat{\mathcal{R}}$ is illustrated for the model problem of a particle of unit mass in the potential $V(x, y) = \text{sech}^2 x + \frac{1}{2}(y - a)^2$ with $a = \frac{1}{2}\text{sech}^2 x$ and $\hbar = 0.05$. The case shown is for an energy $E = 0.95$ just below the threshold value $E = 1$. Here transmission is entirely by tunneling and the Weyl symbol is peaked around the initial condition p_0 for $\gamma_E(t)$.

decays as one moves away from p_0. Although the uncertainty principle forbids us from making an explicit pointwise description of transmission probability in phase space, the Weyl symbol allows us to assign a locally averaged probability, for example (see also the discussion following the uniform approximation [Equation 4.26] in Section 4.5.2). In particular, the maximum transmission probability is achieved when the incident wave coincides with the ground state $|\varphi_0\rangle$ of the generator $\hat{h}$, which, in the quadratic approximation, is a coherent state centered on p_0. A straightforward calculation shows that this maximum transmission probability can be written for the case of a two-dimensional barrier in the form $e^{-\theta}\Lambda^{-1/2}$, where $\Lambda > 1$ is a real eigenvalue of the matrix W. This observation follows from denoting the eigenvalue of $\hat{h}$ corresponding to $|\varphi_0\rangle$ by $\hbar\omega/2$ and then noting that $\Lambda = e^{2\pi\omega}$ is an eigenvalue of W. More generally, if the incident wave coincides with excited eigenstates $|\varphi_k\rangle$ of $\hat{h}$ (with $k = 0, 1, 2, \ldots$), then the transmission probability in quadratic approximation is of the form $e^{-\theta}\Lambda^{-k-1/2}$, decaying geometrically as k increases and the state $|\varphi_k\rangle$ is localized further away from p_0 in phase space.

4.3 APPLICATIONS TO TUNNELING RATES FROM QUASIBOUND STATES

As well as characterizing the scattering matrix for a barrier, the operator $\hat{T}$ is a useful means of calculating tunneling rates from quasibound states in potential wells, particularly when these have chaotic dynamics. In fact, it is in precisely this context that the construction was introduced in Creagh and Whelan (1999) [28]. To describe this application, let us consider Gamow–Siegert states

$$\hat{H}\psi_n = \left(E_n - \frac{i\Gamma_n}{2} \right) \psi_n, \tag{4.8}$$

labeled by a quantum number n, where ψ_n satisfies outward-radiating boundary conditions at infinity. We consider tunneling from a chaotic potential well that is connected by a barrier or barriers to open channels, so that Γ_n is dominated by tunneling and exponentially small and ψ_n is almost a bound state supported in the well. Particularly in cases of chaotic dynamics, we may not have direct semiclassical approximations of the wavefunction ψ_n but we may still use the $\hat{T}$ operator to separate the problem of adding tunneling corrections and computing Γ_n from the problem of characterizing ψ_n (and the quasienergy E_n) within the well. A basis for achieving this is the Herring formula [43] and we outline its derivation here because it provides a useful means of characterizing tunneling in phase space later.

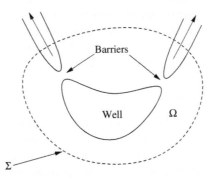

FIGURE 4.6 A schematic representation of potential contours of a well supporting a Gamow–Siegert state is shown. The region Ω contains the well itself and any barriers across which it couples to the continuum. The section Σ is its boundary. The arrows indicate the direction of flux measured by the Herring formula.

4.3.1 HERRING'S FORMULA IN ABSTRACT FORM

Let the section Σ be the boundary of a region Ω containing the potential well and any barriers connecting the well to open channels (see Figure 4.6) and let $\hat{P}_\Omega$ be a Hermitian operator whose classical analog projects onto Ω. An obvious choice is to define $\hat{P}_\Omega$ so that $\hat{P}_\Omega \psi(\mathbf{x}) = \chi_\Omega(\mathbf{x}) \psi(\mathbf{x})$, where $\chi_\Omega(\mathbf{x})$ is the characteristic function of Ω, but it helps later generalization to tunneling in phase space to allow $\hat{P}_\Omega$ to be more general than this in principle. In problems of dynamical tunneling, for example, we might want $\hat{P}_\Omega$ to project onto a region in momentum space, or to project to one side of a generically defined surface in phase space. The essential properties we demand are that $\hat{P}_\Omega \psi_n$ should be an approximate, normalized solution of the Schrödinger equation, so that $(\hat{H} - E_n)\hat{P}_\Omega \psi_n \approx 0$ and $\langle \psi_n | \hat{P}_\Omega | \psi_n \rangle \approx 1$. Then, from Equation 4.8, we have

$$\langle \psi_n | \hat{P}_\Omega \hat{H} | \psi_n \rangle = \left(E_n - \frac{i\Gamma_n}{2} \right) \langle \psi_n | \hat{P}_\Omega | \psi_n \rangle.$$

Subtracting the complex conjugate to extract the imaginary part of this equation gives

$$\langle \psi_n | \hat{P}_\Omega | \psi_n \rangle \Gamma_n = i \langle \psi_n | [\hat{P}_\Omega, \hat{H}] | \psi_n \rangle$$

$$\Rightarrow \quad \Gamma_n \approx \hbar \langle \psi_n //_\Sigma \psi_n \rangle, \tag{4.9}$$

where, for general ψ, the quantity $\langle \psi //_\Sigma \psi \rangle$ is defined by the relation

$$\langle \psi //_\Sigma \psi \rangle \equiv -\frac{1}{i\hbar} \langle \psi | [\hat{P}_\Omega, \hat{H}] | \psi \rangle = -\frac{d}{dt} \langle \psi | \hat{P}_\Omega | \psi \rangle. \tag{4.10}$$

This is a physically natural result, relating Γ_n to the outward flux from the well, but it is also technically appealing because it provides a compact and practically useful basis on which to compute escape rates. A second interpretation of $\langle \psi //_\Sigma \psi \rangle$ is as a "sectional overlap," providing a means by which we can define inner products restricted to the section Σ, related to the formalism underlying the Bogomolny transfer operator of mappings on quantized Poincaré sections [13]. To see this it helps to specialize to the case where $\hat{P}_\Omega \psi(\mathbf{x}) = \chi_\Omega(\mathbf{x}) \psi(\mathbf{x})$, in which case it is not hard to show that (allowing more generally for nondiagonal overlaps)

$$\langle \psi //_\Sigma \varphi \rangle = \frac{\hbar}{2im} \int_\Sigma \left(\psi^* \frac{\partial \varphi}{\partial n} - \frac{\partial \psi^*}{\partial n} \varphi \right) ds,$$

where s denotes arc length (or a higher-dimensional variant of it as appropriate) on Σ and the normal n points away from Ω. The abstract definition above provides us with a means of generalizing such sectional integrals to surfaces Σ which may be defined in phase space and which do not necessarily project simply to surfaces in configuration space. This feature is particularly valuable for a phase-space treatment of barrier penetration in the next section.

4.3.2 TUNNELING FROM CHAOTIC STATES

In order to exploit the Herring formula, we must first know ψ_n on Σ. A useful approach suggested by Auerbach and Kivelson (1985) [2] and by Wilkinson and Hannay (1987) [83] is first to determine ψ_n in or near the well and then to apply Green functions, which may be approximated in terms of complex trajectories (of the sort underlying t_{PR} in the previous section and corresponding to the upper contour in Figure 4.4), to continue the wavefunction across the barrier and out to the section Σ. The attraction of this procedure is that the Green function incorporates tunneling effects while the calculation of ψ_n inside the well may be based on approximations in which tunneling effects are not explicitly included. Although for chaotic problems in particular we may not explicitly know ψ_n

even inside the well, standard techniques of quantum chaos such as periodic orbit theory or Random Matrix Theory may be applied to characterize these wavefunctions collectively.

Qualitatively, while the instanton orbit on its own sets a dominant mean scale for tunneling rates within a given small window of energy (or any other parameter), there are significant fluctuations between one state and the next which are determined by the weight of the wavefunction ψ_n on the orbit $\gamma_E(t)$ [6,12,25–29,36,62]. It was shown in Creagh and Whelan (1999) [28] that the quantitative relation between the wavefunction and the orbit $\gamma_E(t)$ can be compactly represented as a sectional overlap on a section running through the potential well. For simplicity, assume that a single barrier provides a dominant route for tunneling from the well. Then the result can be formally written as a matrix element of the operator $\hat{T}$ (this time given directly by Equation 4.4, without the conjugation in Equation 4.5),

$$\Gamma_n = \langle \tilde{\psi}_n | \hat{T} | \tilde{\psi}_n \rangle, \tag{4.11}$$

where $\tilde{\psi}_n$ is a reduced wavefunction (in the sense of Bogomolny [13]) on the Poincaré section. This provides us with a means of interpreting the escape rates of individual states in terms of the morphology of the wavefunction in the well. As in the case of a simple isolated barrier, the complex orbit $\gamma_E(t)$ has a real segment extending into the potential well. Let p_0 be the point at which this real segment intersects the Poincaré section for which Equation 4.11 is written. Then, in view of the strong localization of phase-space representations of $\hat{T}$ around p_0 as illustrated in Figure 4.5, Equation 4.11 simply represents Γ_n as a measure of the strength of the wavefunction around $\gamma_E(t)$ (in real space). The real segment of $\gamma_E(t)$ runs up to the most efficient route through the barrier (i.e., to the barrier-crossing orbit with the least imaginary action), so this result is unsurprising in qualitative terms. It should be emphasized, however, that this result also provides an explicit qualitative estimate relating Γ_n to the detailed structure of the wavefunction around $\gamma_E(t)$.

A very similar calculation allows us to treat the case of energy-level splittings in multidimensional, symmetric double-well potentials. The splitting ΔE_n of a doublet with levels $E_n^{\pm} = E_n \mp \Delta E_n/2$ can be expressed in the same form,

$$\Delta E_n = 2 \langle \tilde{\psi}_n | \hat{T} | \tilde{\psi}_n \rangle, \tag{4.12}$$

except that the boundary conditions of the complex orbit underlying $\hat{T}$ are different. Here the orbit starts in one well, crosses the barrier once, and ends in the other well at the image under the underlying symmetry operation of the initial condition. This differs from the orbit underlying Γ_n for a Gamow–Siegert state because it has traversed the barrier only once, whereas the orbit underlying Γ_n traverses the barrier twice, once in each direction. Otherwise, however, the tunneling operators underlying splittings and resonances are formally almost the same.

In problems where the well dynamics are chaotic, so that wavefunctions are generically uniformly spread over the corresponding energy-level surface, the matrix element (Equation 4.11) will be dominated by the immediate neighborhood of p_0 on a Poincaré section (in two dimensions, within an area of $O(\hbar)$). In this case, the generating form for $\hat{T}$ in Equation 4.7, with $\hat{h}$ truncated at quadratic order, suffices. Then $\hat{T}$ can be constructed as the quantization of the symplectic matrix W linearizing the map F around p_0 and its Weyl symbol, for example, is a simple Gaussian straightforwardly constructed from the matrix elements of W. The matrix elements (Equation 4.11 or 4.12) then provide explicit quantitative estimates for the tunneling rates of individual states in terms of a Gaussian-weighted integral of a Wigner function of the state over a small region containing p_0 [28]. Although for such chaotic problems we do not have a direct analytical representation of individual wavefunctions, we can often model eigenstates collectively, using periodic orbit theory [6,26,27] or statistically [12,29,36,62], for example, and the operator $\hat{T}$ allows us to make these connections transparently.

(a)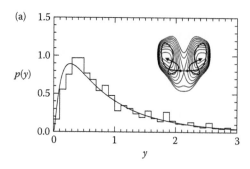

(b)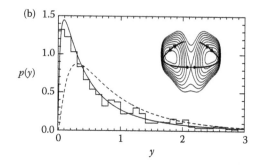

FIGURE 4.7 Distributions of the fluctuations of tunneling rates about the mean are modeled statistically by assuming the wavefunction is distributed according to a Gaussian distribution. A generic case is shown in (a). In (b), the distribution is for a special case where the real extension of the tunneling orbit is periodic. In this case, the distribution deviates from the generic distribution (dashed curve) and must be modified to account for the effects of scarring (solid curve). Note that in this case scarring has a leading order effect on the distribution and that this remains the case in the limit $\hbar \to 0$. (Adapted from S. C. Creagh, S.-Y. Lee, and N. D. Whelan. *Ann. Phys.*, 295:194, 2002.)

As an example of this sort of approach we highlight in particular predictions for the statistics of tunneling rates which can be deduced as a consequence of the Berry–Voros conjecture [7,80], which in this context means that the quasirandom fluctuations in the wavefunction from one level to the next can be modeled as a Gaussian random distribution when the underlying dynamics are ergodic. Furthermore, this calculation offers a very stark, physically measurable, manifestation of the phenomenon of *scarring* [42], because the statistics of tunneling rates may deviate strongly from those of generic models in problems where the real segment of $\gamma_E(t)$ is a periodic orbit of the real, in-well dynamics (Figure 4.7) [12,25,29,36]. Scarring here refers to the fact that wavefunction statistics along (real) periodic orbits are anomalous [46,47]. Furthermore, although one typically thinks of scarring as an effect that becomes weaker in the semiclassical limit, where Schnirelmann's theorem [33,66] guarantees that almost all wavefunctions are uniformly distributed, scarring-induced anomalies in tunneling statistics are seen to persist as $\hbar \to 0$.

It is helpful to separate fluctuations in tunneling rates from their average behavior by using the decomposition

$$\Gamma_n = y_n \bar{\Gamma}(E)$$

(with an obvious analog for splittings ΔE_n), where the fluctuating factors y_n are normalized so that

$$\langle y_n \rangle = 1.$$

The mean tunneling rate $\bar{\Gamma}(E)$ changes monotonically with system parameters such as energy and can be approximated in terms of the action and stability matrix W of the periodic orbit $\gamma_E(t)$ [28]. While $\bar{\Gamma}(E)$ sets the dominant overall scale for tunneling rates in a given parameter window, one finds that there are also fluctuations from one state to the next which are characterized by y_n, whose variance is $O(1)$ for chaotic problems. By combining the matrix element (Equation 4.11) with the $\hbar$-generated form for $\hat{T}$ in Equation 4.7, one can deduce that (for two-dimensional problems)

$$y_n \propto \sum_{k=0}^{\infty} \Lambda^{-k} |\langle \varphi_k | \tilde{\psi}_n \rangle|^2,$$

where $\Lambda > 1$ is a (real) eigenvalue of the stability matrix W and $|\varphi_k\rangle$, $k = 0, 1, 2, \ldots$ are the eigenvectors of the generator $\hat{h}$. By assuming that overlaps $x_k = \langle \varphi_k | \tilde{\psi}_n \rangle$ can be modeled as identically distributed, statistically independent random variables, in line with the Berry–Voros conjecture, probability distributions can be derived for y_n [29] which work well in generic problems such as the case illustrated in Figure 4.7a (which, although showing splittings rather than resonance widths, is modeled in exactly the same way).

This assumption regarding the statistics of x_k is not valid, however, when the states $|\varphi_k\rangle$ are localized in phase space on a periodic orbit of the real, in-well dynamics. It has been shown by Kaplan and coworkers [46,47] that significant deviations occur in the overlap statistics in that case that can be accounted for by classic constructions in the theory of scarring. That approach can be adapted to predict anomalous distributions for the tunneling-rate fluctuations y_n also [25], as illustrated in Figure 4.7b. It is interesting to note that, because the footprints of the states $|\varphi_k\rangle$ shrink as $\hbar \to 0$, such deviations persist in the semiclassical limit, without violating Schnirelman's theorem.

4.4 UNIFORM APPROXIMATION AND BARRIER PENETRATION IN PHASE SPACE

We have so far treated barrier penetration essentially as an extension of Equation 4.2, in which the transmitted wave is obtained using a simple spatial continuation of the incident wave along complex trajectories and across the barrier. While this is a usefully direct way of approximating tunneling in a variety of multidimensional tunneling problems, there are at least two contexts in which it becomes unsatisfactory. First, as the energy approaches and rises above threshold, where classical transmission becomes possible, we can no longer treat caustics or turning points on the two sides of the barrier as being distinct and must turn instead to uniform approximations which treat them jointly. Second, especially in problems of dynamical tunneling, we may need to generalize the picture we use to allow for tunneling in momentum rather than in configuration space, or even tunneling across more general divisions of phase space. For such calculations it is useful to allow for more general phase-space–based descriptions of tunneling, such as we outline in this section. We motivate the treatment initially by considering the problem of one-dimensional barrier penetration as the energy approaches threshold.

4.4.1 DIRECT APPROACH TO UNIFORM TRANSMISSION

The primitive, one-dimensional transmission amplitude in Equation 4.3 has the following generalization to the case where the energy is near, or even above, the threshold energy of the barrier [9]:

$$t \approx \frac{e^{-\theta/2+i\delta}}{\sqrt{1+e^{-\theta}}}, \tag{4.13}$$

where

$$\delta = \frac{\theta}{2\pi} \ln \left| \frac{\theta}{2\pi e} \right| \arg \Gamma \left(\frac{1}{2} - \frac{i\theta}{2\pi} \right).$$

Note that by writing the action θ as a contour integral

$$\theta = \frac{1}{i\hbar} \oint_C p \, dx$$

in the complex x-plane, around a closed contour C enclosing the turning points (x_1, x_2), we see that the exponent θ varies smoothly as a function of energy across threshold even though the turning points collide and move up the imaginary action (Figure 4.8). This feature generalizes to the action of a periodic orbit in the multidimensional case, as illustrated in Figure 4.2. Conventions are such that θ is positive below threshold and changes sign at threshold, becoming negative above.

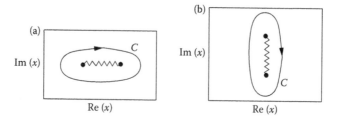

FIGURE 4.8 A schematic illustration is given of the contour C used to define the tunneling action in (a) the below-threshold case and (b) the above-threshold case. Although the turning points collide at threshold and move onto the imaginary axis, the action depends smoothly on energy across this transition.

The corresponding transmission probability $T = |t|^2$ is simpler than the amplitude itself:

$$T = \frac{e^{-\theta}}{1 + e^{-\theta}}, \tag{4.14}$$

rising from being very small below threshold to being close to unity above, in a smooth version of the abrupt classical transition. This relative simplicity of the transmission probability is understandable in view of the corresponding complex orbits, which can be associated with time contours of the form shown in Figure 4.4. An orbit underlying the amplitude t approaches the barrier in positive real time, crosses it in imaginary time, and then carries on on the other side of the barrier with further positive real-time evolution (as for the top contour in Figure 4.4). As energy approaches the barrier top, the real segments of this evolution get ever longer and there is a resulting singularity in the classical dynamics. By contrast, the concatenated orbit underlying T follows this evolution with a retracing of the real segments in negative time (as for the lower contour in Figure 4.4), leaving a net imaginary time of evolution which remains finite as energy approaches threshold. The resulting classical dynamics are then perfectly regular, a feature which carries through to their multidimensional generalization as the dynamics of the map F of Section 4.2, which can be smoothly continued above threshold. This will later be used as the basis of a multidimensional generalization of Equation 4.14.

The most direct derivation of Equation 4.13 is achieved by seeking a change of coordinate $x \to s(x)$ (and a corresponding norm-preserving transformation of the wavefunction $\psi(x)$ to $\varphi(s) = x'(s)^{1/2}\psi(x(s))$) which transforms the Schrödinger equation to a standard, solvable form [9,54]. For barrier penetration, this new form

$$-\frac{\hbar^2}{2}\varphi''(s) + U(s)\varphi(s) = \mathcal{E}\varphi(s)$$

should be the Schrödinger equation for an inverted parabola potential $U(s) = -s^2/2$. It is not hard to see that a semiclassical expansion which achieves this can be written, which begins at leading order with a coordinate change such that action is preserved:

$$S(x, x_0) = \int_{s_0}^{s} \sqrt{2(\mathcal{E} - U(s'))}\,ds'.$$

The resulting solutions can be applied globally (on the real line) as long as a one-to-one matching can be made between turning points of the original problem and turning points of the transformed problem. In the case of a barrier, the Schrödinger equation for the inverted parabola potential can be solved in terms of known functions and the transmission amplitude in Equation 4.13 can then be deduced from their asymptotics. This *method of comparison equations* is described in detail in the classic review article of Berry and Mount (1972) [9], for example, and, like the simple continuation of the wavefunction around turning points implicit in Equation 4.2, is an appealingly direct means of treating near-threshold transmission problems.

4.4.2 UNIFORM TRANSMISSION BY PHASE-SPACE TRANSFORMATION

The method of comparison equations cannot be applied to higher-dimensional problems, however, and nor does it generalize easily to Hamiltonians that are not of kinetic-plus-potential type, or to situations where tunneling occurs across more general divisions of phase space. It is more powerful in such cases to consider canonical transformations which act on the entire phase space rather than on configuration space alone, as happens with the method of comparison equations. The goal of such a transformation is to change a general Hamiltonian into, say, some standard normal form which is more amenable to direct analytical investigation, and in particular to use such a classical transformation to guide a corresponding quantum unitary change of basis. A semiclassical description of the quantization of canonical transformations into quantum unitary operators has been provided by Miller (1974) [55]. This can be used as the beginning of a more systematic transformation at arbitrary order in $\hbar$ to a desired normal form, as described in Cargo et al. (2005) [18], for example. Alternatively, direct quantum analogs of classical normal form expansions can be developed which again generate a desired normal form for the Hamiltonian operator (see Ali (1985) [1], Crehan (1990) [32], and Eckhardt (1986) [39] for early approaches to bound states and Waalkens et al. (2008) [81] for a more recent treatment directed explicitly at barrier problems), although these latter approaches necessarily apply locally in phase space whereas Miller's semiclassical approach is in principle capable of producing global transformations. If one is interested only in calculating invariants such as resonant energies or fluxes, however, it is not necessary to know in detail how such transformations work, only to know that in principle they can be constructed and then one can calculate the required invariants directly in the transformed basis. This is the approach we take here. Transmission probabilities and related higher-dimensional constructions can be defined so that they are independent of the choice of basis. We therefore simply assert that a unitary basis change can be found which transforms any given Hamiltonian operator into a quantum analog of a classical normal form and compute fluxes for the transformed problem.

The uniform transmission probability in Equation 4.14 is an example of a quantity that can be calculated in this way. We first assume that the original one-dimensional barrier problem can be transformed to one for which the classical Hamiltonian is

$$H_0 = f(I), \quad I = \frac{1}{2}\left(p_0^2 - q_0^2\right),$$

where (q_0, p_0) are the new canonical coordinates and I is an action variable appropriate to the hyperbolic dynamics around the unstable equilibrium at the barrier top. This form for the Hamiltonian is a natural product of normal form calculations. An alternative representation of the action variable I is obtained by a further $\pi/4$ rotation in phase space (Figure 4.9), leading us to

$$I = -QP,$$

where (Q, P) are the rotated canonical coordinates. The sign of I is chosen here so that in later calculations it is positive above threshold.

This latter form enables us very easily to characterize the complex periodic orbit and to compute its action. Hamilton's equations immediately yield the solution

$$Q(t) = e^{-\lambda t}Q(0), \quad P(t) = e^{\lambda t}P(0), \tag{4.15}$$

where $\lambda = f'(I)$ is constant on each orbit, and the complex periodic orbit is trivially found in this representation since these solutions admit the imaginary period

$$Q(t - i\tau_0) = Q(t) \quad \text{and} \quad P(t - i\tau_0) = P(t),$$

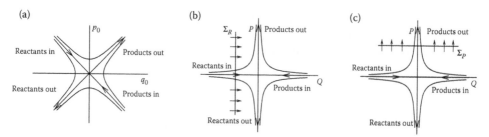

FIGURE 4.9 Illustration of the phase space rotation and directions of flux used to calculate uniform transmission probabilities. (a) Illustrates the original coordinates while (b) and (c) illustrate the rotated coordinates, with the section Σ_R shown in (b) and the section Σ_P shown in (c).

where $\tau_0 = 2\pi/\lambda$. Furthermore, an elementary calculation gives

$$\theta = \frac{1}{i\hbar} \oint_{\gamma_E(t)} P dQ = -\frac{2\pi I}{\hbar} \tag{4.16}$$

for the imaginary action of one of these periodic orbits.

The transformed and rotated Hamiltonian therefore trivialises the calculation of the complex periodic orbit underlying the barrier penetration problem. This process has an even starker simplifying effect on the quantum calculation. Let us suppose that an analogous unitary transformation of the quantum Hamiltonian puts it in the form $\hat{H} = f(\hat{I})$, where $\hat{I} = (\hat{p}_0^2 - \hat{q}_0^2)/2$. A more detailed investigation reveals that a quantization of the $\pi/4$ rotation that followed in the classical case results in the symmetrized action operator

$$\hat{I} = -\frac{1}{2} \left(\hat{Q}\hat{P} + \hat{P}\hat{Q} \right). \tag{4.17}$$

This final form of the action is desirable because, in a representation of the wavefunction where $\hat{Q}\psi(Q) = Q\psi(Q)$, the corresponding eigenvalue equation $\hat{I}\psi = \mathcal{I}\psi$ is a first-order differential equation and easily solved without recourse to special functions. In fact, a simple calculation leads to the two fundamental solutions [34,35,60]

$$\psi_+(Q) = \Theta(Q)Q^{-1/2-i\mathcal{I}/\hbar}, \quad \psi_-(Q) = \psi_+(-Q),$$

where $\Theta(Q)$ denotes the Heaviside step function. With the sign conventions used (see Figure 4.9), the solution $\psi_-(Q)$ represents a wave incident on the barrier of the original problem from the left or, in chemical language, from the reactant side. The solution $\psi_+(Q)$ represents a solution incident from the right, or product side.

Probabilities of reflection and transmission for such solutions can be reduced to a calculation of fluxes. These can be defined invariantly and can therefore be calculated directly in the transformed Hamiltonian without having to supply details of the transformation used to get to the representation $\psi(Q)$. The key to this calculation is the generalized definition of flux in Equation 4.10. It can be shown that [21], for states ψ which satisfy $\hat{I}\psi = \mathcal{I}\psi$, and for a projection operator of the form $\hat{P}_\Omega \psi(Q) = \Theta(Q - Q_0)\psi(Q)$, the generalized flux in Equation 4.10 takes the specialized form

$$\langle \psi /\!/_{\Sigma_R} \psi \rangle = \lambda Q_0 |\psi(Q_0)|^2, \tag{4.18}$$

where here we denote $\lambda = f'(\mathcal{I})$. In this case, Σ_R is a section defined by $Q = Q_0 < 0$ and sign conventions are such that flux is measured from left to right across Σ_R, as illustrated in Figure 4.9b. The solution

$$\psi_{\mathcal{I}}(Q) = \lambda^{-1/2}\psi_-(Q)$$

is then normalized to have unit flux across Σ_R:

$$\langle \psi_{\mathcal{I}} // _{\Sigma_R} \psi_{\mathcal{I}} \rangle = 1.$$

In the original problem illustrated in Figure 4.9a, this corresponds to a wave incident from the reactant side of the barrier and normalized to have unit flux coming toward the barrier.

The advantage of the representation of the Hamiltonian in Equation 4.17 is that fluxes moving away from the barrier are calculated just as easily and from an almost identical calculation (reflecting the symmetry in the representation between Q and P). Fluxes away from the barrier can be measured across sections Σ_P of the form $P = P_0$, as in the illustration in Figure 4.9c. Working with a momentum-space representation

$$\varphi_{\mathcal{I}}(P) = \frac{1}{\sqrt{2\pi\hbar}} \int e^{-iQP/\hbar} \psi_{\mathcal{I}}(Q)\,\mathrm{d}Q,$$

flux across such a section is calculated using a direct analog of Equation 4.18

$$\langle \varphi_{\mathcal{I}} // _{\Sigma_P} \varphi_{\mathcal{I}} \rangle = \lambda P_0 |\varphi(P_0)|^2. \tag{4.19}$$

If the flux incoming on the reactant side of the barrier is unity, this outgoing flux is a probability of transmission to the product side where $P > P_0 > 0$ in our sign conventions. For the solution $\psi_{\mathcal{I}}(Q)$, this calculation can be completed straightforwardly [21] and leads to

$$T \equiv \langle \varphi_{\mathcal{I}} // _{\Sigma} \varphi_{\mathcal{I}} \rangle = \frac{e^{\pi\mathcal{I}/\hbar}}{e^{\pi\mathcal{I}/\hbar} + e^{-\pi\mathcal{I}/\hbar}}, \tag{4.20}$$

which reproduces Equation 4.14 with the identification $\theta = -\mathcal{I}/(2\pi\hbar)$. Note that this calculation is exact for any Hamiltonian of the form $\hat{H} = f(\hat{I})$. Semiclassical errors in Equation 4.14 arise entirely from the transformation which puts the original Hamiltonian in this form in the first place.

4.5 UNIFORM MULTIDIMENSIONAL BARRIER CROSSING

Besides offering an elegant derivation of the uniform approximation (Equation 4.14) for one-dimensional tunneling, phase-space transformations also allow us to treat multidimensional barrier penetration. Again, we will adopt the terminology of chemical reaction and base our treatment on a normal form Hamiltonian.

4.5.1 CLASSICAL PICTURE

Canonical coordinates $(Q, P, q, p) = (Q, P, q_1, \ldots, q_d, p_1, \ldots, p_d)$ are chosen so that the Hamiltonian begins with the quadratic terms

$$H(Q, P, q, p) = \lambda_0 I + \sum_{i=1}^{d} \omega_i(q_i^2 + p_i^2) + \cdots,$$

where, as in the previous section, $I = -QP$. At higher order, these coordinates can be chosen so that the Hamiltonian still depends on the coordinates (Q, P) only through I, so that

$$H(Q, P, q, p) = g(q, p, I). \tag{4.21}$$

Let us further denote by

$$f(I) = g(q_e, p_e, I)$$

the minimum value of $g(q,p,I)$ obtained by varying the transverse coordinates (q,p) while keeping I fixed. This minimum occurs at $(q,p) = (q_e(I), p_e(I))$, where $q_e(0) = p_e(0) = 0$.

Separate from its use here to characterize tunneling, this normal form construction has recently emerged as a useful tool in the application of classical transition state theory to the calculation of chemical reaction rates (see Uzer (2002) [79] and Waalkens et al. (2008) [81] and references therein). Before describing its relevance to tunneling, let us summarize some key features of the classical geometry of such problems (illustrated schematically in Figure 4.10). These can be constructed immediately once the normal form is known, although they do not rely on the normal form for their definition. The set $Q = P = 0$ describes a normally hyperbolic invariant manifold (or NHIM), which generalizes to higher dimensions and to a phase-space setting the periodic orbit dividing surface PODS construction used in Pechukas (1976) [61] and Pollak et al. (1980) [65] to divide reactants from products in configuration space. For us, the important feature of this construction is that the stable and unstable manifolds of the NHIM, which satisfy the condition $I = 0$, form a boundary in phase space between the reacting and nonreacting sets of trajectories. We focus here especially on the stable manifold W^s of the NHIM which, in our conventions (see Figure 4.9), can be defined by the condition $P = 0$. Above threshold the set of reacting trajectories forms a cylinder bounded by W^s, as illustrated in Figure 4.10 (with, as we will see, the instanton orbit $\gamma_E(t)$ running through its center). We denote by V the set of initial conditions on a section Σ_R for these reacting trajectories. As total energy decreases to its threshold value, this set of reacting trajectories shrinks and collapses on the real segment of $\gamma_E(t)$ exactly at threshold. Below threshold, there is no longer a classically reacting region but we will see that, in a sense, a quantum ghost of the departed reacting region remains behind in the form of the orbit $\gamma_E(t)$ and its immediate neighborhood (cf. Figures 4.5 and 4.11).

As in the one-dimensional case, the normal form construction allows us explicitly to construct the orbit $\gamma_E(t)$ in the coordinates (Q,P,q,p). Substitution in Hamilton's equations shows that $\gamma_E(t)$ corresponds to the following extension of Equation 4.15,

$$Q(t) = e^{-\lambda t}Q(0), \quad P(t) = e^{\lambda t}P(0), \quad q(t) = q_e, \quad p(t) = p_e, \qquad (4.22)$$

where $\lambda = f'(I)$, I is determined by the condition $f(I) = E$ and t varies over a contour of the form illustrated in Figure 4.4 with a net imaginary time of $-i\tau_0 \equiv -2\pi i/\lambda$. The action is obtained by an extension of Equation 4.16

$$\theta = \frac{1}{i\hbar} \oint_{\gamma_E(t)} P dQ + p dq = -\frac{2\pi I}{\hbar}. \qquad (4.23)$$

One also sees that, corresponding to a minimum of $g(q,p,I)$ at fixed I, the coordinates (q_e, p_e) of $\gamma_E(t)$ in the transverse degrees of freedom correspond to a point in the middle of the reacting region as illustrated schematically in Figure 4.10.

The instanton orbit and its above-threshold extension can be used as the basis for an approximation of cumulative reaction probability [56] by evaluating its contribution to the trace

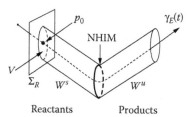

FIGURE 4.10 Schematic representation of the geometry of the normally hyperbolic invariant manifold (NHIM), its stable manifold W^s and the real segment of the instanton orbit $\gamma_E(t)$. The stable manifold defines the boundary of the reacting subset V of the section Σ_R and the initial condition p_0 for $\gamma_E(t)$ on Σ_R is at the centre of V.

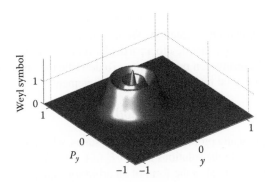

FIGURE 4.11 Weyl symbol of $\hat{\mathcal{R}}$ for the same system as in Figure 4.5 but now with an energy $E = 1.25$ above threshold. This value of energy is such that about five states of area h fit in the classically reacting region. The Weyl symbol in this case undergoes a sharp but smooth transition across the boundary of V from oscillation to exponential decay.

formula [15,41]. A quantum version of the normal form can also be used, however, to determine also how that transmission probability is distributed over phase space.

4.5.2 QUANTUM PICTURE

To tackle the quantum tunneling problem, we next assume that a similar transformation can be achieved for the quantum Hamiltonian which leaves it in an analogous form $\hat{H} = g(\hat{q}, \hat{p}, \hat{I})$. With the help of the one-dimensional scattering states

$$\hat{I}\psi_{\mathcal{I}}(Q) = \mathcal{I}\psi_{\mathcal{I}}(Q)$$

described in the previous section, we can solve for the stationary states of this transformed problem explicitly [21,81]. We simply substitute separated scattering states

$$\Psi_{\mathcal{I},k}(Q,q) = \psi_{\mathcal{I}}(Q)\varphi_{\mathcal{I},k}(q)$$

and find that these are stationary states satisfying $\hat{H}\Psi_{\mathcal{I},k}(Q,q) = E_k\Psi_{\mathcal{I},k}(Q,q)$ provided the transverse part satisfies an eigenvalue equation

$$H(\hat{q}, \hat{p}, \mathcal{I})\varphi_{\mathcal{I},k}(q) = E_k(\mathcal{I})\varphi_{\mathcal{I},k}(q),$$

where the operator $H(\hat{q}, \hat{p}, \mathcal{I})$ represents a partial symbol of $\hat{H} = g(\hat{q}, \hat{p}, \hat{I})$ in which the operator $\hat{I}$ is replaced with the eigenvalue $\mathcal{I}$. We assume that this transverse problem has a discrete spectrum labeled by the quantum number k and note that these eigenvalues are parametrized by $\mathcal{I}$ because the transverse Hamiltonian $H(\hat{q}, \hat{p}, \mathcal{I})$ is parametrized by $\mathcal{I}$. Note that k is a quantum number characterizing scattering states in a neighborhood of the barrier where the normal form applies and is distinct from quantum numbers of asymptotic scattering states. The distinction arises because evolution across the barrier region does not commute with evolution from the barrier region to asymptotically far incoming channels.

The punchline of this procedure is that a calculation of generalized flux can be performed in a direct generalization of the calculation in the previous section. The result is an expression

$$T \equiv \langle \Psi_{\mathcal{I},k} // \_{\Sigma}\Psi_{\mathcal{I},k} \rangle = \frac{e^{\pi\mathcal{I}/\hbar}}{e^{\pi\mathcal{I}/\hbar} + e^{-\pi\mathcal{I}/\hbar}}$$

for the transmission probability of the scattering state $\Psi_{\mathcal{I},k}(Q,q)$ which looks identical to the one-dimensional result (Equation 4.20).

This result is not of direct use without knowing explicitly the transformation of the state $\Psi_{\mathcal{I},k}(Q,q)$ back to the original representation. We can, however, use it as the basis of an operator form of Equation 4.14 which is not tied to a particular representation and which can be used for a practical calculation of transmission probabilities. To describe the outcome of this procedure we identify scattering states at a total fixed energy with basis elements

$$\Psi_{\mathcal{I}_k(E),k}(Q,q) \sim |k\rangle$$

of the space $\mathcal{H}_R$ of asymptotically propagating incoming states, which we assume has dimension M. Here $\mathcal{I}_k(E)$, $k = 1,\ldots,M$ denote the solutions in $\mathcal{I}$ of $E_k(\mathcal{I}) = E$. Then, a calculation of transmitted flux for an incoming state which takes the form of an arbitrary superposition $|\chi\rangle = \sum_k a_k|k\rangle$ can be shown [21] to be of the form

$$T = \langle\chi|\hat{\mathcal{R}}|\chi\rangle, \tag{4.24}$$

where

$$\hat{\mathcal{R}} = \sum_{k=1}^{M} \frac{e^{2\pi\mathcal{I}_k(E)/\hbar}}{1 + e^{2\pi\mathcal{I}_k(E)/\hbar}} |k\rangle\langle k| \tag{4.25}$$

and where the states $|k\rangle$ form an orthonormal basis for $\mathcal{H}_R$ with respect to the inner product $\langle\Psi_{\mathcal{I}_k(E),k} \,//_{\Sigma_R}\, \Psi_{\mathcal{I}_{k'}(E),k'}\rangle \equiv \langle k|k'\rangle = \delta_{kk'}$. By defining

$$\hat{T}(E) = \sum_{k=1}^{M} e^{2\pi\mathcal{I}_k(E)/\hbar} |k\rangle\langle k|,$$

this can be written in the form

$$\hat{\mathcal{R}} = \frac{\hat{T}(E)}{1 + \hat{T}(E)}. \tag{4.26}$$

By writing

$$\hat{T}(E) = e^{-\theta}e^{-2\pi\hat{h}/\hbar}, \quad \text{where} \quad \hat{h} = -\sum_{k=1}^{M} \left(\mathcal{I}_k(E) + \frac{\hbar\theta}{2\pi}\right) |k\rangle\langle k|,$$

the operator $\hat{T}$ is expressed formally very much like the operator of the same symbol defined in Section 4.2. In fact, a more detailed investigation reveals that $\hat{h}$ is indeed a quantum analog of the classical generator $h(q,p)$ of a complex Poincaré map $F : \Sigma_R \to \Sigma_R$. The only difference is that the section Σ_R is defined using the new canonical coordinates (Q,P,q,p). Once this identification has been made, we now simply observe that the unitary conjugation needed to express this operator in the original representation of the scattering problem is achieved within semiclassical approximation by performing an analogous conjugation of the corresponding classical Poincaré maps. The result is that, in the original coordinates, $\hat{T}$ is constructed as described in the discussion following Equation 4.5.

Once $\hat{T}$ is interpreted as a quantization of a Poincaré map in this way, we no longer require explicit consideration of the normal form in order to exploit the uniform approximation in Equation 4.26. The operator $\hat{\mathcal{R}}$ can in principle be calculated following approximations of $\hat{T}$ of the sort given in Equation 4.4, for example. To interpret the resulting semiclassical approximation it is instructive to consider a phase-space representation such as the Weyl symbol defined in Equation 4.6. For energies below barrier, where $\hat{T}$ is small and we can use the primitive approximation $\hat{\mathcal{R}} \approx \hat{T}$, the Weyl symbol $\mathcal{W}_{\hat{\mathcal{R}}}(y,p_y)$ is a localized Gaussian as discussed in Section 4.2 and illustrated in Figure 4.5a. As energy passes through and rises above threshold, on the other hand, it becomes important to use the full uniform expression (Equation 4.26). The qualitative nature of $\hat{\mathcal{R}}$ changes significantly during this transition, reflecting the fact that a classically reacting region emerges in phase space. Significantly above threshold and away from the boundary of the reacting subset V of Σ_R, and in the limit $\hbar \to 0$, the Weyl symbol of $\hat{\mathcal{R}}$ approaches the classical transmission probability

$$\mathcal{W}_{\hat{\mathcal{R}}}(y, p_y) \sim \chi_V(y, p_y).$$

Here, $\chi_V(y, p_y)$ denotes the characteristic function of V. For finite values of $\hbar$, however, the uniform approximation (Equation 4.26) captures interesting deviations from this classical limit due to tunneling. These are especially significant outside V, near its boundary, or when energy is close to threshold.

A specific numerical example is shown in Figure 4.11. The energy has been chosen in this case so that the classically reacting region is small in semiclassical terms, having an area of about five units of Planck's constant h. Therefore, even though the classical footprint of the reacting region is clearly evident in this figure—the sharp decay of the Weyl symbol in Figure 4.11 begins at the boundary of V—there are also clear signatures of quantum structure in the oscillations inside the reacting region and in the smooth decay across the boundary of V. Explicit calculation shows that this sort of quantum structure can be very well reproduced by inserting semiclassical approximations of the sort (Equation 4.4) in Equation 4.26 [23,37,38]. Although the inversion of $1 + \hat{T}$ in Equation 4.26 has been performed numerically in Drew et al. (2005) [38], fully semiclassical implementation of Equation 4.26 can be based on generating forms for $\hat{T}$ such as given in Equation 4.7, for example [23,37].

A phase-space representation of $\hat{\mathcal{R}}$ such as $\mathcal{W}_{\hat{\mathcal{R}}}(y, p_y)$ comes close to generalizing Equation 4.14 to a transmission probability that is distributed as a function of position in phase space, although the uncertainty principle prevents us from making this identification in a simple pointwise manner. The oscillations in the classically reacting region in Figure 4.11, for example, have at least as much to do with the nature of the Weyl representation as they do with the structure of $\hat{\mathcal{R}}$. One can, however, make an explicit identification in an averaged sense. The probability of transmission expressed in Equation 4.24 can be expressed in the Wigner–Weyl formalism in the form

$$T = \int_{\Sigma_R} W(y, p_y) \mathcal{W}_{\hat{\mathcal{R}}}(y, p_y) \, dy \, dp_y,$$

where $W(y, p_y)$ is the Wigner function on Σ_R of the incoming state $|\chi\rangle$ (defined as the Weyl symbol of $|\chi\rangle\langle\chi|/(2\pi\hbar)^d$). This provides a natural quantum analog of the corresponding classical expression

$$T_{cl} = \int_{\Sigma_R} f(y, p_y) \chi_V(y, p_y) \, dy \, dp_y$$

for the reacting fraction of a classical ensemble described by a density $f(y, p_y)$. The closest we can come in the quantum case to a pointwise specification of transmission probability in phase space is when the Wigner function $W(y, p_y)$ is that of a minimum-uncertainty wavepacket, in which case the transmission probability is an average of $\mathcal{W}_{\hat{\mathcal{R}}}(y, p_y)$ over a region of area $O(h)$ in Σ_R.

4.6 TUNNELING BEYOND THE INSTANTON

A description of tunneling which is based on the instanton trajectory $\gamma_E(t)$ (and the map F it induces on a section Σ_R) captures the dominant rates of transmission across a barrier when the energy is near or below threshold. There are important exceptions, however, where more complicated complex dynamics taking place further away from $\gamma_E(t)$ take over and the simple instanton-based description given in the previous section is no longer sufficient.

A particularly important case is that of scattering problems in which the incoming wave is localized in phase space away from $\gamma_E(t)$ (or any classically reacting region surrounding it), or in which the total energy is significantly higher than the barrier energy. It has been shown by Takahashi and Ikeda [72–76] and by Levkov and coworkers [10,11,49–51] that new complex orbits can then dominate tunneling. These new orbits begin on the Lagrangian manifold λ^{in} underlying the incoming state and approach the NHIM along its stable manifold W^s, before falling into the product channel along its unstable manifold W^u. Geometrically, initial conditions for them can be found on the section Σ_R near the complex intersections of λ^{in} with W^s and by beginning near the stable manifold the imaginary action is minimized. One finds associated plateau structures in this regime in the spectrum of transmission probability to outgoing modes, where the imaginary action depends only weakly on the quantum numbers of the outgoing mode. This mechanism is described in more detail in an article by Takahashi in this volume.

A second scenario where the instantonic description of tunneling may be inadequate arises in the context described in Section 4.3, where we would like to find tunnel splittings or resonance widths of states supported in multidimensional wells. If the wavefunctions in the well do not have significant support around $\gamma_E(t)$, then the basic results (Equations 4.11 and 4.12), while still formally correct, need to be treated more carefully. This may happen particularly where the real segment of $\gamma_E(t)$ is in a region of phase space where dynamics is regular rather than chaotic. Most states in this case will be localized in phase space either on invariant tori or in chaotic regions which do not cover $\gamma_E(t)$ and, due to effects of dynamical tunneling, will already be in a state of exponential decay around $\gamma_E(t)$.

A treatment which combines this decay due to dynamical tunneling with decay along the instanton is achieved by using saddle point approximation to evaluate the relevant matrix element. To fix ideas, let us consider the case of a two-dimensional Gamow–Siegert state localized on a quantized invariant torus. When the problem is far enough from being completely integrable—where mechanisms such as resonance-assisted and chaos-assisted tunneling dominate, for example—it may be impractical or even impossible to characterise the evanescent decay of the wavefunction over $\gamma_E(t)$ sufficiently well to carry this program successfully to its conclusion. Close enough to integrability, however, it seems natural to base such a calculation on the standard WKB form,

$$\psi \approx \sum_{\text{branches}} \sqrt{\rho}\, e^{iS/\hbar}, \qquad (4.27)$$

where the density ρ and action S are obtained as a by-product of the construction of action angle variables for the torus, which we denote by (I, J, θ, ϕ). A steepest-descents approximation of Equation 4.11 for such a state receives contributions from initial conditions for the map F which are complex, lying deeper in complex space the further the torus is localized from the real segment of $\gamma_E(t)$. The matrix element may be approximated in terms of the geometry of the intersection in a section Σ_R of the (complexified) torus with its image under the map F. In practice, however, natural boundaries may prevent analytic continuation of the torus from ever reaching such an intersection as discussed in more detail below.

We describe this situation further using an alternative approach [82] in which equivalent results are obtained following direct substitution of the WKB form (Equation 4.27) in Herring's formula (Equation 4.9). In this picture we are required to extend the WKB form (Equation 4.27) as far as a section Σ outside the well of the sort illustrated in Figure 4.6. This approach also allows us to treat problems in which escape is entirely by dynamical tunneling rather than guided by the instanton, such as in the scalar optical problem illustrated in Figure 4.12. Assuming that the classical data required by Equation 4.27 can be found, steepest-descents evaluation of Herring's formula receives dominant contributions from escaping trajectories that are real. The examples illustrated in Figure 4.12 are consistent with this picture, in which with the directions of peak escaping intensity occur along such real escaping trajectories.

In evaluating Herring's integral, one finds that there is a striking qualitative difference between problems which are exactly integrable and those which are merely nearly integrable (or KAM-like [52]). In exactly integrable problems, the invariants (such as angular momentum in a circular cavity) which define the underlying invariant tori are global and, furthermore, real-valued when evaluated on real phase space. This means that the escaping orbits which have tunnelled outside the well define a continuously real family as they escape to infinity. The imaginary part of the exponent

(a) 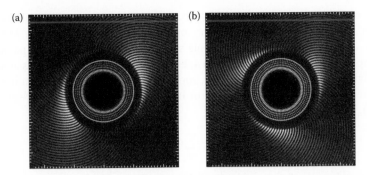 (b)

FIGURE 4.12 The emitted wave is illustrated for Gamow–Siegert states in slightly deformed circular cavities (with the solution outside the cavity being exaggerated graphically to make it visible). The model used is that of a scalar Helmholtz equation with refractive index $n = 2$ inside the cavity and $n = 1$ outside, which can alternatively be mapped to a step-potential quantum problem. The deformation in (a) is elliptical whereas in (b) the radial perturbation is proportional to $\cos 3\theta$. The directions of strongest escaping intensity in these examples are consistent with the WKB form (Equation 4.27) around (isolated) real escaping rays, although a detailed analysis of the case in (b) shows that natural boundaries will often have intervened before the underlying classical data can be reached.

is therefore stationary along the entire range of integration in the Herring formula and the result is an expression for the resonance width which looks one-dimensional:

$$\Gamma^{\text{int}} = \frac{\hbar\omega}{2\pi} e^{-K_0/\hbar}, \tag{4.28}$$

where iK_0 is an imaginary action connecting the real torus inside the well to the real escaping family and ω is a frequency of motion. For merely near-integrable problems, on the other hand, there is no such symmetry (under complex conjugation) and the escaping rays are generically complex. An invariant J for which the escaping family forms a level surface will be complex-valued when evaluated on real escaping rays. In addition, the real escaping rays around which saddle-point contributions are obtained in the Herring formula are isolated. In this case, the Herring formula yields a version of Wilkinson's formula [19,71,82],

$$\Gamma^{\text{ni}} = \sum_{\text{real escaping rays}} \left(\frac{\hbar}{2\pi}\right)^{1/2} \frac{\omega e^{-K_0/\hbar}}{\sqrt{i\{J, J^*\}}}, \tag{4.29}$$

where J is an extension of an interior action variable to the escaping family and ω is the frequency in the nonconjugate angle variable. (Note that a similar result has been given in Levkov et al. (2007) [50] for instanton-guided transmission probabilities of waves incident on a barrier.) If a perturbation strength is decreased so that a near-integrable problem approaches an integrable one, Equation 4.29 diverges but can be replaced by a uniform approximation joining it to the integrable limit (Equation 4.28) [70]. We remark finally that, besides changing the nature of the resonance splitting, the breaking of exact integrability has a dramatic effect on emission patterns of the sort illustrated in Figure 4.12. Whereas in near-integrable systems emission intensity peaks strongly along isolated real trajectories, all escaping trajectories are real in exactly integrable problems and there are no particularly preferred emission directions. Any directionality of emission is due to the amplitude term in Equation 4.27 and is qualitatively weaker [22,30].

Although the preceding picture is qualitatively consistent with emission patterns and tunneling rates observed in weakly perturbed integrable systems, it ignores an important problem, however. There do exist problems where Equation 4.29 and its variants give a complete and quantitative description of tunneling rates, such as splittings between states supported in multiple circular cavities [24,69] or lifetimes of whispering gallery modes in elliptic cavities [31]. More commonly, however, even very weakly perturbed integrable systems are found to develop natural boundaries before the complex data required for Equation 4.29, or for the calculation of emission patterns such as shown in Figure 4.12, can be reached. The impact of this problem is most immediately seen in an explicit representation of the underlying torus in the form of multiple Fourier series in the angle variables (θ, ϕ), such as

$$q(\theta, \phi) = \sum_{mn} q_{mn} e^{mi\theta + ni\phi}$$

for a position coordinate q, for example. Such Fourier series are often very convenient means of representing the classical data in numerical calculations and allow us to obtain complexified data simply by substituting complex values for (θ, φ) when these angles are close enough to the real plane. Deep enough into the complex angle plane, however, the exponentials $e^{mi\theta + ni\phi}$ will grow so quickly with (m, n) as to overwhelm the decay in the coefficients q_{mn} and the series diverges. For integrable systems, the band of convergence of the Fourier series is determined by isolated singularities in the angle variables and the series can in principle be resummed and continued beyond this limit. For KAM-like systems, however, one finds that the breakdown of the Fourier series coincides with a breakdown of the function itself. The series diverges along a *natural boundary*, a curve dense with singularities and beyond which continuation is meaningless for any physical purpose. The ubiquity of this natural boundary for KAM-type systems was first pointed out by Percival and Greene [63,64] and now represents a fundamental barrier to our ability to treat tunneling in KAM systems [19,30].

The question of whether semiclassical treatments of tunneling can be achieved in the regime of natural boundaries is an important open problem. The closest that we currently come to a systematic treatment of such structure in semiclassical approximation is in the work of Shudo and coworkers [67,68], who find that Julia sets, of which the natural boundary forms a part, have an important role to play in the dynamical evolution of wavefunctions. A Julia set consisting of trajectories in the complexified system which remain bounded indefinitely in forward iteration of the dynamics approaches in long-time evolution the set of orbits contributing with smallest imaginary action to iteration of a wavefunction. Dealing with problems of evolution in time that work does not directly answer the questions posed here regarding the properties of energy eigenstates. In principle, however, an understanding of the time development of wavefunctions can be used to characterize energy eigenstates, although this connection has not yet been made concretely in the context of Julia sets and natural boundaries.

Some progress can be made toward calculating energy eigenstates when the system is sufficiently close to being integrable. In explicit solutions of the quantum or wave problem one frequently finds behavior that seems qualitatively consistent with WKB approximations of the form (Equation 4.27) even when natural boundaries are known to intervene in the exact classical data. This is the case in the emitted wave of Figure 4.12b, for example. This suggests that an ansatz of the form (Equation 4.27) may successfully describe the wave solution even in the presence of natural boundaries. One means of realizing this is to allow the action S to satisfy the resulting eikonal equation only approximately, so that it is smooth enough to be continued beyond the natural boundary of the exact solution, and yet accurate enough to provide a faithful description of the wave solution. For perturbations of an integrable problem in which the perturbation parameter is of order $\varepsilon = O(\hbar)$, this can be achieved simply by beginning a classical perturbation series

$$S = S_0 + \varepsilon S_1 + \cdots$$

for the action, whose development is a classical textbook problem [52]. When $\varepsilon = O(\hbar)$, it suffices to truncate this series at the first-order term in order to develop a consistent semiclassical expansion and this approach has proved to be a successful strategy in quantum [22,70] and optical [30] problems. It can be used to explain emission patterns of the sort illustrated in Figure 4.12, for example (in which the perturbation is too large to apply Rayleigh–Schrödinger perturbation directly to the wave problem itself).

If the perturbation parameter is larger than $O(\hbar)$, but still classically small, one might in principle be able to make progress by developing the classical perturbation series to higher order. It is not clear, however, how far this procedure can be pushed before the natural boundary presents an impenetrable obstacle to direct semiclassical approximation. It is also known that, eventually, new regimes take over where the tunneling mechanism involves more than just a simple extension of the torus to complex space. As the perturbation parameter increases, one first encounters resonance-assisted tunneling [3,16,17,40,48], in which resonant island chains of the classical problem mediate coupling of the state to other tori which may have a faster tunneling rate, or be themselves resonantly coupled to further tori with yet faster tunneling rates, and so on. At larger perturbations still, coupling to chaotic seas can also take over, in the regime of chaos-assisted tunneling [14,78]. Considerable progress has been made in achieving quantitative analyses of tunneling in these regimes, but important gaps remain in our understanding of even simple aspects of tunneling from regular states. Even the regime of direct tunneling from and between states supported on KAM-like tori in which direct WKB approximations (Equation 4.27) seem like the most appropriate route to a quantitative description is not fully understood. Until these simpler problems are understood, we cannot hope to arrive at a complete semiclassical understanding of dynamically richer regimes such as those of resonance-assisted and chaos-assisted tunneling. The outstanding problems of understanding tunneling from stationary states in the presence of natural boundaries and finding connections (if any) with the role played by Julia sets in time-dependent problems are therefore worthy targets of attention in future investigation.

4.7 CONCLUSIONS

All semiclassical treatments of tunneling have at their heart problems in the complexified dynamics of the underlying classical limit. These complex dynamical calculations are sometimes technically challenging and there are consequently many problems in tunneling whose complete solution still seems rather far away. The examples explored in this article hopefully demonstrate, however, that one can also sometimes make significant progress on the basis of relatively simple dynamical constructions.

The most basic tunneling problem of all is one-dimensional barrier penetration and we have argued that a solution of this problem already contains the nucleus of a treatment of tunneling in a variety of more interesting multidimensional scenarios. The discussion is centered on the instanton orbit, and dynamics in a neighborhood of it in the multidimensional case. Although dynamics around the instanton cannot explain everything about the multidimensional problem [10,11,49–51,72–76], we can use it to describe the dominant tunneling rates across barriers in the time-independent problem. On the basis of a quantized version of a Poincaré map constructed around the instanton, we can, for example, develop phase-space pictures of the below-threshold scattering matrix or understand tunneling from states trapped in chaotic potential wells. As a particularly stark illustration of the relevance of the instanton geometry to nontrivial tunneling problems, we have highlighted the fact that the scarring phenomenon has an important influence on escape-rate statistics when the real extension of the instanton is a periodic orbit.

The instanton also provides the basis of a phase-space representation of scattering from a multidimensional barrier which unifies the picture above threshold of a nonempty classically reacting region with the picture below threshold of a classically small region of maximum tunneling probability. Our treatment has taken advantage of recent emphasis on use of normal forms to treat the classical problem [79,81] although it should be emphasized that the final result (Equation 4.26) can in principle be used without explicit reference to the normal form at all.

There is more to tunneling than the instanton, however, and we have concluded by describing some particular instances where alternative approaches are needed. This discussion focuses particularly on tunneling from near-integrable states, primarily because some explicit analytic progress is possible for that problem. Even in this restricted regime, where more complicated mechanisms such as resonance-assisted and chaos-assisted tunneling have yet to take over, a direct calculation of tunneling is thwarted by the emergence of natural boundaries which prevent continuation of the classical data sufficiently far into the complex plane. The tunneling wavefunctions observed for such problems are nevertheless qualitatively consistent with what one expects from a straightforward WKB ansatz, so it seems natural that some direct description of tunneling should be possible. For small enough deviations from exact integrability, a limited solution is outlined which proceeds on the basis of perturbation theory applied to the classical solutions (rather than directly to the wave problem itself). The broader problems, however, of how to approach direct tunneling for larger deviations from integrability, where the limits of direct tunneling lie, and the question of how to characterize the boundary between this regime and those of resonance-assisted and chaos-assisted tunneling still present significant theoretical challenges.

ACKNOWLEDGMENTS

The author is indebted to Niall Whelan, in collaboration with whom the tunneling operator picture described in Section 4.2 was developed and from which the uniform approximation of barrier penetration described in Section 4.4 subsequently emerged. He is also grateful to Soo-Young Lee, Graeme Smith, Chris Drew, and Richard Tew for collaborations whose outcomes have also been described in this article. The author acknowledges the support of the EPSRC under grant number EP/F036574/1.

REFERENCES

1. M. K. Ali. The quantum normal form and its equivalents. *J. Math. Phys.*, 26:2565, 1985.
2. A. Auerbach and S. Kivelson. The path decomposition expansion and multidimensional tunneling. *Nuc. Phys. B*, 257:799, 1985.
3. A. Bäcker, R. Ketzmerick, S. Löck, and L. Schilling. Regular-to-chaotic tunneling rates using a fictitious integrable system. *Phys. Rev. Lett.*, 100:104101, 2008.
4. T. Banks and C. M. Bender. Coupled anharmonic oscillators. II.+ Unequal-mass case. *Phys. Rev. D*, 8:3366, 1973.
5. T. Banks, C. M. Bender, and T. T. Wu. Coupled anharmonic oscillators. I. Equal-mass case. *Phys. Rev. D*, 8:3346, 1973.
6. M. W. Beims, V. Konratovich, and J. B. Delos. Semiclassical representation of width-weighted spectra. *Phys. Rev. Lett.*, 81:4537, 1998.
7. M. V. Berry. Regular and irregular semiclassical wavefunctions. *J. Phys. A*, 10:2083, 1977.
8. M. V. Berry. Quantum scars of classical closed orbits in phase space. *Proc. Roy. Soc. A*, 423:219, 1989.
9. M. V. Berry and K. E. Mount. Semiclassical approximations in wave mechanics. *Rep. Prog. Phys.*, 35:315, 1972.
10. F. Bezrukov and D. G. Levkov. Transmission through a potential barrier in quantum mechanics of multiple degrees of freedom: Complex way to the top. arXiv:quant-phys0301022v1, 2003.
11. F. Bezrukov and D. G. Levkov. Dynamical tunneling of bound systems through a potential barrier: Complex way to the top. *JETP*, 98:820, 2004.
12. W. E. Bies, L. Kaplan, and E. J. Heller. Scarring effects on tunneling in chaotic double-well potentials. *Phys. Rev. E*, 64:061204, 2001.
13. E. Bogomolny. Semiclassical quantization of multidimensional systems. *Nonlinearity*, 5:85, 1992.
14. O. Bohigas, S. Tomsovic, and D. Ullmo. Manifestations of classical phase space structures in quantum mechanics. *Phys. Rep.*, 223:45, 1992.
15. M. Brack and R. K. Bhaduri. *Semiclassical Physics*. Westview, Boulder, 2003.
16. O. Brodier, P. Schlageck, and D. Ullmo. Resonance-assisted tunneling. *Phys. Rev. Lett.*, 87:064101, 2001.

17. O. Brodier, P. Schlagheck, and D. Ullmo. Resonance-assisted tunneling. *Ann. Phys.*, 300:88, 2002.

18. M. Cargo, A. Gracia-Saz, R. G. Littlejohn, M. W. Reinsch, and P. de M Rios. Quantum normal forms, Moyal star product and Bohr-Sommerfeld approximation. *J. Phys. A*, 38:1977, 2005.

19. S. C. Creagh. Tunnelling in two dimensions. In S. Tomsovic (Ed.), *Tunnelling in Complex Systems*, pp. 35–100. World Scientific, Singapore, 1998.

20. S. C. Creagh. Classical transition states in quantum theory. *Nonlinearity*, 17:1261, 2004.

21. S. C. Creagh. Semiclassical transmission across transition states. *Nonlinearity*, 18:2089, 2005.

22. S. C. Creagh. Directional emission from weakly eccentric resonators. *Phys. Rev. Lett.*, 98:153901-1, 2007.

23. S. C. Creagh, C. Drew, and R. H. Tew. Uniform approximation of barrier transmission from normal forms. in preparation, 2010.

24. S. C. Creagh and M. D. Finn. Evanescent coupling between disks: A model for near-integrable tunneling. *J. Phys. A*, 34:3701, 2001.

25. S. C. Creagh, S.-Y. Lee, and N. D. Whelan. Scarring and the statistics of tunneling. *Ann. Phys.*, 295:194, 2002.

26. S. C. Creagh and N. D. Whelan. Complex periodic orbits and tunneling in chaotic potentials. *Phys. Rev. Lett.*, 77:4975, 1996.

27. S. C. Creagh and N. D. Whelan. Homoclinic structure controls chaotic tunneling. *Phys. Rev. Lett.*, 82:5237, 1999.

28. S. C. Creagh and N. D. Whelan. A matrix element for chaotic tunneling rates and scarring intensities. *Ann. Phys.*, 272:196, 1999.

29. S. C. Creagh and N. D. Whelan. Statistics of chaotic tunneling. *Ann. Phys.*, 272:196, 1999.

30. S. C. Creagh and M. White. Evanescent waves outside eccentric optical cavities. In A. V. Kudryashov, A. H. Paxton, and V. S. Ilchenko (Eds.), *Laser Resonators and Control XII*, Vol. 7579 of *Proc. SPIE*, pp. 716–731, 2010.

31. S. C. Creagh and M. M. White. Evanescent escape from the dielectric ellipse. *J. Phys. A*, 43:456102, 2010.

32. P. Crehan. The proper quantum analogue of the Birkhoff–Gustavson method of normal forms. *J. Phys. A*, 23:5815, 1990.

33. Y. Colin de Verdière. Ergodicité et fonctions propres du laplacien. *Commun. Math. Phys.*, 102:497, 1985.

34. Y. Colin de Verdière and B. Parisse. Equilibre instable en régime semiclassique I. *Commun. Math. Part. Diff. Eqns.*, 19:1535, 1994.

35. Y. Colin de Verdière and B. Parisse. Equilibre instable en régime semiclassique II. *Ann. Inst. H. Poincaré (Physique Théorique)*, 61:347, 1994.

36. D. Delande and J. Zakrzewski. Experimentally attainable example of chaotic tunneling: The hydrogen atom in parallel static electric and magnetic fields. *Phys. Rev. E*, 68:062110, 2003.

37. C. S. Drew. *Approaches to Transmission across Multidimensional Barriers*. PhD thesis, University of Nottingham, 2006.

38. C. S. Drew, S. C. Creagh, and R. H. Tew. Uniform approximation of barrier penetration in phase space. *Phys. Rev. A*, 72:062501, 2005.

39. B. Eckhardt. Birkhoff–Gustavson normal form in classical and quantum mechanics. *J. Phys. A*, 19:2961, 1986.

40. C. Eltschka and P. Schlagheck. Resonance- and chaos-assisted tunneling in mixed regular-chaotic systems. *Phys. Rev. Lett.*, 94:014101, 2005.

41. F. Haake. *Quantum Signatures of Chaos*. Springer, New York, NY, 2001.

42. E. J. Heller. Bound-state eigenfunctions of classically chaotic Hamiltonian systems: Scars of periodic orbits. *Phys. Rev. Lett.*, 53:1515, 1984.

43. C. Herring. Spin coupling at large distances. *Rev. Mod. Phys.*, 34:631, 1962.

44. C. Jaffé, D. Farrelly, and T. Uzer. Transition state in atomic physics. *Phys. Rev. A*, 60:3833, 1999.

45. C. Jaffé, D. Farrelly, and T. Uzer. Transition state without time-reversal symmetry: chaotic ionization of the hydrogen atom. *Phys. Rev. Lett.*, 84:610, 2000.

46. L. Kaplan. Wavefunction intensity statistics from unstable periodic orbits. *Phys. Rev. Lett.*, 80:2582, 1998.

47. L. Kaplan and E. J. Heller. Linear and nonlinear theory of eigenfunction scars. *Ann. Phys.*, 264:171, 1998.

48. S. Keshavamurthy. Resonance-assisted tunneling in three degrees of freedom without discrete symmetry. *Phys. Rev. E*, 72:045203R, 2005.

49. D. G. Levkov, A. G. Panin, and S. M. Sibiryakov. Complex trajectories in chaotic dynamical tunneling. *Phys. Rev. E*, 76:046209, 2007.

50. D. G. Levkov, A. G. Panin, and S. M. Sibiryakov. Unstable semilcassical trajectories in tunneling. *Phys. Rev. Lett.*, 99:170407, 2007.

51. D. G. Levkov, A. G. Panin, and S. M. Sibiryakov. Signatures of unstable semiclassical trajectories in tunneling. *J. Phys. A*, 42:205102, 2009.

52. A. J. Lichtenberg and M. A. Lieberman. *Regular and Chaotic Dynamics*. Springer, New York, NY, April 1992.

53. R. S. Mackay. Flux over a saddle. *Phys. Lett. A*, 145:425, 1990.

54. S. C. Miller and R. H. Good. A WKB-type approximation to the Schrödinger equation. *Phys. Rev.*, 91:174, 1953.

55. W. H. Miller. Classical-limit quantum mechanics and the theory of molecular collisions. *Adv. Chem. Phys.*, 25:69, 1974.

56. W. H. Miller. Semiclassical limit of quantum mechanical transition state theory for nonseparable systems. *J. Chem. Phys.*, 62:1899, 1975.

57. W. H. Miller. The classical S-matrix in molecular collisions. *Adv. Chem. Phys.*, 30:77, 1976.

58. W. H. Miller and T. F. George. Analytic continuation of classical mechanics for classically forbidden collision processes. *J. Chem. Phys.*, 56:5668, 1972.

59. W. H. Miller and T. F. George. Classical s-matrix theory of reactive tunneling: linear $h + h_2$ collisions. *J. Chem. Phys.*, 57:2458, 1972.

60. S. Nonnenmacher and A. Voros. Eigenstate structures around a hyperbolic point. *J, Phys. A*, 30:295, 1997.

61. P. Pechukas. Statistical approximations in collision theory. In W. H. Miller (Ed.), *Dynamics of Molecular Collisions*, pp. 269–322. Plenum, New York, NY, 1976.

62. L. M. Pecora, H. Lee, and D.-H. Wo. Regularization of tunneling rates with quantum chaos. Preprint, 2010.

63. I. C. Percival. Chaotic boundary of a hamiltonian map. *Physica D: Nonlinear Phenomena*, 6:67–77, 1982.

64. I. C. Percival and J. M. Greene. Hamiltonian maps in the complex plane. *Physica D: Nonlinear Phenomena*, 3:530–548, 1981.

65. E. Pollak, M. S. Child, and P. Pechukas. Classical transition state theory: A lower bound to the reaction probability. *J. Chem. Phys.*, 72:1669, 1980.

66. A. I. Schnirelmann. Ergodic properties of eigenfunctions. *Usp. Mater. Nauk.*, 29:181, 1974.

67. A. Shudo, Y. Ishii, and K. S. Ikeda. Julia sets and chaotic tunneling: I. *J, Phys. A*, 42:265101, 2009.

68. A. Shudo, Y. Ishii, and K. S. Ikeda. Julia sets and chaotic tunneling: II. *J, Phys. A*, 42:265102, 2009.

69. G. C. Smith. *Multidimensional Tunneling in Regular Systems, with Applications to Electromagnetic Problems*. PhD thesis, University of Nottingham, 2004.

70. G. C. Smith and S. C. Creagh. Tunnelling in near-integrable systems. *J. Phys. A*, 39:8283, 2006.

71. S. Takada. Multidimensional tunneling in terms of complex classical mechanics: Wave functions, energy splittings and decay rates in nonintegrable systems. *J. Chem. Phys.*, 104:3742, 1996.

72. K. Takahashi and K. S. Ikeda. Complex-domain semiclassical theory: Application to time-dependent barrier tunneling problems. *Found. Phys.*, 231:177, 2001.

73. K. Takahashi and K. S. Ikeda. An intrinsic multidimensional mechanism of barrier tunneling. *Europhys. Lett.*, 71:193, 2005.

74. K. Takahashi and K. S. Ikeda. Anomalously long passage through a rounded-off-step potential do to a new mechanism of multidimensional tunneling. *Phys. Rev. Lett.*, 71:240403, 2006.

75. K. Takahashi and K. S. Ikeda. A plateau structure in the tunneling spectrum as a manifestation of a new tunneling mechanism in multidimensional barrier systems. *J. Phys. A*, 41:095101, 2008.

76. K. Takahashi and K. S. Ikeda. Spectroscopic signature of the transition in a tunneling mechanism from the instanton path to a complexified stable–unstable manifold. *Phys. Rev. A*, 79:052114, 2009.

77. M. Toller, G. Jacucci, G. DeLorenzi, and C. P. Flynn. Theory of classical diffusion jumps in solids. *Phys. Rev. B*, 32:2082, 1985.

78. S. Tomsovic and D. Ullmo. Chaos-assisted tunneling. *Phys. Rev. E*, 50:145, 1994.

79. T. Uzer, C. Jaffé, J. Palacián, P. Yanguas, and S. Wiggins. The geometry of reaction dynamics. *Nonlinearity*, 15:957, 2002.

80. A. Voros. Semiclassical ergodicity of quantum eigenstates in the Wigner representation. In G. Casati and J. Ford (Eds.), *Stochastic Behavior in Classical and Quantum Hamiltonian Systems*, p. 326. Springer, Berlin, 1979.
81. H. Waalkens, R. Schubert, and S. Wiggins. Wigner's dynamical transition state theory in phase space: Classical and quantum. *Nonlinearity*, 21:R1, 2008.
82. M. Wilkinson. Tunnelling between tori in phase space. *Physica D*, 21:341, 1986.
83. M. Wilkinson and J. H. Hannay. Multidimensional tunneling between excited states. *Physica D*, 27:201, 1987.

5 Semiclassical Analysis of Multidimensional Barrier Tunneling

Kin'ya Takahashi

CONTENTS

5.1 INTRODUCTION

Tunneling for multidimensional barrier systems is the essential and long-standing problem in the fields of quantum physics and molecular science. It is considered as one of the minimum models for study of tunneling phenomena in multidimensional systems.

To the author's knowledge, the first pioneer work applying (complex) semiclassical method to the multidimensional scattering problem was done by Miller and his coworkers [1,2]. They calculated with the complex semiclassical S-matrix (the so-called Miller's classical S-matrix) a tunneling reaction path in collinear collision between an atom and a diatomic molecule. What they found was a novel tunneling path which is not predicted by a 1D adiabatic approximation: it is not a path along the reaction coordinate defined by the adiabatic approximation, but takes a short cut across the corner of a curved gully potential.

This tunneling path is really a product of the multidimensional geometry of the potential, but still forms a 1D-like trajectory as an instanton, which works even for multidimensional systems which are classically integrable or nearly integrable [3].

Tunneling peculiar multidimensional systems is observed for classically nonintegrable systems. Except for a few cases, most cases in which essentially multidimensional tunneling is observed and has been studied in recent works are not of energy barrier tunneling but of dynamical tunneling [4–6]. A typical situation of the multidimensional tunneling is such that an initial quantum state put in a classically regular region penetrates into a chaotic sea nearby it or into another regular region related with the original one by some geometrical symmetry, for example, tunneling between twin tori. Chaos- and resonance-assisted tunneling are good examples of this type of tunneling [7–12].

In multidimensional barrier tunneling,* this type of tunneling, namely dynamical barrier tunneling [14], is observed for periodically perturbed 1D barrier potentials [15–21] and for 2D and more than 2D barrier potentials into which an incident particle with a total energy more than the saddle of potential and with a properly chosen channel state is injected [22,23]. In the latter case, there exists a dividing surface above the saddle separating the reactant side from the product side, which is formed by an unstable periodic orbit for 2D systems or a normally hyperbolic invariant manifold (NHIM) for more than 2D systems [24], and acts as a dynamical barrier for a proper choice of initial channel state.

The question is what is the carrier of quantum tunneling transportation if the instanton method or 1D reductive tunneling path along the reaction coordinate is not available to explain a tunneling phenomenon observed. Unfortunately, we do not still have a complete answer for this question, which allows us to understand the basic mechanism inherent in multidimensional tunneling. One of the candidates which clarify the basic mechanism of tunneling from the semiclassical point of view was introduced by Shudo, Ishii, and Ikeda [25,26]. They applied a fully complexified semiclassical method, according to Miller's guideline for complexification of semiclassical operator [1,2], to the tunneling problem for quantum kicked rotors or quantum maps, and they found that the Julia set, a sort of invariant set in the complex phase space, plays a key role to make tunneling transportation through the multidimensional complex phase space. Roughly speaking, forward and backward Julia sets correspond to complexified stable and unstable manifolds, respectively, and guide tunneling trajectories.

The essentially same mechanism was also found by Takahashi and Ikeda for multidimensional barrier tunneling [16–21,27]. They studied tunneling for periodically perturbed 1D barrier potentials with a complex semiclassical method improved based on Miller's classical S-matrix [1,2,28], and found that complexified stable and unstable manifolds of the unstable periodic orbit above the potential barrier play a key role for tunneling transportation when the perturbation strength is strong enough. Since the stable and unstable manifolds are still robust invariant manifolds in the complex phase space, it is quite natural to consider from the semiclassical point of view that they guide complexified classical flows carrying quantum tunneling probability in classically forbidden regions. Let us call the new tunneling mechanism as Stable–Unstable Manifold Guided Tunneling (SUMGT). As products of SUMGT, several novel tunneling phenomena are found, for example, the characteristic waveform of scattering eigenstate called fringed tunneling [16,17] and the characteristic tunneling spectrum called plateau spectrum [18,19]. Note that the same mechanism was also reported in later works with a different terminology, the so-called "sphaleron" [23].

In this review, we will discuss transition of semiclassical mechanism from the instanton tunneling suitable for integrable systems to SUMGT inherent in classically nonintegrable systems with change of perturbation strength [20]. To do this, we need to study continuous time systems instead of discretized time systems, namely mappings, though it is easier to handle complex-domain dynamics

---

* Most of multidimensional scattering systems with a nonlinear coupling potential are classically nonintegrable by the Painlevé test [13].

for discretized time systems than for continuous time systems. This is because the discretized time systems have no analytical continuation to complex time, namely a δ kick as a distribution is not an analytical function and has no analytical extension to the complex plane. On the other hand, continuous-time systems are naturally extended into the complex phase space involving complex time evolution. To discuss instantons created with imaginary time evolution [3], it is crucial to analytically continue the system under consideration to a complex time domain. This is the reason why we treat continuous-time systems.

5.2 CHARACTERISTIC OF MULTIDIMENSIONAL BARRIER TUNNELING

5.2.1 1.5D MODEL SYSTEM AND S-MATRIX

The minimum model system suitable for our purpose is an oscillating 1D barrier potential. Here we choose an oscillating Eckart barrier, whose Hamiltonian is written by

$$\hat{H}(Q, \hat{P}, \omega t) = \frac{1}{2}\hat{P}^2 + (1 + \varepsilon \sin \omega t)\text{sech}^2(Q). \tag{5.1}$$

Let us assume that an incident wave is coming from $+Q$ infinity with a constant momentum $P_1(<0)$. If the incident energy $E_1(= P_1^2/2)$ is taken small enough such that a classical particle with an arbitrary phase with respect to the applied force is always reflected by the oscillating potential, then the quantum probability observed in the transmissive side is of quantum tunneling. The tunneling spectrum is calculated with the S-matrix.

An expression of the S-matrix suited for reduction to a semiclassical formula was first introduced by Miller et al., who wrote it in terms of Green function [1,2]. Miller's S-matrix modified to be fit for periodically perturbed 1D systems is given by [19,28]

$$S(E_2, E_1) = \frac{\sqrt{|P_1||P_2|}}{2\pi\hbar} \lim_{|Q_1|, |Q_2| \to \infty} e^{(-iP_2 Q_2 + iP_1 Q_1)/\hbar} \int_{-\infty}^{\infty} dt_2 \, e^{i(E_2 - E_1)t_2/\hbar}$$
$$\times \int_0^{\infty} ds \, < Q_2|\hat{U}(\omega t_2 : \omega t_2 - \omega s)|Q_1 > e^{iE_1 s/\hbar}, \tag{5.2}$$

where subscripts 1 and 2 of dynamical variables stand for input and output, respectively, and $\hat{U}(\theta + \omega t : \theta)$ denotes the time propagator of the system under consideration:

$$\hat{U}(\theta + \omega t : \theta) = \mathcal{T} \exp\left\{-\frac{i}{\hbar} \int_0^t ds \, \hat{H}(\theta + \omega s)\right\}, \tag{5.3}$$

where $\mathcal{T}$ denotes the time ordering operator. The initial point Q_1 is put in the positive asymptotic side $Q_1 \to \infty$, and the observation point Q_2 is set in a negative or positive asymptotic side depending on the dynamical variable observed, for example, $Q_2 \to -\infty$ for tunneling. The S-matrix does not depend on choice of initial and observation points if they are in asymptotic ranges, since the phase term $e^{(-iP_2 Q_2 + iP_1 Q_1)/\hbar}$ balance the initial and final phases out.

5.2.2 TUNNELING SPECTRUM

The energy spectrum given by the S-matrix (Equation 5.2) can be calculated numerically by the scheme in Takahashi and Ikeda (1997) [29]. Figure 5.1 shows tunneling spectra obtained numerically, where the input energy is put at $E_1 = 0.5$ [19,20]. The tunneling spectrum consists of delta spikes with the interval $\hbar\omega$ due to the periodicity of the perturbation. For a strong perturbation at $\varepsilon = 0.4$, the spectrum envelop forms a plateau spread over a wide range of energy. As shown later, this characteristic spectrum is the result of SUMGT and its width, which is roughly estimated as $1 - \varepsilon < E_2 < 1 + \varepsilon$, corresponds to the oscillating range of the real unstable manifold W_{uR} at an

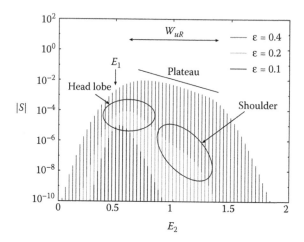

FIGURE 5.1 Tunneling spectra (absolute value of the S-matrix) at three representative values of the perturbation strength at $\varepsilon = 0.1$, $\varepsilon = 0.2$, and $\varepsilon = 0.4$. Other parameters are taken as $E_1 = 0.5$, $\omega = 0.3$, and $\hbar = 1000/(3\pi \times 2^{10}) \sim 0.1036$. The spectrum is normalized such that $\sum_n |S(E_2 = E_1 + n\hbar\omega, E_1)|^2$ gives the total transmissive probability.

asymptotic side ($|Q| \gg 1$). On the other hand, for a weak perturbation at $\varepsilon = 0.1$, the spectrum is localized around E_1, for which a perturbed instanton theory must be available. The interesting case appears for an intermediate strength at $\varepsilon = 0.2$. The spectrum seems to be constructed by the superposition of two characteristic spectra: a head lobe will be explained by the perturbed instanton theory, while a shoulder over an upper range of energy will be formed by SUMGT. Thus, the two tunneling mechanisms will coexist in this case and so we explore what kind of change occurs with ε from the semiclassical point of view in the following sections.

5.3   SEMICLASSICAL FORMULA

We introduce the semiclassical formula of the S-matrix. Applying the saddle-point approximation to the integrals in Equation 5.2 gives the semiclassical expression of the S-matrix for periodically perturbed systems [19,28],

$$S(E_2, E_1) \sim \lim_{|Q_1|,|Q_2| \to \infty} \sum_{\text{c.t.}} \frac{\sqrt{|P_2||P_1|}}{\sqrt{2\pi i \hbar P_1 P_2}} \sqrt{-\frac{\partial^2 S_S}{\partial E_1 \partial E_2}} e^{-i(P_2 Q_2 - P_1 Q_1)/\hbar} e^{i S_S(Q_2, E_2, Q_1, E_1)/\hbar}, \qquad (5.4)$$

where the classical action S_S is defined by

$$S_S(Q_2, E_2, Q_1, E_1) = \int_{Q_1}^{Q_2} P dQ - \int_{t_1}^{t_2} H(Q, P, \omega t) dt + E_2 t_2 - E_1 t_1, \qquad (5.5)$$

where $H(Q, P, \omega t)$ denotes the classical counterpart of Hamiltonian (Equation 5.1). The summation $\sum_{\text{c.t.}}$ is taken over all the contributing trajectories satisfying the boundary condition, which are calculated by the set of canonical equations,

$$\dot{Q} = \frac{\partial H}{\partial P} = P,$$

$$\dot{P} = -\frac{\partial H}{\partial Q} = -\frac{\partial V(Q, \omega t)}{\partial Q}, \qquad (5.6)$$

where $V(Q, \omega t)$ is the potential of the system under consideration. To calculate complex trajectories contributing to tunneling, Equation 5.6 is extended to the complex domain.

Let us consider the boundary condition of complex trajectories contributing to tunneling. The initial and final sets of dynamical variables deciding the classical action, namely (Q_1, E_1) and (Q_2, E_2) in this case, are quantum observables specifying the initial and final states. They must be assigned real numbers. On the other hand, t_1 and t_2 are canonically conjugate to E_1 and E_2, respectively and are not observed quantum mechanically. In the complex semiclassical method initiated by Miller et al. [1,2], it is assumed that classical dynamical variables not observed quantum mechanically are able to take any complex numbers. In this case, t_1 and t_2 can be assigned any complex values (and the lapse time $s = t_2 - t_1$ may take a complex number). If the initial and final sets of dynamical variables, (Q_1, E_1) and (Q_2, E_2), are fixed at certain real values, contributing complex trajectories are each determined to start at a point on the initial complex plane

$$\mathcal{I} \equiv \left\{ (t_1, Q, P) | t_1 \in \mathbf{C}, Q = Q_1, P = P_1 (= -\sqrt{2E_1}) \right\}, \tag{5.7}$$

and to end at a point on the final complex plane

$$\mathcal{F} \equiv \left\{ (t_2, Q, P) | t_2 \in \mathbf{C}, Q = Q_2, P = P_2 (= \pm\sqrt{2E_2}) \right\}. \tag{5.8}$$

Those trajectories are isolated to each other, except for degenerated ones reaching a caustic. In general, not a single but plural trajectories simultaneously contribute to the semiclassical calculation for fixed values of the observables. In practical calculation, we can regard (Q, P) as functions of the lapse time $s \equiv t - t_1 \ (\in \mathbf{C})$, the initial time $t_1 \ (\in \mathbf{C})$, and the set of fixed initial values $(Q_1, P_1) \ (\in \mathbf{R}^2, Q_1 > 0)$ so that t_1 is taken as a complex search parameter to find trajectories satisfying the output boundary condition, that is, $Q = Q_2$(fixed) and $E_2 \in \mathbf{R}^+$(positive real) [19,28].

In order to obtain the energy spectrum as a function of E_2, E_2 is scanned along the positive real axis with the coordinate Q_2 fixed at a positive or negative large real number, then the search parameter t_1 will trace a 1D set on the complex plane, that is, 1D curves. Let us call those 1D curves "complex branches." A convenient expression of the set of branches on $\mathcal{I}$ is given by

$$\mathcal{M}_S = \left\{ t_1 \in \mathbf{C} \mid \mathrm{Im}\{P(t_2 - t_1, t_1, P_1, Q_1)\} = 0, \ Q(t_2 - t_1, t_1, P_1) = Q_2(\ll -1) \right\}, \tag{5.9}$$

which is called $\mathcal{M}$-set [25]. The $\mathcal{M}$-set enables us to visualize the structure of the set of initial points of contributing trajectories on the search plane. We also introduce the Lagrange manifold, so-called $\mathcal{L}$-set [25], which is defined by

$$\mathcal{L}_S = \{(Q', P') \mid Q' = Q(t - t_1, t_1, P_1, Q_1), \ P' = P(t - t_1, t_1, P_1, Q_1), \ t_1 \in \mathcal{M}_S, \ t \in \mathbf{R}\}. \tag{5.10}$$

It is the set of points (Q', P') of the trajectories which start from $\mathcal{M}_S$ and are observed at a given real time t. Note that the $\mathcal{L}$-set of the S-matrix, that is, $\mathcal{L}_S$, is not the set of the end points of contributing trajectories, that is, the output branches on the final plane $\mathcal{F}$, but it is a snapshot of contributing trajectories at the given time t. This definition of the $\mathcal{L}$-set is convenient to see the geometrical structure of contributing branches in the phase space.

Owing to the periodicity of the system, complex branches appear periodically with the period $T(= 2\pi/\omega)$ in the initial plane $\mathcal{I}$. Then it is possible to reproduce the tunneling spectrum by $\mathcal{M}_S$ in a unit interval with the semiclassical S-matrix rewritten as [19,28]

$$S(E_2, E_1) \sim \lim_{|Q_1|, |Q_2| \to \infty} \sum_n \hbar\omega\delta(E_2 - E_1 - n\hbar\omega) \sum_{c.t. \in \mathcal{I}^*} \frac{1}{\sqrt{2\pi i\hbar}} \frac{\sqrt{|P_2||P_1|}}{\sqrt{P_1 P_2}} \sqrt{-\frac{\partial^2 S_S}{\partial E_1 \partial E_2}}$$
$$\times \ e^{-i(P_2 Q_2 - P_1 Q_1)/\hbar} \ e^{iS_S(Q_2, E_2, Q_1, E_1)/\hbar}, \tag{5.11}$$

where $\mathcal{I}^* = \{t_1| -T < \mathrm{Re}\, t_1 \leq 0\}$ denotes a unit of $\mathcal{I}$. The periodicity creates comb spikes with the interval $\hbar\omega$, which is represented as $\sum_n \hbar\omega\delta(E_2 - E_1 - n\hbar\omega)$. In numerical calculation, $\delta(E_2 - E_1 - n\hbar\omega)$ is, for normalization, replaced by $\delta_{n,(E_2-E_1)/\hbar\omega}$.

In practical implementation of the complexified semiclassical method, we always face a problem which integration path among many candidates is relevant to reproducing physically meaningful tunneling probability, because a classical solution extended into the complex domain normally has singularities, which cause (infinite) many branches of the solution. In next subsection, we consider this problem for the unperturbed system ($\varepsilon = 0$) as a simple case.

5.4 UNPERTURBED SYSTEM AND INSTANTON

5.4.1 COMPLEXIFIED CLASSICAL SOLUTION OF THE UNPERTURBED SYSTEM

Let us consider the unperturbed system. The classical solution in the range ($0 < E_1 < 1$) is given by [2]

$$Q(t) = \sinh^{-1}\left(\lambda \cosh\left(\sqrt{2E_1}(t - t_0)\right)\right), \tag{5.12}$$

where $\lambda \equiv \sqrt{1/E_1 - 1}$ and t_0 is the time at which the trajectory hits the turning point at $Q_{turn} = \pm\log\left[\lambda + \sqrt{\lambda^2 + 1}\right]$ [2,28]. For a given initial condition ($Q = Q_1(\gg 1), P = P_1(= -\sqrt{2E_1} < 0)$) at $t = t_1$, the interval between t_0 and t_1 is determined by $t_{01} \equiv t_0 - t_1 = (Q_1 - \log\lambda)/\sqrt{2E_1}$.

The solution has singularities in the lapse time plane s, that is, $s = t - t_1$, at

$$Sg_n^{\pm} = (Q_1 - \log\lambda)\Big/\sqrt{2E_1} \pm \frac{1}{\sqrt{2E_1}}\sinh^{-1}(1/\lambda)$$
$$+ i(-n + 1/2)\Delta t_I/2, \quad \left(\Delta t_I \equiv 2\pi\Big/\sqrt{2E_1}\right). \tag{5.13}$$

The singularities correspond, but not one-to-one, to the singularities of the potential $V_0 = \mathrm{sech}^2(Q)$ at $Q_m = i(2m + 1)\pi/2$, and around a singularity the solution has a form, $Q - Q_m \propto \sqrt{s - Sg_n^{\mp}}$, which is a branch point of it.

Figure 5.2a shows the singularities in the lapse time plane s. There are two types of singularities, namely entrance singularities Sg_n^- and exit singularities Sg_n^+. Representative integration paths C_n with different topologies with respect to the singularities are also shown in this picture. Figure 5.2b shows trajectories corresponding to the integration paths for the case $E < 1$ in the complex phase space projected on the space ($\mathrm{Re}\, Q, \mathrm{Re}\, P, \mathrm{Im}\, P$). Note that there is another possibility of taking integration paths with positive imaginary time evolution. However, they always make unphysical contributions, that is, diverging tunneling tails, and should be ignored. Actually they can be removed by the proper treatment of the Stokes phenomenon [30].

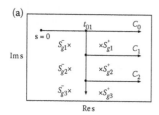 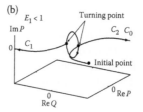 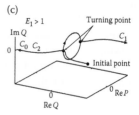

FIGURE 5.2 Singularities, integration paths, and complexified trajectories for the unperturbed system. (a) Singularities and representative integration paths on the lapse time plane. (b) Complex trajectories for $E < 1$. (c) Complex trajectories for $E > 1$.

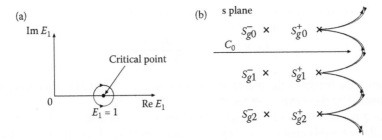

FIGURE 5.3 Critical point and movement of the singularities. (a) Critical point at $E_1 = 1$ on the complex energy plane and two topologically different contours. (b) Movement of the singularities Sg_n^+ along the contours in (a) and integration path C_0 on the lapse time plane.

As shown in Figure 5.2b, the trajectory starting at an initial point in an asymptotic side hits the turning point at $s = t_{01}$ and goes around a cycle in the classically forbidden region with imaginary time evolution along the vertical line in Figure 5.2a. If it takes an odd integration path C_{2n+1}, it reaches the opposite turning point after a half-integer times rotation and goes toward the transmitted side with real time evolution. But it, for an even path C_{2n}, goes back to the same turning point after nth rounds and is scattered to the reflective side.

Thus only trajectories along odd integration paths contribute to reproducing the tunneling weight and the trajectory along C_1 makes the dominant contribution among them, because it undergoes the shortest imaginary time evolution giving rise to the smallest imaginary component of the classical action. The complex trajectory with imaginary time evolution along the vertical part of C_1 is called "instanton" [3] and so the imaginary depth of an instanton is determined by [20,21]

$$t_{\text{inst}}(E_1) = -\Delta t_I/2 = -\pi/\sqrt{2E_1}. \tag{5.14}$$

For the case $E_1 > 1$, the location of singularities in the lapse time plane is similar to that for $E_1 < 1$, but the topology of integration paths with respect to the singularities changes as shown in Figure 5.2c: trajectories along odd integration paths go back to the reflective side, though those for even ones reach the transmissive side. That topological change of integration paths is caused by the movement of the singularities Sg_n^+ with change of the energy E_1 [16,17]. Indeed, as shown in Figure 5.3, the singularities Sg_n^+ diverge logarithmically at $E_1 = 1$, say critical point, as

$$Sg_n^+ \sim \frac{1}{\sqrt{2E_1}} \log(E_1/(1 - E_1)) + \mathrm{i}(-n + 1/2)\Delta t_I/2 + \text{const.}, \tag{5.15}$$

though the singularities Sg_n^- remain in finite ranges. The logarithmic divergence of Sg_n^+ at $E_1 = 1$ induces their shift by the interval $\Delta t_I/2$ along the imaginary axis, taking an up or down shift depending on choice of contour on the complex E_1 plane. As a result, the topology of an integration path changes from transmissive one to reflective one, and vice versa.

That logarithmic divergence of some group of singularities of the classical solution is an important nature of stable and unstable manifolds extended to the complex domain, for this case, the solution at $E_1 = 1$, and it also occurs for periodically perturbed systems for which a critical point appears on initial plane $\mathcal{I}$ as an intersection with the complexified stable manifold as shown later.

5.4.2 WEIGHT OF INSTANTON

The tunneling rate for the unperturbed system is estimated by the classical action of the trajectory with C_1, say instanton. Indeed, the imaginary time evolution in the classical forbidden region by

$it_{inst}(E_1)$ gives rise to the action with a positive imaginary value,

$$\mathrm{Im}\, S_{I0} = \sqrt{2E_1}\, \frac{a_0 - E_1}{E_1 + \sqrt{a_0 E_1}}, \tag{5.16}$$

where a_0 denotes the height of the potential, for our case $a_0 = 1$ [20,21]. Therefore, the tunneling weight is estimated by $W_I = \exp(-\mathrm{Im}\, S_{I0}/\hbar)$.

Even when a nonzero but small enough perturbation is applied to the system, the topology of integration paths is roughly characterized by that of the unperturbed system [17,19]. In the case that $\omega \ll 1$ and $\varepsilon \ll 1$, the adiabatic approximation based on the instanton method is available. Since $\omega \ll 1$, then the movement of potential can be ignored, while the tunneling particle goes through it, and it is replaced by the instantaneous one. So, the effective imaginary time evolution of the instanton seems to be given by $t_{inst}(E) \sim -\pi/\sqrt{2E}$ at $E \sim E_1 + \varepsilon$; that is, the shortest imaginary path during the period of perturbation. However, the tunneling weight is more precisely estimated by the time average of instantaneous instanton weights [20,21]:

$$W_{\mathrm{av}} = \frac{1}{T} \int_0^T \exp\left(-\frac{1}{\hbar} \mathrm{Im}\, S_I\right) dt, \tag{5.17}$$

where $\mathrm{Im}\, S_I$ is given by

$$\mathrm{Im}\, S_I = \sqrt{2E_1}\, \frac{a(t) - E_1}{E_1 + \sqrt{a(t)E_1}}, \tag{5.18}$$

with $a(t) = 1 + \varepsilon \sin \omega t$, the time-dependent height of the oscillating barrier.

5.5 STABLE–UNSTABLE MANIFOLD GUIDED TUNNELING

5.5.1 Brief Sketch of Stable–Unstable Manifold Guided Tunneling

As shown by the quantum calculation in Section 5.2.2, when the perturbation becomes relatively large, a novel tunneling phenomenon, which cannot be explained by the instanton method, is observed. This is the case that a different tunneling mechanism, namely SUMGT, dominates the tunneling process [16,17,19,27]. First we briefly explain this new tunneling mechanism SUMGT.

Figure 5.4 is a schematic picture of the Poincaré maps of periodically perturbed barrier systems, in which stable and unstable manifolds of a saddle are depicted by W_s and W_u—that is, the red and blue lines, respectively. As discussed in Section 5.3, in order to apply the semiclassical method to the analysis of tunneling, we need to prepare a set of initial points of classical trajectories as a classical counterpart of the quantum initial condition, which forms a hypersurface $\mathcal{I}$ in the complex

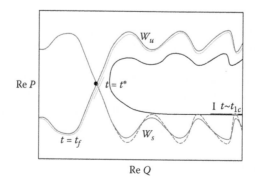

FIGURE 5.4 **(See color insert.)** Brief sketch of SUMGT. Color version available online.

phase space. Note that $\mathcal{I}$ in Figure 5.4—that is, a piece of thick black line—is not that $\mathcal{I}$ determined by Equation 5.7, but is the map of $\mathcal{I}$ on the Poincaré section with time evolution, and corresponds to one period, $\mathcal{I}^* = \{t_1 | -T < \mathrm{Re}\, t_1 \leq 0\}$, which is enough for semiclassical calculation due to the periodicity of the system.

For the situation of tunneling, $\mathrm{Re}\, \mathcal{I}$ does not touch the real stable manifold W_s and mappings of $\mathrm{Re}\, \mathcal{I}$ form an invariant manifold drawn by a thin black curve, which never goes to the product side. However, the oscillation of the stable manifold is usually amplified in the complex domain as shown by a broken red curve and so the intersection between W_s and $\mathcal{I}$ occurs in the complex domain forming an isolated point at $t_1 = t_{1c}$, called *critical point* [16,17,19]. The SUMGT trajectories are those trajectories which start from a small neighborhood of a critical point, evolve being first guided by the complexified stable manifold, pass close to the saddle, and are finally scattered along the unstable manifold—as shown in Figure 5.4, a neighborhood of t_{1c}, indicated by a small piece of green line superposed on $\mathcal{I}$, reaches one close to the saddle at $t = t^*$ and is extended along W_u at $t = t_f$. Particularly, the trajectories scattered along the real unstable manifold make dominant contributions, since they have smaller imaginary actions than others scattered along purely complex parts of the unstable manifold. As a result, the tunneling probability affected by the structure of the unstable manifold is observed. Therefore, SUMGT is essentially different from the instanton-type tunneling and dominates it, if the imaginary part of the initial time of the SUMGT trajectory is nearly equal to or less than that of the instanton, namely $\mathrm{Im}\, t_{1c} \sim |t_{\mathrm{inst}}|$ or $\mathrm{Im}\, t_{1c} < |t_{\mathrm{inst}}|$, because SUMGT breaks the major contributing instantons and the imaginary action of the SUMGT trajectory becomes less than that of the instanton [19–21].

5.5.2 MELNIKOV METHOD: EXISTENCE OF CRITICAL POINTS

The existence of the critical point can be proved by using the Melnikov method [17,31]. Actually, the energy of a trajectory on the stable manifold at a given initial time t_1 can be evaluated by

$$H(Q_s(t), P_s(t), \omega t) = H(Q_{\mathrm{ups}}(t), P_{\mathrm{ups}}(t), \omega t) + \int_{\infty}^{t} \left\{ \frac{\partial V}{\partial t'}(Q_s(t'), \omega t') - \frac{\partial V}{\partial t'}(Q_{\mathrm{ups}}(t'), \omega t') \right\} dt', \tag{5.19}$$

where $(Q_{\mathrm{ups}}, P_{\mathrm{ups}})$ denotes the unstable periodic orbit and (Q_s, P_s) a trajectory on the stable manifold. At the lowest-order approximation, namely the Melnikov method [31], the solution $Q_s(t)$ in Equation 5.19 is replaced by the unperturbed solution Q_{s0} on the stable manifold [17],

$$Q_{s0}(t, \mu) = \sinh^{-1}\left(e^{-\sqrt{2}(t-\mu)}\right), \tag{5.20}$$

where the parameter μ indicates the initial phase or initial time of the solution, and Equation 5.19 is integrated analytically. As a result, the energy at the initial time t_1 is given as a function of μ [17],

$$H(t_1) \sim 1 + \varepsilon(1 - \chi(\omega)) \sin \omega \mu, \tag{5.21}$$

where $\chi(\omega)$ is defined by

$$\chi(\omega) \equiv 2\omega \int_0^{\infty} \frac{\sin \omega s}{1 + e^{2\sqrt{2}s}} ds, \tag{5.22}$$

and μ is related with t_1 as $\mu = t_1 + (Q_1 - \log 2)/\sqrt{2}$.

Figure 5.5 shows the reconstructed stable manifold as a function of complexified μ together with the initial plane at $E = E_1$ in the complex space $(\mathrm{Re}\, E, \mathrm{Re}\, \mu, \mathrm{Im}\, \mu)$. The bottom wavy curve drawn by a thick line is the real stable manifold and the upper part is purely complex. Since the initial energy E_1 is fixed at a certain real number, the intersection μ_c between the stable manifold and the initial plane at $E = E_1$ is determined by

$$E_1 = P_1^2/2 = 1 + \varepsilon(1 - \chi(\omega)) \sin \omega \mu_c. \tag{5.23}$$

In the case that $1 - \varepsilon(1 - \chi(\omega)) \leq E_1 \leq 1 + \varepsilon(1 - \chi(\omega))$, the intersection occurs in the real domain and forms two isolated points on the real stable manifold every period of the perturbation, but the points degenerate at $E_1 = 1 \pm \varepsilon(1 - \chi(\omega))$.

As shown in Figure 5.5, the stable manifold in the complex domain takes real values in energy E only on the gentlest ascent lines drawn by thick lines, and so the intersection for $E_1 < 1 - \varepsilon(1 - \chi(\omega))$—that is, tunneling case—is an isolated point determined by

$$\text{Re}\,\mu_c = (3\pi/2 + 2n\pi)/\omega, \tag{5.24}$$

$$\text{Im}\,\mu_c = \frac{1}{\omega}\cosh^{-1}(\varepsilon^{-1}(1 - E_1)/(1 - \chi(\omega))). \tag{5.25}$$

Then the imaginary depth of the critical point $\text{Im}\,t_{1c} \sim \text{Im}\,\mu_c$ is given by Equation 5.25. Note that the critical point appears periodically by the period $T(= 2\pi/\omega)$ due to the periodicity of the perturbation (see Equation 5.24).

From Equation 5.25, $\text{Im}\,t_{1c}$ changes with the strength of the perturbation ε. Actually it increases with decreasing ε. If ε is small enough such that $\text{Im}\,t_{1c}$ is much larger than the imaginary depth of the instanton $|t_{\text{inst}}|$, no instanton branches are disturbed by the critical point t_{1c}. This is the case of weak perturbation regime. However, with an increase of ε, the critical point goes down at the same level of the instanton branches and SUMGT takes the place of the instanton tunneling, namely the strong perturbation regime.

The critical strength of perturbation ε_c, at which the tunneling mechanism changes from the instanton tunneling to SUMGT, and vice versa, is defined by [20,21]

$$\text{Im}\,t_{1c}(\varepsilon_c) = \frac{1}{\omega}\cosh^{-1}(\varepsilon_c^{-1}(1 - E_1)/(1 - \chi(\omega))) = 1.5|t_{\text{inst}}|. \tag{5.26}$$

The threshold line at $t_1 = 1.5|t_{\text{inst}}|$ is set at the middle of the integration paths C_1 and C_2. As shown in Table 5.1, the evaluation of $\text{Im}\,t_{1c}$ by the Melnikov method provides good agreement with the numerical ones for three values of ε, that is, $\varepsilon = 0.1,\ 0.2,\ 0.4$. Table 5.2 shows imaginary depths of the integration paths C_1, C_2, and C_3. Comparison of $\text{Im}\,t_{1c}$ with the imaginary depths of the integration paths indicates that SUMGT works if ε is larger than 0.2, which shows good agreement with the quantum calculation.

5.5.3 Weight of SUMGT

Weights of SUMGT trajectories are evaluated from the classical action determined by Equation 5.5 [20,21]. Actually, if the contribution of the amplitude factor $\sqrt{-\partial^2 S_S/\partial E_1 \partial E_2}$ in Equation 5.11 is

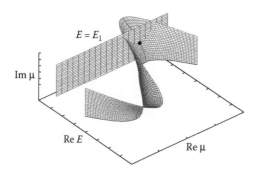

FIGURE 5.5 Stable manifold reproduced by the Melnikov method. An initial plane at $E = E_1$ is also depicted. The bullet denotes the intersection between the stable manifold and the initial energy plane in the complex domain.

TABLE 5.1

Depth of Critical Point $\omega \text{Im} \, t_{1c}$

ϵ	Numerical	Melnikov
0.1	2.3570	2.3112
0.2	1.6226	1.5869
0.4	0.7698	0.7234

ignored, the weight of SUMGT, W_S, is evaluated by

$$W_S = |S(E_1, E_2)| \sim \frac{\hbar \omega}{\sqrt{2\pi\hbar}} \exp\left(-\frac{1}{\hbar} \text{Im} \, S_S\right), \tag{5.27}$$

where the coefficient $\hbar\omega$ arises from the spike density of the spectrum (see Equation 5.11).

Note that the amplitude factor $\sqrt{-\partial^2 S_S / \partial E_1 \partial E_2}$ is roughly estimated as $\sqrt{1/(\varepsilon\omega)}$ in a low-frequency range ($\omega < 1$) by a more precise analysis based on an adiabatic solution combined with the Melnikov method [19] and it takes larger values for smaller εs. Small but visible differences in weight between the full-semiclassical calculations of SUMGT trajectories and W_S in Figures 5.6, 5.8, and 5.9, which increase with a decrease of ε, come from ignoring the amplitude factor. We do not have any simple evaluation in a high-frequency range ($\omega > 1$), but similar discrepancies between them are observed [21].

The problem is how to calculate the action S_S for SUMGT trajectories. The values of S_S for SUMGT trajectories are well approximated by that of the critical trajectory starting at t_{1c} which guides them until being close to the saddle. This is because the major contributing trajectories of SUMGT that are scattered along the real unstable manifold each gain a negligibly small imaginary part of S_S after passing close to the saddle and Im S_S of those trajectories are almost equal to that of the critical trajectory. It can be evaluated by using the Melnikov method again [20,21]. First, the classical action is rewritten as

$$S_S(Q_2, E_2, Q_1, E_1) = \int_{t_{1c}}^{t_2} H(Q, P, \omega t) \, dt - \int_{t_{1c}}^{t_2} 2V(Q, \omega t) + E_2 t_2 - E_1 t_{1c}, \tag{5.28}$$

where $V(Q, \omega t) = (1 + \varepsilon \sin \omega t) \, \text{sech}^2 Q$. The time evolution of the Hamiltonian $H(Q, P, \omega t)$ is given by Equation 5.19.

According to the usage of the Melnikov method [31], classical variables $Q(t)$ and $P(t)$ are replaced by those of the unperturbed trajectory at $\mu = \mu_c$ on the stable manifold along an integration path, $t_{1c}(\notin \mathbf{R}) \to t_2'(\to t_2(\in \mathbf{R})$, where Im $t_2' =$ Im $t_{1c} > 0$ and Re $t_2' = t_2 \to \infty$. After some

TABLE 5.2

Depth of Integration Path $\omega n \Delta t_I / 2$

Path	Depth
C_3	2.8274
C_2	1.8850
C_1	0.9425

calculations, we finally obtain the expression of S_S,

$$S_S \sim 2E_1(\mu_c - t_1) + \mu_c(1 - E_1) + t_2(E_2 - 1) + \frac{\varepsilon}{\omega}\cos\omega t_2 - i\frac{1 - E_1}{\omega}\frac{\sinh(\omega\mathrm{Im}\,\mu_c)}{\cosh(\omega\mathrm{Im}\,\mu_c)}$$

$$+ i\sqrt{2}\sinh(\omega\mathrm{Im}\,\mu_c)\int_\infty^0 dx \int_\infty^x ds\frac{\varepsilon\sin\omega s}{\cosh^2(\sqrt{2}s)}. \tag{5.29}$$

Taking into account the boundary condition that $E_1, E_2 \in \mathbf{R}$, $\mathrm{Im}\,t_2 \sim 0$, and $\mathrm{Im}\,t_{1c} \sim \mathrm{Im}\,\mu_c$, the imaginary part of S_S is evaluated by [20,21]

$$\mathrm{Im}\,S_S \sim \mathrm{Im}\,\mu_c(1 - E_1) - \frac{1 - E_1}{\omega}\frac{\sinh(\omega\mathrm{Im}\,\mu_c)}{\cosh(\omega\mathrm{Im}\,\mu_c)} + \sqrt{2}\sinh(\omega\mathrm{Im}\,\mu_c)\int_\infty^0 dx \int_\infty^x ds\frac{\varepsilon\sin\omega s}{\cosh^2(\sqrt{2}s)},$$

$$\tag{5.30}$$

which allows us to calculate the tunneling weight through Equation 5.27.

5.6 SUMGT VERSUS INSTANTON

In this section, we see to what extent the semiclassical method can reproduce tunneling spectra, and study how the competition between the two characteristic semiclassical mechanisms, SUMGT and instanton tunneling is reflected in the formation of the tunneling spectrum and what change occurs with an increase or decrease of the perturbation strength [18–20].

5.6.1 NUMERICAL RESULTS: STRONG PERTURBATION REGIME

First, we see the nature of spectra in the strong perturbation regime, in which SUMGT dominates the tunneling process [18,19]. Figure 5.6 shows numerical results in the strong perturbation regime at $\varepsilon = 0.4$. In Figure 5.6a, a typical example of the $\mathcal{M}$-set in the strong perturbation regime is depicted. On the t_1-plane, there is a critical point indicated by X. Due to the periodicity of the perturbation, critical points appear periodically with the period of perturbation T and the structure of branches in the $\mathcal{M}$-set is also repeated periodically. Then, we show the $\mathcal{M}$-set of the unit period in this picture.

It should be first noted that the parts of branches drawn by broken lines indicate noncontributing parts which make unphysical contributions and can be removed by the proper treatment of the Stokes phenomenon [30].

The black horizontal line **r** indicates the initial set of real trajectories contributing to reflection. There are several complex branches drawn by color lines and the structure of those branches is complicated. However, only a few branches located in a neighborhood of the critical point make major contributions to tunneling, namely the branches **1**, **2**, and **3**.

In the inserted box, a blown-up picture near the critical point is drawn. Parts of branches are located along a straight line passing through the critical point and others are along a circle with its center at the critical point. Then, they are recategorized into two groups: the parts along the straight line, called $\mathcal{M}_\mathcal{R}$, and the other parts along the circle, called $\mathcal{M}_C$ [18,19]. $\mathcal{M}_C$ is separated into paired semicircles by cuts transversely across it. They make different types of contributions to the S-matrix and go into different Riemann sheets beyond the cuts. That is, trajectories starting from the branches in the upper-half plane above the cuts reach the reflective side, while trajectories in the lower-half plane go to the transmissive side. So the upper and lower parts of $\mathcal{M}_\mathcal{R}$ above and below the critical point must contribute to reflection and transmission, respectively.

Figure 5.6b shows the corresponding Lagrange manifold $\mathcal{L}$-set projected on the real phase space. The Lagrange manifold $\mathcal{L}_R$ corresponding to $\mathcal{M}_\mathcal{R}$ is extended from the reflective side to the transmissive side as $\mathbf{1} \to \mathbf{2} \to \mathbf{3}$ and the major part of it follows the real unstable manifold indicated by W_{uR}. It means that the destination of the trajectory changes as the initial time t_1 passes through the critical point t_{1c} along $\mathcal{M}_\mathcal{R}$. Therefore, the branch $\mathcal{M}_\mathcal{R}$ is regarded as a *merged* object composed

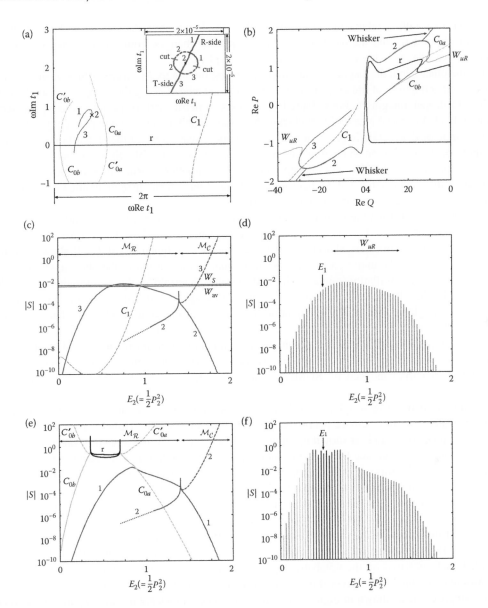

FIGURE 5.6 (See color insert.) Semiclassical results for the strong perturbation regime at $\varepsilon = 0.4$. (a) $\mathcal{M}$-set. (b) $\mathcal{L}$-set. (c) Weights of tunneling branches. (d) Tunneling spectrum. (e) Weights of reflective branches. (f) Reflection spectrum. Color version available online.

of the tunneling and reflective branches of the unperturbed system and so the integration path of the trajectory switches from C_1 to C_2, and vice versa, as t_1 passes through t_{1c}, namely switching of the path topology happens. This fact can be explained by the divergence movement of the singularities Sg_n^+ at the critical point as [16,17,19]

$$Sg_n^+ \sim -\frac{1}{\sqrt{2}}\log(t_1 - t_{1c}) - i\frac{\pi}{\sqrt{2}}\left(n - \frac{1}{2}\right) + \text{const.}, \qquad (5.31)$$

which can be proved by the more detailed analysis combining the adiabatic approximation in a low-frequency range with the Melnikov method [17,19]. Since Equation 5.31 corresponds to

Equation 5.15 with t_1 taking the place of E_1, then the logarithmic divergence of Sg_n^+ occurs at the critical point t_{1c} and the destination of the trajectory changes from the reflective side to the transmissive side, and vice versa, when t_1 passes through t_{1c}.

On the other hand, the Lagrange manifold $\mathcal{L}_C$ corresponding to $\mathcal{M}_C$ forms whiskers going from apexes and antiapexes of the oscillating $\mathcal{L}_R$, which are extended along the complex unstable manifold into deeper imaginary sides. Paired branches symmetric with respect to $\mathcal{M}_R$ in the same semicircle of $\mathcal{M}_C$ make a pair of branches in the $\mathcal{L}$-set, which are almost complex conjugate, then they seem to be degenerate in projection on the real phase space. One of the paired branches makes an unphysical contribution and is removed by the proper treatment of the Stokes phenomenon [30].

Note that the branches C_{0a} and C_{0b} with the integration path C_0 form whiskers emanating from caustics of the real branch $\mathbf{r}$ in the reflective side of the $\mathcal{L}$-set, which make tunneling tails appearing in the reflection spectrum as shown in Figure 5.6e.

In order to reconstruct the plateau spectrum semiclassically, we need to take into account only the contributions from $\mathcal{M}_R$ and $\mathcal{M}_C$, namely the branches **2** and **3** in the original picture. Figure 5.6c indicates semiclassical weights of those branches as functions of E_2. It shows that the branches $\mathcal{M}_R$ form the major part of the spectrum, that is, the flat top of the plateau whose width corresponds to the oscillating range of the real unstable manifold in the asymptotic region. The weight of SUMGT W_S given by Equation 5.27 shows good agreement with the height of the plateau, though the weight of instant W_{av} in Equation 5.17 is also at the same level. Note that a tiny difference between the height of the plateau and W_S is due to the omission of $\sqrt{-\partial^2 S_S/\partial E_1 \partial E_2}$ as discussed in Section 5.5.3. Figure 5.6d shows the tunneling spectrum reconstructed by the semiclassical method. Its agreement with the purely quantum result in Figure 5.1 is very good. Hence, the plateau spectrum induced by SUMGT reflects the classical structure of the unstable manifold.

Figure 5.6e and f show weights of the branches contributing to reflection and the reflection spectrum reconstructed from them, respectively. In Figure 5.6e the black lines labeled $\mathbf{r}$ are contributions of the real trajectories and, as shown in Figure 5.6f, form the top of the reflection spectrum, which oscillates due to interference between the folded parts of the branch $\mathbf{r}$. The branches C_{0a} and C_{0b} form tunneling tails dropping from the top, while the blue ones are contributions of SUMGT, namely the branches **1** and **2**, which form a shoulder of the reflection spectrum. The semiclassical reflection spectrum shows good agreement with the quantum calculation, though it is not shown in a picture.

The characteristic of SUMGT is in the existence of the branches $\mathcal{M}_R$ and $\mathcal{M}_C$ in a small neighborhood of the critical point so that the contributing trajectories starting from those branches are guided by the stable manifold. The existence of $\mathcal{M}_R$ and $\mathcal{M}_C$ can be proved through the more detailed analysis based on the adiabatic approximation in a low-frequency range combined with the Melnikov method [19]. According to the analysis, $\mathcal{M}_C$ forms a set of concentric circles rather than a single circle as shown in Figure 5.7a. However, inner circles are usually exponentially small compared with the outermost circle, then only the outermost is observed in our numerical calculation. Figure 5.7a and b schematically show the relation between the characteristic $\mathcal{M}$-set and the corresponding Lagrange manifold $\mathcal{L}$-set predicted by the detail analysis. Like the numerical results, the straight line $\mathcal{M}_R$ corresponds to the part following the real unstable manifold in the $\mathcal{L}$-set, while the circles $\mathcal{M}_C$ correspond to the whiskers extending along the complexified unstable manifold.

5.6.2 Numerical Results: Intermediate Perturbation Regime

In this subsection, we see results of the middle perturbation regime [20]. In Figure 5.8a, the $\mathcal{M}$-set is drawn and there is a critical point accompanied by the characteristic branches of SUMGT labeled "**1, 2, 3**." The inserted box shows a blow-up picture near the critical point, in which the characteristic branches $\mathcal{M}_R$ and $\mathcal{M}_C$ form a straight line and a circle around it, respectively. In this case, the critical point and the characteristic branches **1–3** are in the level of the integration path C_2, then

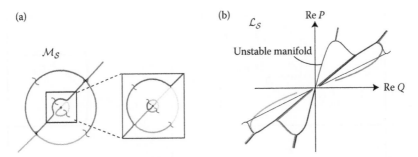

FIGURE 5.7 **(See color insert.)** Correspondence between (a) $\mathcal{M}$-set and (b) $\mathcal{L}$-set. The same color specifies the corresponding parts of $\mathcal{M}_S$ and $\mathcal{L}_S$. In (b), the thin black lines, which closely overlap with colored branches of $\mathcal{L}_S$, indicate the unstable manifold W_u: the wavy curve is the real part W_{uR}, while the straight lines are parts of the purely complex unstable manifold with real energy values and guide $\mathcal{L}_C$. Color version available online.

it is possible to find a different branch with the integration path C_1. Indeed, C_{1b} is a branch of the instanton tunneling with C_1.

In the $\mathcal{L}$-set drawn in Figure 5.8b, the characteristic branches of SUMGT well follow the real unstable manifold W_{uR} from the reflective side to the transmissive side like $\mathbf{1} \to \mathbf{2} \to \mathbf{3}$. However, the instanton branch C_{1b} locates in the transmissive side and its contributing part drawn by a solid line takes relatively lower values in energy compared with W_{uR} in the asymptotic region.

Figure 5.8c shows weights of the branches contributing to tunneling. The active part of the branch C_{1b} forms the head lobe of the spectrum, whose height is well approximated by the averaged instanton weight W_{av}. On the other hand, the branches $\mathbf{2}$ and $\mathbf{3}$ contribute to forming the side shoulder and its height is also well estimated by W_S, where a small discrepancy between the height of the shoulder and W_S is due to the omission of $\sqrt{-\partial^2 S_S / \partial E_1 \partial E_2}$. It means that the instant tunneling and SUMGT simultaneously contribute to the tunneling process. As shown in Figure 5.8d, the tunneling spectrum reproduced by the semiclassical method shows good agreement with the quantum calculation in Figure 5.1.

Figure 5.8e and f show weights of the branches contributing to reflection and the spectrum reconstructed from them, respectively. Like the results of strong perturbation regime, the real branches form the oscillating top of the reflection spectrum accompanied by tunneling tails reproduced by the branches C_{0a} and C_{0b} with the integration path C_0. The SUMGT branches $\mathbf{1}$ and $\mathbf{2}$ form a shoulder of the reflection spectrum. The semiclassical reflection spectrum shows good agreement with the quantum calculation, though it is not shown here.

5.6.3 Numerical Results: Weak Perturbation Regime

In this subsection, we see results in the weak perturbation regime, in which the instanton tunneling dominates the tunneling process [20]. Figure 5.9a shows the initial set in t_1-plane, in which we still find a critical point accompanied by the characteristic branches of SUMGT, whose structure is confirmed in the inserted box. This fact means that the intersection of the initial set $\mathcal{I}$ with the stable manifold W_s still occurs in a deep complex domain even for the weak perturbation regime. In this case, the branches of SUMGT are in the level of the integration path C_3, then different types of branches with the path C_1 or C_2 are able to exist. Actually, C_{1b} and C_{2b} with the integration path C_1 and C_2 contribute to tunneling and reflection, respectively. Those branches are explained by the perturbed instanton theory.

In the $\mathcal{L}$-set shown in Figure 5.9b, the main part of SUMGT branches still follows the real unstable manifold W_{uR} from the reflective side to the transmissive side like $\mathbf{1} \to \mathbf{2} \to \mathbf{3}$. However, the branches C_{1b} and C_{2b} exist in the transmissive side and in the reflective side, respectively.

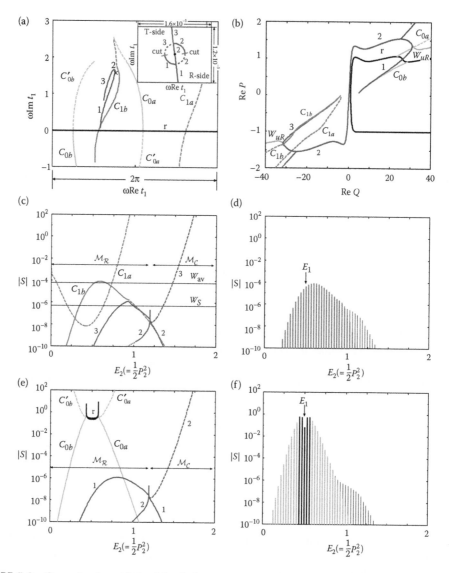

FIGURE 5.8 (See color insert.) Semiclassical results for the intermediate perturbation regime at $\varepsilon = 0.2$. (a) $\mathcal{M}$-set. (b) $\mathcal{L}$-set. (c) Weights of tunneling branches. (d) Tunneling spectrum. (e) Weights of reflective branches. (f) Reflection spectrum. Color version available online.

Figure 5.9c and d show weights of the branches contributing to tunneling and the tunneling spectrum reproduced semiclassically, respectively. These pictures are extended into a lower range up to 10^{-14} compared with the quantum tunneling spectrum given in Figure 5.1. In Figure 5.9c, the major part of the spectrum, namely the head lobe, is reproduced by the instanton branch C_{1b}, but a small shoulder created by the SUMGT branches **2** and **3** appears in a range less than 10^{-10}. The blue thin vertical lines denote the spectrum given by the quantum calculation, which confirms that the small shoulder is not an artificial fact by semiclassics, but really exists. W_{av} and W_S also well approximate the heights of the head lobe and shoulder, respectively. Note that a small discrepancy between the height of the shoulder and W_S originates from ignoring $\sqrt{-\partial^2 S_S/\partial E_1 \partial E_2}$. The reconstructed spectrum in Figure 5.9d shows good agreement with quantum one.

Figure 5.9e and f show weights of the branches contributing to reflection and the spectrum reconstructed from them, respectively. As the cases of strong and intermediate perturbation regimes, the real branches **r** and the complex branches C_{0a} and C_{0b} with the path C_0 form the major part of

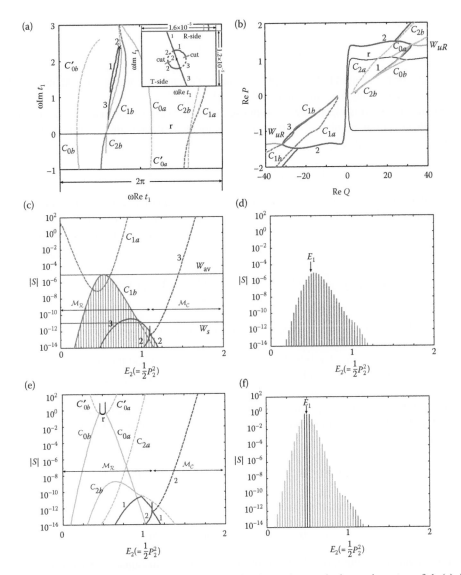

FIGURE 5.9 **(See color insert.)** Semiclassical results for the weak perturbation regime at $\varepsilon = 0.1$. (a) $\mathcal{M}$-set. (b) $\mathcal{L}$-set. (c) Weights of tunneling branches. (d) Tunneling spectrum. (e) Weights of reflective branches. (f) Reflection spectrum. Color version available online.

the reflection spectrum, namely the oscillating top accompanied by tunneling tails. However, the branch C_{2b} creates a second reflection component caused by an instanton going around the full circle with the path C_2. Indeed, the branch C_{2b} forms the head lobe of the second reflection component, which is accompanied by a small shoulder reproduced by the SUMGT branches **1** and **2**. The semiclassical reflection spectrum shows good agreement with the quantum calculation, though it is not shown here.

5.7 BARRIER TUNNELING OF A 2D SYSTEM

In this section, we study tunneling for a purely 2D barrier system. We start with a model system of 2D barrier tunneling, which is given by the following Hamiltonian [19]:

$$\hat{H}_{\text{tot}}(Q,\hat{P},q,\hat{p}) = \frac{1}{2}\hat{P}^2 + (1+\beta q)\operatorname{sech}^2 Q + \hat{H}_{\text{ch}}(q,\hat{p}), \qquad (5.32)$$

where $\hat{H}_{ch}$ denotes the Hamiltonian representing an unperturbed channel motion given by

$$\hat{H}_{ch}(q,p) = \tfrac{1}{2}\hat{p}^2 + \tfrac{1}{2}\omega^2 q^2. \tag{5.33}$$

This model can be considered to be a simplified model of chemical reactions, for example, collinear collision between an atom and a diatomic molecule [1,2].

To construct a scattering eigenstate, we suppose that an incident wave excited at a channel eigenstate w_{n_1} with a quantum number $n_1 (\geq 0)$ is propagating along the reaction coordinate Q from $+\infty$ with a constant momentum $P_1 (< 0)$:

$$\Psi_{in} \propto \exp\{iP_1 Q/\hbar\} w_{n_1}, \quad Q \to +\infty. \tag{5.34}$$

To consider the relation to the 1.5D system given by Equation 5.1, we introduce the classical action and angle coordinates for the channel degree of freedom, namely $I = H_{ch}/\omega$ and $\theta = \arctan(\omega q/p)$. Using the action and angle, (I, θ), the classical counterpart of Hamiltonian (Equation 5.32) is written by

$$H_{tot}(Q, P, q, p) = \frac{1}{2}P^2 + \left(1 + \beta\sqrt{\frac{2I}{\omega}}\sin\theta\right)\text{sech}^2 Q + \omega I. \tag{5.35}$$

Since the initial classical action I_1 corresponds to the quantum number n_1 as $I_1 = \hbar(n_1 + 1/2)$, then the total energy is given by $E_{tot} = P_1^2/2 + \hbar\omega(n_1 + 1/2)$.

Under the assumption that the initial channel is highly excited and the coupling strength is not very strong, so that the variation of I in the scattering process is much less than its initial value $I_1 = (n_1 + 1/2)\hbar$: the classical action I is well approximated by its initial value I_1 and the classical angle θ is written by $\theta = \omega t + \theta_0$. Thus, the action variable I in the coupling term in Equation 5.35 can be replaced by the constant c-number I_1 and the Hamiltonian of the 2D model (Equation 5.35) is very well approximated as

$$H_{tot}(Q, P, \theta, I) \sim \frac{1}{2}P^2 + (1 + \varepsilon\sin\theta)\,\text{sech}^2(Q) + \omega I, \tag{5.36}$$

where ε denotes the effective perturbation strength defined by

$$\varepsilon \equiv \beta\sqrt{2I_1/\omega} = \beta\sqrt{(2n_1 + 1)\hbar/\omega}. \tag{5.37}$$

This Hamiltonian is nothing more than that of the 1.5D system in Equation 5.1 and it is expected that similar tunneling phenomena are observed for proper choices of the couping strength through Equation 5.37.

Figure 5.10 shows tunneling spectra for the 2D system at $\varepsilon = 0.1$, 0.2, and 0.4 [19], which are calculated by Miller's S-matrix [1,2]. Initial conditions and other parameters are set up to match with corresponding ones in the 1.5D system (see the figure caption in Figure 5.10). Each spectrum is represented as a function of the kinetic energy $E_2 = P_2^2/2$ along the reaction coordinate observed in the asymptotic region of the product side. The spectra observed are almost the same as those of the 1.5D system in the weak, middle and strong coupling regimes, respectively. Indeed, a localized spectrum at $\varepsilon = 0.1$ changes to a plateau spectrum at $\varepsilon = 0.4$ through an intermediate shape with a side shoulder at $\varepsilon = 0.2$. Therefore, the same change of semiclassical mechanism should occur with an increase of ε: the instanton-type tunneling dominates for the weak coupling regime but SUMGT takes the place of it for the strong coupling regime. In the middle, both mechanisms simultaneously contribute to forming the characteristic spectrum.

Let us consider the situation for which SUMGT contributes to tunneling. As will be shown below, this situation is not of energy barrier tunneling, but is regarded as a sort of dynamical barrier tunneling which is observed when the total energy is larger than the potential saddle. Indeed, in

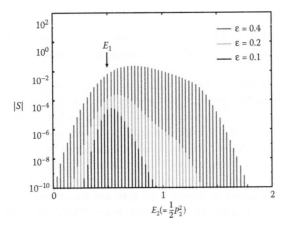

FIGURE 5.10 Tunneling spectra (absolute value of the S-matrix) of the 2D model for three representative values of the effective perturbation strength at $\varepsilon = 0.1, 0.2$, and 0.4. $E_1 = P_1^2/2 = 0.5$, $\omega = 0.3$, $n_1 = 128$, and $\hbar = 1000/(3\pi \times 2^{10}) \sim 0.1036$.

order for SUMGT to work, a real unstable periodic orbit should exist on the energy surface given by the total energy as $E_{tot} = H_{tot}$, such that E_{tot} is larger than the potential energy at its saddle point $(Q,q) = (0, -\beta/\omega^2)$ and an *unstable periodic orbit* O exists above the saddle of the barrier potential at $Q = 0$, that is, a harmonic vibration described by the Hamiltonian $H_{ch}(q,p) + \beta q$ with the energy $E_{tot} - 1$. As shown in Figure 5.11a, the unstable periodic orbit O forms the transition state separating the product side from the reactant side [24] and is accompanied by the two characteristic sets, the stable manifold W_s and the unstable manifold W_u, which form tubes extended into the asymptotic regions, reactant and product sides, respectively.

Classical particles inside the tube W_s go through the unstable periodic orbit O and reach the product side, however, particles outside it are reflected back to the reactant side by the effective dynamical barrier. If we choose the initial channel eigenstate outside W_s on the initial surface ΣQ_1 at $Q = Q_1 \gg 1$, namely the ellipse defined by $h_e = \{(q,p)|\frac{1}{2}p^2 + \frac{1}{2}\omega^2 q^2 = \omega l_1\}$ encircles W_s on ΣQ_1 without any intersections as shown in Figure 5.11b, then any quantum probabilities observed in the product side are caused by tunneling.

In this situation, any intersections of W_s with h_e, that is, the critical points, do not occur in the real plane. However, if extension of h_e into the complex domain is achieved by complexifying the angle variable $\theta = \arctan(\omega q/p)$ keeping the initial channel energy at a fixed real number as $e_{ch} = \omega l_1$,

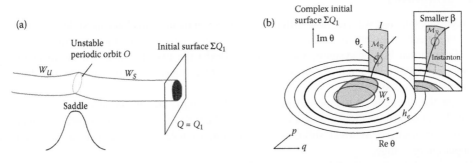

FIGURE 5.11 SUMGT mechanism for 2D-barrier tunneling. (a) Schematic illustration of the unstable periodic orbit O accompanied by the tubes formed by W_s and W_u. (b) Intersection of $\mathcal{I}$ with W_s on ΣQ_1 and the branches $\mathcal{M}_\mathcal{R}$ and $\mathcal{M}_\mathcal{C}$ on $\mathcal{I}$. The inset box shows the case for a smaller value of the coupling constant β.

then as is depicted in Figure 5.11b, there emerges a complex critical point $\theta_c(e_{ch})$ as the intersection between the complexified W_s and h_e. In a small vicinity of the critical point θ_c, the characteristic branches $\mathcal{M_R}$ and $\mathcal{M_C}$ exist and complex trajectories starting from them make contributions to tunneling as SUMGT.

When the coupling constant β is large enough such that the imaginary part of θ_c is in the same level as that of instant branches, the SUMGT branches take the place of the instanton branches and dominate the tunneling process. However, as shown in the inset box in Figure 5.11b, for a smaller value of the coupling constant β, the critical point shifts deeper into the imaginary side and so do $\mathcal{M_R}$ and $\mathcal{M_C}$, which means that the actions along the trajectories from them gain larger imaginary parts and contribute less to the S-matrix. In this case, instanton branches appear in a lower imaginary side of the complexified h_{ch} and become the major carrier of tunneling probability. The change of the tunneling spectrum shown by Figure 5.10 indicates the change of the underlying classical mechanism making the major tunneling contribution, from SUMGT to the instanton tunneling with decrease of β.

Note that the analysis based on the Melnikov method developed in Section 5.5 is also available for the 2D system. This is because the unperturbed 2D system with $\beta = 0$ in Equation 5.35 is separable so that the solution of the reaction coordinate is the same as that of the unperturbed 1D system, while the channel motion is represented by the harmonic oscillator for which $\theta = \omega t$ plays the role of time t. Therefore all the results of the Melnikov method, that is, the imaginary depth of the critical point Im t_{1c} and the weight of SUMGT W_S, are the same as those for the 1.5D system. Furthermore, the same analysis based on the instanton method in Section 5.4 is also available. Therefore, judging the winner of the competition between SUMGT and the instanton tunneling as well as estimating the tunneling weight are done by the same ways as for the 1.5D system.

Recently, Creagh and coworkers developed a semiclassical theory of multidimensional tunneling, in a situation similar to that considered in this review, by introducing the reaction operator, which is convenient for visualizing the transmission probability on Σ_Q by using the Weyl symbol [22]. Their semiclassical formalism taking into account the contribution to tunneling from the outside of the tube W_s seems to well reproduce the fully quantum results, although their tunneling trajectories relevant for the reaction operator formula are not the same as those we treated in this review. Whether or not the novel tunneling mechanism SUMGT plays any role in their results is not clear yet. According to the treatment of the reaction operator, contributions from all the initial states of the channel at a given total energy E_{tot} are summed up and so the contributions from the branches $\mathcal{M_R}$ and $\mathcal{M_C}$ may be concealed by the averaging effect and/or by the ambiguity inherent in the phase-space representation of the quantum state. It is desired that the two approaches are unified to give a comprehensive view of multidimensional barrier tunneling.

5.8 SUMMARY

Through this chapter, we saw how different the novel tunneling mechanism SUMGT is from the tunneling mechanism based on instantons and how it contributes to the tunneling process for multidimensional barrier systems. Indeed, there coexist the two mechanisms, the instanton-type tunneling and SUMGT, and the competition between them determines which mechanism dominates the tunneling process.

When the strength of the perturbation is small enough, the instanton tunneling overcomes SUMGT and contributes to forming a major part of the tunneling spectrum, namely the head lobe, but even for this case, SUMGT still exists contributing to make a shoulder part of the spectrum in an upper energy range.

However, if the strength of the perturbation becomes stronger beyond a threshold value, SUMGT takes the place of the instanton tunneling and the characteristic tunneling spectrum called "plateau spectrum" is created by SUMGT. The threshold value is given by the comparison of the imaginary part of the critical point, the intersection between the stable manifold and the initial plane in the complex domain, with the imaginary time depth of the instanton. The existence of the critical point for any strength of the perturbation is proved with the Melnikov method.

In the strong perturbation regime, the imaginary part of the critical point is in the same level as or smaller than the imaginary depth of the instanton, so the instanton branches are broken by the critical point. Instead, the characteristic branches $\mathcal{M}_{\mathcal{R}}$ and $\mathcal{M}_{\mathcal{C}}$ appear in a small neighborhood of the critical point in the initial plane and contribute to forming the plateau spectrum, which is considered as a manifestation of the structure of the unstable manifold. It can be proved that $\mathcal{M}_{\mathcal{R}}$ and $\mathcal{M}_{\mathcal{C}}$ exist in the small neighborhood of the critical point by using a certain approximate solution which works in a low-frequency regime [19]. Note that even for the weak perturbation regime, $\mathcal{M}_{\mathcal{R}}$ and $\mathcal{M}_{\mathcal{C}}$ exist near the critical point in a deeper imaginary side and contribute to forming the shoulder part of the spectrum.

The tunneling weight can be evaluated with calculating the classical action. The weight of the instanton is simply estimated by the imaginary part of the classical action along the instanton with imaginary time evolution in the classically forbidden region, but averaging over the period is necessary for periodically perturbed systems. The weight of SUMGT is estimated by the imaginary part of the classical action along the critical trajectory starting at the critical point. This is because the imaginary parts of classical actions of that $\mathcal{M}_{\mathcal{R}}$, which forms a major part of the plateau, are almost as same as that of the critical trajectory. Judging the dominant mechanism with comparing the imaginary depth of the critical point with that of the instanton and using the weight of SUMGT or the averaged weight of the instantons, we can simply estimate the tunneling weight for multi-dimensional barrier systems. To reconstruct more detailed information of tunneling, for example, the shape of the spectrum, we need full semiclassical calculations implemented in Section 5.6.

It is worthwhile to mention the problem how the transition of underlying tunneling mechanism occurs with the change of the perturbation frequency from a nearly zero to an extremely large value. Actually, there exist significant transitions with the change of the frequency, in which SUMGT dominates in a middle range, while the time-averaged instanton and a single instanton work in low- and high-frequency limits, respectively [21].

The systems treated in this review, the 1.5D and 2D barrier potentials, are in appearance classically integrable, since there is no chaotic behavior in the real scattering process. However, it is judged that those are nonintegrable by the Painlevé test or singularity analysis [13]. Thus, there should exist a chaotic motion created by the entanglement between stable and unstable manifolds in a deeper complex domain, although it is very hard to find it numerically. Note that an entanglement between stable and unstable manifolds is observed for a free-motion system kicked periodically by a barrier potential, though it is not a continuous time system [32]. In some sense, the fact that the complexified channel energy surface h_e intersects with the stable manifold W_s for the 2D barrier system, though it may be a much simpler case, corresponds to the fact that the KAM torus meets a natural boundary in the complex domain for usual nonintegrable systems, giving rise to the instanton-type tunneling breaking down [10,33]. Hence the mechanism of SUMGT must be universal in multidimensional systems and should provide a unified theoretical foundation for understanding various novel tunneling phenomena peculiar to multidimensional systems, including chaos- and resonance-assisted tunnelings.

REFERENCES

1. W.H. Miller, Semiclassical theory of atom–diatom collisions: path integrals and the classical S matrix, *J. Chem. Phys.*, vol. 53, pp. 1949–1959, 1970; W.H. Miller and T.F. George, Analytic continuation of classical mechanics for classically forbidden collision processes, *J. Chem. Phys.*, vol. 56, pp. 5668–5681, 1972; T.F. George and W.H. Miller, Classical S-matrix theory of reactive tunneling: Linear H+H2 collisions, *J. Chem. Phys.*, vol. 57, pp. 2458–2467, 1972.

2. W.H. Miller, Classical-limit quantum mechanics and the theory of molecular collisions, *Adv. Chem. Phys.*, vol. 25 pp. 69–177, 1974.
3. L.S. Schulman, *Techniques and applications of path integration*, Wiley, Inc., New York, NY, 1981.
4. S. Tomsovic (Ed.), *Tunneling in Complex Systems*, World Scientific, Inc., Singapore, 1998.
5. J. Ankerhold, *Quantum Tunneling in Complex Systems: The Semiclassical Approach*, Springer-Verlag, Inc., Berlin, 2007.
6. S. Keshavamurthy, Dynamical tunnelling in molecules: Quantum routes to energy flow, *Int. Rev. Phys. Chem.*, vol. 26, pp. 521–584, 2007.
7. O. Bohigas, S. Tomsovic, and D. Ullmo, Classical transport effects on chaotic levels, *Phys. Rev. Lett.*, vol. 65, pp. 5–8, 1990; O. Bohigas, S. Tomsovic, and D. Ullmo, Manifestations of classical phase-space structures in quantum-mechanics, *Phys. Rep.*, vol. 223, pp. 43–133, 1993.
8. W.K. Hensinger, H. Haffer, A. Browaeys, N.R. Heckenberg, K. Helmerson, C. McKenzie, G.J. Milburn, et al., Dynamical tunnelling of ultracold atoms, *Nature*, vol. 412, pp. 52–55, 2001; D.A. Steck, W.H. Oskay, and M.G. Raizen, Observation of chaos-assisted tunneling between islands of stability, *Science*, vol. 293, pp. 274–278, 2001; C. Dembowski, H.D. Graf, A. Heine, R. Hofferbert, F. Rehfeld, and A. Richter, First experimental evidence for chaos-assisted tunneling in a microwave annular billiard, *Phys. Rev. Lett.* vol. 84, pp. 867–870, 2000.
9. O. Brodier, P. Schlagheck, and D. Ullmo, Resonance-assisted tunneling in near-integrable systems, *Phys. Rev. Lett.*, vol. 87, 064101, 2001.
10. O. Brodier, P. Schlagheck, and D. Ullmo, Resonance-assisted tunneling, *Ann. Phys.(NY)*, vol. 300, pp. 88–136, 2002.
11. C. Eltschka and P. Schlagheck, Resonance-and chaos-assisted tunneling in mixed regular-chaotic systems, *Phys. Rev. Lett.*, vol. 94, 014101, 2005.
12. A. Bäcker, R. Ketzmerick, S. Löck, and L. Schilling, Regular-to-chaotic tunneling rates using a fictitious integrable system, *Phys. Rev. Lett.*, vol. 100, 104101, 2008; A. Bäcker, R. Ketzmerick, S. Lock, M. Robnik, G. Vidmar, R. Höhmann, U. Kuhl, and H.-J. Stöckmann, Dynamical tunneling in mushroom billiards, *Phys. Rev. Lett.*, vol. 100, 174103, 2008; A. Bäcker, R. Ketzmerick, S. Löck, J. Wiersig, and M. Hentschel, Quality factors and dynamical tunneling in annular microcavities, *Phys. Rev.*, Ser. A, vol. 79, 063804, 2009.
13. M.J. Ablowitz, A. Ramani, and H. Segur, A connection between non-linear evolution-equations and ordinary differential-equations of P-type. 1., *J. Math. Phys.*, vol. 21, pp. 715–721, 1980; A. Ramani, B. Grammaticos, and T. Bountis, The painlevé property and singularity analysis of integrable and non-integrable systems, *Phys. Rep.*, vol. 180, pp. 159–245, 1989.
14. M.J. Davis and E.J. Heller, Quantum dynamical tunneling in bound states, *J. Chem. Phys.* vol. 75, pp. 246–254, 1981.
15. M. Büttiker and R. Landauer, Traversal time for tunneling, *Phys. Rev. Lett.*, vol. 49, pp. 1739–1742, 1982; M. Büttiker and R. Landauer, Traversal time for tunneling, *Physica Scripta*, vol. 32, pp. 429–434, 1985.
16. K. Takahashi, A. Yoshimoto, and K.S. Ikeda, Movable singularities, complex-domain heteroclinicity, and fringed tunneling in multi-dimensional systems, *Phys. Lett.*, Ser. A, vol. 297, pp. 370–375, 2002.
17. K. Takahashi and K.S. Ikeda, Complex-classical mechanism of the tunnelling process in strongly coupled 1.5-dimensional barrier systems, *J. Phys.* Ser. A, vol. 36, pp. 7953–7987, 2003.
18. K. Takahashi and K.S. Ikeda, An intrinsic multi-dimensional mechanism of barrier tunneling, *Europhys. Lett.*, vol. 71, pp. 193–199, 2005 (Erratum: *Europhys. Lett.*, vol. 75, p. 355, 2006).
19. K. Takahashi and K.S. Ikeda, A plateau structure in the tunnelling spectrum as a manifestation of a new tunnelling mechanism in multi-dimensional barrier systems, *J. Phys.*, Ser. A, vol. 41, 095101, 2008.
20. K. Takahashi and K.S. Ikeda, Spectroscopic signature of the transition in a tunneling mechanism from the instanton path to a complexified stable–unstable manifold, *Phys. Rev.*, Ser. A, vol. 79, 052114, 2009.
21. K. Takahashi and K.S. Ikeda, Structural change of tunneling spectrum with perturbation frequency, *J. Phys.*, Ser. A, vol. 43, 192001, 2010.
22. S.C. Creagh, Classical transition states in quantum theory, *Nonlinearity*, vol. 17, pp. 1261–1303, 2004; S.C. Creagh, Semiclassical transmission across transition states, *Nonlinearity*, vol. 18, pp. 2089–2110, 2005; C.S. Drew, S.C. Creagh, and R.H. Tew, Uniform approximation of barrier penetration in phase space, *Phys. Rev.*, Ser. A vol. 72, 062501, 2005.

23. F. Bezrukov and D. Levkov, Dynamical tunneling of bound systems through a potential barrier: Complex way to the top, *J. Exp. Theor. Phys.*, vol. 98, pp. 820–836, 2004; D.G. Levkov, A.G. Panin, and S.M. Sibiryakov, Unstable semiclassical trajectories in tunneling, *Phys. Rev. Lett.* vol. 99, 170407, 2007; D.G. Levkov, A.G. Panin, and S.M. Sibiryakov, Complex trajectories in chaotic dynamical tunneling, *Phys. Rev.*, Ser. E vol. 76, 046209, 2007; D.G. Levkov, A.G. Panin, and S.M. Sibiryakov, Signatures of unstable semiclassical trajectories in tunneling, *J. Phys.* Ser. A, vol. 42, 205102, 2009.

24. S. Wiggins, L. Wiesenfeld, C. Jaffe, and T. Uzer, Impenetrable barriers in phase space, *Phys. Rev. Lett.*, vol. 86, pp. 5478–5481, 2001; T. Uzer, C. Jaffe, J. Palacian, P. Yanguas, and S. Wiggins, The geometry of reaction dynamics, *Nonlinearity*, vol. 15, pp. 957–992, 2002.

25. A. Shudo and K.S. Ikeda, Complex classical trajectories and chaotic tunneling, *Phys. Rev. Lett.*, vol. 74, pp. 682–685, 1995; A. Shudo and K.S. Ikeda, Chaotic tunneling: A remarkable manifestation of complex classical dynamics in non-integrable quantum phenomena, *Phys.*, Ser. D, vol. 115, pp. 234–292, 1998.

26. A. Shudo, Y. Ishii, and K.S. Ikeda, Julia set describes quantum tunnelling in the presence of chaos, *J. Phys.*, Ser. A, vol. 35, pp. L225–L231, 2002; A. Shudo, Y. Ishii, and K.S. Ikeda, Chaos attracts tunneling trajectories: A universal mechanism of chaotic tunneling, *Europhys. Lett.*, vol. 81, 50003, 2008; A. Shudo, Y. Ishii, and K.S. Ikeda, Julia sets and chaotic tunneling: I, *J. Phys.*, Ser. A, vol. 42, 265101, 2009; A. Shudo, Y. Ishii, and K.S. Ikeda, Julia sets and chaotic tunneling: II, *J. Phys.*, Ser. A, vol. 42, 265102, 2009.

27. K. Takahashi and K.S. Ikeda, Anomalously long passage through a rounded-off-step potential due to a new mechanism of multidimensional tunneling, *Phys. Rev. Lett.*, vol. 97, 240403, 2006.

28. K. Takahashi and K.S. Ikeda, Complex semiclassical description of scattering problem in systems with 1.5 degrees of freedom, *Ann. Phys. (NY)*, vol. 283, pp. 94–140, 2000.

29. K. Takahashi and K.S. Ikeda, Application of symplectic integrator to stationary reactive-scattering problems: Inhomogeneous Schrödinger equation approach, *J. Chem. Phys.*, vol. 106, pp. 4463–4480, 1997.

30. A. Shudo and K.S. Ikeda, Stokes geometry for the quantum Hénon map, *Nonlinearity*, vol. 21, pp. 1831–1880, 2008 and references therein.

31. S. Wiggins, *Introduction to Applied Nonlinear Dynamical Systems and Chaos*, Springer-Verlag, Inc., New York, NY, 1990.

32. T. Onishi, A. Shudo, K.S. Ikeda, and K. Takahashi, Tunneling mechanism due to chaos in a complex phase space, *Phys. Rev.*, Ser. E, vol. 64, 025201(R), 2001; T. Onishi, A. Shudo, K.S. Ikeda, and K. Takahashi, Semiclassical study on tunneling processes via complex-domain chaos, *Phys. Rev.*, Ser. E, vol. 68, 056211, 2003.

33. J.M. Greene and I.C. Percival, Hamiltonian maps in the complex plane, *Physica*, Ser. D, vol. 3, pp. 530–548, 1981.

6 Direct Regular-to-Chaotic Tunneling Rates Using the Fictitious Integrable System Approach

Arnd Bäcker, Roland Ketzmerick, and Steffen Löck

CONTENTS

6.1 INTRODUCTION

Tunneling is one of the most fundamental manifestations of quantum mechanics. For 1D systems, the theoretical description is well established by the Wentzel–Kramers–Brillouin (WKB) method and related approaches [1,2]. However, for higher dimensional systems no such simple description exists. In these systems, typically regular and chaotic motion coexists and in the two-dimensional case regular tori are absolute barriers to the motion. Quantum mechanically, the eigenfunctions either concentrate within the regular islands or in the chaotic sea, as expected from the semiclassical eigenfunction hypothesis [3–5]. These eigenfunctions are coupled by so-called dynamical tunneling [6] through the dynamically generated barriers in phase space. In particular, this leads to a substantial enhancement of tunneling rates between phase-space regions of regular motion due to the presence of chaotic motion, which was termed chaos-assisted tunneling [7–9]. Such dynamical tunneling processes are ubiquitous in molecular physics and were realized with microwave cavities [10] and cold atoms in periodically modulated optical lattices [11,12].

In the quantum regime, $h_{\text{eff}} \lesssim A$, where the effective Planck constant h_{eff} is smaller but still comparable to the area of the regular island A, the process of tunneling into the chaotic region is dominated by a *direct regular-to-chaotic tunneling* mechanism [13–16]. For the prediction of tunneling rates, the fictitious integrable system approach was introduced recently [16]. It relies on a fictitious integrable system [8,14] that resembles the regular dynamics within the island under consideration. The approach has been applied to quantum maps [16], billiard systems [17], and the annular microcavity [18]. In the semiclassical regime, $h_{\text{eff}} \ll A$, however, the direct tunneling contribution alone is typically not sufficient to describe the observed tunneling rates, because nonlinear resonance chains within the regular island lead to resonance-assisted tunneling [19–22]. Recently, a combined prediction of dynamical tunneling rates from regular to chaotic phase-space regions was derived [23], which combines the direct regular-to-chaotic tunneling mechanism in the quantum regime with an improved resonance-assisted tunneling theory in the semiclassical regime, see the contributions of Schlagheck, Mouchet, and Ullmo in this book. In this text, we concentrate on the fictitious integrable system approach for the theoretical description of the direct regular-to-chaotic tunneling mechanism and how it can be applied analytically, semiclassically, and numerically to quantum maps.

This text is organized as follows: we start by defining different convenient kicked systems, their quantization, and the numerical determination of tunneling rates. In Section 6.3, we describe the approach to determine dynamical tunneling rates by using a fictitious integrable system. In Section 6.4, we apply this theory to the previously introduced systems and compare its results with numerical data. We conclude with a brief summary.

6.2 KICKED SYSTEMS

Kicked systems are particularly suited to study classical and quantum effects appearing in a mixed phase space as they can be easily treated analytically and numerically. In contrast to time-independent systems, where at least two degrees of freedom are necessary to break integrability, one-dimensional-driven systems can show chaotic motion. We consider time-periodic one-dimensional kicked systems

$$H(q,p,t) = T(p) + V(q) \sum_{n \in \mathbb{Z}} \tau \delta(t - n\tau), \tag{6.1}$$

which are described by the kinetic energy $T(p)$ and the potential $V(q)$ which is applied once per kick period $\tau = 1$. The classical dynamics of a kicked system is given by its stroboscopic mapping, for example, evaluated just after each kick

$$\begin{aligned} q_{n+1} &= q_n + T'(p_n), \\ p_{n+1} &= p_n - V'(q_{n+1}). \end{aligned} \tag{6.2}$$

It maps the phase-space coordinates after the nth kick to those after the $(n + 1)$th kick. For the stroboscopic mapping, we consider a compact phase space with periodic boundary conditions for $q \in [-1/2, 1/2]$ and $p \in [-1/2, 1/2]$. The phase space generally consists of regions with regular motion surrounded by chaotic dynamics, see Figure 6.1b. We focus on the situation of just one regular island embedded in the chaotic sea, Figure 6.2a and b. At the center of the island one has an elliptic fixed point, which is surrounded by invariant regular tori. Classically, a particle within a regular island will never enter into the chaotic region and vice versa, that is, tori form absolute barriers to the motion.

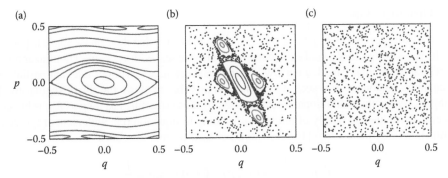

FIGURE 6.1 Phase space of the standard map for (a) $\kappa = 0.3$, (b) $\kappa = 2.4$, and (c) $\kappa = 8$. In (a), the system is nearly integrable. Most of the phase space is filled by invariant tori (lines). The second figure (b) shows a mixed phase space with regular motion (lines) and chaotic motion (dots). The four small islands surrounding the central island form a nonlinear resonance chain. In the chaotic sea partial barriers exist, which is reflected by the different densities of chaotic points. For large values of κ, as in (c), the phase space seems macroscopically chaotic.

6.2.1 STANDARD MAP

The paradigmatic model of an area preserving map is the standard map, defined by Equation 6.2 with the functions

$$T'(p) = p, \tag{6.3}$$

$$V'(q) = \frac{\kappa}{2\pi} \sin(2\pi q). \tag{6.4}$$

For $\kappa = 0$, the system is integrable and the dynamics takes place on invariant tori with constant momentum. For small $\kappa > 0$, see Figure 6.1a, the system is no longer integrable. However, as stated by the Kolmogorov–Arnold–Moser (KAM) theorem [24], many invariant tori persist. In between these tori sequences of stable and unstable fixed points emerge, which are described by the Poincaré–Birkhoff theorem [25]. The stable fixed points of such a sequence form the so-called nonlinear resonance chains, see Figure 6.1b, while in the vicinity of the unstable fixed points typically chaotic dynamics is observed. For larger κ, more and more regular tori break up and the chaotic dynamics occupies a larger area of phase space. At $\kappa \approx 3$, one large regular region ("regular island") is surrounded by a region of chaotic dynamics ("chaotic sea"). At the border of the regular island, a hierarchical region exists, where self-similar structures of regular and chaotic dynamics are found on all scales. Also in the chaotic sea additional structures can be found. Here, partial barriers limit the classical flux between different regions of chaotic motion caused by the stable and unstable manifolds of unstable periodic orbits or caused by cantori, which are the remnants of regular tori [26]. For even larger κ, the regular islands disappear and the motion seems macroscopically chaotic, see Figure 6.1c.

All these structures in phase space, such as nonlinear resonances, the hierarchical regular-to-chaotic transition region, and partial barriers within the chaotic sea, have an influence on the dynamical tunneling process from the regular island to the chaotic sea. In order to quantitatively understand dynamical tunneling in systems with a mixed phase space, it is helpful to first consider simpler model systems. By designing maps such that the phase-space structures can be switched on and off one by one, it is possible to study their influence on the dynamical tunneling process separately. Afterwards, the results can be tested using the generic example of the standard map.

6.2.2 DESIGNED MAPS

Our aim is to introduce kicked systems which are designed such that their phase space is particularly simple. It shows one regular island embedded in the chaotic sea, with very small nonlinear resonance chains within the regular island, a narrow hierarchical region, and without relevant partial barriers in the chaotic component. For such a system, it is possible to study the direct regular-to-chaotic tunneling process without additional effects caused by these structures as long as h_{eff} is big enough. To this end, we define the family of maps $\mathcal{D}$, according to Equation 6.2, with an appropriate choice of the functions $T'(p)$ and $V'(q)$ [16,23,27,28]. For this, we first introduce

$$t'(p) = \frac{1}{2} \pm (1 - 2p) \quad \text{for } 0 < \pm p < \frac{1}{2}, \tag{6.5}$$

$$v'(q) = -rq + Rq^2 \quad \text{for } -\frac{1}{2} < q < \frac{1}{2} \tag{6.6}$$

with $0 < r < 2$ and $R \geq 0$. Considering periodic boundary conditions the functions $t'(p)$ and $v'(q)$ show discontinuities at $p = 0, \pm 1/2$ and $q = \pm 1/2$, respectively. In order to avoid these discontinuities, we smooth the periodically extended functions $v'(q)$ and $t'(q)$ with a Gaussian

$$G(z) = \frac{1}{\sqrt{2\pi\varepsilon^2}} \exp\left(-\frac{z^2}{2\varepsilon^2}\right), \tag{6.7}$$

resulting in analytic functions

$$T'(p) = \int dz\, t'(z) G(p - z), \tag{6.8}$$

$$V'(q) = \int dz\, v'(z) G(q - z), \tag{6.9}$$

which are periodic with respect to the phase-space unit cell. With this we obtain the map $\mathcal{D}$ depending on the parameters r, R, and the smoothing strength ε. The smoothing ε determines the size of the hierarchical region at the border of the regular island. Tuning the parameters r and R, one can find situations where all nonlinear resonance chains inside the regular island are small.

For $R = 0$, both functions $v'(q)$ and $t'(p)$ are linear in q and p, respectively. In this case, we find a harmonic oscillator-like regular island with elliptic invariant tori and constant rotation number. We choose the parameters $r = 0.46$, $R = 0$, and $\varepsilon = 0.005$ for which the phase space of the resulting map $\mathcal{D}_{\text{ho}}$ is shown in Figure 6.2a.

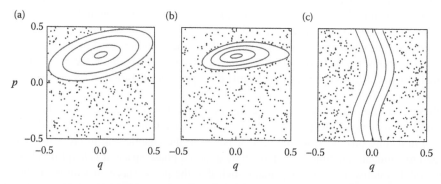

FIGURE 6.2 Phase space of (a) the designed map $\mathcal{D}_{\text{ho}}$ with a harmonic oscillator-like regular island, (b) $\mathcal{D}_{\text{d}}$ with a deformed regular island, and (c) $\mathcal{D}_{\text{rs}}$ with a regular stripe. For parameters see text.

In typical systems, the rotation number of regular tori changes from the center of the regular region to its border which typically has a nonelliptic shape. Such a situation can be achieved for the family of maps $\mathcal{D}$ with the parameter $R \neq 0$. For most combinations of the parameters, r and R resonance structures appear inside the regular island. They limit the h_{eff}-regime in which the direct regular-to-chaotic tunneling process dominates. Hence, we choose a situation in which the nonlinear resonances are small such that their influence on the tunneling process is expected only at small h_{eff}. The phase space of the map $\mathcal{D}_{\mathrm{d}}$, obtained for $r = 0.26$, $R = 0.4$, and $\varepsilon = 0.005$, is shown in Figure 6.2b.

Another designed kicked system was introduced in Ishikawa et al. (2007) [29], Shudo and Ikeda (1995) [35], and Shudo and Ikeda (1998) [36]. Here the regular region consists of a stripe in phase space, see Figure 6.2c. In our notation, the mapping, Equation 6.2, is specified by the functions

$$V'(q) = -\frac{1}{2\pi}\left(8\pi aq + d_1 - d_2 + \frac{1}{2}[8\pi aq - \omega + d_1]\tanh[b(8\pi q - q_d)]\right.$$

$$\left. + \frac{1}{2}[-8\pi aq + \omega + d_2]\tanh[b(8\pi q + q_d)]\right) \tag{6.10}$$

$$T'(p) = -\frac{\kappa}{8\pi}\sin(2\pi p). \tag{6.11}$$

The kinetic energy $T(p)$ is periodic with respect to the phase-space unit cell. We label the resulting map with a regular stripe by $\mathcal{D}_{\mathrm{rs}}$ using the parameters $a = 5$, $b = 100$, $d_1 = -24$, $d_2 = -26$, $\omega = 1$, $q_d = 5$, and $\kappa = 3$.

The map $\mathcal{D}_{\mathrm{rs}}$ is similar to the system $\mathcal{D}_{\mathrm{ho}}$ as it also destroys the integrable region by smoothly changing the function $V'(q)$, here at $|q| = q_d/(8\pi)$. For $|q| < q_d/(8\pi)$ the potential term is almost linear while it tends to the standard map for $|q| > q_d/(8\pi)$. The parameter b determines the width of the transition region.

6.2.3 QUANTIZATION

Quantum mechanically kicked systems are described by the unitary time-evolution operator U [30]. It can be written as $U = U_V U_T$ with

$$U_V = \mathrm{e}^{-iV(q)/\hbar_{\mathrm{eff}}}, \tag{6.12}$$

$$U_T = \mathrm{e}^{-iT(p)/\hbar_{\mathrm{eff}}}. \tag{6.13}$$

Its eigenstates $|\psi_n\rangle$ and quasienergies φ_n are determined by

$$U|\psi_n\rangle = \mathrm{e}^{i\varphi_n}|\psi_n\rangle. \tag{6.14}$$

We consider a compact phase space with periodic boundary conditions in q and p. This implies that the effective Planck constant can take only the discrete values [31]

$$h_{\mathrm{eff}} = \frac{1}{N}, \tag{6.15}$$

where N is the dimension of the Hilbert space. In addition the position and momentum coordinates are quantized according to

$$q_k = \frac{k}{N} - \frac{1}{2}, \tag{6.16}$$

$$p_l = \frac{l}{N} - \frac{1}{2} \tag{6.17}$$

with $k, l = 0, \ldots, N-1$.

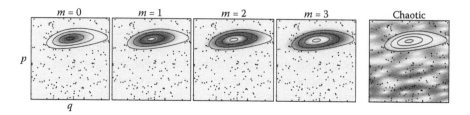

FIGURE 6.3 Husimi representation of eigenfunctions of the map $\mathcal{D}_d$ for $h_{\text{eff}} = 1/50$. The regular ground state $m = 0$ and the first three excited states $m = 1, 2, 3$ as well as a chaotic state are shown.

For systems with a mixed phase space, in the semiclassical limit ($h_{\text{eff}} \to 0$), the semiclassical eigenfunction hypothesis [3–5] implies that the eigenstates $|\psi_n\rangle$ can be classified as either regular or chaotic according to the phase-space region on which they concentrate, see Figure 6.3. In order to understand the behavior of eigenstates away from the semiclassical limit, that is, at finite values of the effective Planck constant h_{eff}, one has to compare the size of phase-space structures with h_{eff}.

The so-called regular states are concentrated on tori within the regular island and fulfill the Bohr–Sommerfeld-type quantization condition

$$\oint p \, dq = h_{\text{eff}} \left(m + \frac{1}{2} \right), \quad m = 0, \dots, N_{\text{reg}} - 1. \tag{6.18}$$

For a given value of h_{eff} there exist N_{reg} of such regular states, where $N_{\text{reg}} = \lfloor A/h_{\text{eff}} + 1/2 \rfloor$ and A is the area of the regular island. The $N_{\text{ch}} = N - N_{\text{reg}}$ chaotic states extend over the chaotic sea. Due to dynamical tunneling, however, the regular and chaotic eigenfunctions of U always have a small component in the other region of phase space, respectively. This is most clearly seen for hybrid states which even have the same weight in each of the components as they are involved in an avoided level crossing.

6.2.4 Numerical Determination of Tunneling Rates

The structure of the considered phase space, with one regular island surrounded by the chaotic sea, allows for the determination of tunneling rates by introducing absorption somewhere in the chaotic region of phase space. For quantum maps, this can, for example, be realized by using a nonunitary open quantum map [32]

$$U^o = PUP, \tag{6.19}$$

where P is a projection operator onto the complement of the absorbing region.

While the eigenvalues of U are located on the unit circle the eigenvalues of U^o are inside the unit circle as U^o is subunitary. The eigenequation of U^o reads

$$U^o |\psi_n^s\rangle = z_n |\psi_n^s\rangle \tag{6.20}$$

with eigenvalues

$$z_n = e^{i\left(\varphi_n + i\frac{\gamma_n}{2}\right)}. \tag{6.21}$$

The decay rates are characterized by the imaginary part of the quasienergies in Equation 6.21 and one has

$$\gamma_m = -2\log|z_m| \approx 2(1 - |z_m|). \tag{6.22}$$

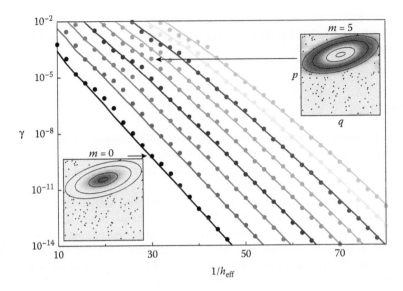

FIGURE 6.4 Numerical data (dots) for $m \leq 8$ for the map $\mathcal{D}_{\text{ho}}$ with a harmonic oscillator-like island. Comparison with Equation 6.27 (lines), see Section 6.3.1 for its derivation. The insets show Husimi functions of the regular states for the quantum numbers $m = 0$ and 5 at $1/h_{\text{eff}} = 30$.

If the chaotic region does not contain partial barriers and shows no dynamical localization, it is justified to assume that the rate of escaping the regular island is equal to the rate of leaving through the absorbing regions located in the chaotic sea. Then, the location of the absorbing regions in the chaotic part of phase space has no effect on the tunneling rates.

In generic systems, however, partial barriers will appear in the chaotic region of phase space. The additional transition through these structures further limits the quantum transport such that the calculated decay through the absorbing region occurs slower than the decay from the regular island to the neighboring chaotic sea. Similarly, dynamical localization in the chaotic region may slow down the decay. The influence of partial barriers and dynamical localization on the tunneling process is an open problem. Here, we will suppress their influence, if necessary, by moving the absorbing regions closer to the regular island.

In Figure 6.4 tunneling rates are determined numerically by opening the system for the map $\mathcal{D}_{\text{ho}}$ (dots). These agree very well with an analytical prediction (lines), Equation 6.27, which is derived in the next section.

6.3 TUNNELING USING A FICTITIOUS INTEGRABLE SYSTEM

Dynamical tunneling in systems with a mixed phase space couples the regular island and the chaotic sea, which are classically separated. This coupling can be quantified by tunneling rates γ_m, which describe the decay of regular states to the chaotic sea. To define these tunneling rates one can consider a wave packet started on the mth quantized torus in the regular island coupled to a continuum of chaotic states, as in the case for an infinite chaotic sea or in the presence of an absorbing region somewhere in the chaotic sea. Its decay $e^{-\gamma_m t}$ is described by a tunneling rate γ_m. For systems with a finite phase space this exponential decay occurs at most up to the Heisenberg time $\tau_H = h_{\text{eff}}/\Delta_{\text{ch}}$, where Δ_{ch} is the mean level spacing of the chaotic states.

In the quantum regime, $h_{\text{eff}} \lesssim A$, where h_{eff} is smaller but comparable to the area A of the regular island, the rates γ_m are dominated by the direct regular-to-chaotic tunneling mechanism, while contributions from resonance-assisted tunneling are negligible. We concentrate on situations where additional phase-space structures within the chaotic sea are not relevant for tunneling. In the following, we derive a prediction for the direct regular-to-chaotic tunneling rates using the fictitious integrable system approach [16].

6.3.1 Derivation

In order to find a prediction for the direct regular-to-chaotic tunneling rates, we decompose the Hilbert space of the quantum map U into two parts, which correspond to the regular and chaotic regions. While classically such a decomposition is unique (neglecting tiny phase-space structures), quantum mechanically this is not the case due to the uncertainty principle. We find a decomposition by introducing a fictitious integrable system U_{reg} (a related idea was presented in Bohigas et al. (1993) [8], Podolskiy and Narimanov (2003) [14], and Sheinman et al. (2006) [15]). It has to be chosen such that its dynamics resembles the classical motion corresponding to U within the regular island as closely as possible and continues this regular dynamics beyond the regular island of U, see Figure 6.5a and d. The eigenstates $|\psi_{\text{reg}}^m\rangle$ of U_{reg} are purely regular in the sense that they are localized on the mth quantized torus of the regular region and continue to decay beyond this regular region, see Figure 6.5e. This is the decisive property of $|\psi_{\text{reg}}^m\rangle$ which has no chaotic admixture, in contrast to the predominantly regular eigenstates of U, see Figure 6.5b. The explicit construction of U_{reg} is discussed in Section 6.3.2.

With the eigenstates $|\psi_{\text{reg}}^m\rangle$ of U_{reg}, $U_{\text{reg}}|\psi_{\text{reg}}^m\rangle = \mathrm{e}^{\mathrm{i}\varphi_{\text{reg}}^m}|\psi_{\text{reg}}^m\rangle$, we define a projection operator

$$P_{\text{reg}} := \sum_{m=0}^{N_{\text{reg}}-1} |\psi_{\text{reg}}^m\rangle\langle\psi_{\text{reg}}^m|, \tag{6.23}$$

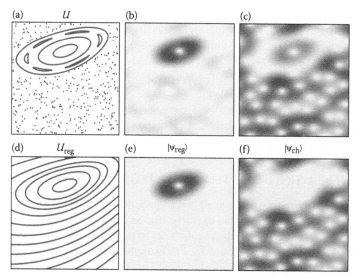

FIGURE 6.5 (a) Illustration of the classical phase space corresponding to some mixed quantum map U together with the Husimi representation of (b) a regular and (c) a chaotic eigenstate of U which both have a small component in the other region. (d) Illustration of the classical phase space of the fictitious integrable system U_{reg}. (e) Eigenstates $|\psi_{\text{reg}}\rangle$ of U_{reg} are purely regular while (f) the basis states $|\psi_{\text{ch}}\rangle$ are localized in the chaotic region of phase space.

using the first N_{reg} regular states of U_{reg} which approximately projects onto the regular island corresponding to U. The orthogonal projector

$$P_{ch} := \mathbb{1} - P_{reg} \qquad (6.24)$$

approximately projects onto the chaotic phase-space region. These projectors, P_{reg} and P_{ch}, define our decomposition of the Hilbert space into a regular and a chaotic subspace.

Introducing an orthonormal basis $|\psi_{ch}\rangle$ in the chaotic subspace we can write

$$P_{ch} = \sum_{ch} |\psi_{ch}\rangle\langle\psi_{ch}|. \qquad (6.25)$$

The tunneling rate is obtained using a dimensionless version of Fermi's golden rule,

$$\gamma_m = \sum_{ch} |v_{ch,m}|^2, \qquad (6.26)$$

where the sum is over all chaotic basis states $|\psi_{ch}\rangle$ and thus averages the modulus squared of the fluctuating matrix elements $v_{ch,m}$. Here, we apply Fermi's golden rule in the case of a discrete spectrum, which is possible if one considers the decay $e^{-\gamma_m t}$ up to the Heisenberg time $\tau_H = h_{eff}/\Delta_{ch}$ only. Inserting Equation 6.25 we obtain

$$\gamma_m = \| P_{ch} U |\psi_{reg}^m\rangle \|^2 = \| (\mathbb{1} - P_{reg}) U |\psi_{reg}^m\rangle \|^2 \qquad (6.27)$$

as the basis of all our following investigations. It allows for the prediction of tunneling rates from a regular state localized on the mth quantized torus to the chaotic sea. Equation 6.27 confirms the intuition that the tunneling rates are determined by the amount of probability of $|\psi_{reg}^m\rangle$ that is transferred to the chaotic region after one application of the time evolution operator U. In Equation 6.27 properties of the fictitious integrable system U_{reg} and the chaotic projector $P_{ch} = \mathbb{1} - P_{reg}$ enter, which rely on the chosen decomposition of Hilbert space.

In cases where one finds a fictitious integrable system U_{reg} which resembles the dynamics within the regular island of U with very high accuracy, Equation 6.27 can be approximated as

$$\gamma_m \approx \| (U - U_{reg}) |\psi_{reg}^m\rangle \|^2, \qquad (6.28)$$

using $P_{reg} U |\psi_{reg}^m\rangle \approx P_{reg} U_{reg} |\psi_{reg}^m\rangle = U_{reg} |\psi_{reg}^m\rangle$. Instead of the operator product $P_{ch} U$ in Equation 6.27, the difference $U - U_{reg}$ enters in Equation 6.28. It allows for further derivations, which are presented in Section 6.3.3.

6.3.2 FICTITIOUS INTEGRABLE SYSTEM

The most difficult step in the application of Equation 6.27 to a given system is the determination of the fictitious integrable system U_{reg}. On the one hand its dynamics should resemble the classical motion of the considered mixed system within the regular island as closely as possible. As a result the contour lines of the corresponding integrable Hamiltonian H_{reg}, Figure 6.5d, approximate the KAM curves of the classically mixed system, Figure 6.5a, in phase space. This resemblance is not possible with arbitrary precision as the integrable approximation does not contain, for example, nonlinear resonance chains and small embedded chaotic regions. Moreover, it cannot account for the hierarchical regular-to-chaotic transition region at the border of the regular island. Similar problems appear for the analytic continuation of a regular torus into complex space due to the existence of natural boundaries [19–21,33–37]. However, for not too small h_{eff}, where these small structures are not yet resolved, an integrable approximation with finite accuracy turns out to be sufficient for a prediction of the tunneling rates.

On the other hand the integrable dynamics of H_{reg} should extrapolate smoothly beyond the regular island of H. This is essential for the quantum eigenstates of H_{reg} to have correctly decaying tunneling tails which are according to Equation 6.27 relevant for the determination of the tunneling rates. While typically tunneling from the regular island occurs to regions within the chaotic sea close to the border of the regular island, there exist other cases, where it occurs to regions deeper inside the chaotic sea, as studied in Sheinman et al. (2006) [15]. Here H_{reg} has to be constructed such that its eigenstates have the correct tunneling tails up to this region.

For quantum maps we determine the fictitious integrable system in the following way: We employ classical methods, see below, to obtain a one-dimensional time-independent Hamiltonian $H_{reg}(q, p)$ which is integrable by definition and resembles the classically regular motion corresponding to the mixed system. After its quantization we obtain the regular quantum map $U_{reg} = e^{-iH_{reg}/\hbar_{eff}}$ which has the same eigenfunctions $|\psi_{reg}^m\rangle$ as H_{reg}. For the numerical evaluation of Equation 6.27 we use $P_{ch} = \mathbb{1} - P_{reg} = \mathbb{1} - \sum |\psi_{reg}^m\rangle\langle\psi_{reg}^m|$, according to Equations 6.23 and 6.24, where the sum extends over $m = 0, 1, \ldots, N_{reg} - 1$.

Two examples for the explicit construction of H_{reg} will be mentioned below. Note, that also other methods, for example based on the normal-form analysis [38,39] or on the Campbell–Baker–Hausdorff formula [40] can be employed in order to find H_{reg}. For the example systems considered in this chapter, however, they show less good agreement.

One possible choice for the determination of the fictitious integrable system for quantum maps is the Lie-transformation method [41]. It determines a classical Hamilton function as a power series in q and p

$$H_{reg}^K(q, p) = \sum_{l=1}^{K} h_l(q, p). \tag{6.29}$$

with expansion coefficients $h_{k,l}$, see Brodier et al. (2002) [20] and Figure 6.6a for examples. Typically, the order of the expansion K can be increased up to 20 within reasonable numerical effort. The Lie-transformation method provides a regular approximation H_{reg} which interpolates the dynamics inside the regular region and gives a smooth continuation into the chaotic sea. At some order K the series should diverge due to the nonlinear resonances inside the regular island. For strongly perturbed systems, such as the standard map at $\kappa > 2.5$, the Lie-transformation method may not be

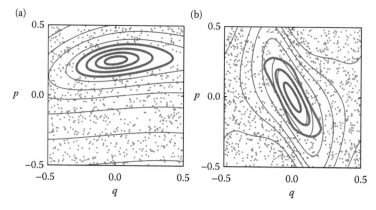

FIGURE 6.6 Application of the Lie transformation method: (a) Orbits (gray) of the map $\mathcal{D}_d$ and of the corresponding integrable system (thin lines) of order $K = 15$. Here H_{reg} accurately resembles the regular dynamics of U. (b) Orbits of the standard map for $\kappa = 2.9$ (gray) and of the corresponding integrable system (thin lines) of order $K = 7$. Here H_{reg} does not accurately resemble the regular dynamics of H.

able to reproduce the regular dynamics of U, see Figure 6.6b. Here we use a method [16] based on the frequency map analysis [42].

An important question is whether the direct tunneling rates obtained using Equation 6.27 depend on the actual choice of H_{reg} and how these results converge depending on the order K of its perturbation series. Ideally, one would like to use classical measures, which describe the deviations of the regular system H_{reg} from the originally mixed one, to predict the error of Equation 6.27 for the tunneling rates. However, these classical measures can only account for the deviations within the regular region but not for the quality of the continuation of H_{reg} beyond the regular island of U. Currently, the quality of an integrable system can be estimated *a posteriori* by comparison of the predicted tunneling rates with numerical data. It remains an open question how to obtain a direct connection between the error on the classical side and the one for the tunneling rates.

6.3.3   APPROXIMATE FICTITIOUS INTEGRABLE SYSTEM

For an analytical evaluation of the result, Equation 6.27, of the direct regular-to-chaotic tunneling rates, we approximate the fictitious integrable system U_{reg} by a kicked system, $\widetilde{U}_{\text{reg}} = U_{\widetilde{V}}U_T$ or $\widetilde{U}_{\text{reg}} = U_V U_{\widetilde{T}}$ with

$$U_{\widetilde{V}} = \mathrm{e}^{-i\widetilde{V}(q)/\hbar_{\text{eff}}}, \tag{6.30}$$

$$U_{\widetilde{T}} = \mathrm{e}^{-i\widetilde{T}(p)/\hbar_{\text{eff}}}. \tag{6.31}$$

Here, the functions $\widetilde{V}(q)$ and $\widetilde{T}(p)$ are a low order Taylor expansion of $V(q)$ and $T(p)$, respectively, around the center of the regular island. Note, that the classical dynamics corresponding to $\widetilde{U}_{\text{reg}}$ is typically not completely regular. Still the following derivation is applicable if $\widetilde{U}_{\text{reg}}$ has the following properties: (1) Within the regular island it has an almost identical classical dynamics as U, including nonlinear resonances and small embedded chaotic regions. (2) It shows mainly regular dynamics for a sufficiently wide region beyond the border of the regular island of U.

Now we consider the specific case $\widetilde{U}_{\text{reg}} = U_{\widetilde{V}}U_T$ and assume that both properties (1) and (2) are fulfilled. As the dynamics of U and $\widetilde{U}_{\text{reg}}$ are almost identical within the regular island of U, the approximate result, Equation 6.28, can be applied with U_{reg} replaced by $\widetilde{U}_{\text{reg}}$, giving

$$\gamma_m \approx \|(U - U_{\widetilde{V}}U_T)|\psi_{\text{reg}}^m\rangle\|^2 \tag{6.32}$$

$$= \|(UU_T^{\dagger}U_{\widetilde{V}}^{\dagger} - \mathbb{1})U_{\widetilde{V}}U_T|\psi_{\text{reg}}^m\rangle\|^2. \tag{6.33}$$

We now use that $|\psi_{\text{reg}}^m\rangle$, which is an eigenstate of the exact U_{reg} and H_{reg}, is an approximate eigenstate of $U_{\widetilde{V}}U_T$, leading to $U_{\widetilde{V}}U_T|\psi_{\text{reg}}^m\rangle \approx \mathrm{e}^{i\varphi_{\text{reg}}^m}|\psi_{\text{reg}}^m\rangle$. With this we obtain

$$\gamma_m \approx \|(U_V U_{\widetilde{V}}^{\dagger} - \mathbb{1})|\psi_{\text{reg}}^m\rangle\|^2. \tag{6.34}$$

In position representation this reads

$$\gamma_m \approx 2\sum_{k=0}^{N-1}|\psi_{\text{reg}}^m(q_k)|^2\left[1 - \cos\left(\frac{\Delta V(q_k)}{\hbar_{\text{eff}}}\right)\right], \tag{6.35}$$

where $\Delta V(q) := V(q) - \widetilde{V}(q)$ and $q_k = k/N - 1/2$. In the semiclassical limit the sum in Equation 6.35 can be replaced by an integral

$$\gamma_m \approx 2\int_{-1/2}^{1/2} \mathrm{d}q\,|\psi_{\text{reg}}^m(q)|^2\left[1 - \cos\left(\frac{\Delta V(q)}{\hbar_{\text{eff}}}\right)\right], \tag{6.36}$$

where we integrate over the whole position space $q \in [-1/2, 1/2]$. Note, that for the complementary situation, where $U_{\text{reg}} \approx U_V U_{\widetilde{T}}$ is used in Equation 6.28, a similar result can be obtained in momentum representation

$$\gamma_m \approx 2 \int\limits_{-1/2}^{1/2} dp \, |\psi_{\text{reg}}^m(p)|^2 \left[1 - \cos\left(\frac{\Delta T(p)}{\hbar_{\text{eff}}}\right)\right], \tag{6.37}$$

with $\Delta T(p) := T(p) - \widetilde{T}(p)$.

We now use a WKB expression for the regular states $|\psi_{\text{reg}}^m\rangle$. For simplicity, we restrict to the case $H_{\text{reg}} = p^2/2 + W(q)$ leading to

$$\psi_{\text{reg}}^m(q) \approx \sqrt{\frac{\omega}{2\pi|p(q)|}} \exp\left(-\frac{1}{\hbar_{\text{eff}}} \int\limits_{q_m^r}^{q} |p(q')| dq'\right), \tag{6.38}$$

which is valid for $q > q_m^r$. Here q_m^r is the right classical turning point of the mth quantizing torus, ω is the oscillation frequency, and $p(q) = \sqrt{2(E_{\text{reg}}^m - W(q))}$. The eigenstates $\psi_{\text{reg}}^m(q)$ decay exponentially beyond the classical turning point q_m^r. The difference of the potential energies $\Delta V(q) = V(q) - \widetilde{V}(q)$ approximately vanishes within the regular region and increases beyond its border to the chaotic sea. Hence, the most important contribution in Equation 6.36 arises near the left or the right border, q_b^l or q_b^r, of the regular island. For $q > q_b^r$ we rewrite the regular states

$$\psi_{\text{reg}}^m(q) \approx \psi_{\text{reg}}^m(q_b^r) \exp\left(-\frac{1}{\hbar_{\text{eff}}} \int\limits_{q_b^r}^{q} |p(q')| dq'\right) \sqrt{\frac{p(q_b^r)}{p(q)}} \tag{6.39}$$

$$\approx \psi_{\text{reg}}^m(q_b^r) \exp\left(-\frac{1}{\hbar_{\text{eff}}}(q - q_b^r)|p(q_b^r)|\right), \tag{6.40}$$

where in the last step we use $p(q) \approx p(q_b^r)$ in the vicinity of the border.

In order to use Equation 6.36 we split the integration interval in two parts, such that

$$\gamma_m = \gamma_m^l + \gamma_m^r,$$

corresponding to the contributions from the left and the right. We now approximate

$$\Delta V(q) \approx \begin{cases} 0 & , q_m^r \leq q \leq q_b^r \\ c_b(q - q_b^r) & , q > q_b^r \end{cases} \tag{6.41}$$

with some constant c_b and find

$$\gamma_m^r \approx 2\hbar_{\text{eff}} |\psi_{\text{reg}}^m(q_b^r)|^2 \int\limits_{0}^{x_{\text{max}}} e^{-2x|p(q_b^r)|} [1 - \cos(c_b x)] dx \tag{6.42}$$

$$\approx \frac{I\hbar_{\text{eff}}}{\pi} |\psi_{\text{reg}}^m(q_b^r)|^2, \tag{6.43}$$

where $x = (q - q_b^r)/\hbar_{\text{eff}}$ and $x_{\text{max}} = (1/2 - q_b^r)/\hbar_{\text{eff}}$. In the semiclassical limit $x_{\text{max}} \to \infty$ and for fixed quantum number m the integral in Equation 6.42,

$$I = \int\limits_{0}^{x_{\text{max}}} e^{-2x|p(q_b^r)|} [1 - \cos(c_b x)] dx, \tag{6.44}$$

becomes an h_{eff}-independent constant. The tunneling rate γ_m^r is proportional to the square of the modulus of the regular wave function at the right border q_b^r of the regular island. With Equation 6.38 we obtain

$$\gamma_m^r \approx \frac{I\omega h_{\text{eff}}}{2\pi^2|p(q_b^r)|} \exp\left(-\frac{2}{h_{\text{eff}}} \int_{q_m^r}^{q_b^r} |p(q')|\, dq'\right). \tag{6.45}$$

A similar equation holds for γ_m^l.

As an example for the explicit evaluation of Equation 6.45 we now consider the harmonic oscillator $H_{\text{reg}}(p,q) = p^2/2 + \omega^2 q^2/2$, where ω denotes the oscillation frequency and gives the ratio of the two half axes of the elliptic invariant tori. Its classical turning points $q_m^{r,l} = \pm\sqrt{2E_m}/\omega$, the eigenenergies $E_m = h_{\text{eff}}\omega(m + 1/2)$, and the momentum $p(q) = \sqrt{2E_m - q^2\omega^2}$ are explicitly given. Using these expressions in Equation 6.45 and $\gamma_m = 2\gamma_m^r$ we obtain

$$\gamma_m = c\,\frac{h_{\text{eff}}}{\beta_m} \exp\left(-\frac{2A}{h_{\text{eff}}}\left[\beta_m - \alpha_m \ln\left(\frac{1+\beta_m}{\sqrt{\alpha_m}}\right)\right]\right) \tag{6.46}$$

as the semiclassical prediction for the tunneling rate of the mth regular state, where $\alpha_m = (m + 1/2)(A/h_{\text{eff}})^{-1}$, $\beta_m = \sqrt{1 - \alpha_m}$, and A is the area of the regular island. The exponent in Equation 6.46 was also derived in Le Deunff and Mouchet (2010) [43]. The prefactor

$$c = \frac{I}{\pi^2}\sqrt{\frac{\pi\omega}{A}} \tag{6.47}$$

can be estimated semiclassically by solving the integral, Equation 6.44, for $x_{\max} \to \infty$. For a fixed classical torus of energy E one obtains

$$I \approx \frac{1}{2|p(q_b^r)|} - \frac{2|p(q_b^r)|}{4|p(q_b^r)|^2 + c_b^2}. \tag{6.48}$$

With this prefactor, the prediction Equation 6.46 gives excellent agreement with numerically determined data over 10 orders of magnitude in γ (see Figure 6.7). For a fixed quantum number m the energy E_m goes to zero in the semiclassical limit such that one can approximate $|p(q_b^r)| \approx \omega q_b^r$ in Equation 6.48 which does not depend on h_{eff}.

Let us make the following remarks concerning Equation 6.46: The only information about this nongeneric island with constant rotation number is A/h_{eff} as in Podolskiy and Narimanov (2003) [14]. In contrast to Equation 6.27, it does not require further quantum information such as the quantum map U. While the term in square brackets semiclassically approaches one, it is relevant for large h_{eff}. In contrast to Equation 6.36, where the chaotic properties are contained in the difference $V(q) - \widetilde{V}(q)$, they now appear in the prefactor c via the linear approximation of this difference.

In the semiclassical limit the tunneling rates predicted by Equation 6.46 decrease exponentially. For $h_{\text{eff}} \to 0$ the values α_m go to zero and β_m to one, such that $\gamma \sim e^{-2A/h_{\text{eff}}}$ remains which reproduces the qualitative prediction obtained in Hanson et al. (1984) [13]. We find that the non-universal constant in the exponent is 2, which is comparable to the prefactor $3 - \ln 4 \approx 1.61$ derived in Podolskiy and Narimanov (2003) [14] and Sheinman (2005) [44]. However, our result shows more accurate agreement to numerical data and does not require an additional fitting parameter, as will be shown in Section 6.4.

6.4 APPLICATIONS

We study the direct regular-to-chaotic tunneling process in the systems introduced in Section 6.2 starting with the simplest example $\mathcal{D}_{\text{ho}}$ with a harmonic oscillator-like regular island. As further systems, we consider the map $\mathcal{D}_d$ with a deformed regular region, the map $\mathcal{D}_{rs}$ with a regular stripe, and finally the standard map which shows a generic mixed phase space.

The map $\mathcal{D}_{ho}$ has a particularly simple phase-space structure with a harmonic oscillator-like regular island with elliptic invariant tori and constant rotation number, see the insets in Figure 6.4. Numerically, we determine tunneling rates by using absorbing boundary conditions at $|q| \geq 1/2$. Analytically, for the approach derived in Section 6.3, Equation 6.27, we use the Hamiltonian of a harmonic oscillator as the fictitious integrable system U_{reg}. It is squeezed and tilted according to the linearized dynamics in the vicinity of the stable fixed point located at the center of the regular island.

Figure 6.4 shows the prediction of Equation 6.27 compared to numerical data. We find excellent agreement over more than 10 orders of magnitude in γ. In the regime of large tunneling rates γ small deviations occur which can be attributed to the influence of the chaotic sea on the regular states: These states are located on quantizing tori close to the border of the regular island and are affected by the regular-to-chaotic transition region. However, the deviations in this regime are smaller than a factor of two.

In Figure 6.7 we compare the results of the semiclassical prediction, Equation 6.46, to the numerical data. Because of the approximations performed in the derivation of this formula stronger deviations are visible in the regime of large tunneling rates while the agreement in the semiclassical regime is still excellent.

In Podolskiy and Narimanov (2003) [14] and Sheinman (2005) [44] a prediction was derived for the tunneling rate of the regular ground state,

$$\gamma_0 = c \frac{\Gamma(\alpha, 4\alpha)}{\Gamma(\alpha, 0)}, \tag{6.49}$$

where Γ is the incomplete gamma function, $\alpha = A/h_{eff}$, and c is a constant. Equation 6.49 can be approximated semiclassically (see Schlagheck et al. (2006) [22]), leading to

$$\gamma_0 \propto e^{-\frac{A}{h_{eff}}(3-\ln 4)}. \tag{6.50}$$

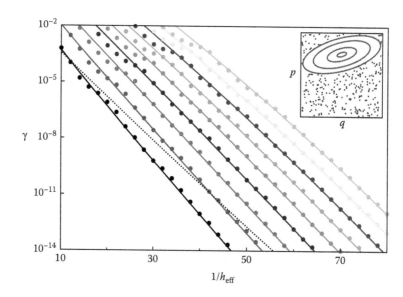

FIGURE 6.7 Numerical data (dots) for $m \leq 8$ for the map $\mathcal{D}_{ho}$ with a harmonic oscillator-like island. Comparison with Equation 6.46 (solid lines). The prediction of Podolskiy and Narimanov (2003) [14] and Sheinman (2005) [44], Equation 6.49, for $m = 0$ with a fitted prefactor is shown (dotted line).

Figure 6.7 shows the comparison of Equation 6.49 (dotted line) to the numerical data for the map $\mathcal{D}_{ho}$. Especially in the semiclassical limit deviations are visible. The factor 2 which appears in the exponent of Equation 6.46 is more accurate than the factor $3 - \ln 4$ in Equation 6.50.

In typical systems the rotation number of regular tori changes from the center of the regular region to its border which typically has a nonelliptic shape. Such a situation can be achieved using the map $\mathcal{D}_d$. Here nonlinear resonances are small such that their influence on the tunneling process is expected only at large $1/h_{eff}$, see the inset in Figure 6.8 for its phase space.

We determine the fictitious integrable system H_{reg} by means of the Lie-transformation method described in Section 6.3.2. It is then quantized and its eigenfunctions are determined numerically. Figure 6.8 shows a comparison of numerically determined tunneling rates (dots) to the prediction of Equation 6.27 (solid lines) yielding excellent agreement for tunneling rates $\gamma \gtrsim 10^{-11}$. For smaller values of γ, deviations occur due to resonance-assisted tunneling which is caused by a small 10:1 resonance chain. Similar to the case of the harmonic oscillator-like island the fictitious integrable system U_{reg} can be approximated by a kicked system $U_{reg} \approx U_{\widetilde{V}} U_T$ using $\widetilde{V}(q) = -rq^2/2 + Rq^3/3$. Hence, Equations 6.35 and 6.36 can be evaluated giving similarly good agreement (not shown). The prediction of Equation 6.49 [14,44] (dotted line) shows large deviations to the numerical data.

We now consider the kicked system $\mathcal{D}_{rs}$, for which the regular region consists of a stripe in phase space, see the inset in Figure 6.9. In Ishikawa et al. (2007) [29], this map is used to study the evolution of a wave packet initially started in the regular region by means of complex paths. Also for this system one can predict tunneling rates by means of Equation 6.27. The fictitious integrable system U_{reg} is determined by continuing the dynamics within $|q| < q_d/(8\pi)$ to the whole phase space. It is given as a kicked system, Equation 6.2, defined by the functions

$$\widetilde{V}'(q) = -\frac{1}{2\pi}\left(\omega + \frac{d_1}{2} - \frac{d_2}{2}\right), \tag{6.51}$$

$$T'(p) = -\frac{\kappa}{8\pi}\sin(2\pi p). \tag{6.52}$$

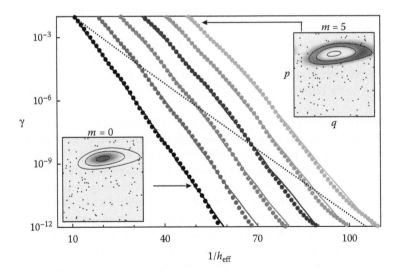

FIGURE 6.8 Dynamical tunneling rates from a regular island to the chaotic sea for the map $\mathcal{D}_d$: Numerical results (dots) and prediction following from Equation 6.27 (lines) versus $1/h_{eff}$ for quantum numbers $m \leq 5$. The insets show Husimi representations of the regular states $m = 0$ and $m = 5$ at $1/h_{eff} = 50$. The prediction of Podolskiy and Narimanov (2003) [14] and Sheinman (2005) [44], Equation 6.49, for $m = 0$ with a fitted prefactor is shown (dotted line).

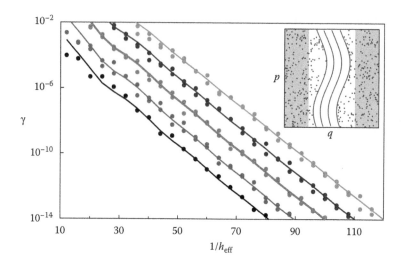

FIGURE 6.9 Dynamical tunneling rates from a regular stripe to the chaotic sea for the map $\mathcal{D}_{rs}$: We compare numerical results (dots) and the prediction following from Equation 6.27 (lines) versus $1/h_{eff}$ for the quantum numbers $|m| \leq 4$. The inset shows the phase space of the system. The numerical data is obtained using an absorbing region at $|q| \geq 1/4$ (gray-shaded area of the inset).

For sufficiently large smoothing parameter b (see Equation 6.10), the dynamics of the mixed map inside the regular region is equivalent to that of the regular map. Thus, Equation 6.27 can be applied and we compare its results to numerically determined data. Absorbing boundary conditions at $|q| \geq 1/2$ lead to strong fluctuations of the numerically determined tunneling rates as a function of h_{eff}, presumably due to dynamical localization. Choosing $|q| \geq 1/4$ for the opening, which is closer to the regular stripe, we find smoothly decaying tunneling rates, see Figure 6.9. The comparison with the theoretical prediction shows quite good agreement. Note that, due to the symmetry of the map, there are always two regular states with comparable tunneling rates except for the ground state $m = 0$. These two states are located symmetrically around the center of the regular stripe. While the theoretical prediction, Equation 6.27, is identical for both of these states, the numerical results differ slightly due to the different chaotic dynamics in the vicinity of the left and right border of the regular region.

The paradigmatic model of an area preserving map is the standard map (see Section 6.2.1). For κ between 2.5 and 3.0 one has a large generic regular island with a relatively small hierarchical region surrounded by a 4:1 resonance chain, see the inset in Figure 6.10. Absorbing boundary conditions at $|q| \geq 1/2$ lead to strong fluctuations of the numerically determined tunneling rates as a function of h_{eff}, presumably caused by partial barriers. Choosing absorbing boundary conditions at $|q| \geq 1/4$, which is closer to the island, we find smoothly decaying tunneling rates (dots in Figure 6.10). Evaluating Equation 6.27 for $\kappa = 2.9$ gives good agreement with these numerical data, see Figure 6.10 (solid lines). Here we determine H_{reg} using a method based on the frequency map analysis, as the Lie transformation is not able to reproduce the dynamics within the regular island of U, see Section 6.3.2. With increasing order of the expansion series of H_{reg}, the tunneling rates following from Equation 6.27 diverge. Hence, for the predictions in Figure 6.10 we choose terms up to second order only. Note that, at such small order the accuracy of H_{reg} within the regular region of U is inferior compared to the examples discussed before. Hence, in Equation 6.27 the state $U|\psi_{reg}^m\rangle$ has small contributions of all purely regular states $|\psi_{reg}^n\rangle$ in the regular island. These contributions are removed by the application of the projector P_{ch}. However, this projector depends on the number of regular states N_{reg}, which grows with $1/h_{eff}$. If N_{reg} increases by one, P_{reg} projects onto a larger

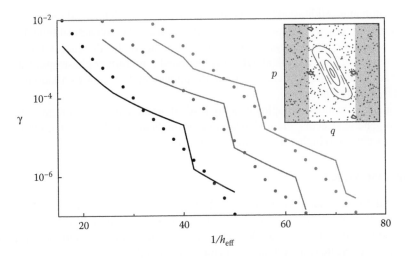

FIGURE 6.10 Tunneling rates for the standard map ($\kappa = 2.9$) for $m \leq 2$. Prediction of Equation 6.27 (lines) and numerical results (dots), obtained using an absorbing region at $|q| \geq 1/4$ (gray-shaded area of the inset).

region in phase space. This explains the steps of the theoretical prediction, Equation 6.27, visible in Figure 6.10.

6.5 SUMMARY

Dynamical tunneling plays an important role in many areas of physics. Therefore, a detailed understanding and quantitative description is of great interest. In this text we have given an overview on determining direct regular-to-chaotic tunneling rates using the fictitious integrable system approach. The direct regular-to-chaotic tunneling mechanism is valid in the regime where Planck's constant is large compared to additional structures in the regular island, such as nonlinear resonances. To include the effect of such resonances, resonance-assisted tunneling [19,21] has to be considered in addition with the direct regular-to-chaotic tunneling contribution. Recently, the two mechanisms have been studied in a combined theory [23], giving a full quantitative description of tunneling from a regular island, including resonances, into the chaotic region, see the contributions of Schlagheck, Mouchet, and Ullmo in this book.

The approach to determine direct regular-to-chaotic tunneling rates can also be generalized to the case of billiard systems. It can be applied if the fictitious integrable system H_{reg} is known, as for example, for the annular or the mushroom billiard. In addition to H_{reg} a description for the chaotic states of the mixed system H is needed, for which we employ random wave models [4] which account for the relevant boundary conditions of the billiard. For the mushroom billiard the fictitious integrable system is easily found as the semicircle billiard and with this an analytical expression for the tunneling rates was derived in Bäcker et al. (2008) [17]. This result has been compared to experimental data obtained from microwave spectra of a mushroom billiard with adjustable stem height and good agreement was found. It was also shown that tunneling rates manifest themselves in exponentially diverging localization lengths of nanowires with one-sided surface roughness in a perpendicular magnetic field [45,46].

The fictitious integrable system approach has been extended to open optical microcavities in Bäcker et al. (2009) [18]. In particular, the annular microcavity was studied, which allows for unidirectional emission of light and shows modes of high-quality factors simultaneously. This is desirable for most applications. In contrast to closed billiards the leakiness of the cavity has to be considered, which leads to the contribution Q_{dir} to the quality factor caused by the direct coupling of the regular mode to the continuum. Additionally, the contribution caused by dynamical tunneling Q_{dyn} is relevant, where $1/Q = 1/Q_{\mathrm{dir}} + 1/Q_{\mathrm{dyn}}$. Q_{dyn} is directly related to the dynamical tunneling rates given by the theory using a fictitious integrable system. The prediction for the quality factors Q has been compared to numerical data. Excellent agreement is found if no further phase-space structures exist in the chaotic sea. If additional structures appear, the numerical data show oscillations, which cannot be explained by the present theory.

Future challenges include a completely semiclassical prediction of direct regular-to-chaotic tunneling rates in generic systems, the understanding of implications of additional phase space structures, and the extension to higher dimensional systems.

ACKNOWLEDGMENTS

We are grateful to S. Creagh, S. Fishman, M. Hentschel, R. Höhmann, A. Köhler, U. Kuhl, A. Mouchet, M. Robnik, P. Schlagheck, A. Shudo, H.-J. Stöckmann, S. Tomsovic, G. Vidmar, and J. Wiersig for stimulating discussions. We further acknowledge financial support through the DFG Forschergruppe 760 "Scattering systems with complex dynamics."

REFERENCES

1. L. D. Landau and E. M. Lifschitz, *Lehrbuch der Theoretischen Physik, Band 3: Quantenmechanik*, Akademie Verlag, Berlin, 1979.
2. E. Gildener and A. Patrascioiu, Pseudoparticle contributions to the energy spectrum of a one-dimensional system, *Phys. Rev. D* 16, 423–430, 1977.
3. I. C. Percival, Regular and irregular spectra, *J. Phys. B* 6, L229–L232, 1973.
4. M. V. Berry, Regular and irregular semiclassical wavefunctions, *J. Phys. A* 10, 2083–2091, 1977.
5. A. Voros, *Stochastic Behavior in Classical and Quantum Hamiltonian Systems*, Springer Verlag, Berlin, 1979.
6. M. J. Davis and E. J. Heller, Quantum dynamical tunneling in bound states, *J. Chem. Phys.* 75, 246–254, 1981.
7. W. A. Lin and L. E. Ballentine, Quantum tunneling and chaos in a driven anharmonic oscillator, *Phys. Rev. Lett.* 65, 2927–2930, 1990.
8. O. Bohigas, S. Tomsovic, and D. Ullmo, Manifestations of classical phase space structures in quantum mechanics, *Phys. Rep.* 223, 43–133, 1993.
9. S. Tomsovic and D. Ullmo, Chaos-assisted tunneling, *Phys. Rev. E* 50, 145–162, 1994.
10. C. Dembowski, H.-D. Gräf, A. Heine, R. Hofferbert, H. Rehfeld, and A. Richter, First experimental evidence for chaos-assisted tunneling in a microwave annular billard, *Phys. Rev. Lett.* 84, 867–870, 2000.
11. D. A. Steck, W. H. Oskay, and M. G. Raizen, Observation of chaos-assisted tunneling between islands of stability, *Science* 293, 274–278, 2001.
12. W. K. Hensinger, H. Häffner, A. Browaeys, N. R. Heckenberg, K. Helmerson, C. McKenzie, G. J. Milburn, et al., Dynamical tunnelling of ultracold atoms, *Nature* 412, 52–55, 2001.
13. J. D. Hanson, E. Ott, and T. M. Antonsen, Influence of finite wavelength on the quantum kicked rotator in the semiclassical regime, *Phys. Rev. A* 29, 819–825, 1984.
14. V. A. Podolskiy and E. E. Narimanov, Semiclassical description of chaos-assisted tunneling, *Phys. Rev. Lett.* 91, 263601, 2003.
15. M. Sheinman, S. Fishman, I. Guarneri, and L. Rebuzzini, Decay of quantum accelerator modes, *Phys. Rev. A* 73, 052110, 2006.
16. A. Bäcker, R. Ketzmerick, S. Löck, and L. Schilling, Regular-to-chaotic tunneling rates using a fictitious integrable system, *Phys. Rev. Lett.* 100, 104101, 2008.
17. A. Bäcker, R. Ketzmerick, S. Löck, M. Robnik, G. Vidmar, R. Höhmann, U. Kuhl, and H.-J. Stöckmann, Dynamical tunneling in mushroom billiards, *Phys. Rev. Lett.* 100, 174103, 2008.
18. A. Bäcker, R. Ketzmerick, S. Löck, J. Wiersig, and M. Hentschel, Quality factors and dynamical tunneling in annular microcavities, *Phys. Rev. A* 79, 063804, 2009.
19. O. Brodier, P. Schlagheck, and D. Ullmo, Resonance-assisted tunneling in near-integrable systems, *Phys. Rev. Lett.* 87, 064101, 2001.

20. O. Brodier, P. Schlagheck, and D. Ullmo, Resonance-assisted tunneling, *Ann. Phys.* 300, 88–136, 2002.
21. C. Eltschka and P. Schlagheck, Resonance- and chaos-assisted tunneling in mixed regular-chaotic systems, *Phys. Rev. Lett.* 94, 014101, 2005.
22. P. Schlagheck, C. Eltschka, and D. Ullmo, Resonance- and chaos-assisted tunneling, in *Progress in Ultrafast Intense Laser Science I*, edited by K. Yamanouchi, S. L. Chin, P. Agostini, and G. Ferrante, Springer, Berlin, 2006.
23. S. Löck, A. Bäcker, R. Ketzmerick, and P. Schlagheck, Regular-to-chaotic tunneling rates: From the quantum to the semiclassical regime, *Phys. Rev. Lett.* 104, 114101, 2010.
24. M. V. Berry, Regular and irregular motion, *Am. Inst. Phys. Conf. Proc.* 46, 16–120, 1978.
25. D. K. Arrowsmith and C. M. Place, *An Introduction to Dynamical Systems*, Cambridge University Press, Cambridge, 1990.
26. R. S. MacKay, J. D. Meiss, and I. C. Percival, Transport in Hamiltonian systems, *Physica D* 13, 55–81, 1984.
27. H. Schanz, M.-F. Otto, R. Ketzmerick, and T. Dittrich, Classical and quantum Hamiltonian ratchets, *Phys. Rev. Lett.* 87, 070601, 2001.
28. A. Bäcker, R. Ketzmerick, and A. G. Monastra, Flooding of chaotic eigenstates into regular phase space islands, *Phys. Rev. Lett.* 94, 054102, 2005.
29. A. Ishikawa, A. Tanaka, and A. Shudo, Quantum suppression of chaotic tunneling, *J. Phys. A.* 40, F397–F405, 2007.
30. M. V. Berry, N. L. Balzas, M. Tabor, and A. Voros, Quantum maps, *Ann. Phys.* 122, 26–63, 1979.
31. S.-J. Chang and K.-J. Shi, Evolution and exact eigenstates of a resonant quantum system, *Phys. Rev. A* 34, 7–22, 1986.
32. H. Schomerus and J. Tworzydło, Quantum-to-classical crossover of quasibound states in open quantum systems, *Phys. Rev. Lett.* 93, 154102, 2004.
33. M. Wilkinson, Tunnelling between tori in phase space, *Physica D* 21, 341–354, 1986.
34. S. C. Creagh, Tunneling in two dimensions, in *Tunneling in Complex Systems*, World Scientific, Singapore, 1998.
35. A. Shudo and K. S. Ikeda, Complex classical trajectories and chaotic tunneling, *Phys. Rev. Lett.* 74, 682–685, 1995.
36. A. Shudo and K. S. Ikeda, Chaotic tunneling: A remarkable manifestation of complex classical dynamics in non-integrable quantum phenomena, *Physica D* 115, 234–292, 1998.
37. T. Onishi, A. Shudo, K. S. Ikeda, and K. Takahashi, Tunneling mechanism due to chaos in a complex phase space, *Phys. Rev. E* 64, 025201, 2001.
38. F. G. Gustavson, On constructing formal integrals of a Hamiltonian system near an equilibrium point, *The Astronomical Journal* 71, 670–686, 1966.
39. A. Bazzani, M. Giovannozzi, G. Servizi, E. Todesco, and G. Turchetti, Resonant normal forms, interpolating Hamiltonians and stability of area preserving maps, *Physica D* 64, 66–97, 1993.
40. R. Scharf, Quantum maps, adiabatic invariance and the semiclassical limit, *J. Phys. A* 21, 4133–4147, 1988.
41. A. J. Lichtenberg and M. A. Lieberman, *Regular and Stochastic Motion*, Springer, New York, 1983.
42. J. Laskar, C. Froeschlé, and A. Celletti, The means of chaos by the numerical analysis of the fundamental frequencies. Application to the standard mapping, *Physica D* 56, 253–269, 1992.
43. J. Le Deunff and A. Mouchet, Instantons revisited: Dynamical tunnelling and resonant tunnelling, *Phys. Rev. E* 81, 046205, 2010.
44. M. Sheinman, *Decay of Quantum Accelerator Modes*, Master Thesis, Technion, Haifa, Israel, 2005.
45. J. Feist, A. Bäcker, R. Ketzmerick, S. Rotter, B. Huckestein, and J. Burgdörfer, Nano-wires with surface disorder: Giant localization lengths and quantum-to-classical crossover, *Phys. Rev. Lett.* 97, 116804, 2006.
46. J. Feist, A. Bäcker, R. Ketzmerick, J. Burgdörfer, and S. Rotter, Nanowires with surface disorder: Giant localization length and dynamical tunneling in the presence of directed chaos, *Phys. Rev. B* 80, 245322, 2009.

7 Complex Semiclassical Approach to Chaotic Tunneling

Akira Shudo and Kensuke S. Ikeda

CONTENTS

7.1   INTRODUCTION

It is a strict requirement that physical observables should be real-valued; the position and momentum of a particle obeying the Newtonian equations of motion are both real, and any dynamical variables derived from them are also real. The wavefunction in quantum mechanics may be complex-valued, but observables obtained averaged over the square modulus of the wavefunction turn out to be real. Occasionally, one examines the nature of singularities in the complex plane in order to judge whether the system under consideration is completely integrable or not, but in almost all cases complex variables appear only implicitly in the description of physical phenomena. Indeed, there is no serious demand or even motivation to investigate dynamics whose variables are complex.

As a result of this fact, although the study of complex dynamics is making significant progress in mathematics [1–5], physicists are almost unconcerned with it except for appreciating beautiful fractal pictures generated by the iteration of one complex variable [6].

There are, however, some subtle issues in physics which may need the help of complex variables, not as a supplementary tool but as a language to understand it, in the border between classical and quantum mechanics. The tunneling phenomenon, which we discuss in this chapter, is peculiar to quantum mechanics and no analog exists in classical mechanics. What is most frustrating is that it is a purely quantum effect, nevertheless a plenty of evidence exists showing that the nature of tunneling is strongly influenced by underlying classical dynamics.

Such a suspensive aspect can be well captured in the short wavelength limit where classical and quantum mechanics are closest with each other, and the Wentzel–Krammers–Brillouin (WKB) or semiclassical method is a technique developed to understand the connection between them. In particular, relieving the frustration especially in the tunneling problem can be achieved by extending the dynamics to the complex plane, more precisely, using the semiclassical approximation taking into account complex orbits.

The semiclassical treatment using complex orbits is actually quite common, rather it goes back almost to the beginning of the WKB theory, even to text book examples of quantum tunneling in one dimension [7]. Further developments follow by introducing the so-called instanton method in which the tunneling probability is evaluated by a classical path moving along imaginary time [8–11]. A generalization to higher-dimensional systems is not straightforward even in completely integrable systems, but as a direct descendant of the instanton several formulations have been elaborated [12,13]. Note that the instanton method has been revisited recently and a possible extension is presented [14]. On the other hand, a rather sophisticated treatment essentially utilizing the complex plane and based on the Borel–Laplace resummation was proposed [15–17] earlier than recent developments of WKB analyses driven by the topics of quantum chaology. This is closely related with another works for the quantic oscillator [18–20], and has driven recent progress on the so-called exact WKB analysis or resurgent theory [21–24].

The situation drastically changes if we turn to nonintegrable systems, especially systems with mixed phase space. What is specific in mixed systems is that phase space is divided into infinitely many invariant components and they are generally intermingled with each other. The transition between different invariant components is forbidden in classical dynamics, and invariant components play the role of dynamical barriers in phase space. In particular, Kolmogorov–Arnold–Moser (KAM) curves confine the orbits in locally restricted regions and form major obstacles to prevent ergodic exploration in the entire phase space.

In quantum mechanics, the transition through dynamical barriers thus formed may become possible as a result of quantum effects. The so-called quasimode [25], each of which is associated with a KAM torus in phase space, can couple with its congruent partner via the quantum transition. *Dynamical tunneling* has originally been understood as quantum tunneling between regular islands [26], in analogy with potential tunneling in the double well, but possible roles of chaos in the tunneling process have been thereafter recognized in several ways [27–32]. Of course there is no reason to limit ourselves to an original setting, where a chaotic state is sandwiched between two regular states, rather any transition which occurs between distinct classical invariant components should be regarded as dynamical tunneling as well. Further important developments have been made by focusing on not only the role of chaos but also nonlinear resonances inside KAM islands [33,34]. It is now possible to evaluate the tunneling probability or energy splitting affected by the presence of nonlinear resonances very accurately based on the resonance-assisted mechanism with further refinements [35,36].

It is natural to suppose that dynamical tunneling, even though it is purely quantum mechanical, should be linked to and driven by the underlying dynamics, but there are no clues in real phase space in order to explore its origin by its very definition. This strongly motivates us to use the complex variables, and an essential difference between completely integrable and nonintegrable cases indeed exist, which is exactly our major concern in this chapter. One important point we should remind the readers is that an explicit quantization rule based on clearly specified invariant manifolds (invariant tori) is known in completely integrable systems as Einstein–Brillouin–Keller (EBK) quantization, whereas no alternatives exist in systems with mixed phase space. We may ask tunneling occurs from which state to which state in integrable situations, but cannot specify it so obviously in mixed phase space.

The question we will specifically ask is *why and how classically disconnected regions are connected.* Typically, dynamical tunneling is discussed between regular states that have supports on invariant KAM curves in the real plane [26,28–30]. Each invariant curve is separated from each other and the real dynamics does not allow the transition from one KAM curve to another. However, as well illustrated in Creagh (1998) [37], as long as the system is integrable, these disconnected curves in the real plane are connected in the complex curve to form a bridge between disjointed regions. This provides a basis for the description of tunneling using classical orbits.

On the other hand, this is not the case in nonintegrable systems. As will be explained in detail in this chapter, *natural boundaries* of invariant curves appear in the complex plane, and the bridge connecting real separated regions is broken [38]. Alternatively stated, the carrier for the tunneling transport does not seem to exist not only in the real but also in the complex space. Our issue is therefore to find the counterparts which *bond* separated invariant curves together, otherwise no dynamical description for dynamical tunneling is available. This is a central subject which we want to discuss here, and we will show that the *complex chaos* plays the role of the bridge between classically disconnected regions.

Here we only consider the discrete map in which the time evolution is given by the iteration of two variables. It is not self-evident whether or not the area-preserving map in the complex plane can be a model of the flow system as the real map models Poincaré map of the flow system [39], and technical issues arise especially in the flow system as to, for example, which variables should be complexified and which should not. We will not go into discussions of this point, but just remark that the key mechanism which drives the tunneling transport is common to both [40–42].

In the discrete dynamics, complexification of dynamical variables is almost straightforward if the iteration rule is expressed by analytic functions. Restricting ourselves further to the polynomial map, we can share special benefits of recent mathematical results on complex dynamics in several variables [43–48]. However, as mentioned above, due to the lack of motivations, such results have not been known so widely in the physicists' community, and at the same time mathematicians do their works without intention to link to any physical problems. Our task here is to translate these remarkable mathematical results, especially recent ones, in this field into the language familiar with physicists as much as possible, and drive intercommunication between each other.

More concretely, we will provide *informal* accounts for our results making full use of recent mathematical results [49], and in doing so we try to convey the essence of recent outcomes on multidimensional complex dynamics. Our previous results were presented in a formal manner, which may blur important points. Our strategy to achieve this is not straightforward however; since rigorous statements for multidimensional complex dynamics are full of unfamiliar terminologies and technicalities inevitable to develop the theory in several dimensions, we think that it is not appropriate to start with multidimensional cases. Instead, we will first provide some elementary expositions on one-dimensional complex dynamics, especially one-dimensional polynomial maps, together with introducing a specific approach which is commonly used and becomes powerful tools especially in the analysis of several dimensions. After that, we present the results for multidimensional polynomial maps as an analogy of one-dimensional case, although losing mathematical rigors in several places. For the readers who are interested in more details, we will cite related references. In the final part, we will discuss how the dominant tunneling orbits are characterized and some prediction necessarily derived from the arguments is provided.

7.2 SETTING FOR DYNAMICAL TUNNELING IN THE DISCRETE MAP

This section is devoted to introducing some fundamentals for semiclassical approaches to the dynamical tunneling problem. Here we consider the time-domain approach, in which the time evolution propagator for given initial and final states is examined. How the complex dynamics comes in to describe classically forbidden processes and how to formulate our problem are explained. We also present the most significant questions we will be interested in.

7.2.1 Area-Preserving Map: Classical and Quantum

We consider the following area-preserving map:

$$F : \begin{pmatrix} p' \\ q' \end{pmatrix} = \begin{pmatrix} p - V'(q) \\ q + T'(p') \end{pmatrix}, \tag{7.1}$$

where $T(p)$ and $V(q)$ are kinetic and potential functions.

A standard procedure to construct the quantum mechanics of the area-preserving map is to express the time evolution unitary operator in the discretized Feynman path integral form. The one-step unitary operator is given as

$$\hat{U} = \exp\left\{ -\frac{i}{\hbar} T(p) \right\} \exp\left\{ -\frac{i}{\hbar} V(q) \right\}. \tag{7.2}$$

General formulation of time-domain semiclassics is possible for arbitrary initial and final states [49]. To be concrete and for simplicity, we proceed by taking the momentum representation hereafter. The

n-step quantum propagator in the momentum representation is expressed as

$$K_n(p,p') = \langle p'|\hat{U}^n|p\rangle = \int \cdots \int \prod_j dq_j \prod_j dp_j \exp\left\{\frac{i}{\hbar}S_n(p,p')\right\},$$ (7.3)

where $S_n(p,p')$ denotes the discrete action functional along each path with fixed ends (p,p'):

$$S_n(p,p') = S(p_0 = p, q_0, p_1, q_1, \ldots, p_n = p', q_n) = \sum_{j=1}^{n}\left[T(p_j) + V(q_{j-1}) + q_{j-1}(p_j - p_{j-1})\right].$$ (7.4)

It is easy to check that the classical mapping rule F is recovered by imposing the variational condition

$$\frac{\partial S_n(p,p')}{\partial q_j} = 0, \quad \frac{\partial S_n(p,p')}{\partial p_j} = 0.$$ (7.5)

7.2.2 Semiclassical Propagator

The semiclassical method we apply is to evaluate the multiple integral $K_n(p,p')$ by the method of stationary phase or saddle-point method. The resulting semiclassical formula is nothing more than a discrete version of the Van Vleck–Gutzwiller propagator

$$K_n^{sc}(p,p') = \sum_{\ell} A_n^{(\ell)}(p,p')\exp\left\{\frac{i}{\hbar}S_n^{(\ell)}(p,p') + i\frac{\pi}{2}\mu^{(\ell)}\right\},$$ (7.6)

where $A_n^{(\ell)}(p,p')$ and $S_n^{(\ell)}(p,p')$ denote the amplitude factor associated with the stability of each orbit ℓ and its classical action, respectively. The semiclassical propagator $K_n^{sc}(p,p')$ is evaluated by summing over classical trajectories ℓ which start with $p = p_a$ and end with $p' = p_b$, where the time step n and initial and final momenta p and p' are given. Both initial and final momenta p_a and p_b should be real-valued since they are observables.

Suppose the initial manifold $p = p_a$ is contained in a regular region and the final manifold $p' = p_b$ is in the chaotic sea (see Figure 7.1). If all the initial points are confined in the regular region, then they cannot go out of the regular region. In such a case, the set of points contributing to the semiclassical propagator $K_n^{sc}(p,p')$ is empty. As long as we evaluate $K_n^{sc}(p,p')$ within real dynamics, the transition amplitude to the chaotic sea turns out to be zero although the tunneling penetration is always present in the time evolution of quantum wavepacket.

However, extending the dynamics to the complex plane secures the situation. Let us define the initial and final manifolds in $\mathbb{C}^2$, which are respectively represented as

$$\mathcal{A} = \{ (q,p) \in \mathbb{C}^2 \mid p = p_a \in \mathbb{R} \}, \quad \mathcal{B} = \{ (q,p) \in \mathbb{C}^2 \mid p = p_b \in \mathbb{R} \}.$$ (7.7)

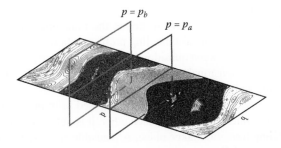

FIGURE 7.1 Initial and final manifolds in the momentum representation (schematic). The initial manifold, $\{p = p_a\} \cap \mathbb{R}^2$, is contained in the regular region, and the final manifold is placed in the chaotic sea.

Sets of points which contribute to the sum in $K_n^{sc}(p, p')$ are then expressed as

$$\mathcal{M}_n^{a,b} = \mathcal{A} \cap F^{-n}(\mathcal{B}), \tag{7.8}$$

on the initial manifold side and

$$\mathcal{L}_n^{a,b} = F^n(\mathcal{A}) \cap \mathcal{B}, \tag{7.9}$$

on the final manifold side, respectively.

7.2.3 QUESTIONS

In this setting, we particularly focus on the following questions: in the case where $\mathcal{A}$ and $\mathcal{B}$ are dynamically separated in $\mathbb{R}^2$, that is, $(\mathcal{A} \cap \mathbb{R}^2) \cap F^{-n}(\mathcal{B} \cap \mathbb{R}^2) = \emptyset$,

1. How are $\mathcal{A}$ and $\mathcal{B}$ in $\mathbb{R}^2$ are connected under the dynamics in $\mathbb{C}^2$?
2. Is it possible to relate the dynamics from $\mathcal{A}$ to $\mathcal{B}$ to some invariant objects in $\mathbb{C}^2$?
3. How does one evaluate, or even define the tunneling probability from $\mathcal{A}$ to $\mathcal{B}$?
4. Does some specific relevant orbit(s) (like the instanton) exclusively control the transition from $\mathcal{A}$ to $\mathcal{B}$, or are there any other principles?

Question 1 is somewhat vague, and a minimum answer follows immediately. Assume, for simplicity, that the kinetic and potential terms $T(p)$ and $V(q)$ are polynomial functions of each variable. For given $p_a, p_b \in \mathbb{R}$, the equation $(p' = p_b, q_n) = F^n(p = p_a, q_0)$ gives simultaneous algebraic equations. Due to fundamental theorem of algebra, there always exist complex roots whose number is exactly the number of degree of algebraic equations in question. Hence, for any pair of $p = p_a$ and $p' = p_b$, even though $(\mathcal{A} \cap \mathbb{R}^2) \cap F^{-n}(\mathcal{B} \cap \mathbb{R}^2) = \emptyset$, there exist *complex trajectories* which connect $\mathcal{A}$ and $\mathcal{B}$. The initial and final manifolds are always connected via dynamics in this sense.

However, only knowing the existence of complex roots does not tell us anything about the nature of dynamics involved in the tunneling process. Recall that an original spirit of the semiclassical method would be to understand quantum phenomena whose classical counterparts exhibit chaos by observing the structure of "backbone" or "skeleton" behind the wave phenomena, as the periodic orbits play the role in the Gutzwiller's trace formula. We therefore ask our question more sharply, which leads to Question 2. Chaos can only be captured in the long time limit, and invariant structures in phase space are most relevant, so seeking the link to canonical objects in dynamical systems would be a natural direction.

Questions 1 and 2 are rather mathematics oriented, whereas Questions 3 and 4 are motivated by interests in physics. Dynamical tunneling is expected to take place quite broadly [50–55], although experimental observations and related theoretical studies are still limited unfortunately. A natural question we should address is, as well as one-dimensional tunneling problems, how to evaluate to the tunneling rate in the dynamical tunneling process. Question 4 is linked not only to specifying "backbone" of the tunneling transport but also to finding explicit analytical or numerical scheme to evaluate the tunneling probability.

To calculate the tunneling rate, we have to define the tunneling rate in advance. This is not a trivial task indeed in case of dynamical tunneling problems. In the one-dimensional double-well problem, the tunneling splitting between quasidoublet states characterizes the timescale of tunneling oscillations. Also in the one-dimensional system with dissociative channel, the imaginary part of the energy defines the life time and the tunneling rate as well.

In generic nonintegrable systems phase space is composed of infinitely many invariant components and they are generally intermingled with each other. It is not easy to specify from which state to which state the tunneling transition takes place. A plausible strategy to surmount it is to design the classical phase space, especially preparing the phase space where regular and chaotic regions are sharply bounded [56,57], and to introduce the tunneling rate under a certain principle. The readers should consult recent advances in this direction [58–61].

The main focus of the present chapter will be put on Questions 1 and 2—that is, qualitative natures of complex dynamics that bridges the initial and final manifolds. We further divide our discussion into two parts: (a) the nature of complex dynamics itself and (b) boundary conditions which need to be imposed for the semiclassical propagator $K_n^{sc}(p, p')$, that is, $p = p_a$ and $p' = p_b$. The latter discussion comes from the fact that the propagator $K_n^{sc}(p, p')$ is an object that evaluated at a fixed time step n. Classical orbits in a finite time interval is not compatible with any canonical objects in dynamical systems. To find a link to certain invariant structures in phase space, we have to take the limit $n \to \infty$ as mentioned. Such a task has actually been performed in Section 2.2 of our second paper [49]. However, this step contains rather technical issues and might blur relevant points. Therefore, although the issues (a) and (b) are not independent of each other of course, our stress in the chapter will be especially put on the issue (a); the complex dynamics especially when dynamical barriers separate the real phase space into disjointed invariant components.

We will also not touch on the Stokes phenomenon which inevitably appears when one applies the semiclassical approximation in the complex plane. Although this offers several novel aspects of the complex semiclassical treatment [62], we will put aside this issue here.

7.3   COMPLEX DYNAMICS IN ONE VARIABLE

In this section, we briefly review one-dimensional complex dynamics. Our minimal model is the area-preserving map (Equation 7.1), so it is necessary for us to examine the complex dynamics at least in two variables. Before going to the two variables case, however, we think it necessary to get used to simpler cases since even the complex dynamics in one variable has not been so much discussed in physics so far. Besides, the complex dynamics in more than one dimension needs many elaborate notions to be developed, some of which appear only in the complex analysis in several variables. To avoid the situation where too much mathematical technicalities make it difficult to gain our intuitive understanding, we begin by providing some expositions of one-dimensional complex dynamics, especially stressing some materials necessary to understand the dynamics in more than one dimension.

7.3.1   JULIA SET

Let us consider one-dimensional polynomial maps with degree d:

$$P : z \mapsto P(z), \tag{7.10}$$

where

$$P(z) = z^d + a_1 z^{d-1} + \cdots + a_d, \quad (d \geq 2). \tag{7.11}$$

Our first task is to classify the orbits according to their behavior as $n \to \infty$. It is natural to define

$$I_P = \{ \, z \in \mathbb{C} \mid \lim_{n \to \infty} P^n(z) = \infty \, \}, \tag{7.12}$$

$$K_P = \{ \, z \in \mathbb{C} \mid \lim_{n \to \infty} P^n(z) \text{ is bounded} \, \}, \tag{7.13}$$

where I_P and K_P are called the *set of escaping points* and the *filled Julia set* of P, respectively. In particular,

$$J_P = \partial K_P \tag{7.14}$$

is called the *Julia set*, and

$$F_P = \mathbb{C} - J_P \tag{7.15}$$

is called the *Fatou set*, which consists of the interiors of I_P and K_P. Note that the Fatou set can be introduced in different ways: one way to define it is to employ the notion of the *normal family*, and another is based on the *equicontinuity*. Note that all these are equivalent (see more details in Devaney (1989) [1] and Beardon (1991) [2]).

7.3.2   SIMPLE EXAMPLES

The simplest possible example which exhibits nontrivial behavior is $P(z) = z^2$. It is easy to show that $I_P = \{|z| > 1\}$, $K_P = \{|z| \leq 1\}$, and $J_P = \{|z| = 1\}$. Since $z = \infty$ is an attracting fixed point of P, the points $z \notin K_P$ converge to $z = 0$ monotonically. Similarly, $z = 0$ is another attracting fixed point of P, and the points $z \in K_P \backslash J_P$ converge to $z = 0$ monotonically. On the other hand, the orbits $z \in J_P$ are chaotic. This can be checked by putting $z = e^{2\pi i \theta}$. The map on J_P is reduced to the Bernoulli shift, $\theta \mapsto 2\theta \pmod 1$.

Another simple case is $P(z) = 2z^2 - 1$. It is also easy to show that $I_P = \mathbb{C}\backslash[-1,1]$ and $K_P = J_P = [-1,1]$. Since $P(\cos\theta) = \cos(2\theta)$, we generally have $P^n(\cos\theta) = \cos(2^n\theta)$. Then the iteration on $z \in [-1,1]$ is again described by the Bernoulli shift, $\theta \mapsto 2\theta \pmod 1$. One can also show that if $z \in \mathbb{C}\backslash[-1,1]$, $P^n(z) \to \infty$ as $n \to \infty$.

In both examples, the Julia set J_P has zero Lebesgue measure in the z-plane, that is, area$(J_P) = 0$, while area$(K_P) \neq 0$ for the first example while area$(K_P) = 0$ for the second example. Here area$(\cdot)$ denotes the Lebesgue measure.

7.3.3   PROPERTIES OF THE JULIA SET

The behavior of the orbits in the Fatou set is simple. Nontrivial behavior of orbits is observed only in the Julia set [1]. In fact, the orbits exhibit chaos behavior only in the Julia set in the above simply "solvable" examples. This holds in general, that is, *P is chaotic in J_P*. The definition of chaos is mathematically given in several ways. For instance, if one follows a definition proposed by Devaney [1], the Julia set satisfies; (1) sensitive dependence on initial conditions, (2) density of repelling periodic orbits, and (3) topological transitivity. The reader can find more precise contents and the proof of claims, for example, in the textbook of Devaney [1].

7.3.4   FILTRATION PROPERTY OF POLYNOMIAL MAPS

Recall that, as a model of Poincaré map, we often use the area-preserving map (Equation 7.1) in which transcendental functions appear in kinetic and/or potential terms. For the standard map, the potential function is given as sinusoidal functions. Sinusoidal functions appear both in kinetic and potential terms in the case of the kicked Harper model, which is also taken as a typical model with mixed phase space. Thus, it must be helpful to mention the nature of complex dynamics with transcendental functions.

In contrast to real dynamics, several important differences in complex dynamics exist between polynomial and transcendental maps. One of the crucial differences is that polynomial maps have the *filtration property*, whereas transcendental maps do not. In the case of the polynomial map, for sufficiently large $R > 0$, one can show that $P(V) \subset V$, where $V = \{|z| > R\}$. This means that once an orbit goes out of a region V in the set of escaping points I_P, it never goes back to the original region V but monotonically tends to infinity. Such a filtration leads

$$I_P = \bigcup_{n=1}^{\infty} P^{-n}(V). \tag{7.16}$$

On the other hand, the orbits for transcendental maps do not diverge in a monotonic way and can go back to a neighborhood of the origin even though it lies close to infinity. Thus, we may say that orbits in general behave much wilder than those of polynomial maps. This difference originates from the degree of singularities at $z = \infty$. In particular, $z = \infty$ is an essential singularity of transcendental functions, which makes the dynamics much more complicated. Rigorous mathematical results for complex dynamics in more than one variable with transcendental functions are limited, but the absence of filtration property is common as explained below.

7.4 POTENTIAL THEORETIC APPROACH IN THE ONE-VARIABLE CASE

As mentioned in Section 7.3.3, chaotic behavior appears in the Julia set J_P. In the remaining region, which is called the Fatou set F_P, orbits behave regularly. Nontrivial dynamics is therefore realized only in the Julia set J_P. This fact has been well established in polynomial maps in one variable. Several approaches exist to prove it (see, e.g., Devaney (1989) [1] and Beardon (1991) [2]). Here, for later reference, we will introduce the so-called *potential theoretic approach*.

7.4.1 DYNAMICS AROUND $z = 0$

Since $z = 0$ and $z = \infty$ are both attracting fixed points of the polynomial map (Equation 7.10) with $a_1 \neq 0$, we can expect that the behavior around $z = 0$ and $z = \infty$ is rather simple. Around $z = 0$, the lowest term z governs the dynamics, $P(z) \sim z$, and the dynamics follows that of the top term z^d around $z = \infty$. A standard approach to capture such situations is to introduce the *conjugation* from the original map F to more simplified coordinates.

First we discuss the case around $z = 0$. If the map is either contracting or expanding around $z = 0$, there is an old theorem on linearization around the origin. Note that the following theorem applies not only to polynomial maps but the map with a Taylor expansion around $z = 0$.

Theorem (Koenigs)

Let $F(z)$ be holomorphic near $z = 0$ and has a Taylor expansion

$$F(z) = \lambda z + c_2 z^2 + \cdots, \quad (0 < |\lambda| < 1). \tag{7.17}$$

Then there exists a conformal map $\psi : U \to \mathbb{C}$ which satisfies the functional equation (Schröder equation)

$$\psi\big(F(z)\big) = \lambda \psi(z), \quad (z \in U), \tag{7.18}$$

where U denotes a neighborhood of $z = 0$.

The dynamics around $z = 0$ is thus described by the conjugating function $\psi(z)$. The proof is straightforward: first assume a formal expansion as $\psi(z) = \sum_{\ell=0}^{\infty} a_\ell z^\ell$. By inserting it into the functional equation 7.18, we get $a_\ell = K_\ell(c_2, \ldots, c_\ell, a_2, \ldots, a_{\ell-1})/(\lambda^\ell - \lambda)$, where $K_\ell(c_2, \ldots, c_\ell, c_2, \ldots, a_{\ell-1})$ are a polynomial function of $(c_2, \ldots, c_\ell, a_2, \ldots, a_{\ell-1})$. The final step is to prove the convergence of the expansion.

It is rather easy to construct a conjugating function around $z = 0$ for the case $|\lambda| \neq 1$ (in case $|\lambda| > 1$, one can show the same assertion by considering the inverse function). On the other hand, the case with $|\lambda| = 1$ is subtle. This is because denominators of coefficients a_ℓ for the formal solution diverges and the same method cannot apply. Linearization around a *neutral fixed point* has been a long-standing problem and the following theorem is a cornerstone of this issue:

Theorem (Siegel–Moser)

Let $F(z)$ be holomorphic near $z = 0$ and has the Taylor expansion of the type

$$F(z) = \lambda z + c_2 z^2 + \cdots, \quad (\lambda = e^{2\pi i \alpha}, \ \alpha : \text{irrational}). \tag{7.19}$$

Suppose that there exist $a, b > 0$ such that $\left| \alpha - \dfrac{p}{q} \right| > \dfrac{a}{q^b}$ for all $p, q \in \mathbb{Z}$. Then there is a neighborhood U of $z = 0$ on which $F(z)$ is analytically conjugate to the irrational rotation—that is, $z \mapsto \lambda z$.

This is a sufficient condition to have a convergent conjugating function, and necessary and sufficient condition was especially presented for the quadratic map $F(z) = \lambda z + c_2 z^2$ ($\lambda = e^{2\pi i \alpha}, \alpha :$ irrational) [63].

An analogous small denominator problem arises in constructing the KAM circles in the area-preserving map. The latter is much more subtle than linearization around a neutral fixed point in $\mathbb{C}$, but several aspects are common. We will discuss the KAM case later.

In conjunction with the linearization problem, the so-called *Siegel disk* appears as a result of linearization [64]. The theorem of Siegel is that around a neutral fixed point, $F(z)$ has a region $\mathcal{D}$ which is conjugate to an irrational rotation. The region $\mathcal{D}$ is called the Siegel disk. Note that if there exists a Siegel disk $\mathcal{D}$, then not only a single Siegel disk but also a bundle of them appear in $\mathbb{C}$. As a result, we have area $(K_P) > 0$. We will discuss this point in Section 7.5.5, the area (or volume) of rotation domains in the complex plane, which will become one of the key issues.

7.4.2 Dynamics around $z = \infty$

A similar strategy is taken to analyze the complex dynamics around $z = \infty$. The existence of the conjugating function is guaranteed by the following theorem.

Theorem (Böttcher)

For a sufficiently large R, there exists a conformal map $\varphi(z)$ of $V = \{ |z| > R \}$ into $\mathbb{C}$ which has the form

$$\varphi(z) = z + b_0 + \frac{b_1}{z} + \cdots \tag{7.20}$$

and satisfies

$$\varphi(P(z)) = \big(\varphi(z)\big)^d. \tag{7.21}$$

In particular, $\varphi(z)$ is called the *Böttcher function*.

The Böttcher function $\varphi(z)$ is constructed explicitly in the following manner. First, consider

$$\psi(z) = \log \frac{P(z)}{z^d},$$

which is a single-valued holomorphic function satisfying $\lim\limits_{z \to \infty} \psi(z) = 0$. Since $P(z) = z^d \exp \psi(z)$, we have $P^2(z) = z^{d^2} \exp\big(d\psi(z) + \psi(P(z))\big)$. Inductively, we have

$$P^n(z) = z^{d^n} \exp\big(d^{n-1} \psi(z) + d^{n-2} \psi(P(z)) + \cdots + \psi(P^{n-1}(z)) \big).$$

Now we define $\varphi_n(z) \equiv \big(P^n(z)\big)^{d^{-n}}$, then we have

$$\varphi_n(z) = z \exp\left(\frac{1}{d} \psi(z) + \frac{1}{d^2} \psi(P(z)) + \cdots + \frac{1}{d^n} \psi(P^{n-1}(z)) \right). \tag{7.22}$$

$\sum_{j=1}^{\infty} \frac{1}{d^j} \psi\left(P^{j-1}(z)\right)$ is uniformly convergent, hence

$$\varphi(z) = \lim_{n \to \infty} \varphi_n(z) = z \exp\left(\frac{1}{d}\psi(z) + \frac{1}{d^2}\psi(P(z)) + \cdots\right) \tag{7.23}$$

does so. It is easy to check that $\varphi(z)$ satisfies the desired functional relation (Equation 7.21).

7.4.3 Green Function and Potential Theory in $\mathbb{C}$

The Böttcher function $\varphi(z)$ describes the behavior around $z = \infty$, and the Green function $G(z)$, which is a key object in potential theory, is defined using $\varphi(z)$ as

$$G(z) \equiv \log|\varphi(z)|. \tag{7.24}$$

Initially, the Böttcher function $\varphi(z)$ is introduced around $z = \infty$, thus the Green function $G(z)$ is only defined in the same region. However, it is possible to show that $G(z)$ can be extended to the the set of escaping set I_P as a harmonic function, that is, $\Delta G(z) = 0$. For $K_P = \mathbb{C}\backslash F_P$, we define $G(z) = 0$. Then one can prove that $G(z)$ is continuous and subharmonic in $\mathbb{C}$. Here, a real-valued function $u(z)$ defined on a domain $U \subset \mathbb{C}$ is called *subharmonic* if it satisfies the two conditions:

i. $u(z)$ is continuous (rigorously upper semicontinuous).
ii. Consider an arbitrary disk D in U. If we take any harmonic function $h(z)$ which equals to $u(z)$ on the boundary of D, then $u(z) \leq h(z)$ in D.

This is a two-variable ($\operatorname{Re} z$ and $\operatorname{Im} z$) analog of a one-variable convex function $d^2 u(x)/dx^2 \geq 0$. Indeed, a subharmonic function satisfies the relation $\triangle u \geq 0$. A significance of this property is that $\triangle u$ describes a measure or a distribution on U. A well-known example is the electrostatic potential, where $\triangle u$ gives the charge distribution. Note also that a subharmonic function $f(z)$ satisfies

$$f(z) \leq \frac{1}{2\pi}\int_0^{2\pi} f(z + re^{i\theta})\,d\theta. \tag{7.25}$$

Recall that the equality holds for the harmonic function.

Using Equations 7.22 and 7.23 for the Böttcher function, we have a more explicit expression

$$G(z) = \lim_{n \to \infty}\frac{1}{d^n}\log^+\left|P^n(z)\right|, \tag{7.26}$$

where $\log^+ t \equiv \max\{\log t, 0\}$. To check that this definition is consistent with Equation 7.24, it is suffice to recall $\varphi_n(z) = \left(P^n(z)\right)^{d^{-n}}$ and take "log" and $n \to \infty$ in both sides. Note that the definition $G(z) = 0$ for $z \in K_P$ is consistent with the fact $|P^n(z)| < \infty$ in K_P.

7.4.4 Complex Equilibrium Measure

Next we introduce $\mu(z)$ by

$$\mu(z) = \frac{1}{2\pi}\Delta G(z). \tag{7.27}$$

The meaning of the Green function can be understood by the analogy of the electrostatic potential of a perfect conductor. The charge on a perfect (earthed) conductor is distributed over the surface. Once the potential function is known, the surface distribution of charge is given by applying the Laplacian to the electrostatic potential, and we can identify the surface of the conductor as the support of the charge distribution. The Green function just corresponds to the potential function.

The following theorem justifies that $\mu(z)$ is certainly a complex equilibrium measure [65].

Theorem (Brolin)

1. $\mu_n(z) = \dfrac{1}{d^n} \displaystyle\sum_{z_0 \in P^{-n}(a)} \delta(z - z_0) \to \mu(z)$ *for arbitrary* $z = a$
2. $\operatorname{supp} \mu(z) = J_P$
3. *The map* P *preserves the measure* μ, *and is strongly mixing.*

For arbitrary fixed $a \in I_P$, we define a measure μ_n as the sum of point masses with weight d^{-n} supported by d^n points of $(P^n)^{-1}(a)$, which takes into account multiplicity. Then, the statement 1 claims that μ_n converges to μ (more precisely in the sense of weak convergence). The statement 2 shows that the equilibrium charge is localized on the boundary of K_P. The statement 3 concerns the nature of dynamics. Due to general theory of dynamical system, ergodicity of μ immediately follows.

Sketch of the proof.

1. Recall that $g(z) = \log|z|$ is a fundamental solution for $(1/2\pi)\Delta g(z) = \delta(z)$. Therefore $(1/2\pi)\Delta \log|P^n(z) - a| = \sum_{P^n(z_0)=a} \delta(z - z_0)$. We can prove $\lim\limits_{n \to \infty}(1/d^n)\log|P^n(z) - a| = G(z)$, where $G(z) = \lim\limits_{n \to \infty}(1/d^n)\log^+|P^n(z)|$. This is because in case $z \in I_P$, $P_n(z) \to \infty$, thus $|P^n(z) - a| \sim |P^n(z)|$, also in case $z \in K_p$, $P_n(z)$ is bounded, so $\lim\limits_{n \to \infty}(1/d^n)\log|P^n(z) - a| = 0$. Finally, we apply "$(1/2\pi)\Delta$" to both sides.

2. We first show that $\operatorname{supp} \mu \subset J_P$ holds.

   The Green function $G(z) = \log|\varphi(z)| = \lim\limits_{n \to \infty}(1/d^n)\log^+|P^n(z)|$ is harmonic on I_P, which implies $\mu = 0$ on I_P. Recall $G = 0$ on K_P (definition of G). Thus, $\operatorname{supp} \mu \subset J_P$ follows.

   Next we show that $\operatorname{supp} \mu \supset J_P$ holds.

   Suppose that there exists a point $z \in J_P$ and its neighborhood U such that $\operatorname{supp} \mu \cap U = \emptyset$. This implies $\Delta G = 0$ on U (i.e., G is harmonic on U). On the other hand, $G \equiv 0$ in $(U \cap K_P)$ (by definition of G) and $G \geq 0$ on $\mathbb{C}$, thus $G \equiv 0$ on the whole U due to the principle of minimum values (since G is harmonic). This contradicts that $G > 0$ on $U \cap I_P$ (G is positive on I_P).

3. In order to prove P is mixing, for arbitrary L^2-functions f and g, it is suffice to show

$$\lim_{n \to \infty} \int_{J_P} f(P^n(z)) g(z) \mathrm{d}\mu(z) = \int_{J_P} f(z)\mathrm{d}\mu(z) \cdot \int_{J_P} g(z)\mathrm{d}\mu(z). \qquad (7.28)$$

First, consider the mass distribution $\{\mu_n(\cdot, w)\}$ produced by a starting point w. If we allow w to be a function of n, we obtain a sequence $\{\mu_n(\cdot, w_n)\}$. $\mu_n(z) \to \mu(z)$ (statement 1) implies that $\mu_n(\cdot, w_n) \to \mu(\cdot)$. Next, let $\{Q_j\}_{j=1}^k$ be a finite number of boxes which cover J_P, then we can prove $\mu_n(Q_j, w_n) \to \mu(Q_j)(1 \leq j \leq k)$. For any function $g(z)$ which is constant on each box Q_j, then from the above result we have

$$\lim_{n \to \infty} \sum_{\nu=1}^{d^\nu} \frac{1}{d^n} g(\zeta_{-n}^{(\nu)}) = \int_{J_P} g(z)\,\mathrm{d}\mu(z), \qquad (7.29)$$

where $\zeta \in J_P$ and $\{\zeta_{-n}^{(v)}\}$ are preimages of ζ of order n. Finally, for any function $f(z), g(z)$ which is constant on each box Q_j,

$$
\begin{aligned}
\lim_{n \to \infty} \int_{J_P} f\big(P^n(z)\big) g(z) d\mu(z) &= \lim_{n \to \infty} \lim_{m \to \infty} \sum \frac{1}{d^{n+m}} f(\zeta_{-m}^{(v)}) g(\zeta_{-(m+n)}^{(v)}) \\
&= \lim_{n \to \infty} \lim_{m \to \infty} \sum \frac{1}{d^m} f(\zeta_{-m}^{(v)}) \cdot \sum_{\zeta_{-m}^{(v)} \text{ fixed}} \frac{1}{d^n} g(\zeta_{-(m+n)}^{(v)}) \\
&= \int_{J_P} f(z) d\mu(z) \cdot \int_{J_P} g(z) d\mu(z).
\end{aligned}
\tag{7.30}
$$

7.4.5 CONJUGATING FUNCTIONS AND THEIR NATURAL BOUNDARIES

As has been shown above, the conjugating functions $\psi(z)$ and $\varphi(z)$, each of which respectively transforms the original map P into simpler motions around $z = 0$ and $z = \infty$. Both functions have positive radii of convergence and so analytic around each point. However, if $\varphi(z)$ is holomorphic in the whole $\mathbb{C}$, it turns out $\Delta G(z) = 0$ in $\mathbb{C}$, which is not consistent with the Brolin's theorem. Therefore, $\varphi(z)$ should have some singularities and limited domains of analyticity. Likewise we may expect that $\psi(z)$ does so. $\Delta G(z)$ is not identically zero in the whole Julia set J_P, which implies the presence of logarithmic singularities of $G(z) = \log |\varphi(z)|$ as indicated in the proof of Brolin's theorem. In reality, both $\psi(z)$ and $\varphi(z)$ have been shown to have *natural boundaries*. More precisely, for $|\lambda| < 1$ it was shown for example in Costin and Kruskal (2005) [66] that

the domain of analyticity of $\psi(z)$ is K_P, and $J_P = \partial K_P$ is a singularity barrier (= natural boundary) of $\psi(z)$.

and also shown in Costin and Huang (2009) [67] that

the domain of analyticity of the Böttcher function $\varphi(z)$ is I_P and $J_P = \partial K_P$ is a singularity barrier (= natural boundary) of $\varphi(z)$.

Here, following the proof in Costin and Kruskal (2005) [66], we present a sketch of the proof only for the case of $\psi(z)$. First, recall the Schröder equation 7.18, from which we have

$$
\psi\big(F^n(z)\big) = \lambda^n \psi(z).
\tag{7.31}
$$

We then use $J_P = \overline{\{\text{repelling fixed points}\}}$. As mentioned in Section 7.3.3, the Julia set satisfies the condition of chaos (in the sense of Devaney, for example). The densely distributed repelling fixed points in J_P was one of the conditions, so let us assume that z_0 is a repelling fixed point of $F(z)$ of period n, and is a point of analyticity of $\psi(z)$. Relation (Equation 7.31) implies $\psi(z_0) = 0$, since $|\lambda| < 1$. Now, we have $(F^n)'(z_0) \psi'(z_0) = \lambda^n \psi'(z_0)$, but since $|(F^n)'(z_0)| > 1$ and $|\lambda| < 1$, this implies $\psi'(z_0) = 0$. Inductively, we have $\psi^{(m)}(z_0) = 0$ for all m. We have assumed that $\psi(z)$ is analytic, and this entails $\psi(z) \equiv 0$. This contradicts the fact that $\psi(z)$ is a conjugation function.

Note that the above proof essentially uses the condition $|\lambda| < 1$. To the authors' knowledge, the situation $\lambda = e^{2\pi i \alpha}$, where α is, for example, a Diophantine number, is not clear enough, while such a case is closer to the issue we will discuss hereafter.

7.5 COMPLEX DYNAMICS IN TWO VARIABLES

For the complex map in one variable, we emphasized that a crucial difference exists between the polynomial and transcendental map since the former has the filtration property whereas the latter does not. The nonexistence of wandering domains also characterizes polynomial maps [2,68,69], and in many respects the behavior of the polynomial map is much simpler than the transcendental map. Technical aspects are also important: In the definition of the the Green function (Equation 7.24), the factor $1/d^n$ appears as a normalization factor, which reflects that the modulus of variables grows as z^{d^n} as $|z| \to \infty$. In the transcendental map, an analogous construction of the Green function cannot be made because the modulus around $z = \infty$ is not normalizable. This is also related to the fact that the behavior around $z = \infty$ of the transcendental map is much more complicated than in the polynomial map. As shown below, the same difficulty arises in several dimensional cases as well. Since, at present, potential theory is a unique tool to analyze the higher dimensional complex map, it is crucial whether or not we can introduce the Green function. Mainly for this reason, as is the case of one variable, we consider polynomial complex maps in what follows.

7.5.1 HÉNON MAP AND ITS PARAMETER DEPENDENCE

The polynomial map in one variable is simply expressed in the form of Equation 7.11. There is no ambiguity to write down its standard form. For two-dimensional maps, we first give a canonical form of polynomial maps. Fortunately, there is a classification theorem by Friedland and Milnor [70] which has shown that two-dimensional polynomial diffeormorphisms are conjugate either to (1) elementary map, (2) affine map, and (3) generalized Hénon maps. The former two are simple and generate no chaos. Only the generalized Hénon map is nontrivial and chaos appears. The generalized Hénon map is a composition of well-known Hénon maps. Therefore the Hénon map, a standard form of which is given as

$$P : \begin{pmatrix} x' \\ y' \end{pmatrix} = \begin{pmatrix} y \\ y^2 - bx + a \end{pmatrix} \qquad (7.32)$$

can be considered as the simplest possible two-dimensional polynomial maps which we should first analyze. We note that the Hénon map with $b = 1$ is transformed into the form of the area-preserving map by an affine change of variable $(p,q) = (y-x, y-1)$, which yields the cubic potential map

$$F : \begin{pmatrix} p' \\ q' \end{pmatrix} = \begin{pmatrix} p - V'(q) \\ q + p' \end{pmatrix}, \qquad (7.33)$$

where $V(q) = (1/3)q^3 + cq$ and $c = 1 - a$. As $|b| \to 0$, the Hénon map is reduced to the one-dimensional Logistic map, $y' = y^2 + a$. In the area-preserving case $|b| = 1$, the Hénon map is genuinely two dimensional.

Depending on the nonlinear parameter a, the Hénon map shows a variety of dynamical behaviors: For $a \gg 1$, the horseshoe condition is satisfied and the map is conjugate to the symbolic dynamics with binary full shift (see Figure 7.2). In this parameter regime, all the stable and unstable manifolds for periodic orbits intersect transversally, and the system is uniformly hyperbolic.

At a certain critical value a_f at which the horseshoe is broken. What is unique in the Hénon map is that even after the horseshoe is destroyed there exist many (presumably infinitely many) parameter intervals on which hyperbolicity of the mapping is recovered [71,72]. Figure 7.3 shows stable and unstable manifolds before and after the first tangency. Especially in Figure 7.3b, note that all the stable and unstable manifolds intersect transversally even after the first tangency.

The Hénon map is one of the most extensively studied two-dimensional maps concerning the *pruning* of the invariant set. An idea of the *pruning front* has been proposed as a natural extension of the kneading theory, which provides us with a precise recipe to specify all topological natures of a family of unimodal maps [73]. Pruning front theory aims at giving an analogous border in the two-dimensional symbol plane, which determines admissible and nonadmissible orbits [74–77]. In Figure 7.4, we sketch the parameter space of the area-preserving Hénon map, showing that the Hénon map covers from horseshoe to KAM situations depending on the parameter a.

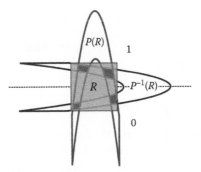

FIGURE 7.2 The forward and backward iteration of the Hénon map. A generating partition is naturally introduced which admits the symbolic dynamics with binary coding 0 and 1.

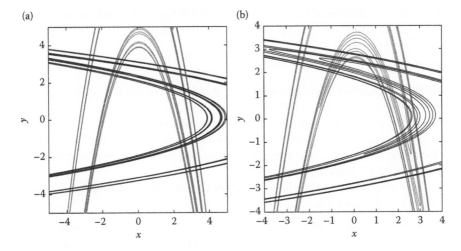

FIGURE 7.3 The stable and unstable manifolds for the Hénon map for (a) before the first tangency $(a > a_f)$ and (b) after the first tangency $(a < a_f)$.

7.5.2 STABLE AND UNSTABLE SETS IN THE HORSESHOE LIMIT

Before going to the Hénon map in the complex plane, we overview the Hénon map in the real plane especially when the horseshoe condition holds. In contrast to one-dimensional maps, the Hénon map is two-dimensional diffeomorphism, so it is invertible. We therefore consider the limiting sets

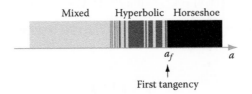

FIGURE 7.4 The parameter dependence of the Hénon map (schematic).

not only in the forward but also in the backward direction. We define stable (resp. unstable) set as

$$K^{\pm} = \{ (x,y) \in \mathbb{R}^2 \mid \|P^n(x,y)\| \text{ is bounded } (n \to \pm\infty) \}, \tag{7.34}$$

$$K = K^+ \cap K^-, \tag{7.35}$$

and their boundaries as

$$J^{\pm} = \partial K^{\pm}, \quad J = J^+ \cap J^-. \tag{7.36}$$

Below, we introduce the filled Julia set and the Julia set in $\mathbb{C}^2$ using the same notation, but for the moment we confine ourselves to the map defined in $\mathbb{R}^2$.

In the horseshoe parameter locus, the symbolic dynamics represented by the binary full shift exists and dynamical behaviors in $\mathbb{R}^2$ can be analyzed in the symbolic space. For later reference, we particularly focus on the following properties. Any of them can be easily proved using the symbolic dynamics.

1. $\overline{\{\text{periodic orbits}\}} = K$.
2. For any unstable periodic orbit p, $\overline{W^s(p)} = J^+$ and $\overline{W^u(p)} = J^-$ hold, where $W^s(p)$ (resp. $W^u(p)$) represents the stable (resp. unstable) manifold of p.
3. $K^{\pm} = J^{\pm}$ and $K = J$.

The sketch of the proof is as follows:

1. Since each periodic orbit is expressed in a symbolic sequence as $p = (s_0 s_1 \ldots s_{n-1})^\infty$. Such orbits are dense in the whole symbolic space.
2. Orbits in $W^s(p)$ are represented as $s = (* \ldots * s_0 \ldots s_{n-1} s_0 \ldots s_{n-1} \ldots)$. By substituting an arbitrary sequence $s_1' s_2' \ldots s_m'$ into $* \ldots *$, s can mimic any sequence (i.e., any point in K^+) including the string $s_1' s_2' \ldots s_m'$, which means that the orbit s can pass arbitrarily close to any element of K^+ before approaching p.
3. Uniformly contracting in the stable direction and uniformly expanding in the unstable direction lead the assertion.

In this way, binary symbolic dynamics is powerful machinery to derive various ergodic properties. However, in the parameter regime where the mixed phase is realized, we can no more expect any symbolic description. Not only that, any of the above statements do not hold:

1. $\overline{\{\text{periodic orbits}\}} \neq K$.
2. $\overline{W^s(p)} \neq K^+$, $\overline{W^u(p)} \neq K^-$.
3. $K^{\pm} \neq J^{\pm}$, $K \neq J$.

All these are due to the existence of KAM curves in $\mathbb{R}^2$.

7.5.3 JULIA SET IN $\mathbb{C}^2$

In the same way as in one-dimensional dynamics in $\mathbb{C}$, we classify the orbits in $\mathbb{C}^2$ according to the behavior of $n \to \infty$:

$$I^{\pm} = \{(x,y) \in \mathbb{C}^2 \mid \lim_{n \to \infty} P^{\pm n}(x,y) \to \infty \, (n \to \infty) \}, \tag{7.37}$$

$$K^{\pm} = \{(x,y) \in \mathbb{C}^2 \mid \lim_{n \to \infty} P^{\pm n}(x,y) \text{ is bounded in } \mathbb{C}^2 \}. \tag{7.38}$$

In particular,

$$K = K^+ \cap K^-, \tag{7.39}$$

$$J^{\pm} = \partial K^{\pm}, \tag{7.40}$$

$$J = J^+ \cap J^-. \tag{7.41}$$

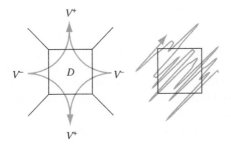

FIGURE 7.5 Filtration property in the polynomial map (left). A wild behavior in the transcendental map (right).

Here $K, J^{\pm}$, and J are respectively called the filled Julia set, forward (resp. backward) Julia set, and the Julia set. We note that for $a > a_f$ where the Hénon map satisfies the horseshoe condition, then the filled Julia set has no interior and it is contained in the real plane, that is, $K = J \subset \mathbb{R}^2$.

Also as in case of one-dimensional maps, the Hénon map has the filtration property. Suppose

$$V^+ = \{ (x,y) \in \mathbb{C}^2 \mid |y| > R, \ |y| > |x| \}, \tag{7.42}$$

$$V^- = \{ (x,y) \in \mathbb{C}^2 \mid |x| > R, \ |x| > |y| \}, \tag{7.43}$$

then for a sufficiently large $R > 0$ it can be shown that $P(V^+) \subset V^+$ and $P^{-1}(V^-) \subset V^-$. Furthermore, as shown in Figure 7.5, once an orbit drops into V^+ (resp. V^-), it diverges to infinity by the forward (resp. backward) iterations monotonically. In analogy with Equation 7.16, we have

$$\bigcup_{n=0}^{\infty} P^{-n}(V^+) = \mathbb{C}^2 \setminus K^+, \quad \bigcup_{n=0}^{\infty} P^{+n}(V^-) = \mathbb{C}^2 \setminus K^-, \tag{7.44}$$

namely, if $z \in V^+$ (resp. V^-), then $P^n(z) \in V^+$ (resp. $P^{-n}(z) \in V^-$).

7.5.4 GREEN FUNCTION AND PLURIPOTENTIAL THEORY FOR $\mathbb{C}^2$

7.5.4.1 Green Function

The Green function for the Hénon map is defined in the same way as in the one-dimensional polynomial map. For the map in $\mathbb{C}$, we have introduced in Equation 7.24 the Green function $G(z)$ via the Böttcher function $\varphi(z)$, where $\varphi(z)$ is constructed to satisfy the functional relation $\varphi(P(z)) = \{\varphi(z)\}^d$. Similarly, we can construct an analog of the Böttcher function in $\mathbb{C}^2$.

Theorem [4]

For the Hénon map P, there exists a holomorphic function $\varphi^+(x,y)$ (resp. $\varphi^-(x,y)$) on V^+ (resp. V^-), which satisfies the functional equation $\varphi(P(x,y)) = \{\varphi^+(x,y)\}^2$ (resp. $\varphi(P^{-1}(x,y)) = \{\varphi^-(x,y)\}^2$.

We then introduce the Green function via the Böttcher function $\varphi^{\pm}(x,y)$ as $G^{\pm}(x,y) = \log |\varphi^{\pm}(x,y)|$. Here we directly give an expression for the Green function using the map $P(x,y)$:

$$G^{\pm}(x,y) \equiv \lim_{n \to +\infty} \frac{1}{2^n} \log^+ \left\| P^{\pm n}(x,y) \right\|, \tag{7.45}$$

where $\log^+ t \equiv \max\{\log t, 0\}$.

7.5.4.2   Plurisubharmonic Function

It is possible to show that $G^{\pm}(x,y)$ is continuous and plurisubharmonic on $\mathbb{C}^2$ [4]. We can briefly say that a plurisubharmonic function is multidimensional analog of a subharmonic function in $\mathbb{C}$, whose definition is presented in Section 7.4.3. More precisely, a real-valued function defined on a domain in $\mathbb{C}^2$ is called *plurisubharmonic* if (i) u is continuous (precisely upper semicontinuous) and (ii) along any one-dimensional line in $\mathbb{C}^2$ u is a subharmonic function in the sense of $\mathbb{C}$.

Next we want to consider the analog of Equation 7.27. In order to proceed in parallel with discussions on the dynamics in $\mathbb{C}$ case, we need to prepare several notions which are specific in $\mathbb{C}^2$.

7.5.4.3   Currents

As a generalization of the measure, we next introduce the *currents*. To this end, first denote the set of differential (p,q)-forms whose coefficients are contained in C_0^{∞} by

$$\mathcal{D}^{p,q}(\Omega) \equiv \left\{ \sum_{|I|=p,|J|=q} u_{I,J}\, dz_{i_1} \wedge \cdots \wedge dz_{i_p} \wedge d\bar{z}_{j_1} \wedge \cdots \wedge dz_{j_q} \;\middle|\; u_{I,J} \in C_0^{\infty}(\Omega) \right\}, \qquad (7.46)$$

where $C_0^{\infty}(\Omega)$ represents the distributions, that is, continuous linear functionals defined on a certain nonempty domain Ω. Here we have used multi-index notations: $I = (i_1,\dots,i_p)$ and $J = (j_1,\dots,j_q)$. $\mathcal{D}^{p,q}(\Omega)$ is often called the *test forms*. Note also that $\mathcal{D}^{2-p,2-q}(\Omega) = C_0^{\infty}(\Omega)$.

Linear functionals on $\mathcal{D}^{2-p,2-q}(\Omega)$, which are elements of the dual space $(\mathcal{D}_0^{p,q}(\Omega))'$, are called (p,q)-*currents*. Here the prime $'$ denotes the dual space. In other words, currents T act on differential forms $\varphi \in \mathcal{D}^{2-p,2-q}(\Omega)$. More concretely, for $T \in (\mathcal{D}^{0,0}(\Omega))'$ and $\varphi \in \mathcal{D}^{0,0}(\Omega)$, the duality is denoted by $T(\varphi)$ or $\int_{\Omega} T \wedge \varphi$. The duality for other bidegrees (p,q) can be similarly described so that we can regard a current T as a *differential form with the distribution coefficients*. Note that if we replace $C_0^{\infty}(\Omega)$ by $C_0^0(\Omega)$, then a current T becomes a differential form with the *measure coefficients*.

7.5.4.4 dd^c Operator and Induced Currents

In order to consider the counterpart of Equation 7.27, we introduce the counterpart of the Laplacian for $\mathbb{C}^2$ as

$$dd^c \equiv 2i \sum_{j,k=1}^{2} \frac{\partial^2}{\partial z_j \partial \bar{z}_k} dz_j \wedge d\bar{z}_k, \qquad (7.47)$$

and we call the dd^c *operator*. A generalization of dd^2 operator to $\mathbb{C}^2$, $(dd^c)^n = dd^c \wedge \cdots \wedge dd^c$ is called the *complex Monge Ampére operator*, which plays a central role in pluripotential theory [78]. We refers to an important result:

Proposition [78]

If u is a plurisubharmonic function, then $dd^c u$ operator is a positive $(1,1)$-current with measure coefficients.

We specify positivity of currents as follows: A (p,p)-current $T \in (\mathcal{D}^{2-p,2-p}(\Omega)')$ is said to be positive if $T(\varphi) \geq 0$ for all $\varphi \in \mathcal{D}^{2-p,2-p}(\Omega)$ with $\varphi \geq 0$. Here a (p,p)-form $\varphi \in \mathcal{D}^{2-p,2-p}(\Omega)$ is called positive if $\varphi(z) \geq 0$ at each $z \in \Omega$.

Note that this proposition is a generalization of the fact that applying the Laplacian in $\mathbb{C}$ to the subharmonic function induces a measure.

Therefore, by identifying $(x,y) = (z_1, z_2)$, we can apply the dd^c-operator to $G^{\pm}(x,y)$ to get the $(1,1)$-currents $\mu^{\pm}$:

$$\mu^{\pm} \equiv \frac{1}{2\pi} dd^c G^{\pm}. \tag{7.48}$$

We notice that Equation 7.48 again is in the form of the Poisson equation and so it can be regarded as a multidimensional version of Equation 7.27. For these $\mu^{\pm}$, we have the following result:

Theorem (Bedford–Smillie)

$$\operatorname{supp}\mu^{\pm} = J^{\pm} \tag{7.49}$$

where $J^{\pm}$ is the forward (resp. backward) Julia set.

Sketch of the proof.

We first show $\operatorname{supp}\mu^+ \subset J^+$.

G^+ is pluriharmonic on F^+, that is, $dd^c G^+ = 0$, which implies $\mu^+ = 0$ on F^+. Recall that $G^+ = 0$ on K^+ (by definition of G^+). Thus, $\operatorname{supp}\mu^+ \subset J^+$.

Next we show $\operatorname{supp}\mu^+ \supset J^+$.

Suppose that there exist a point $z \in J^+$ and its neighborhood W such that $\operatorname{supp}\mu \cap W = \emptyset$. This implies $dd^c G^+ = 0$ on W (i.e., G^+ is pluriharmonic on W). On the other hand, $G^+ \equiv 0$ in $(W \cap K^+)$ (by definition of G^+) and $G^+ \geq 0$ on $\mathbb{C}^2$, thus $G^+ \equiv 0$ on the whole W due to the principle of minimum values (since G^+ is pluriharmonic). This contradicts that $G^+ > 0$ on $W \cap F^+$ (G^+ is positive on K^+).

7.5.4.5 Complex Equilibrium Measure

The above theorem of Bedford and Smillie looks a two-dimensional analog of the second statement of the Brolin's theorem, $\operatorname{supp}\mu(z) = J_P$. However, $\mu(z)$ in $\mathbb{C}$ is a measure while $(1,1)$-currents $\mu^{\pm}$ are not. A counterpart of the equilibrium measure $\mu(z)$ is given by taking the wedge product of μ^+ and μ^-:

$$\mu = \mu^+ \wedge \mu^-. \tag{7.50}$$

Here $\mu^+ \wedge \mu^-$ is defined for continuous plurisubharmonic functions u_1 and u_2 by

$$\langle \mu^+ \wedge \mu^-, \chi \rangle = \langle dd^c u_1 \wedge dd^c u_2, \chi \rangle \equiv \langle dd^c u_1, u_2 dd^c \chi \rangle, \tag{7.51}$$

where $\chi \in \mathcal{D}^{0,0}(U)$ and U denotes the domain of definition of functions u_1 and u_2. There is a general theorem stating that μ thus defined becomes a measure [78]. Therefore, the two-dimensional analog of the Brolin's measure $\mu(z)$ is $\mu = \mu^+ \wedge \mu^-$. Below we denote the support of this measure μ by

$$J^* = \operatorname{supp}\mu. \tag{7.52}$$

7.5.4.6 Convergence Theorem of Currents

We now present a key theorem from which a variety of properties of the measure μ are derived. This has been established by Bedford and Smilie [44,45] and Fornæss and Sibony [79]. It relates the repeated iteration of the current supported by a certain class of complex manifolds and the currents $\mu^{\pm}$:

Theorem

Let M be an algebraic variety, then there is a constant $c > 0$ such that

$$\lim_{n \to \infty} \frac{1}{2^n} [P^{\mp n} M] = c\mu^{\pm} \tag{7.53}$$

in the sense of current, where $[M]$ is the current of integration of M.

An algebraic variety is defined as the zero set of polynomials. A line ($z_1 + z_2 - 1 = 0$) and a sphere ($z_1^2 + z_2^2 - 1 = 0$) are examples of algebraic varieties. The initial and final states $p = p_a$ and $p' = p_b$, where $p_a, p_b \in \mathbb{R}$ in our tunneling problems also provide algebraic varieties.

The definition of the *current of integration* is as follows. Let Ω be an open set in $\mathbb{C}^2$ and let M be an analytic subset of Ω of pure dimension 1. Then M defines the *current of integration* $[M]$ in the following way: let $\varphi \in \mathcal{D}^{1,1}(\Omega)$ and let $\overline{M}$ be the set of regular points of M. Then, $[M]$ is defined by

$$\int_{\Omega} [M] \wedge \varphi = \int_{\overline{M}} \varphi.$$

The current of integration can be rephrased using the Poincaré Lelong formula: let h be a holomorphic function $\mathbb{C}^2 \to \mathbb{C}$, and if M is the set defined as the zeros (with no multiplicity) of h, that is, $M = \{z = (z_1, z_2) \, | \, h(z) = 0\}$, then the current of integration is given by the so-called Poincaré Lelong formula as

$$[M] = \frac{1}{2\pi} dd^c \log |h|. \tag{7.54}$$

In the case of straight line $h(z_1, z_2) = az_1 + bz_2$, for example, taking the new coordinate Z perpendicular to the line, we immediately obtain $dd^c \log |h| = v dZ \wedge d\overline{Z}$, where $v = (\partial^2/\partial Z \partial \overline{Z}) \log |Z| = \delta(Z)\delta(\overline{Z})$ and so the current is distributed along $M = \{(z_1, z_2) \, | \, az_1 + bz_2 = 0\}$.

In the one-variable case, for a given polynomial function $q(z) : \mathbb{C} \to \mathbb{C}$ with only simple roots $R = \{z \, | \, q(z) = 0\}$, the current of integration is given as a current acting on 0-forms, and the following holds:

$$[R] = \frac{1}{2\pi} \Delta \log |q|. \tag{7.55}$$

The above theorem of Bedford and Smillie is called the *convergence theorem of currents*.

7.5.4.7 Some Properties Derived from Convergence Theorem of Currents

Before discussing what this theorem implies, we present some important properties which will be used to derive natures of tunneling orbits.

Theorem (Bedford–Smillie)

1. *For any unstable periodic orbit p, $\overline{W^s(p)} = J^+$ and $\overline{W^u(p)} = J^-$ hold.*
2. *The measure μ is mixing and a hyperbolic measure.*
3. *$\{Unstable\ periodic\ points\} = J^*$.*

Here, the measure μ is said to be the hyperbolic measure if characteristic exponents (Lyapunov exponents) satisfy $\Lambda_1 > 0 > \Lambda_2$. As proved in Section 7.5.2, for an arbitrary unstable periodic orbit p, $\overline{W^s(p)} = J^+$ and $\overline{W^u(p)} = J^-$ hold only when the system satisfies the horseshoe condition. The above statement asserts that the same is true in the complex Hénon map *irrespective of the nonlinear parameter a*. That is, even if the system has mixed phase space in which KAM invariant circles and chaotic sea coexist in the real plane, beautiful rules $\overline{W^s(p)} = J^+$ and $\overline{W^u(p)} = J^-$ always hold.

Combining the fact that $\mathrm{supp}\,\mu^\pm = J^\pm$ and $\overline{W^s(p)} = J^+$ (or $\overline{W^u(p)} = J^-$), we have an intuitive interpretation for the convergent theorem, $\lim_{n\to\infty}(1/2^n)[P^{+n}M] = c\mu^-$: it tells us that any algebraic manifold M tends to $J^- = \overline{W^u(p)}$ as $n \to \infty$. To be more concrete, choose an unstable periodic orbit p in the real chaotic sea, and denote the unstable manifold of p confined in $\mathbb{R}^2$ by $W^u(p)|_{\mathbb{R}^2}$. Since $J^- = \overline{W^u(p)}$, $W^u(p)|_{\mathbb{R}^2}$ is a subset of J^-. We further assume that $M|_{\mathbb{R}^2}$ is confined in the KAM region and cannot escape from the regular region within the real dynamics. Even though $M|_{\mathbb{R}^2}$ is trapped in the regular region, the complex manifold M can get out and approach $W^u(p)|_{\mathbb{R}^2}$, which fill the real chaotic sea. The convergent theorem thus suggests that dynamical processes connecting the regular and chaotic regions always exist even though dynamical barriers exist in the real plane. Below, we will discuss this point more closely by examining the nature of regular regions in the complex plane.

We comment on the relation between the Julia set $J = J^+ \cap J^-$ and the potential theoretic Julia set J^*. The former was naively introduced in analogy with the one-dimensional complex map, and the latter is defined through the relation $J^* = \mathrm{supp}\,\mu$. It is known that support J^* coincides with the Julia set J if the Hénon map is hyperbolic [44]. On the other hand, the best known result in the nonhyperbolic case is that

$$J^* \subset J. \tag{7.56}$$

7.5.5 Interior Points of Filled Julia Set

As explained in Section 7.4.1, for one-dimensional dynamics, if a neutral fixed point is linearizable, the Siegel disk appears around the neutral fixed point, and a set of Siegel disks occupies a positive area in $\mathbb{C}$. In this case, the filled Julia set has interior points. We now ask whether or not the filled Julia sets $K^\pm$ have interior points in the case of the area-preserving Hénon map. As for the non-area-preserving case, we have the following rigorous results [4]:

   i. if $|b| > 1$, then $\mathrm{vol}(K^+) = 0$.
   ii. if $|b| < 1$, then $\mathrm{vol}(K^+) = 0$ or ∞.

Recall that $|b|$ denotes the Jacobian determinant of the map P, and $\mathrm{vol}(K^+)$ the four-dimensional volume (Lebesgue measure) of K^+. The inverse of the Hénon map is also the Hénon map with Jacobian determinant $1/|b|$, thereby the above results apply for the four-dimensional volume of K^- with Jacobian determinant $1/|b|$. For $|b| = 1$, we can say $\mathrm{vol}(K^+) < \infty$ at most. It is not known whether $\mathrm{vol}(K^+) > 0$ or not. We will develop some arguments which imply $\mathrm{vol}(K^\pm) = 0$ thus $\mathrm{vol}(K) = \mathrm{vol}(K^+ \cap K^-) = 0$. $\mathrm{vol}(K) = 0$ means that the filled Julia set K has no interior points.

7.5.5.1 Linearization around a Fixed Point

If linearization is possible around a fixed point, namely, if one can find an analytic conjugation map transforming the dynamics in a neighborhood of an elliptic fixed point into its linearized version, then two-dimensional analogs of Siegel disks appear and $\mathrm{vol}(K^\pm)$ turns out to be nonzero. However, it is obvious that this type of linearization cannot be allowed in the area-preserving map. This is because we always have a pair of eigenvalues $e^{i\kappa}$ and $e^{-i\kappa}$ with $\kappa \in \mathbb{R}$, for the linearized map, which breaks the *nonresonant condition*. Here we say that the linearized matrix satisfies the nonresonant condition if the eigenvalues of the matrix κ_1 and κ_2 satisfies $\prod_{i=1}^{2} \kappa_i^{n_i} - \kappa_j \neq 0$ for $j = 1, 2$ and $(n_1, n_2) \in \mathbb{N}^2$ with $|n_1 + n_2| \geq 2$ (see more details in Herman (1987) [64]).

7.5.5.2   KAM Curves in $\mathbb{C}^2$

Analogous but more elaborate invariant circles in the area-preserving map are KAM curves. For the two-dimensional map, the existence of KAM curves can be formulated in the following way. For a given rotation number ω, the motion on the KAM curve $\mathcal{C}_\omega$, is expressed as a constant rotation in a suitable coordinate θ:

$$\sigma : \theta \mapsto \theta + 2\pi\omega \;(\mathrm{mod}\; 2\pi). \tag{7.57}$$

In order to introduce such a coordinate θ, the conjugation function φ has to satisfy a functional equation $\sigma\{\varphi(q, p)\} = \varphi\{P(q, p)\}$. The solutions of this equation can be studied perturbatively by expanding φ in the Taylor series in the strength of perturbation (or the nonlinear parameter) and in Fourier series in θ [38,80]. The resulting series is called the *Lindstedt series*. If such a series has a positive radius of convergence with respect to the complex θ-variable, then the corresponding invariant curve with a given rotation number ω survives in the real plane. Therefore, the existence of KAM curves is linked to the convergency of the expansion thus constructed. Several works have questioned up to which the expansion can be analytically extended—in other words, how singularities of the expansion appear in the complex θ-plane [80,81]. The results, although numerical, strongly suggest that natural boundaries, not isolated singularities but accumulation of singularities, prevent the analytical prolongation of the conjugating function. We will come back to this point later, instead we focus on the other side of the issue: the expansion with positive radii of convergence.

A positive radius of convergence for a given rotation number ω means that the corresponding KAM curve exists not only in the real plane but also in the complex plane. As a result of its construction, the motion on the resulting curve is also expressed as the same relation (Equation 7.57) in the θ-coordinate, and we call such curves *complex KAM curves*. Since KAM curves, either in $\mathbb{R}^2$ or in $\mathbb{C}^2$, are subsets of $K = K^+ \cap K^-$, they may form interior points of K. Here we examine how fat KAM curves are in $\mathbb{C}^2$.

First, for a given rotation number ω, if the radius of convergence of the conjugating function is positive, θ can be complexified as $\theta = \theta' + i\theta''$. Therefore, not only the motion along each KAM curve, we gain an additional one dimension in the imaginary direction. Second, due to the KAM theorem, the measure of ω on the real ω-axis, for which KAM perturbation is convergent, is positive, while KAM curves do not exist for rational ω. The latter also has positive measure (the Hausdorff dimension of KAM curves on the ω-axis lies between 0 and 1). If we add a nonzero imaginary part to ω as $\omega = \omega' + i\omega''$, then the small denominator problem disappears and the Lindstedt series converges. However, since ω is not real, the orbits cannot be quasiperiodic on the invariant structure constructed in such a way, but they spirals up with a speed ω'', hence they do not form KAM curves. In total, the volume of KAM curves amounts at most to 3. (The corresponding Hausdorff dimension is $2 + \alpha$, where $0 < \alpha < 1$.) This implies the contribution from complex KAM curves to the four-dimensional volume of $K = K^+ \cap K^-$ is null.

7.5.5.3   Numerical Evidence

More straightforward verification to see whether $\mathrm{vol}(K) = 0$ or not is to perform numerical calculations. Thanks to the existence of the filtering property of the polynomial map, one can effectively sort out the escaping orbits from the remaining ones by checking whether or not the trajectories are out of an appropriately chosen finite region. As presented in Figure 7.6, the number of remaining orbits gradually decreases. The rate is not exponentially fast but algebraically slow reflecting that the orbits move in the very close vicinity of complex KAM curves for a long while. Although the Hausdorff dimension of total complex KAM curves is less than 3, they densely fill a certain

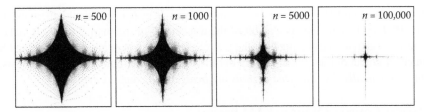

FIGURE 7.6 The slice of nonescaping points by $\operatorname{Re} q$ versus $\operatorname{Im} q$ plane. The parameter of the Hénon map is chosen as $a = 0.9$ for which KAM curves encircle an elliptic fixed point. We plot the initial points whose forward orbits are bounded up to 500, 1000, 5000, 100,000, respectively.

finite region close to the real plane. Note that, however, arbitrary chosen orbits eventually diverge to infinity.

The above speculations 1–3 are consistent with each other and imply that the filled Julia sets of the area-preserving Hénon map have null four-dimensional volume in $\mathbb{C}^2 = \mathbb{R}^4$:

$$\operatorname{vol}(K^+) = \operatorname{vol}(K^-) = \operatorname{vol}(K) = 0,$$

which leads us the following *vacant interior conjecture* [49]:

Conjecture [49]

The filled Julia sets of the area-preserving Hénon map have no interior points:

$$J^\pm = K^\pm, \quad J = K. \tag{7.58}$$

7.5.6 Fundamental Working Hypothesis

Almost all the results on the Hénon map have been derived based on potential theory, and much is known for the potential theoretic Julia set J^*, but not for J. As stated, although $J^* = J$ holds for the hyperbolic case, the best known result obtained so far is $J^* \subset J$ in generic cases. The question of whether or not $J^* = J$ also holds even in the nonhyperbolic case is entirely open. Numerical experiments cannot provide any signatures to prove or disprove it.[†] Nevertheless, in order to proceed further, we hereafter assume $J^* = J$ even in generic mixed cases. Our position is therefore to hypothesize the following two statements, and then to explore what could be deduced from them particularly on the complex orbits connecting the regular and chaotic regions. The two assumptions we put forward are

1. $J^\pm = K^\pm$ and $J = K$.
2. $J^* = J$.

The first ansatz has several supports discussed above, but the latter is genuinely mathematical.

We explain what this hypothesis implies in the context of our tunneling problem. If the hypotheses 1 and 2 are validated, we can conclude $K = J^*$. Since KAM curves are contained in K, whether they are either real or complex, KAM curves are subset of J^*. Further recall theorems claiming that $J^* = \operatorname{supp} \mu$ and the measure μ is mixing. These imply that any close neighborhood of complex KAM curves are connected with each other via some orbits on K. More precisely, for arbitrary

---

[†] Note, however, that the numerical result demonstrated in Figure 7.7 is consistent with the conjecture $J^* = J$.

neighborhoods $U(z_1)$ and $U(z_2)$ of any two points z_1 and z_2 in K respectively, there exists n such that $U(z_1) \cap P^n(U(z_2)) \neq \emptyset$. This is a result of the transitivity of the map P in $\mathbb{C}^2$. It makes a very sharp contrast to the dynamics confined in the real phase space, where any orbits confined by KAM curves can never go out of the regular region. In other words, under our working hypothesis, we can say that *any dynamically disconnected regions are bridged via complex orbits*.

7.5.7 Typical Complex Orbits in KAM Region

In Figure 7.7 we show a typical behavior of orbits which are sandwiched between complex KAM curves. An orbit launched very close to a real KAM curve follows the motion of nearby complex KAM curves initially. But, if the orbit does not lie exactly on some complex KAM curve, it gradually shifts in the imaginary direction. Here we put an initial point (q_0, p_0) such that $(\mathrm{Re}\, q_0, \mathrm{Re}\, p_0)$ is on a certain KAM circle but has very small imaginary component $|\mathrm{Im}\, q_0|, |\mathrm{Im}\, p_0| \ll 1$. An orbit so located rotates along a complex KAM curve that is closest to the initial condition but it gradually leaves the real plane, and spirals up in complex phase space. In almost all choices of initial points, however, orbits moving up in such a way diverge to infinity finally. More precisely, until a certain border, which must be associated with natural boundaries of the conjugating function, orbits rotate along the "cylindrical wall" of complex KAM curves, but once they reach the border, they quickly fly away to infinity. The fact that typical orbits behave in this way is consistent with the fact that $K^\pm$ has no interior points because arbitrarily chosen initial points in the complex plane should belong to the set of escaping points $I^\pm$.

On the other hand, if the initial point is chosen deliberately, we can find orbits such that they initially leave the real plane and tend to borders in a similar way as above, but again go back and approach the real plane. Figure 7.7 exactly demonstrates such an itineracy. An important observation is that, as shown in Figure 7.7b, once the orbit goes back close to the real plane, it rotates along the KAM curve that is different from the initially located one. Several different circles observed in Figure 7.7b is projection of an orbit belonging to different time intervals. This numerical experiment strongly suggests that KAM curves on the real domain are indeed bridged via complex orbits.

A question we should ask as the issue of dynamical tunneling would be *does the tunneling process from regular to chaotic regions utilize the transition mechanism shown here?* Before examin-

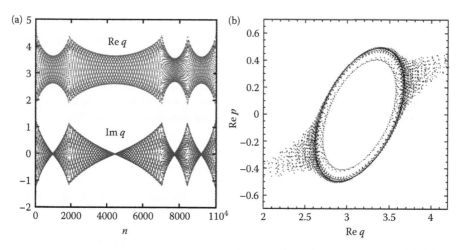

FIGURE 7.7 (a) A typical behavior of an orbit which is placed close to the real plane. $(\mathrm{Re}\, q_0, \mathrm{Re}\, p_0)$ is placed on a KAM curve on the real plane. In the itinerary, when $\mathrm{Im}\, q_n$ is almost zero, $\mathrm{Im}\, p_n$ takes also almost zero. Thus, the orbit is very close to the real plane each time focusing of $\mathrm{Im}\, q_n$ occurs. (b) Projection of an orbit on to (q_n, p_n) plane.

ing this question, we present some of our rigorous results showing that some appropriately defined tunneling orbits, with incorporating the initial and final conditions of the semiclassical propagator, are indeed closely linked to the Julia set.

7.6 TUNNELING ORBITS CONTRIBUTING TO SEMICLASSICAL PROPAGATOR

So far, we have concentrated on characteristics of the complex dynamics itself, not paying any special attentions to the boundary conditions that have to be imposed when we evaluate the semiclassical propagator. Our original question was, however, under given initial and final conditions, which were specified respectively as the manifolds $\mathcal{A}$ and $\mathcal{B}$ Section 7.2.2, to ask how the complex trajectories contributing to the semiclassical propagator $K_n^{sc}(p, p')$ can be related to some invariant objects in the complex plane. As stressed in Section 7.2.3, the orbits in a finite time interval is not compatible with any mathematically canonical objects so that we have to take the limit $n \to \infty$ to explore a link to certain invariant structures in phase space.

7.6.1 NECESSARY CONDITIONS FOR TUNNELING ORBITS

We will introduce "tunneling orbits" by taking the following two steps. Recall, first, that the set $\mathcal{M}_n^{a,b}$ represents the orbits satisfying initial and final conditions imposed on the semiclassical propagator. We now consider an asymptotic limit of $\mathcal{M}_n^{a,b}$. Note that this step itself automatically filters out irrelevant complex orbits [49]. Second, we require that relevant tunneling orbits should satisfy a necessary condition as contributors to the semiclassical sum (Equation 7.3) in the asymptotic limit. These two steps are strongly motivated by our extensive numerical investigations performed in the first paper of Shudo et al. (2009) [49]. The tunneling orbits thus introduced can be related with invariant objects whose natures are discussed in the previous sections. Since we need to take several technical substeps to follow actual procedures, which may make essential points obscure, we here present a brief sketch of the idea and main results with some related remarks. The readers can find the details in Shudo et al. (2009) [49].

The first step is to consider, instead of $\mathcal{M}_n^{a,b}$, the sequence of hypersurface

$$\mathcal{M}_n^{*,b} \equiv \{(p,q) \in \mathbb{C}^2 \mid p_n = p_b\}, \tag{7.59}$$

where $(p_n, q_n) = F^n(p, q)$. We introduce a "limit" of the set $\mathcal{M}_n^{*,b}$ in an appropriate topology (more precisely in the Hausdorff topology[†]) as

$$\mathcal{M}_\infty^b \equiv \lim_{n \to \infty} \mathcal{M}_n^{*,b} \quad \text{and} \quad \mathcal{M}_\infty \equiv \bigcup_{b \in \mathbb{R}} \mathcal{M}_\infty^b. \tag{7.60}$$

With the help of convergence theorem of currents due to Bedford–Smillie, we could prove that $\mathcal{M}_\infty$ has a definite relation with the Julia sets

$$J^+ \subset \mathcal{M}_\infty \subset K^+. \tag{7.61}$$

If the vacant interior conjecture is true, it can be rewritten simply as $\mathcal{M}_\infty = J^+$.

[†] To consider a "limit" of a sequence of sets such as $\{\mathcal{M}_n^{*,b}\}_n$ the Hausdorff distance is introduced as follows: let (X, d) be a metric space and $\mathcal{C}(X)$ be the set of all nonempty compacts in X. For A and B in $\mathcal{C}(X)$ we set

$$\delta(A, B) \equiv \inf\{\varepsilon > 0 : [A]_\varepsilon \supset B \quad \text{and} \quad [B]_\varepsilon \supset A\}$$

and call it the Hausdorff distance between A and B, where $[A]_\varepsilon$ means the ε neighborhood of A:

$$[A]_\varepsilon \equiv \{y \in X : d(x, y) < \varepsilon \text{ for some } x \in A\}.$$

The second requirement comes from the condition for the tunneling orbits to contribute to the sum (Equation 7.3) in the asymptotic limit. Note that the modulus of each contribution in the sum (Equation 7.6) is written as $|A_n^{(\ell)}|\exp\{-\operatorname{Im}S_n^{(\ell)}/\hbar\}$. The amplitude factor $|A_n^{(\ell)}|$ measures the degree of instability of each trajectory ℓ. As long as real trajectories are concerned, $S_n^{(\ell)}$ has no imaginary component, so only $|A_n^{(\ell)}|$ controls the weight of contribution. For the complex trajectories, the imaginary action $\operatorname{Im}S_n^{(\ell)}$ usually plays the major role since only a slight difference of $\operatorname{Im}S_n^{(\ell)}/\hbar$ gives rise to an exponential difference. It turns out that behavior of $\operatorname{Im}S_n^{(\ell)}$ has to be primarily taken into account.

When the imaginary action $\operatorname{Im}S_n^{(\ell)}$ diverges as $n \to \infty$, there are two possibilities: $\operatorname{Im}S_n^{(\ell)} \to +\infty$ or $\operatorname{Im}S_n^{(\ell)} \to -\infty$. The former type of orbits is negligible in amplitude, but the latter cannot be excluded. However, the divergence of imaginary action is obviously unphysical, and the orbits with $\operatorname{Im}S_n^{(\ell)} \to -\infty$ should not be contained in the semiclassical sum. Expositions about the Stokes phenomenon are necessary to explain why one can drop these contributions. Here we only mention that the Stokes phenomenon occurs in general when apply the saddle-points method, and the orbits of the latter type are to be removed as a result of the Stokes phenomenon.

In any case, both types of orbits with divergent imaginary actions do not contribute to the semiclassical propagator, and it is natural to require that "tunneling orbits" are the orbits whose imaginary actions are asymptotically finite:

$$\mathcal{C}_{\text{Laputa}} \equiv \big\{(q,p) \in \mathcal{M}_\infty \mid \operatorname{Im}S_n(q,p) \text{ converges absolutely at } (q,p)\big\}. \qquad (7.62)$$

Here the name "Laputa" originates in our previous paper [32] in which we discovered that tunneling is controlled by a particular subset in $\mathcal{M}_n^{*,b}$ which we called "Laputa chain." The "Laputa chain" float in the imaginary domain just like the Laputa island in Swift's *Gullivers' Travels*.

Important remarks are in order. The condition of absolute convergence of $\operatorname{Im}S_n$ only specifies a necessary condition for tunneling orbits. The most important issue would be to single out dominant tunneling orbit(s) with *minimal imaginary action*, which substantially controls the tunneling process. This issue will be a central matter of concern in the next section.

Our final step is then to find a connection between the tunneling orbits defined as (7.62) with the Julia set. We reached the following assertion [49].

Theorem

For the Hénon map P,

   i. *If P is hyperbolic and $h_{\text{top}}(P|_{\mathbb{R}^2}) = \log 2$, then $\mathcal{C}_{\text{Laputa}} = J^+$.*
   ii. *If P is hyperbolic and $h_{\text{top}}(P|_{\mathbb{R}^2}) > 0$, then $\overline{\mathcal{C}_{\text{Laputa}}} = J^+$.*
   iii. *If $h_{\text{top}}(P|_{\mathbb{R}^2}) > 0$, then $J^+ \subset \overline{\mathcal{C}_{\text{Laputa}}} \subset K^+$.*

Here h_{top} denotes topological entropy and $\overline{X}$ indicates the closure of X. The results obtained by Bedford and Smillie are essential to prove the theorem.

We add some remarks:

1. If the vacant interior conjecture (i.e., $J^\pm = K^\pm$, $J = K$) is true, then $\overline{\mathcal{C}_{\text{Laputa}}} = J^+$ holds in the generic case as well. Thus, we can say that tunneling orbits are dense in the forward Julia set J^+.

2. Note that $\overline{\mathcal{C}_{\text{Laputa}}} = J^+$ holds in hyperbolic (or generic) cases, whereas $\mathcal{C}_{\text{Laputa}} = J^+$ in the horseshoe situation. There indeed exist exponentially many orbits contained in $J^+ \backslash \mathcal{C}_{\text{Laputa}}$ in hyperbolic (or generic) cases. They itinerate in the complex space and do not have

convergent imaginary action. The possible role of chaos in the complex plane is discussed in Section 3.2 of Shudo et al. (2009) [49].

3. It looks somewhat unusual to consider the cases (i) and (ii) where the system is purely hyperbolic and thus no KAM circles forming the tunneling barrier exist in phase space. However, they provide an idealized situation which contains an essential ingredient of complicated aspects observed in the mixed phase space [82].

Numerical results and phenomenological arguments developed so far are all consistent with the claims stated above. In this sense, we can say that, in contrast to one-dimensional tunneling in which instanton controls the tunneling process, the orbits in the Julia set are carriers of tunneling, so the underlying dynamical mechanism is entirely different.

However, as remarked above, there is still an unsettled important issue: how to specify the most dominant complex orbit(s) out of the set $\mathcal{C}_{\text{Laputa}}$. We have to narrow down and much more sharply specify the candidates orbits and characterize them.

Before discussing this point, we will point out the importance of the boundary conditions in the time-domain propagator, choosing initial and final states from another viewpoint. The following observation is possibly helpful in order to be convinced that tunneling in nonintegrable systems differs essentially from that in the integrable one.

7.6.2 Tunneling in Integrable Map

As a special limit of the area-preserving map (Equation 7.1), let us consider the following completely integrable map:

$$F : \begin{pmatrix} p' \\ q' \end{pmatrix} = \begin{pmatrix} p + K \sin q \\ q + \omega \end{pmatrix}. \tag{7.63}$$

As shown in Figure 7.8a, the whole phase space is foliated with smooth invariant curves. We apply the same formulation of quantum unitary operator (Equation 7.3) and its semiclassical approximation (Equation 7.6).

We first see what happens if we take the momentum representation. After n time iteration, the initial manifold $\mathcal{A} \cap \mathbb{R}^2$ is evolved into $F^n(\mathcal{A} \cap \mathbb{R}^2)$ to form bended curves in the phase space. Here, $\mathcal{A} = \{(q,p) \in \mathbb{C}^2 \,|\, p = p_a \in \mathbb{R}\}$ and $\mathcal{B} = \{(q,p) \in \mathbb{C}^2 \,|\, p = p_b \in \mathbb{R}\}$. Since the phase space

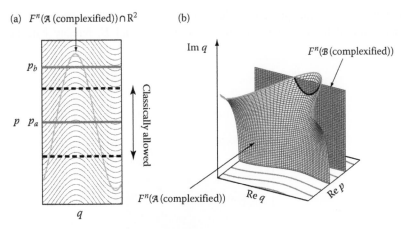

FIGURE 7.8 Invariant curves for the integrable map (Equation 7.63) (a). The upper and lower borders up to which the time-evolved classical manifold $F^n(\mathcal{A} \cap \mathbb{R}^2)$ can reach are drawn as dotted horizontal lines. A bended curve is the projection of $F^n(\mathcal{A})$ onto $\mathbb{R}^2$. The intersection between the time-evolved $F^n(\mathcal{A})$ and $\mathcal{B}$ (b). Here, $\mathcal{A} \equiv \{(q,p) \in \mathbb{C}^2 \,|\, p = p_a \in \mathbb{R}\}$ and $\mathcal{B} \equiv \{(q,p) \in \mathbb{C}^2 \,|\, p = p_b \in \mathbb{R}\}$.

is filled with invariant curves, there exist upper and lower limits in the p-direction up to which $F^n(\mathcal{A} \cap \mathbb{R}^2)$ can reach. The time-evolved manifold $F^n(\mathcal{A} \cap \mathbb{R}^2)$ cannot go beyond these borders (dotted horizontal lines in Figure 7.8). In other words, $(\mathcal{A} \cap \mathbb{R}^2) \cap F^{-n}(\mathcal{B} \cap \mathbb{R}^2) = \emptyset$ for $\forall n \in \mathbb{Z}$ if $\mathcal{B} \cap \mathbb{R}^2$ is chosen outside the classically allowed region.

On the other hand, if we extend the initial and final manifolds to the complex plane, then we can go outside classical allowed regions. As illustrated in Figure 7.8a, the projection of the manifold $F^n(\mathcal{A})$ onto $\mathbb{R}^2$ can go beyond the classical borders.

Since the map is completely solvable, not only in the real domain but also in the complex domain, we can evaluate the semiclassical propagator (Equation 7.6) analytically [32]. It turns out that the tunneling contribution in the semiclassical sum (Equation 7.6) is approximately evaluated as

$$\exp \frac{i}{\hbar} \left[S_n(p_a, p_b) \right] \approx \left(2(p_b - p_a) \right)^{\pm K_n(p_b - p_a)/\hbar}, \tag{7.64}$$

where $K_n = K \sin(\omega n/2) / \sin(\omega/2)$. This shows that for a fixed time step n, the modulus of wave-function outside the classically allowed region decays exponentially, which is an expected behavior in the classical forbidden region. The decay rate is controlled by the factor K_n, which oscillates with a frequency $\omega/2$. The oscillation reflects the underlying oscillation of the classical manifold $F^n(\mathcal{A})$. The plus and minus sign in front of the exponent means that the one solution gives the exponentially decaying contribution, and the other is exponentially divergent. The latter is to be removed as a result of the Stokes phenomenon.

Here we notice that the origin of the exponential decaying behavior as a function of $(p_b - p_a)$ is somewhat fake because the initial (resp. final) state $|p\rangle$ (resp. $|p'\rangle$) is not an eigenstate of this system but a superposition of eigenstates. In the initial state, $|p\rangle$, contains *exponentially small* components of eigenstates for classical forbidden regions. The exponential decaying behavior seen in Equation 7.64 is a result of this fact.

Our system is integrable and it has a globally analytic conserved quantity, which is nothing more than the action $I(q, p)$ (see the caption of Figure 7.9 for its analytical representation). In order to make the issue more transparent, instead of taking the momentum representation, we alternatively take two quantum states classically corresponding to the invariant curves

$$\mathcal{A} = \{ (p, q) \in \mathbb{C}^2 \,|\, I(p, q) = a \}, \quad \mathcal{B} = \{ (p, q) \in \mathbb{C}^2 \,|\, I(p, q) = b \}. \tag{7.65}$$

as initial and final states: the propagator is now $K(a, b) = \langle b | \hat{U}^n | a \rangle$. The semiclassical evaluation is then reduced to finding the orbits connecting $\mathcal{A}$ and $\mathcal{B}$, namely calculating the set

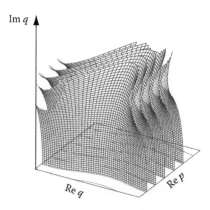

FIGURE 7.9 Complex invariant curves for the integrable model. The map (Equation 7.63) has an integral of motion of the form $M(q, p) = \omega p + (\omega K/2 \sin \omega/2) \sin(q - \omega/2)$.

$\mathcal{M}_n^{a,b} = \mathcal{A} \cap F^{-n}(\mathcal{B})$. Now, since $I(p,q)$ is classically invariant in the whole $\mathbb{C}^2$- plane, there exist no complex classical orbits connecting $\mathcal{A}$ and $\mathcal{B}$. This immediately leads to the null transition probability $K(a,b) = 0$. This is a classical interpretation to the fact that no quantum transition takes place between the different sectors of the conserved observable.

The above argument is generalized as follows. Let $K_n(\alpha, \beta) = \langle \beta | \hat{U}^n | \alpha \rangle$ be the n-step propagator using arbitrary representations $|\alpha\rangle$ and $|\beta\rangle$, and denote the corresponding classical counterparts by $\mathcal{A}$ and $\mathcal{B}$:

$$\mathcal{A} = \{ (p,q) \in \mathbb{C}^2 \mid \alpha(p,q) = a \}, \quad \mathcal{B} = \{ (p,q) \in \mathbb{C}^2 \mid \beta(p,q) = b \}. \tag{7.66}$$

Semiclassically, the support of the time-evolved wavefunction initially put on $\alpha(p,q) = a$ is given in the β-representation as

$$\operatorname{supp} K_n(\alpha, \beta) = \{ b \in \mathbb{R} \mid F^n(\mathcal{A}) \cap \mathcal{B} \neq \emptyset \}. \tag{7.67}$$

Now, we regard $\{ b \in \mathbb{R} \mid F^n(\mathcal{A} \cap \mathbb{R}^2) \cap \mathcal{B} \neq \emptyset \}$ and the remaining part as "classical" and "tunneling" components, respectively. For the completely integrable map, as shown above, there exist smooth classical manifolds which provide supports of eigenfunctions. In the time-domain picture, this is equivalent to the fact that there exist smooth or analytic initial and final manifolds that are not connected even by complex classical orbits. We may alternatively say that, in the integrable map, we can eliminate "tunneling" components to be zero if we take appropriately chosen initial and final states. Namely, there exist classical supports between which even the tunneling transition is forbidden.

In contrast, such smooth classical manifolds no more exist in the nonintegrable case. As seen below, under any representations, classical orbits inevitably connect the initial and final manifolds, and thus tunneling exists between the states whose classical counterparts are smooth in the phase space, which means that the tunneling transition is always allowed between any classical supports. The object always connecting initial and final manifolds is nothing but the Julia set.

7.7 NATURAL BOUNDARIES AND DOMINANT TUNNELING ORBITS

As mentioned in Section 7.5.5, even if KAM curves can be analytically extended to the complex plane, the domains of analyticity are limited in the nonintegrable system. Natural boundaries appear in the complex plane, and complex KAM curves can be extended only up to these boundaries. While the invariant curves are globally analytic in the completely integrable map, there appears the border

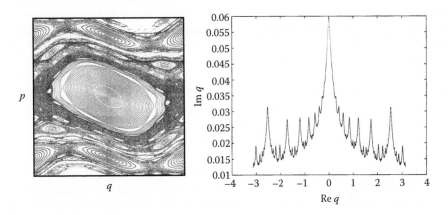

FIGURE 7.10 Typical mixed phase space for the Chirikov–Taylor standard map (left) and an example of the natural boundary (right).

of analyticity, which usually has a fractal structure in the nonintegrable system. Figure 7.10 depicts an example of the natural boundary for the Chirikov–Taylor standard map. As is analyzed using the Páde approximation, singularities of the conjugating function emerge densely to generally form a fractal curve in the complex plane [80].

The existence of natural boundaries is incompatible with a complex extension of the EBK-type torus quantization. The description of dynamical tunneling based on the EBK quantization is only possible in the completely integrable system [12,13] where the constants of motion are globally analytic. Natural boundaries make it impossible any more to construct the whole eigenstate including the tunneling part on the partly destroyed KAM invariant curve. There are reasons to believe that natural boundaries become key objects for our central question addressed in Section 7.2.3.

7.7.1 Natural Boundaries of KAM Curves and Julia Set

We first discuss a possible relation between natural boundaries and the Julia set. For the one-dimensional map, we introduced in Section 7.5.5 a conjugating function $\psi(z)$ which describes the dynamics around $z = 0$, and also another conjugating function, called the Böttcher function $\varphi(z)$, which transforms the dynamics into a simple one around $z = \infty$. Of particular importance in the present context was, as stated in Section 7.4.5, that both conjugating functions $\psi(z)$ and $\varphi(z)$ have natural boundaries at the Julia set J_P.

In the following, we assume that an analogous correspondence holds between the natural boundary of KAM curves and the Julia set in the two-dimensional case. There is some underlying reason to justify it; If the vacant interior conjecture, $J^{\pm} = K^{\pm}$ and therefore $J = K$ are true, KAM curves either in real or complex are contained in J since the orbits on KAM curves are bounded both under forward and backward iterations. Hence, we can at least say that natural boundaries of KAM curves are contained in the Julia set J. If we further admit $J = J^*$ (fundamental working hypothesis), then we may conjecture

$$\{\text{Natural boundaries of KAM curves}\} \subset J^*. \tag{7.68}$$

This implies that even though KAM curves are limited by borders of complex analyticity, the orbits can go out to the chaotic sea using the transitive property of the Julia set J^*.

7.7.2 Pseudoextension of KAM Curves

Under the working hypothesis introduced in Section 7.5.6, we can say that any dynamically regions are bridged via complex orbits in the Julia set. This means that KAM curves do not work as dynamical barriers if we extend the dynamics to the complex space. As presented in Section 7.6.1, the tunneling orbits with finite imaginary action Im S_n are shown to be dense in the Julia set.

This certainly provides a new general principle for the tunneling transport between classically disconnected regions. It is however noted that the tunneling orbits, denoted by $\mathcal{C}_{\text{Laputa}}$, were still candidates and we have not yet clarified whether the whole $\mathcal{C}_{\text{Laputa}}$ equally relevant or only some of them substantially dominates the others. As shown in Figure 7.7, typical orbits put close to the real KAM curves repeat spiral-up and spiral-down motions which usually take much time to get out to the outside chaotic sea. It is questionable whether such orbits are dominant contributors.

As mentioned above, except for the completely integrable map, any quantum and semiclassical outputs are inevitably representation dependent. We here proceed our argument by assuming *pseudoextension* of KAM curves. What we mean by pseudoextension of KAM curves is to construct an analytical curve mimicking a KAM curve as well as possible in such a way that it approximates the KAM curve below the natural boundary and can be extended analytically beyond it. This is in practice possible since KAM curves are singular only in a very thin region close to the natural boundary. In fact the Páde approximation applied to the Fourier series indicates that the singularities may be weak enough and the invariant curve can formally be extended beyond the natural boundary in the nearly integrable regime [81]. The pseudo-KAM curves can be also constructed systematically by fictitious integrable systems which well approximate the actual system in the real phase space in a similar way proposed in Bäcker (2008) [59]. But we do not enter into details about the construction of pseudo-KAM curves since our current aim to introduce them is merely to explain how "dominant tunneling orbits" are related to natural boundaries and the Julia set. It should be emphasized that we may assume that the natural boundaries of original KAM curves are close enough to such pseudoextended KAM curves.

7.7.3 Family of Dominant Tunneling Orbits

Assuming a pseudoextended KAM curve as an initial condition, we hereafter discuss what nature the dominant tunneling orbits should have. For the moment, we will not consider the final state.

A plausible principle to select out the orbits with the largest weight is reduced to finding the orbits with *smallest imaginary action*. We may expect that the longer the period of itinerary is, the larger the imaginary action of the orbit grows. If an orbit spends long time in the complex plane, it must gain larger imaginary actions. (This of course assumes no phase coherence that cancels the gained imaginary action.) If this is actually the case, the orbits, which are initially placed very close to the real plane and go out straight to the chaotic sea without any roundabout itinerary in the complex plane, would gain the minimal imaginary action. However, such "direct" paths cannot exist, but the orbits close to the real KAM curves always undergo spiral motions as illustrated in Figure 7.7.

The second best strategy to save imaginary action would be

i. To start at an edge of complex KAM curves to minimize the initial imaginary depth
ii. To go down to the real plane as fast as possible to minimize imaginary action gained in the itinerary

In the following, on the basis of mathematical results for the Hénon map, which was explained in Sections 7.5.4, we show the existence, density, and characters of the *optimal tunneling orbits*, which indeed satisfy the above conditions. As for the existence the optimal orbits,

1. There exist complex orbits which leave the natural boundary of the KAM curve and tend to the real plane exponentially fast.

The reasoning is as follows. If we admit the fundamental hypothesis, $\overline{\{\text{natural boundaries}\}}$ $\subset J^*$ holds as in Section 7.7.1. Furthermore, we have $\overline{\{\text{unstable periodic orbits}\}} = J^*$ due to the theorem of Bedford–Smillie. These imply that there exists an unstable periodic orbit P that is contained in the natural boundary of a given KAM curve. For any neighborhood of U of P, there exists an unstable periodic orbit $R \in \mathbb{R}^2$ such that $U \cap W^s(R) \neq \emptyset$. Here we have used the statement $\overline{W^s(R)} = J^+$. Such orbits on $W^s(R)$ therefore leave at an arbitrary close neighborhood of the natural boundary and tends to the real plane. Since the orbits are on the stable manifolds, they are attracted exponentially by the real unstable periodic orbits, which saves the time of flight in the complex plane and thus makes the gain of imaginary action smallest.

The orbits on the stable manifold for real unstable periodic orbits indeed satisfy the above optimal conditions (i) and (ii). However, there are many unstable periodic orbits, say R_1, $R_2, R_3 \ldots$, in the real plane in general. It is not clear whether only a single orbit with the smallest imaginary action dominates or many orbits with the same or comparable imaginary action coexist. If imaginary action for some special orbit is sufficiently small as compared to the others, it gives the dominant tunneling contribution. On the other hand, there appear the complex orbits with almost the same weights, all these should be taken into account in the semiclassical sum. We should therefore ask the density of orbits with comparable imaginary action. In this regard, we can show the following:

2. There exist exponentially many optimal orbits with comparable imaginary action.

In order to show this, take an optimal path γ, which starts at the natural boundary of a KAM curve and tends to an unstable periodic orbit, say R, on the real plane. Since $\gamma \in J^+$ due to our hypothesis and $\overline{W^s(p)} = J^+$ for any unstable periodic orbits p including R, exponentially many accompanying orbits exist γ', γ'', ..., which are approximate and very close to γ.

This statement tells us that if there exists an orbit γ which starts at a natural boundary and approaches the real plane, there is a *family*, each of which is accompanied by exponentially many complex orbits which are running very close to the reference orbit γ. We note that such a family is exactly what we called the *Laputa chain* in our previous papers [32,41,49]. The Laputa chain is a chained structure appearing in the initial set $\mathcal{M}_n^{a,b} = \mathcal{A} \cap F^{-n}(\mathcal{B})$ and it was shown that all the characteristic structures of tunneling wavefunction are linked to the Laputa chain.

In this way, the optimal orbits are attracted by unstable orbits in the real chaotic region, and approach the real plane. After reaching the real plane, they behave as if they are real orbits:

3. Optimal orbits follow almost real dynamics after reaching the real plane.

This claim is due to the following theorem which was proved in the second paper of Shudo et al. (2009) [49]:

Theorem

Let $0 < h_{\text{top}}(P|_{\mathbb{R}^2}) < \log 2$ and let $C_i (1 \leq i \leq N)$ and R be unstable periodic points in $\mathbb{C}^2 \setminus \mathbb{R}^2$ and a real unstable periodic point, respectively. Take any positive integers k_i and arbitrary neighborhood U_i of C_i $(1 \leq i \leq N)$. Then, there exists a point $P \in W^s(R)$ in $\mathbb{C}^2 \setminus \mathbb{R}^2$ such that its orbit stays in U_i at least k_i-times iterates.

To prove this theorem the fact that W^u of C_i and W^s of C_j (or R) intersects densely on J (see Bedford–Lyubich–Smillie [47]) is used. The above theorem holds also in the case where the complex saddles $C_i(1 \leq i \leq N)$ are all replaced by the real saddles $R_i(1 \leq i \leq N)$. Since we may assume that unstable periodic orbits are densely distributed in the real chaotic sea, the theorem implies the existence of an orbit on $W^s(R)$ which, after reaching the real plane, almost follows the real dynamics wandering over real chaotic sea.

Characters of the dominant tunneling orbits are illustrated in Figure 7.11 and summarized as follows:

1. Go down from the edge of KAM curve to the chaotic region along the stable manifolds
2. Attracted by the real chaotic sea (unstable periodic orbits on $\mathbb{R}^2$) exponentially
3. Move as if they are real orbits after reaching the real chaotic sea

Closing this section, we present numerical evidence for the scenario explained here. Figure 7.12 shows the projection of pseudoextesion of a KAM curve onto the real phase space and the natural boundary associated with the KAM curve. We also draw the projection of the set of the initial manifold $\mathcal{M}_n^{a,b}$, which is taken on the pseudoextension of the KAM curve, onto the real plane. As is seen, initial points of complex orbits $\mathcal{M}_n^{a,b}$ approach the natural boundary with an increase of the time step n. This reveals that the mechanism explained above works well. Notice also that the most dominant complex paths, which have the smallest imaginary action, are not necessarily

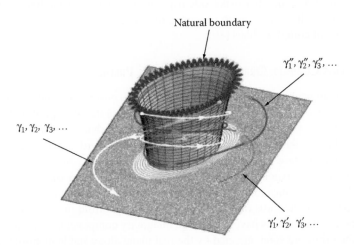

FIGURE 7.11 Families of optimal tunneling orbits (schematic). Each family, denoted by $\gamma_i', \gamma_i'', \gamma_i''' \ldots$ ($i = 1, 2, 3 \ldots$), leaves at the natural boundary and tends to each unstable orbit. After reaching the real plane, they behave as almost real orbits. We represent a complex KAM curve with a natural boundary by the cylinder.

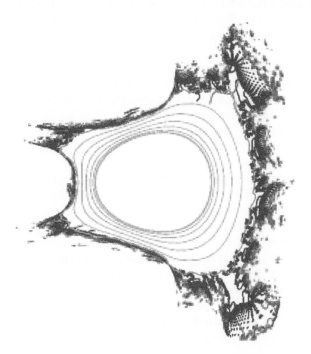

FIGURE 7.12 **(See color insert.)** Projection onto (q, p)-plane of a complexified KAM curve (yellows) and the corresponding natural boundary (watery blue) which are projected onto the real (q, p)-plane. (The yellows are the slices of complexified KAM curve at several $\mathrm{Im}\,\theta = \mathrm{const.}$) The Hénon map with $T(p) = p^2/2$ and $V(q) = \varepsilon(2q^2 + q^3/3)$ ($\varepsilon = 0.4$) is used. The sets $\mathcal{M}_n^{a,b}$ with $n = 40$ (red) and $n = 200$ (blue) are also presented. With increase of the time step n, $\mathcal{M}_n^{a,b}$ approaches the natural boundary from the outside. Yellow-green dots represent initial conditions which give the smallest imaginary action, that is, Laputa chains. The intersections between the pseudo-KAM curve and the stable manifolds along which the dominant tunneling paths approach the real plane also appear in the same positions. Color version available online.

located very close to the natural boundary. The orbits launched close to the natural boundary do not go straight to the real plane, but make side trips in the complex plane before approaching the real plane. This implies that the scenario presented here needs to be further refined by taking into account the presence of complex chaos [49].

7.7.4 AMPHIBIOUS COMPLEX ORBITS: WHAT CAN WE PREDICT?

The mixing and topological transitivity in the complex plane ensure the transition between dynamically disjointed regions in the real plane. A remarkable fact is that the spiral motion inside KAM regions, as schematically shown in Figure 7.11, and almost real dynamics in the chaotic sea is realized in the same orbit. In other words, the orbits predicted in Bedford and Smillie's theorem assume an *amphibious* character. Here we discuss what we can predict based on this observation.

As summarized above, exponentially many complex orbits, not a single instanton path as in integrable tunneling, contributes to the semiclassical propagator. Such complex orbits start from an initial manifold placed in the torus region and reach a final manifold in the chaotic sea. In the first stage of their itinerary these orbits have some large imaginary components and move in the complex plane, whereas the orbits are soon attracted by the real plane along stable manifolds and then move very close to the real plane in the later stage. The rate of the attraction is exponential, reflecting the fact that the behavior of the orbits is controlled by the stable manifolds of unstable periodic orbits. These observations imply that they act as tunneling orbits when they stay in the torus region and behave as if they are almost real orbits when reaching the chaotic sea. Furthermore, as shown above, exponentially many optimal tunneling orbits with almost equal weights appear in the semiclassical sum. A well-recognized aspect of dynamical tunneling, that is, the enhancement of tunneling, is explained in this way.

It should be noted, however, that their contributions to the tunneling amplitude are merely "potentially existing." This is because the time evolution in the later stage may involve opposite effects on the transport property, especially when *dynamical localization* takes place in the chaotic region. Note that the orbit correlation associated with dynamical localization was actually found in a chaotic system [83]. The amphibious nature of complex orbits predicts that the tunneling penetration through integrable barriers and the dynamics in the chaotic sea are not independent of each other, rather they must be intimately related.

Therefore, we expect that if the spreading of wavefunction is suppressed due to dynamical localization in the chaotic sea, which appears as a result of destructive interference effects [83], the tunneling transport is *simultaneously* suppressed. In order words, potentially existing tunneling orbits are exponentially many, but they interfere with each other to form localization.

Numerical evidence exists showing that drastic enhancement of tunneling occurs if external noise is applied to dynamically localized states [61,84]. That is, potentially existing chaotic tunneling is recovered by attenuating destructive interference generating dynamical localization. The *flooding* of eigenfunction discovered in Hufnagel et al. (2002) [85] and Bäcker (2005) [86] can be therefore interpreted as a consequence of the flooding of potentially existing tunneling orbits. Alternatively stated, *amphibious states* [85,86] appear due to the fact that complex orbits connecting regular and chaotic regions take the amphibious character.

7.8 OUTLOOK

It would not have been easy even to guess in numerical studies that complex dynamics in $\mathbb{C}^2$ has mixing or ergodicity, which was proved in potential theory of $\mathbb{C}^2$. This is not only due to the difficulty of visualization of the orbits in $\mathbb{C}^2$, but also due to the fact that the Julia set does not have positive volume in $\mathbb{C}^2$ (vacant interior conjecture) and arbitrary chosen orbits diverge to infinity. Mathematical settings or tools are somewhat elaborated and not so familiar as compared to the analysis in the real dynamics, but the results are remarkable. It is surprising that one of the most central results in the theory of complex dynamics has an exact counterpart in physical phenomena. We have not yet fully understood whether this is just accidental coincidence or not. We could, however, almost answer the first two questions we have raised in Section 7.2.3 about qualitative natures of complex orbits.

We give some remarks on the rest of questions mainly regarding the quantitative prediction. As explained in Section 7.7.3, optimal tunneling orbits form families $\gamma', \gamma'', \gamma'''$. Each family, represented as γ, is associated with an unstable periodic orbit R on $\mathbb{R}^2$. Since unstable periodic orbits on $\mathbb{R}^2$ are also exponentially many and dense in the real chaotic sea. The orbits contained in each family γ have almost comparable imaginary actions, but it is not clear what if we compare imaginary actions among different families. In order to evaluate the tunneling probability, which we assume to be able to define in some proper manner, it is necessary to know which family has the minimal imaginary action. The potential theoretic approach to complex dynamics is no more useful in this respect. It may happen that several or many families have again almost the same imaginary actions, or only some specific family controls the tunneling probability. So far, we have studied several maps [32,40,62] but not yet obtained a conclusive answer. However, the case of the Hénon map with a dominant island chains [62], and a piecewise linear map which is complexified in some proper manner [82], must offer good testing grounds for this issue.

ACKNOWLEDGMENTS

The present chapter was written based on the lectures given in the international seminar and workshop "Tunneling and Scattering in Complex Systems—From Single to Many Particle Physics" at MPIPKS, Dresden. We thank the organizers of this meeting, especially Arnd Bäcker and Peter Schlagheck for their warm hospitalities and stimulating discussions. We are also grateful to Yutaka Ishii for many helpful comments and stimulating discussions on complex dynamics in several dimensions.

REFERENCES

1. R.L. Devaney, *Introduction to Chaotic Dynamical Systems, 2nd edn.* Addison–Wesley, Redwood City, CA, 1989.
2. A.F. Beardon, *Iteration of Rational Functions: Complex Analytic Dynamical Systems*. Springer, NY, 1991.
3. J. Milnor, *Dynamics in One Complex Variable*. Princeton University Press, Princeton, 2006.
4. S. Morosawa, Y. Nishimura, M. Taniguchi, and T. Ueda, *Holomorphic Dynamics*, Cambridge University Press, Cambridge, 1999.
5. L. Carlenon and T.W. Gamelin, *Complex Dynamics*. Springer, New York, NY, 1993.
6. H.O. Peitgen and P.H. Richter, *The Beauty of Fractals*. Springer, New York, NY, 1986.
7. M.V. Berry and K.E. Mount, Semiclassical approximations in wave mechanics. *Rep. Prog. Phys.* 35, 315, 1972.
8. J.S. Langer, Statistical theory of the decay of metastable states. *Ann. Phys. NY*, 54, 258, 1969.
9. S. Coleman, Fate of the false vacuum: Semiclassical theory. *Phys. Rev. D*, 15, 2929, 1977.
10. W.H. Miller, Semiclassical theory of atom-diatom collisions: Path integrals and the classical S matrix. *J. Chem. Phys.* 53, 1949, 1970; The classical S-matrix in molecular collisions. *Adv. Chem. Phys.* 25, 69, 1974.
11. L.S. Shulman, *Techniques and Applications of Path Integration*, Wiley, New York, NY, 1981.
12. M. Wilkinson, Tunnelling between tori in phase space. *Physica D* 21, 341, 1986; M. Wilkinson and J.H. Hannay, Multidimensional tunnelling between excited states. *Physica D* 27, 201, 1987.
13. S.C. Creagh, Tunnelling in multidimensional systems. *J. Phys. A* 27, 4969, 1994.
14. J. Le Deunff and A. Mouchet, Instantons re-examined: Dynamical tunneling and resonant tunneling. 81, 046205, 2010.
15. R. Balian and C. Bloch, Solution of the Schrödinger equation in terms of classical paths. *Ann. Phys. NY* 85, 514, 1974.

16. A. Voros, The return of the quartic oscillator. The complex WKB method. *Ann. Inst. H. Poincaré A* 39, 211, 1983.

17. J. Écalle, Les Fonctions Résergentes, *Publ. Math.* Université de Paris-Sud. 3 vols, 1981.

18. C.M. Bender and T.T. Wu, Anharmonic oscillator. *Phys. Rev* 184, 1231, 1969.

19. J. Zinn-Justin, Instantons in quantum mechanics: Numerical evidence for a conjecture. *J. Math. Phys.* 25, 549, 1984.

20. H.J. Silverstone, JWKB connection-formula problem revisited via Borel summation. *Phys. Rev. Lett.* 55, 2523, 1985.

21. F. Pham, Resurgence, quantized canonical transformations, and multi-instanton expansions. *Algebraic Analysis* Vol. 2, p. 699, Academic Press, New York, NY, 1988.

22. E. Delabaere, H. Dillinger, and F. Pham, Resurgence de Voros et periodes des courbes hyperelliptiques. *Ann. Inst. Fourier (Grenoble)* 43, 163, 1993.

23. T. Kawai and T. Takei, Secular equations through the exact WKB analysis. *Algebraic Analysis of Singular Perturbation Theory*, Translations of Mathematical Monographs, AMS, 2005.

24. M.V. Berry and C.J. Howls, Hyperasymptotics. *Proc. L. Soc. London A* 430, 653, 1991; Hyperasymptotics for integrals with saddles. *Proc. L. Soc. London A* 434, 657, 1991.

25. V.I. Arnold, *Mathematical Methods of Classical Mechanics*, Springer, New York, NY, 1978.

26. M.J. Davis and E.J. Heller, Quantum dynamical tunneling in bound states. *J. Chem. Phys.* 75, 246, 1981.

27. W.A. Lin and L.E. Ballentine, Quantum tunneling and chaos in a driven anharmonic oscillator. *Phys. Rev. E* 65, 2927, 1990.

28. O. Bohigas, S. Tomsovic, and D. Ullmo, Manifestations of classical phase space structures in quantum mechanics. *Phys. Rep.* 223, 43, 1993.

29. S. Tomsovic and D. Ullmo, Chaos-assisted tunneling. *Phys. Rev. E* 50, 145, 1994.

30. E. Doron and S.D. Frischat, Semiclassical description of tunneling in mixed systems: Case of the annular billiard. *Phys. Rev. Lett.* 75, 3661, 1995; Dynamical tunneling in mixed systems. *Phys. Rev. E* 57, 1421, 1998.

31. S.C. Creagh and N.D. Whelan, Complex periodic orbits and tunneling in chaotic potentials. *Phys. Rev. Lett.* 77, 4975, 1996; Homoclinic structure controls chaotic tunneling. *Phys. Rev. Lett.* 82, 5237, 1999.

32. A. Shudo and K.S. Ikeda, Complex classical trajectories and chaotic tunneling. *Phys. Rev. Lett.* 74, 682, 1995; Chaotic tunneling: A remarkable manifestation of complex classical dynamics in non-integrable quantum phenomena. *Physica D* 115, 234, 1998.

33. O. Brodier, P. Schlagheck, and D. Ullmo, Resonance-assisted tunneling in near-integrable systems. *Phys. Rev. Lett.* 87, 064101, 2001; Resonance-assisted tunneling. *Ann. Phys.* 300, 88, 2002.

34. P. Schlagheck, C. Eltschka, and D. Ullmo, Resonance- and chaos-assisted tunneling, in: K. Yamanouchi, S.L. Chin, P. Agostini, and G. Ferrante (eds), *Progress in Ultrafast Intense Laser Science I*, 279 pp. 107–131. Springer, Berlin, 2006.

35. P. Schlagheck, A. Mouchet, and D. Ullmo, Resonance-assisted tunneling in mixed regular-chaotic systems, in: S. Keshavamurthy and P. Schlagheck (eds), *Dynamical Tunneling: Theory and Experiment*, pp. 177–210. CRC Press, Boca Raton, FL, 2011.

36. S. Löck, A. Bäcker, R. Ketzmerick, and P. Schlagheck, Regular-to-chaotic tunneling rates: From the quantum to the semiclassical regime. *Phys. Rev. Lett.* 104, 114101, 2010.

37. S.C. Creagh, Tunnelling in two dimensions, in: S. Tomsovic (ed.), *Tunneling in Complex Systems*, p. 35. World Scientific, Singapore, 1988.

38. J.M. Greene and I.C. Percival, Hamiltonian maps in the complex plane. *Physica D* 3, 540, 1982; I.C. Percival, Chaotic boundary of a Hamiltonial map. *Physica D* 6, 67, 1982.

39. K. Takahashi and K.S. Ikeda, Complex-domain semiclassical theory: Application to time-dependent barrier tunneling problems. *Found. Phys.* 31, 177, 2001; K. Takahashi, A. Yoshimoto, and K.S. Ikeda, Movable singularities, complex-domain heteroclinicity, and fringed tunneling in multi-dimensional systems. *Phys. Lett.* 297, 370, 2002; K. Takahashi and K.S. Ikeda, Complex-classical mechanism of the tunnelling process in strongly coupled 1.5-dimensional barrier systems. *J. Phys. A* 36, 7953, 2003.

40. T. Onishi, A. Shudo, K.S. Ikeda, and K. Takahashi, Tunneling mechanism due to chaos in a complex phase space. *Phys. Rev. E* 68, 025201(R), 2001; Semiclassical study on tunneling processes via complex-domain chaos. *Phys. Rev. E* 64, 056211, 2003.

41. A. Shudo, Y. Ishii, and K.S. Ikeda, Chaos attracts tunneling trajectories: A universal mechanism of chaotic tunneling. *Europhys. Lett.* 81, 5003, 2008.

42. K. Takahashi and K.S. Ikeda, An intrinsic multi-dimensional mechanism of barrier tunneling. *Europhys. Lett.* 71, 193, 2005.

43. J.H. Hubbard and R.W. Oberste-Vorth, Hénon mappings in the complex domain I: The global topology of dynamical space. *Publ. Math.* 79, 5, 1994.

44. E. Bedford and J. Smillie, Polynomial diffeomorphisms of $\mathbb{C}^2$: Currents, equilibrium measure and hyperbolicity. *Invent. Math.* 103, 69, 1991.

45. E. Bedford and J. Smillie, Polynomial diffeomorphisms of $\mathbb{C}^2$. II: Stable manifolds and recurrence. *J. Am. Math. Soc.* 4, 657, 1991.

46. E. Bedford and J. Smillie, Polynomial diffeomorphisms of $\mathbb{C}^2$. *Math. Ann.* 294, 395, 1992.

47. E. Bedford, M. Lyubich, and J. Smillie, Polynomial diffeomorphisms of $\mathbb{C}^2$. IV: The measure of maximal entropy and laminar currents. *Invent. Math.* 112, 77, 1993.

48. J. Smillie, Complex dynamics in several dimensions. *Flavors of Geometry*, MSRI publications, 31, 117, 1997.

49. A. Shudo, Y. Ishii, and K.S. Ikeda, Julia sets and chaotic tunneling: I. *J. Phys. A: Math. Theor.* 42, 265101, 2009; Julia sets and chaotic tunneling: II. *J. Phys. A: Math. Theor.* 42, 265102, 2009.

50. W.K. Hensinger, H. Haffner, A. Browaeys, N.R. Heckenberg, K. Helmerson, C. McKenzie, G.J. Milburn, W.D. Phillips, S.L. Rolston, H. Rubinsztein-Dunlop, and B. Upcroft, Dynamical tunnelling of ultracold atoms. *Nature* 412, 52, 2001.

51. C. Dembowski, H.-D. Gräf, A. Heine, A. Hofferbert, H. Rehfeld, and A. Richter, First experimental evidence for chaos-assisted tunneling in a microwave annular billiard. *Phys. Rev. Lett.* 84, 867, 2000.

52. D.A. Steck, W.H. Oskay, and M.G. Raizen, Observation of chaos-assisted tunneling between islands of stability. *Science* 293, 274, 2001.

53. A. Mouchet, C. Miniatura, R. Kaiser, B. Grémaud, and D. Delande, Chaos-assisted tunneling with cold atoms. *Phys. Rev. E* 64, 016221, 2001.

54. A. Mouchet and D. Delande, Signatures of chaotic tunneling. *Phys. Rev. E* 67, 046216, 2003.

55. S. Keshavamurthy, Dynamical tunnelling in molecules: Quantum routes to energy flow. *Int. Rev. Phys. Chem.* 26, 521, 2007.

56. M. Wojtkowski, A model problem with the coexistence of stochastic and integrable behaviour. *Comm. Math. Phys.* 80, 453, 1981.

57. L.A. Bunimovich, Mushrooms and other billiards with divided phase space. *Chaos* 11, 802, 2001.

58. A.H. Barnett and T. Betcke, Quantum mushroom billiards. *Chaos* 17, 043125, 2007.

59. A. Bäcker, R. Ketzmerick, S. Löck, and L. Schilling, Regular-to-chaotic tunneling rates using a fictitious integrable system. *Phys. Rev. Lett.* 100, 104101, 2008.

60. A. Bäcker, R. Ketzmerick, S. Löck, M. Robnik, G. Vidmar, R. Höhmann, U. Kuhl, and H.-J. Stöckmann, Dynamical tunneling in mushroom billiards. *Phys. Rev. Lett.* 100, 174103, 2008.

61. A. Ishikawa, A. Tanaka, and A. Shudo, Recovery of chaotic tunneling due to destruction of dynamical localization by external noise. *Phys. Rev. E* 80, 046204, 2009.

62. A. Shudo and K.S. Ikeda, Stokes geometry for the quantum Hénon map. *Nonlinearity*, 21, 1831, 2008.

63. J. Yoccoz, Petits diviseurs en dimension 1. *Astérisque* 231, 3, 1996; Conjugaison différentiable des difféomorphismes du cercle dont le nombre de rotation vérifie une condition diophantienne. *Ann. Sci. Ec. Norm. Sup.* 17 333, 1984; in *Lect. Notes in Math.* 1784, 125, 2002.

64. M.R. Herman, Recent results and some open questions on Siegel's linearization theorem of germs of complex analytic diffeomorphisms of $\mathbb{C}^n$ over a fixed point. in *Proc. VIII Int. Cong. Math. Phys.*, World Scientific, Singapore, 1987, p. 138.

65. H. Brolin, Invariant sets under iteration of rational functions. *Ark. Mat.* 6, 103, 1965.

66. O. Costin and M. Kruskal, Analytic methods for obstruction to integrability in discrete dynamical systems. *Comm. Pure Appl. Math.* 58, 723, 2005.

67. O. Costin and M. Huang, Behavior of lacunary series at the natural boundary. *Adv. Math.* 222, 1370, 2009.

68. D. Sullivan, Quasiconformal homeomorphisms and dynamics I. Solution of the Fatou–Julia problem on wandering domains. *Ann. Math.* 122, 401, 1985.

69. L. Bers, On Sullivan's proof of the finiteness theorem and the eventual periodicity theorem. *Am. J. Math.* 109, 833, 1987.

70. S. Friedland and J. Milnor, Dynamical properties of plane polynomial automorphisms. *Ergod Th. Dynam. Sys.* 9, 67, 1989.

71. M.J. Davis, R.S. MacKay and A. Sannami, Markov shifts in the Hénon family. *Physica D* 52, 171, 1991.

72. Z. Arai, On hyperbolic plateaus of the Hénon map. *Exp. Math.* 16, 181, 2007.

73. J. Milnor and W. Thurston, On iterated maps of the interval, in *Lecture Notes in Math.* 1342, Springer, Berlin, p. 465, 1988.

74. P. Cvitanović, G.H. Gunaratne, and I. Procaccia, Topological and metric properties of Hénon-type strange attractors. *Phys. Rev. A* 38, 1988, 1503; P. Cvitanović, Periodic orbits as the skeleton of classical and quantum chaos. *Physica D* 51, 138, 1991.

75. Y. Ishii, Towards a kneading theory for Lozi mappings I: A solution of the pruning front conjecture and the first tangency problem. *Nonlinearity* 10, 731, 1997.

76. D. Sterling, H.R. Dullin, and J.D. Meiss, Homoclinic bifurcations for the Hénon map. *Physica D* 134, 153, 1999.

77. A. de Carvalho and T. Hall, How to prune a horseshoe. *Nonlinearity*, 15, R19–68, 2002.

78. M. Klimek, *Pluripotential Theory*, London Math. Soc. Monographs. New Series, 6. Oxford Science Publ. Oxford University Press, 1991.

79. J.E. Fornæss and N. Sibony, Complex Hénon mappings in $\mathbb{C}^2$ and Fatou-Bieberbach domains. *Duke Math. J.* 65, 345, 1992.

80. A. Berretti and L. Chierchaia, On the complex analytic structure of the golden invariant curve for the standard map. *Nonlinearity* 3, 39, 1990; A. Berretti, and M Marmi, Standard map at complex rotation numbers: Creation of natural boundaries. *Phys. Rev. Lett.* 68, 1443, 1992; A. Berretti, A. Celletti, L. Chierchaia, and C. Falcolini, Natural boundaries for area-preserving twist maps. *J. Stat. Phys.* 66, 1613, 1992.

81. R. Shiromoto, A. Shudo, and K.S. Ikeda, Natural boundary, Padé mending and chaotic tunneling. In press.

82. A. Shudo and K.S. Ikeda, Chaotic tunneling in an ideal limit. In press.

83. A. Shudo and K. Ikeda, Toward the classical understanding of quantum chaological phenomena—Dynamical localization and chaotic tunneling. *Prog. Theor. Phys.* 116, 283, 1994.

84. A. Ishikawa, A. Tanaka, and A. Shudo, Quantum suppression of chaotic tunnelling. *J. Phys. A* 40, F397, 2007; Dynamical tunneling in many-dimensional chaotic systems. *Phys. Rev. Lett.* 104, 224102, 2010.

85. L. Hufnagel, R. Ketzmerick, M.-F. Otto, and H. Schanz, Eigenstates ignoring regular and chaotic phase-space structures. *Phys. Rev. Lett.* 89, 154101, 2002.

86. A. Bäcker, R. Ketzmerick, and A.G. Monastra, Flooding of chaotic eigenstates into regular phase space islands. *Phys. Rev. Lett.* 94, 054102, 2005.

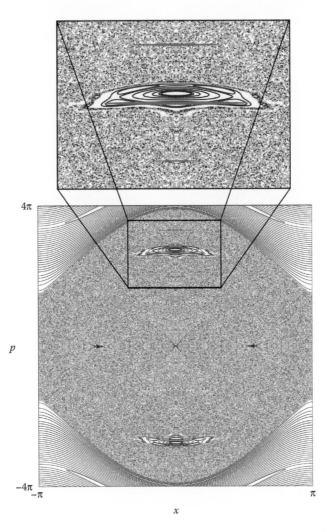

FIGURE 3.3 Phase space corresponding to the experimental conditions for the data in Figure 3.4 ($\alpha = 10.5$). A schematic representation of the atomic initial state is superimposed in red on the upper island ($\bar{k} = 2.08$), showing the subrecoil structure that we expect from the state-preparation procedure. A magnified view of the upper island and initial quantum state is also shown. Color version also available online.

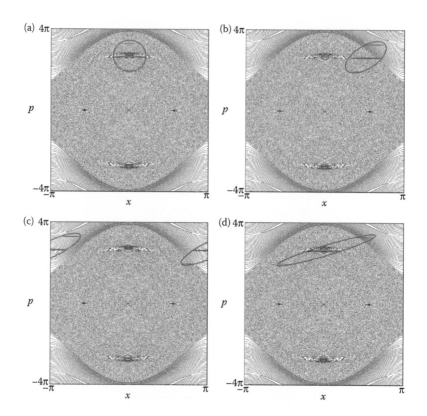

FIGURE 3.11 Initial conditions in phase space for the four time delays (a–d) used in obtaining the data of Figure 3.10. The large ellipse around the three narrow population slices in each case marks the overall profile of the wave packet to guide the eye. Color version also available online.

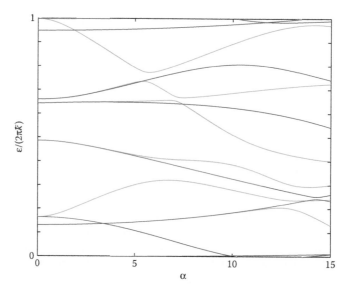

FIGURE 3.22 Calculated quasienergy spectrum for $\bar{k} = 2.077$, corresponding to a 20 μs modulation period. Quasienergies that correspond to states with large momentum (that do not interact with the states shown in this range of α) are suppressed, and the quasienergies shown are for the symmetric momentum ladder (zero quasimomentum). The quasienergies for even-parity Floquet states are shown in green, while the odd-parity states are shown in blue. The even (orange) and odd (red) states with maximal overlap with the outer stable islands are shown, up to the point where the islands bifurcate, as described in the text. The avoided-crossing behavior of the tunneling states is apparent over a broad range of α, where two chaotic states have a clear influence on the tunneling-doublet splitting. Color version also available online.

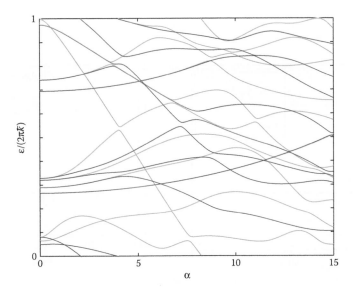

FIGURE 3.23 Calculated quasienergy spectrum for $k = 1.039$, corresponding to a 10 μs modulation period. Quasienergies that correspond to states with large momentum (that do not interact with the states shown in this range of α) are suppressed, and the quasienergies shown are for the symmetric momentum ladder (zero quasimomentum). The quasienergies for even-parity Floquet states are shown in green, while the odd-parity states are shown in blue. The even (orange) and odd (red) states with maximal overlap with the outer stable islands are shown, up to the point where the islands bifurcate, as described in the text. Several avoided crossings of the tunneling doublet with chaotic states are apparent, although the splitting only becomes very large around $\alpha = 10$. Color version also available online.

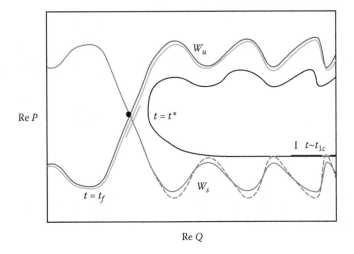

FIGURE 5.4 Brief sketch of SUMGT. Color version also available online.

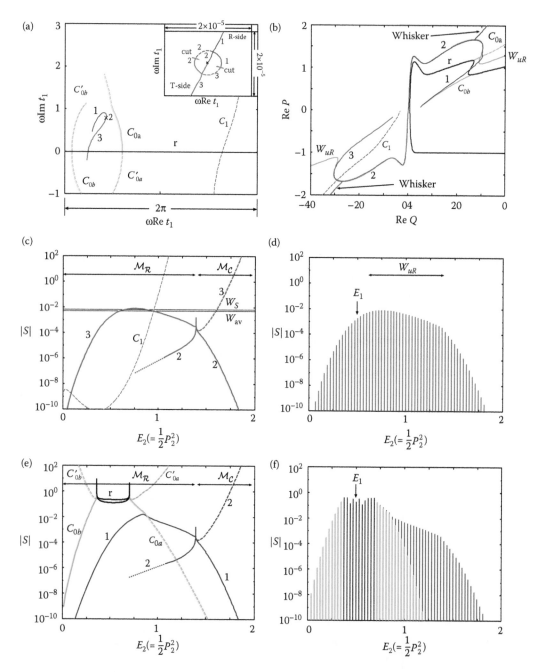

FIGURE 5.6 Semiclassical results for the strong perturbation regime at $\varepsilon = 0.4$. (a) $\mathcal{M}$-set. (b) $\mathcal{L}$-set. (c) Weights of tunneling branches. (d) Tunneling spectrum. (e) Weights of reflective branches. (f) Reflection spectrum. Color version also available online.

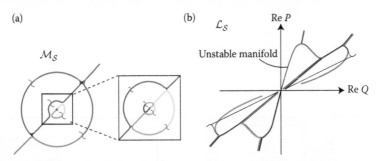

FIGURE 5.7 Correspondence between (a) $\mathcal{M}$-set and (b) $\mathcal{L}$-set. The same color specifies the corresponding parts of $\mathcal{M}_S$ and $\mathcal{L}_S$. In (b), the thin black lines, which closely overlap with colored branches of $\mathcal{L}_S$, indicate the unstable manifold W_u: the wavy curve is the real part W_{uR}, while the straight lines are parts of the purely complex unstable manifold with real energy values and guide $\mathcal{L}_C$. Color version also available online.

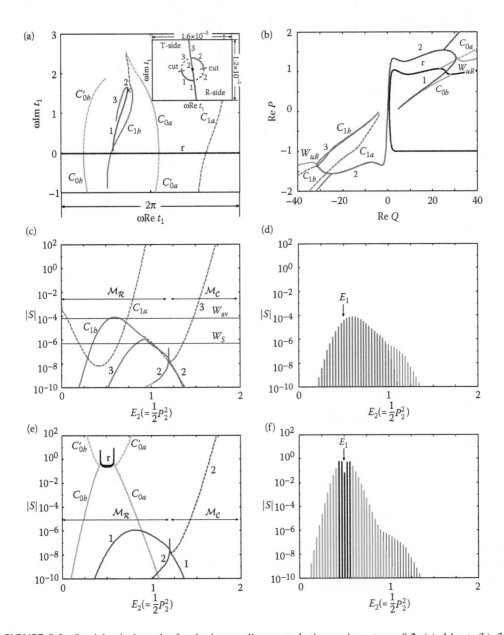

FIGURE 5.8 Semiclassical results for the intermediate perturbation regime at $\varepsilon = 0.2$. (a) $\mathcal{M}$-set. (b) $\mathcal{L}$-set. (c) Weights of tunneling branches. (d) Tunneling spectrum. (e) Weights of reflective branches. (f) Reflection spectrum. Color version also available online.

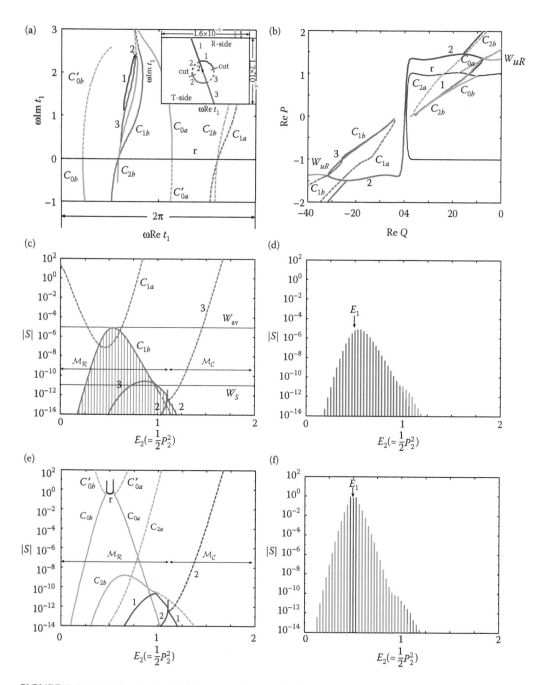

FIGURE 5.9 Semiclassical results for the weak perturbation regime at $\varepsilon = 0.1$. (a) $\mathcal{M}$-set. (b) $\mathcal{L}$-set. (c) Weights of tunneling branches. (d) Tunneling spectrum. (e) Weights of reflective branches. (f) Reflection spectrum. Color version also available online.

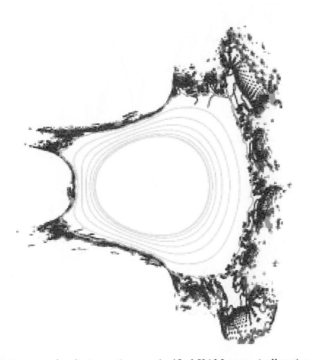

FIGURE 7.12 Projection onto (q, p)-plane of a complexified KAM curve (yellows) and the corresponding natural boundary (watery blue) which are projected onto the real (q, p)-plane. (The yellows are the slices of complexified KAM curve at several $\mathrm{Im}\,\theta$ = const.) The Hénon map with $T(p) = p^2/2$ and $V(q) = \varepsilon(2q^2 + q^3/3)$ ($\varepsilon = 0.4$) is used. The sets $\mathcal{M}_n^{a,b}$ with $n = 40$ (red) and $n = 200$ (blue) are also presented. With an increase of the time step n, $\mathcal{M}_n^{a,b}$ approaches the natural boundary from the outside. Yellow-green dots represent initial conditions which give the smallest imaginary action, that is, Laputa chains. The intersections between the pseudo-KAM curve and the stable manifolds along which the dominant tunneling paths approach the real plane also appear in the same positions. Color version also available online.

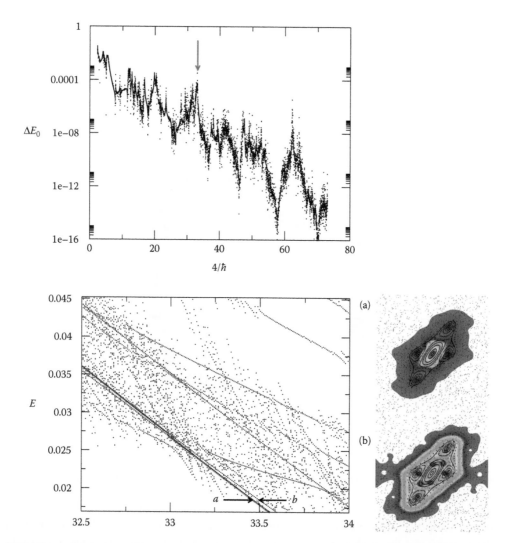

FIGURE 8.6 The upper panel shows the quasienergy splittings in $\Delta E_0 = \hbar\varphi_0$ in the kicked rotor model for $K = 2.28$, corresponding to a 4:1 classical resonance. Here we are concerned with a tunneling in q. The lower panel shows the quasienergy spectrum $E_n^{\pm} = \hbar\varphi_n^{\pm}$ of $\hat{U}$ where only states with a significant overlap with a coherent state localized around $(p,q) = (0,0)$ have been retained. The horizontal arrow (a) marks the central state doublet ($n = 0$, not resolved at that scale) and the arrow (b) indicates the third excited state localized in the island ($n = 4$). Their Husimi distribution superimposed with the Poincaré surface of section are shown in the panels on the right-hand side in order to illustrate the clear correspondence between the classical and the quantum resonance. As $1/\hbar$ increases, the crossing of the doublet by the resonant state provokes the large and wide spike indicated by the vertical arrow in the upper panel. Color version also available online.

8 Resonance-Assisted Tunneling in Mixed Regular–Chaotic Systems

Peter Schlagheck, Amaury Mouchet, and Denis Ullmo

CONTENTS

8.1 INTRODUCTION

8.1.1 TUNNELING IN INTEGRABLE SYSTEMS

Since the early days of quantum mechanics, tunneling has been recognized as one of the hallmarks of the wave character of microscopic physics. The possibility of a quantum particle to penetrate an energetic barrier represents certainly one of the most spectacular implications of quantum theory and has led to various applications in nuclear, atomic, and molecular physics as well as in mesoscopic science. Typical scenarios in which tunneling manifests are the escape route of a quantum particle from a quasibounded region, the transition between two or more symmetry-related, but classically disconnected wells (which we shall focus on in the following), as well as scattering or transport through potential barriers. The spectrum of scenarios becomes even richer when the concept of tunneling is generalized to any kind of classically forbidden transitions in phase space, that is, to transitions that are not necessarily inhibited by static potential barriers but by some other constraints of the underlying classical dynamics (such as integrals of motion). Such "dynamical tunneling" processes arise frequently in molecular systems [1] and were realized with cold atoms propagating in periodically modulated optical lattices [2–4]. Moreover, the electromagnetic analog of dynamical tunneling was also obtained with microwaves in billiards [5].

Despite its genuinely quantal nature, tunneling is strongly influenced by the structure of the underlying classical phase space (see Creagh (1998) [6] for a review). This is best illustrated within the textbook example of a one-dimensional symmetric double-well potential. In this simple case, the eigenvalue problem can be straightforwardly solved with the standard Jeffreys–Wentzel– Kramers–Brillouin (JWKB) ansatz [7]. The eigenstates of this system are, below the barrier height, obtained by the symmetric and antisymmetric linear combination of the local "quasimodes" (i.e., of the wave functions that are semiclassically constructed on the quantized orbits within each well, without taking into account the classically forbidden coupling between the wells), and the splitting of their energies is given by an expression of the form

$$\Delta E = \frac{\hbar \Omega}{\pi} \exp \left[-\frac{1}{\hbar} \int \sqrt{2m(V(x) - E)} \, dx \right]. \tag{8.1}$$

Here E is the mean energy of the doublet, $V(x)$ represents the double-well potential, m is the mass of the particle, Ω denotes the oscillation frequency within each well, and the integral in the exponent is performed over the whole classically forbidden domain, that is, between the inner turning points of the orbits in the two wells. Preparing the initial state as one of the quasimodes (i.e., as the even or odd superposition of the symmetric and the antisymmetric eigenstate), the system will undergo Rabi oscillations between the wells with the frequency $\Delta E/\hbar$. The "tunneling rate" of this system is therefore given by the splitting (Equation 8.1). Keeping all classical parameters fixed, it decreases exponentially with $1/\hbar$, and, in that sense, one can say that tunneling "vanishes" in the classical limit.

8.1.2 Chaos-Assisted Tunneling

The approach presented in the previous section can be generalized to multidimensional, even non-separable systems, as long as their classical dynamics is still integrable [8]. It breaks down, however, as soon as a nonintegrable perturbation is added to the system, for example, if the one-dimensional double-well potential is exposed to a driving that is periodic in time (with period τ, say). In that case, the classical phase space of the system generally becomes a mixture of both regular and chaotic structures.

As visualized by the stroboscopic Poincaré section—which is obtained by retaining the phase space coordinates at every integer multiple of the driving period τ—the phase space typically displays two prominent regions of regular motion, corresponding to the weakly perturbed dynamics within the two wells, and a small (or, for stronger perturbations, large) layer of chaotic dynamics that separates the two regular islands from each other. Numerical calculations of model systems in the early nineties [9,10] have shown that the tunnel splittings in such mixed systems generally become strongly enhanced compared to the integrable limit. Moreover, they do no longer follow a smooth exponential scaling with $1/\hbar$ as expressed by Equation 8.1, but display huge, quasierratic fluctuations when $\hbar$ or any other parameter of the system varies [9,10].

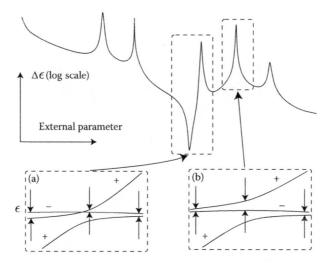

FIGURE 8.1 The two elementary scenarios of huge enhancement (case b) or cancellation (case a) of the tunneling splitting between symmetric (+) and antisymmetric (−) states can be easily understood from the resonant crossing of a third level (here a symmetric one) and the corresponding level repulsion between states of the same symmetry class. The external parameter that triggers the fluctuation of tunneling can be of quantum origin (effective $\hbar$) or classical.

These phenomena are traced back to the specific role that *chaotic* states play in such systems [11–14]. In contrast to the integrable case, the tunnel doublets of the localized quasimodes are, in a mixed regular–chaotic system, no longer isolated in the spectrum, but resonantly interact with states that are associated with the chaotic part of phase space. Due to their delocalized nature in phase space,* such chaotic states typically exhibit a significant overlap with the boundary regions of both regular wells. Therefore, they may provide an efficient coupling mechanism between the quasimodes—which becomes particularly effective whenever one of the chaotic levels is shifted exactly on resonance with the tunnel doublet. As illustrated in Figure 8.1, this coupling mechanism generally enhances the tunneling rate, but may also lead to a complete suppression thereof, arising at specific values of $\hbar$ or other parameters [16].

We point out that this type of resonant tunneling does not necessarily require the presence of classical chaos and may appear also in integrable systems, for instance in a one-dimensional symmetric triple-well potential. This is illustrated in Figure 8.2, which shows the scaling of level splittings associated with the two lateral wells as a function of $1/\hbar$ for such a triple-well potential. On top of an exponential decrease according to Equation 8.1, the splittings display strong spikes occurring whenever the energy of a state localized in the central well becomes quasidegenerate with the energies of the states in the two lateral wells.

For mixed regular–chaotic systems, the validity of this "chaos-assisted" tunneling picture was essentially confirmed by successfully modeling the chaotic part of the quantum dynamics with a random matrix from the Gaussian orthogonal ensemble (GOE) [11,12,17]. Using the fact that the coupling coefficients between the regular states and the chaotic domain are small, this random matrix ansatz yields a truncated Cauchy distribution for the probability density to obtain a level splitting of the size ΔE. Such a distribution is indeed encountered in the exact quantum splittings, which was demonstrated for the two-dimensional quartic oscillator [17] as well as, later on, for the driven pendulum Hamiltonian that describes the tunneling process of cold atoms in periodically modulated

---

* We assume here that the effects of dynamical localization observed in Ishikawa et al. (2009) [15] remain irrelevant.

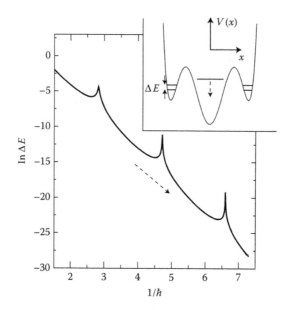

FIGURE 8.2 Resonant tunneling at work for a one-dimensional triple-well potential (here $V(x) = (x^2 - a^2)^2(x^2 - b^2)$ with $a = 1.75$ and $b = .5$). When $1/\hbar$ is increased, tunneling between the two lateral wells is enhanced by several orders of magnitude each time a level in the central well crosses down the doublet.

optical lattices [4,18]. A quantitative prediction of the *average* tunneling rate, however, was not possible in the above-mentioned theoretical works. As we shall argue later on in Section 8.3, this average tunneling rate is directly connected to the coupling matrix element between the regular and the chaotic states, and the strength of this matrix element was unknown and introduced in an *ad hoc* way.

A first step toward this latter problem was undertaken by Podolskiy and Narimanov [19] who proposed an explicit semiclassical expression for the mean tunneling rate in a mixed system by assuming a perfectly clean, harmonic-oscillator like dynamics within the regular island and a structureless chaotic sea outside the outermost invariant torus of the island. This expression turned out to be successful for the reproduction of the level splittings between near-degenerate optical modes that are associated with a pair of symmetric regular islands in a nonintegrable microcavity [19,20]. The application to dynamical tunneling in periodically modulated optical lattices [19], for which splittings between the left- and the right-moving stable eigenmodes were calculated in Mouchet et al. (2001) [4], seems convincing for low and moderate values of $1/\hbar$, but reveals deviations deeper in the semiclassical regime where plateau structures arise in the tunneling rates. Further, and more severe, deviations were encountered in the application of this approach to tunneling processes in other model systems [21].

Bäcker, Ketzmerick, Löck, and coworkers [22,23] recently undertook the effort to derive more rigorously the regular-to-chaotic coupling rate governing chaos-assisted tunneling. Their approach is based on the construction of an integrable approximation for the nonintegrable system, designed to accurately describe the motion within the regular islands under consideration. The coupling rate to the chaotic domain is then determined through the computation of matrix elements of the Hamiltonian of the system within the eigenbasis of this integrable approximation [22]. This results in a smooth exponential-like decay of the average tunneling rate with $1/\hbar$, which was indeed found to be in very good agreement with the exact tunneling rates for quantum maps and billiards [22,23]. Those systems, however, were designed such as to yield a "clean" mixed regular–chaotic phase space, containing a regular island and a chaotic region which both do not exhibit appreciable substructures [22,23].

8.1.3 Role of Nonlinear Resonances

In more generic systems, such as the quantum kicked rotor or the driven pendulum [4], however, even the "average" tunneling rates do not exhibit a smooth monotonous behavior with $1/\hbar$, but display peaks and plateau structures that cannot be accounted for by the above approaches. To understand the origin of such plateaus, it is instructive to step back to the conceptually simpler case of *nearly integrable* dynamics, where the perturbation from the integrable Hamiltonian is sufficiently small such that macroscopically large chaotic layers are not yet developed in the Poincaré surface of section. In such systems, the main classical phase space features due to the perturbation consist in chain-like substructures that surround stable periodic orbits or equilibrium points of the classical motion. Those substructures come from *nonlinear resonances* between the internal degrees of freedom of the system or, for driven systems, between the external driving and the unperturbed oscillations around the central orbit. In a similar way as for the quantum pendulum Hamiltonian, such resonances induce additional tunneling paths in the phase space, which lead to couplings between states that are located near the *same* stable orbit [24,25].

The relevance of this effect for the near-integrable tunneling process between two symmetry-related wells was first pointed out by Bonci et al. [26], who argued that such resonances may lead to a strong enhancement of the tunneling rate, due to couplings between lowly and highly excited states within the well which are permitted by near-degeneracies in the spectrum. In Brodier et al. (2001) [27] and Brodier et al. (2002) [28], a quantitative semiclassical theory of near-integrable tunneling was formulated on the basis of this principal mechanism. This theory allows one to reproduce the exact quantum splittings from purely classical quantities and takes into account high-order effects such as the coupling via a sequence of different resonance chains [27,28]. More recent studies by Keshavamurthy on classically forbidden coupling processes in model systems that mimic the dynamics of simple molecules confirm that the resonance-assisted tunneling scenario prevails not only in one-dimensional systems that are subject to a periodic driving (such as the kicked Harper model studied in Brodier et al. (2001) [27] and Brodier et al. (2002) [28]), but also in autonomous systems with two and even three degrees of freedom [29,30].

In Eltschka and Schlagheck (2005) [31], Schlagheck et al. (2006) [32], Mouchet et al. (2006) [33], and Wimberger et al. (2006) [34] resonance-assisted couplings were incorporated in an approximate manner into the framework of chaos-assisted tunneling in order to provide a quantitative theory for the regular-to-chaotic coupling rate. In this context, it is assumed that the dominant coupling between regular states within and chaotic states outside the island is provided by the presence of a nonlinear resonance within the island. A straightforward implementation of this idea yields good agreement with the exact tunneling rates as far as their average decay with $1/\hbar$ in the deep semiclassical limit is concerned. Moreover, individual plateaus and peak structures could be traced back to the influence of specific nonlinear resonances, not only for double-well-type tunneling in closed or periodic systems [31–33], but also for tunneling-induced decay in open systems [34]. However, the predictive power of this method was still rather limited, insofar as individual tunneling rates at given system parameters could be over- or underestimated by many orders of magnitude. In particular, resonance-assisted tunneling seemed inapplicable in the "quantum" limit of large $\hbar$, where direct regular-to chaotic tunneling proved successful [22,23].

A major advance in this context was achieved by improving the semiclassical evaluation of resonance-induced coupling processes in mixed systems, and by combining it with "direct" regular-to-chaotic tunneling [35]. This combination resulted, for the first time, in a semiclassical prediction of tunneling rates in generic mixed regular–chaotic systems that can be compared with the exact quantum rates on the level of individual peak structures [35]. This confirms the expectation that nonlinear resonances do indeed form the "backbone" behind nonmonotonous substructures in tunneling rates. It furthermore suggests that those rates could, also in systems with more degrees of freedom, possibly be estimated in a quantitatively satisfactory manner via simple classical computations, based on the most prominent nonlinear resonances that are manifested within the regular island.

It is in the spirit of this latter expectation that this contribution has been written. Our aim is not to formulate a formal semiclassical theory of tunneling in mixed systems, which still represents an open problem that would rather have to be solved on the basis of complex classical orbits [36–38]. Instead, we want to provide a simple, easy-to-implement, yet effective prescription on how to compute the rates and time scales associated with tunneling processes solely on the basis of the classical dynamics of the system, without performing any diagonalization (not even any application) of the quantum Hamiltonian or of the time evolution operator. This prescription is based on chaos- and resonance-assisted tunneling in its improved form [35]. The main part of this contribution is, therefore, devoted to a detailed description of resonance-assisted tunneling and its combination with chaos-assisted tunneling in the Sections 8.2 and 8.3, respectively. We present in Section 8.4 the application of this method to tunneling processes in the quantum kicked rotor, and discuss possible limitations and future prospects in the conclusion in Section 8.5.

8.2 THEORY OF RESONANCE-ASSISTED TUNNELING

8.2.1 SECULAR PERTURBATION THEORY

For our study, we restrict ourselves to systems with one degree of freedom that evolve under a periodically time-dependent Hamiltonian $H(p,q,t) = H(p,q,t + \tau)$. We suppose that, for a suitable choice of parameters, the classical phase space of H is mixed regular–chaotic and exhibits two symmetry-related regular islands that are embedded within the chaotic sea. This phase-space structure is most conveniently visualized by a stroboscopic Poincaré section, where p and q are plotted at the times $t = n\tau (n \in \mathbb{Z})$. Such a Poincaré section typically reveals the presence of chain-like substructures within the regular islands, which arise due to nonlinear resonances between the external driving and the internal oscillation around the island's center. Before considering the general situation for which many resonances may come into play in the tunneling process, we start with the simpler case where the two islands exhibit a prominent $r{:}s$ resonance, that is, a nonlinear resonance where s internal oscillation periods match r driving periods and r subislands are visible in the stroboscopic section.

The classical motion in the vicinity of the $r{:}s$ resonance is approximately integrated by secular perturbation theory [39] (see also Brodier et al. (2002) [28]). For this purpose, we formally introduce a time-independent Hamiltonian $H_0(p,q)$ that approximately reproduces the regular motion in the islands and preserves the discrete symmetry of H. In some circumstances, as for instance if H is in the nearly integrable regime, $H_0(p,q)$ can be explicitly computed within some approximation scheme (using for instance the Lie transformation method [39]). We stress though that this will not always be necessary. Assuming the existence of such a H_0, the phase space generated by this integrable Hamiltonian consequently exhibits two symmetric wells that are separated by a dynamical barrier and "embed" the two islands of H. In terms of the action-angle variables (I, θ) describing the dynamics within each of the wells, the total Hamiltonian can be written as

$$H(I, \theta, t) = H_0(I) + V(I, \theta, t), \tag{8.2}$$

where V would represent a weak perturbation in the center of the island.*

The nonlinear $r{:}s$ resonance occurs at the action variable $I_{r{:}s}$ that satisfies the condition

$$r\Omega_{r{:}s} = s\omega, \tag{8.3}$$

with $\omega = 2\pi/\tau$ and

$$\Omega_{r{:}s} \equiv \left. \frac{dH_0}{dI} \right|_{I=I_{r{:}s}}. \tag{8.4}$$

---

* In order not to overload the notation, we use the same symbol H for the Hamiltonian in the original phase-space variables (p,q) and in the action-angle variables (I,θ).

We now perform a canonical transformation to the frame that corotates with this resonance. This is done by leaving I invariant and modifying θ according to

$$\theta \mapsto \vartheta = \theta - \Omega_{r:s}t. \tag{8.5}$$

This time-dependent shift is accompanied by the transformation $H \mapsto \mathcal{H} = H - \Omega_{r:s}I$ in order to ensure that the new corotating angle variable ϑ is conjugate to I. The motion of I and ϑ is therefore described by the new Hamiltonian

$$\mathcal{H}(I,\vartheta,t) = \mathcal{H}_0(I) + \mathcal{V}(I,\vartheta,t) \tag{8.6}$$

with

$$\mathcal{H}_0(I) = H_0(I) - \Omega_{r:s}I, \tag{8.7}$$
$$\mathcal{V}(I,\vartheta,t) = V(I,\vartheta + \Omega_{r:s}t,t). \tag{8.8}$$

The expansion of $\mathcal{H}_0$ in powers of $I - I_{r:s}$ yields

$$\mathcal{H}_0(I) \simeq \mathcal{H}_0^{(0)} + \frac{(I-I_{r:s})^2}{2m_{r:s}} + \mathcal{O}\left[(I-I_{r:s})^3\right] \tag{8.9}$$

with a constant $\mathcal{H}_0^{(0)} \equiv H_0(I_{r:s}) - \Omega_{r:s}I_{r:s}$ and a quadratic term that is characterized by the effective "mass" parameter $m_{r:s} \equiv [d^2H_0/dI^2(I_{r:s})]^{-1}$. Hence, $d\mathcal{H}_0/dI$ is comparatively small for $I \simeq I_{r:s}$, which implies that the corotating angle ϑ varies slowly in time near the resonance. This justifies the application of adiabatic perturbation theory [39], which effectively amounts, in first order, to replacing $\mathcal{V}(I,\vartheta,t)$ by its time average over r periods of the driving (using the fact that $\mathcal{V}$ is periodic in t with the period $r\tau$).[†] We therefore, obtain, after this transformation, the time-independent Hamiltonian

$$\mathcal{H}(I,\vartheta) = \mathcal{H}_0(I) + \mathcal{V}(I,\vartheta) \tag{8.10}$$

with

$$\mathcal{V}(I,\vartheta) \equiv \frac{1}{r\tau}\int_0^{r\tau} \mathcal{V}(I,\vartheta,t)dt. \tag{8.11}$$

By expanding $V(I,\theta,t)$ in a Fourier series in both θ and t—that is,

$$V(I,\theta,t) = \sum_{l,m=-\infty}^{\infty} V_{l,m}(I)e^{il\theta}e^{im\omega t} \tag{8.12}$$

with $V_{l,m}(I) = [V_{-l,-m}(I)]^*$, one can straightforwardly derive

$$\mathcal{V}(I,\vartheta) = V_{0,0}(I) + \sum_{k=1}^{\infty} 2V_k(I)\cos(kr\vartheta + \phi_k), \tag{8.13}$$

defining

$$V_k(I)e^{i\phi_k} \equiv V_{rk,-sk}(I); \tag{8.14}$$

that is, the resulting time-independent perturbation term is $(2\pi/r)$-periodic in ϑ.

For the sake of clarity, we start discussing the resulting effective Hamiltonian neglecting the action dependence of the Fourier coefficients of $\mathcal{V}(I,\vartheta)$. We stress that this dependence can be implemented in a relatively straightforward way using Birkhoff–Gustavson normal-form

---

[†] This step involves, strictly speaking, another time-dependent canonical transformation $(I,\vartheta) \mapsto (\tilde{I},\tilde{\vartheta})$ which slightly modifies I and ϑ (see also Brodier et al. (2002) [28]).

coordinates (cf. Section 8.2.3); it is actually important in order to obtain a good quantitative accuracy. For now, however, we replace $V_k(I)$ by $V_k \equiv V_k(I = I_{r:s})$ in Equation 8.13. Neglecting furthermore the term $V_{0,0}(I)$, we obtain the effective integrable Hamiltonian

$$H_{\text{res}}(I, \vartheta) = H_0(I) - \Omega_{r:s}I + \sum_{k=1}^{\infty} 2V_k \cos(kr\vartheta + \phi_k) \tag{8.15}$$

for the description of the classical dynamics in the vicinity of the resonance. We shall see in Section 8.2.2 that the parameters of H_{res} relevant to the tunneling process can be extracted directly from the classical dynamics of $H(t)$, which is making Equation 8.15 particularly valuable.

8.2.2 PENDULUM APPROXIMATION

The quantum implications due to the presence of this nonlinear resonance can be straightforwardly inferred from the direct semiclassical quantization of H_{res}, given by

$$\hat{H}_{\text{res}} = H_0(\hat{I}) - \Omega_{r:s}\hat{I} + \sum_{k=1}^{\infty} 2V_k \cos(kr\hat{\vartheta} + \phi_k). \tag{8.16}$$

Here we introduce the action operator $\hat{I} \equiv -i\hbar\partial/\partial\vartheta$ and assume antiperiodic boundary conditions in ϑ in order to properly account for the Maslov index in the original phase space [24]. In accordance with our assumption that the effect of the resonance is rather weak, we can now apply quantum perturbation theory to the Hamiltonian (Equation 8.16), treating the $\hat{I}$-dependent "kinetic" terms as unperturbed part and the $\hat{\vartheta}$-dependent series as perturbation. The unperturbed eigenstates are then given by the (anti-periodic) eigenfunctions $\langle\vartheta|n\rangle = (2\pi)^{-1/2}\exp[i(n+1/2)\vartheta]$ ($n \geq 0$) of the action operator $\hat{I}$ with the eigenvalues

$$I_n = \hbar(n+1/2). \tag{8.17}$$

As is straightforwardly evaluated, the presence of the perturbation induces couplings between the states $|n\rangle$ and $|n+kr\rangle$ with the matrix elements

$$\langle n+kr|\hat{H}_{\text{res}}|n\rangle = V_k e^{i\phi_k} \tag{8.18}$$

for strictly positive integer k. As a consequence, the "true" eigenstates $|\psi_n\rangle$ of $\hat{H}_{\text{res}}$ contain admixtures from unperturbed modes $|n'\rangle$ that satisfy the selection rule $|n'-n| = kr$ with integer k. They are approximated by the expression

$$|\psi_n\rangle = |n\rangle + \sum_k \frac{\langle n+kr|\hat{H}_{\text{res}}|n\rangle}{E_n - E_{n+kr} + ks\hbar\omega}|n+kr\rangle$$
$$+ \sum_{k,k'} \frac{\langle n+kr|\hat{H}_{\text{res}}|n+k'r\rangle}{E_n - E_{n+kr} + ks\hbar\omega} \frac{\langle n+k'r|\hat{H}_{\text{res}}|n\rangle}{E_n - E_{n+k'r} + k's\hbar\omega}|n+kr\rangle + \cdots, \tag{8.19}$$

where $E_n \equiv H_0(I_n)$ denote the unperturbed eigenenergies of H_0 and the resonance condition (Equation 8.3) is used. The summations in Equation 8.19 are generally finite due to the finiteness of the phase- space area covered by the regular region.

Within the quadratic approximation of $H_0(I)$ around $I_{r:s}$, we obtain from Equations 8.7 and 8.9

$$E_n \simeq H_0(I_{r:s}) + \Omega_{r:s}(I_n - I_{r:s}) + \frac{1}{2m_{r:s}}(I_n - I_{r:s})^2. \tag{8.20}$$

This results in the energy differences

$$E_n - E_{n+kr} + ks\hbar\omega \simeq \frac{1}{2m_{r:s}}(I_n - I_{n+kr})(I_n + I_{n+kr} - 2I_{r:s}). \tag{8.21}$$

From this expression, we see that the admixture between $|n\rangle$ and $|n'\rangle$ becomes particularly strong if the $r{:}s$ resonance is symmetrically located between the two tori that are associated with the actions I_n and $I_{n'}$—that is, if $I_n + I_{n'} \simeq 2I_{r{:}s}$. The presence of a significant nonlinear resonance within a region of regular motion provides therefore an efficient mechanism to couple the local "ground state"—that is, the state that is semiclassically localized in the center of that region (with action variable $I_0 < I_{r{:}s}$)—to a highly excited state (with action variable $I_{kr} > I_{r{:}s}$).

It is instructive to realize that the Fourier coefficients V_k of the perturbation operator decrease rather rapidly with increasing k. Indeed, one can derive under quite general circumstances the asymptotic scaling law

$$V_k \sim (kr)^\gamma V_0 \exp[-kr\Omega_{r{:}s}t_{\mathrm{im}}(I_{r{:}s})] \tag{8.22}$$

for large k, which is based on the presence of singularities of the complexified tori of the integrable approximation $H_0(I)$ (see Equation 66 in Brodier et al. (2002) [28]). Here, $t_{\mathrm{im}}(I)$ denotes the imaginary time that elapses from the (real) torus with action I to the nearest singularity in complex phase space, γ corresponds to the degree of the singularity, and V_0 contains information about the corresponding residue near the singularity as well as the strength of the perturbation. The expression (Equation 8.22) is of little practical relevance as far as the concrete determination of the coefficients V_k is concerned. It permits, however, to estimate the relative importance of different perturbative pathways connecting the states $|n\rangle$ and $|n + kr\rangle$ in Equation 8.19. Comparing, for example, the amplitude $\mathcal{A}_2$ associated with a single step from $|n\rangle$ to $|n + 2r\rangle$ via V_2 and the amplitude $\mathcal{A}_1$ associated with two steps from $|n\rangle$ to $|n + 2r\rangle$ via V_1, we obtain from Equations 8.21 and 8.22 the ratio

$$\mathcal{A}_2/\mathcal{A}_1 \simeq \frac{2^{\gamma-1}r^{2-\gamma}\hbar^2}{m_{r{:}s}V_0}e^{i(\phi_2-2\phi_1)} \tag{8.23}$$

under the assumption that the resonance is symmetrically located in between the corresponding two tori (in which case we would have $I_{n+r} \simeq I_{r{:}s}$). Since V_0 can be assumed to be finite in mixed regular–chaotic systems, we infer that the second-order process via the stronger coefficient V_1 will more dominantly contribute to the coupling between $|n\rangle$ and $|n + 2r\rangle$ in the semiclassical limit $\hbar \to 0$.

A similar result is obtained from a comparison of the one-step process via V_k with the k-step process via V_1, where we again find that the latter more dominantly contributes to the coupling between $|n\rangle$ and $|n + kr\rangle$ in the limit $\hbar \to 0$. We therefore conclude that in mixed regular–chaotic systems the semiclassical tunneling process is adequately described by the lowest nonvanishing term of the sum over the V_k contributions, which in general is given by $V_1 \cos(r\vartheta + \phi_1)$.* Neglecting all higher Fourier components V_k with $k > 1$ and making the quadratic approximation of H_0 around $I = I_{r{:}s}$, we finally obtain an effective pendulum-like Hamiltonian

$$H_{\mathrm{res}}(I,\vartheta) \simeq \frac{(I - I_{r{:}s})^2}{2m_{r{:}s}} + 2V_{r{:}s}\cos(r\vartheta + \phi_1) \tag{8.24}$$

with $V_{r{:}s} \equiv V_1$ [31].

This simple form of the effective Hamiltonian allows us to determine the parameters $I_{r{:}s}$, $m_{r{:}s}$, and $V_{r{:}s}$ from the Poincaré map of the classical dynamics, without explicitly using the transformation to the action-angle variables of H_0. To this end, we numerically calculate the monodromy matrix $M_{r{:}s} \equiv \partial(p_f, q_f)/\partial(p_i, q_i)$ of a stable periodic point of the resonance (which involves r iterations of the stroboscopic map) as well as the phase space areas $S_{r{:}s}^+$ and $S_{r{:}s}^-$ that are enclosed by the outer and inner separatrices of the resonance, respectively (see also Figure 8.3). Using the fact that the trace

---

* Exceptions from this general rule typically arise in the presence of discrete symmetries that, for example, forbid the formation of resonance chains with an odd number of subislands and therefore lead to $V_1 = 0$ for an $r{:}s$ resonance with an odd r.

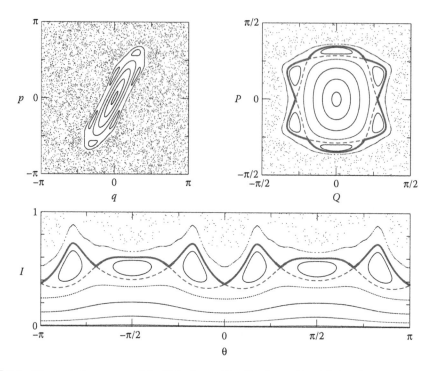

FIGURE 8.3 Classical phase space of the kicked rotor Hamiltonian at $K = 3.5$ showing a regular island with an embedded 6:2 resonance. The phase space is plotted in the original (p,q) coordinates (upper left panel), in approximate normal-form coordinates (P,Q) defined by Equations 8.28 and 8.29 (upper right panel), and in approximate action-angle variables (I,ϑ) (lower panel). The thick solid and dashed lines represent the "outer" and "inner" separatrix of the resonance, respectively.

of $M_{r:s}$ as well as the phase space areas $S_{r:s}^{\pm}$ remain invariant under the canonical transformation to (I,ϑ), we infer

$$I_{r:s} = \frac{1}{4\pi}(S_{r:s}^{+} + S_{r:s}^{-}), \tag{8.25}$$

$$\sqrt{2m_{r:s}V_{r:s}} = \frac{1}{16}(S_{r:s}^{+} - S_{r:s}^{-}), \tag{8.26}$$

$$\sqrt{\frac{2V_{r:s}}{m_{r:s}}} = \frac{1}{r^2\tau}\arccos(\mathrm{tr}M_{r:s}/2) \tag{8.27}$$

from the integration of the dynamics generated by H_{res} [40]. Equations 8.25 through 8.27 make it possible to derive the final expressions for the tunneling rates directly from the dynamics of $H(t)$, without explicitly having to construct the integrable approximation H_0 and making the Fourier analysis of $V(p,q,t) = H(p,q,t) - H_0(p,q)$. As this construction of the integrable approximation may turn out to be highly nontrivial in the mixed regime, avoiding this step is actually an essential ingredient to make the approach we are following practical. We note though that improving the quadratic approximation in Equation 8.24 for H_0 is sometimes necessary, but this does not present any fundamental difficulty.

8.2.3 ACTION DEPENDENCE OF COUPLING COEFFICIENTS

Up to now, and in our previous publications [27,28,31–33], we completely neglected the action dependence of the coupling coefficients $V_k(I)$. This approximation should be justified in the semiclassical limit of extremely small $\hbar$, where resonance-assisted tunneling generally involves multiple coupling processes [28] and transitions across individual resonance chains are therefore expected to take place in their immediate vicinity in action space. For finite $\hbar$, however, the replacement $V_k(I) \mapsto V_k(I_{r:s})$, permitting the direct quantization in action-angle space, is, in general, not sufficient to obtain an accurate reproduction of the quantum tunneling rates. We show now how this can be improved.

To this end, we make the general assumption that the classical Hamiltonian $H(p,q,t)$ of our system is analytic in p and q in the vicinity of the regular islands under consideration. It is then possible to define an analytical canonical transformation from (p,q) to Birkhoff–Gustavson normal-form coordinates (P,Q) [41,42] that satisfy

$$P = -\sqrt{2I}\sin\theta, \tag{8.28}$$

$$Q = \sqrt{2I}\cos\theta, \tag{8.29}$$

and that can be represented in power series in p and q. The "unperturbed" integrable Hamiltonian H_0 therefore depends only on $I = (P^2 + Q^2)/2$.

Writing

$$e^{\pm il\theta} = \left(\frac{Q \mp iP}{\sqrt{2I}}\right)^l \tag{8.30}$$

for positive l, we obtain, from Equation 8.12, the series

$$V(I,\theta,t) = \sum_{m=-\infty}^{\infty} \left\{ V_{0,m}(I) + \sum_{l=1}^{\infty} \frac{1}{\sqrt{2I}^l} \left[V_{l,m}(I)(Q-iP)^l + V_{-l,m}(I)(Q+iP)^l \right] \right\} e^{im\omega t} \tag{8.31}$$

for the perturbation. Using the fact that $V(I,\theta,t)$ is analytic in P and Q, we infer that $V_{l,m}(I)$ must scale at least as $I^{l/2}$. By virtue of Equation 8.14, this implies the scaling $V_k(I) \propto I^{rk/2}$ for the Fourier coefficients of the time-independent perturbation term that is associated with the $r:s$ resonance. Making the ansatz $V_k(I) \equiv I^{rk/2}\tilde{v}_k$ (and neglecting the residual action dependence of $\tilde{v}_k$), we rewrite Equation 8.13 as[*]

$$\mathcal{V}(I,\vartheta) = V_{0,0}(I) + \sum_{k=1}^{\infty} \frac{\tilde{v}_k}{2^{kr/2}} \left[(Q-iP)^{kr}e^{i\phi_k} + (Q+iP)^{kr}e^{-i\phi_k} \right]. \tag{8.32}$$

Each term in the sum is given by the well-known Birkhoff normal form generically describing a $r \geq 3$ resonance when the bifurcation of the stable periodic orbit is controlled by one single parameter [see, e.g., Equation 4.70 in Chapter 4 of Ozorio de Almeida (1988) [43] or Equations 3.3.17 and 3.3.18 in Leboeuf and Mouchet (1999) [44] for another simple derivation of the action dependence of $V_k(I)$]. The term $V_{0,0}(I)$ is neglected in the following as it does not lead to any coupling between different unperturbed eigenstates in the quantum system.

One comment is in order here. The small parameter in the adiabatic approximation (Equation 8.10) is the difference $(I - I_{r:s})$, while in the derivation of Equation 8.32 we have neglected higher powers of I and thus assumed the action I itself to be small (strictly speaking we should work near one bifurcation only). We thus mix a development near the resonant torus with one near the center of the island. This may eventually become problematic if (1) the resonance chain is located far away from the center of the island, in which case the associated coupling strength may contain a nonnegligible relative error when being computed via the assumption $V_k(I) \equiv I^{rk/2}\tilde{v}_k$ with constant $\tilde{v}_k$, and if (2) that coupling strength happens to appear rather often in the main perturbative chain

---

[*] This involves, strictly speaking, another canonical transformation of P and Q to the frame that co-rotates with the resonance.

that connects the quasimodes of the island to the chaotic sea, which generally would be the case for low-order resonances with relatively small r.* Otherwise, we expect that this inconsistency in the definition of the regimes of validity of our perturbative approach does not lead to a significant impact on the numerical values of the semiclassical tunneling rates, which indeed seems to be confirmed by numerical evidence to be discussed below.

This being said, the quantization of the resulting classical Hamiltonian can now be carried out in terms of the "harmonic oscillator" variables P and Q and amounts to introducing the standard ladder operators $\hat{a}$ and $\hat{a}^\dagger$ according to

$$\hat{a} = \frac{1}{\sqrt{2\hbar}}(\hat{Q} + i\hat{P}), \tag{8.33}$$

$$\hat{a}^\dagger = \frac{1}{\sqrt{2\hbar}}(\hat{Q} - i\hat{P}). \tag{8.34}$$

This yields the quantum Hamiltonian

$$\hat{H}_{\text{res}} = H_0(\hat{I}) - \Omega_{r:s}\hat{I} + \sum_{k=1}^{\infty} \tilde{v}_k \hbar^{kr/2} \left[\hat{a}^{kr} e^{-i\phi_k} + (\hat{a}^\dagger)^{kr} e^{i\phi_k} \right] \tag{8.35}$$

with $\hat{I} \equiv \hbar(\hat{a}^\dagger \hat{a} + 1/2)$. As for Equation 8.16, perturbative couplings are introduced only between unperturbed eigenstates $|n\rangle$ and $|n'\rangle$ that exhibit the selection rule $|n' - n| = kr$ with integer k. The associated coupling matrix elements are, however, different from Equation 8.18 and read

$$\langle n + kr | \hat{H}_{\text{res}} | n \rangle = \tilde{v}_k \sqrt{\hbar}^{kr} e^{i\phi_k} \sqrt{\frac{(n+kr)!}{n!}}$$

$$= V_k(I_{r:s}) e^{i\phi_k} \left(\frac{\hbar}{I_{r:s}} \right)^{kr/2} \sqrt{\frac{(n+kr)!}{n!}} \tag{8.36}$$

for strictly positive k. Close to the resonance, that is, more formally, taking the semiclassical limit $n \to \infty$ keeping k and $\delta = (I_{r:s}/\hbar - n)$ fixed, and making use of the Stirling formula $n! \simeq \sqrt{2\pi n}(n/e)^n$, Equation 8.36 reduces to $V_k(I_{r:s})e^{i\phi_k}$. The difference becomes, on the other hand, particularly pronounced if the $r{:}s$ resonance is, in phase space, rather asymmetrically located in between the invariant tori that correspond to the states $|n\rangle$ and $|n + kr\rangle$—that is, if $I_{r:s}$ is rather close to I_n or to I_{n+kr}. In that case, Equation 8.18 may, respectively, strongly over- or underestimate the coupling strength between these states.

8.2.4 MULTIRESONANCE PROCESSES

Up to this point, we considered the couplings between quasimodes generated by a given resonance. In general, however, several of them may play a role for the coupling to the chaotic sea, giving rise to multiresonance transitions across subsequent resonance chains in phase space [27,28]. As was argued in the context of near-integrable systems [28], such multiresonance processes are indeed expected to dominate over couplings involving only one single resonance in the deep semiclassical limit $\hbar \to 0$.

The description of the coupling process across several consecutive resonances requires a generalization of Equation 8.19 describing the modified eigenstate due to resonance-induced admixtures. We restrict ourselves, for this purpose, to including only the first-order matrix elements $\langle n + r | \hat{H}_{\text{res}}^{(r:s)} | n \rangle$ for each resonance (i.e., only the matrix elements with $k = 1$ in Equation 8.36).

---

* Corrections to the form (Equation 8.32) should also arise in the presence of prominent *secondary* resonances, which occur when primary resonances start to overlap and create chains of sub-islands nested inside the primary island chains.

For the sake of clarity, we further consider the particular case of coupling processes that start in the lowest locally quantized eigenmode with node number $n = 0$ (the generalization to initial $n \neq 0$ being straightforward). The prescription we use is to consider that, although the approximation (Equation 8.15) is valid for only one resonance at a time, it is possible to sum the contributions obtained from different resonances. Considering a sequence of consecutive $r{:}s$, $r'{:}s'$, $r''{:}s''$ ... resonances, we obtain in this way

$$
|\psi_0\rangle = |0\rangle + \sum_{k>0} \left(\prod_{l=1}^{k} \frac{\langle lr|\hat{H}_{\text{res}}^{(r:s)}|(l-1)r\rangle}{E_0 - E_{lr} + ls\hbar\omega} \right)
$$

$$
\times \left\{ |kr\rangle + \sum_{k'>0} \left(\prod_{l'=1}^{k'} \frac{\langle kr + l'r'|\hat{H}_{\text{res}}^{(r':s')}|kr + (l'-1)r'\rangle}{E_0 - E_{kr+l'r'} + (ks + l's')\hbar\omega} \right) \right.
$$

$$
\times \left[|kr + k'r'\rangle + \sum_{k''>0} \left(\prod_{l''=1}^{k''} \frac{\langle kr + k'r' + l''r''|\hat{H}_{\text{res}}^{(r'':s'')}|kr + k'r' + (l''-1)r''\rangle}{E_0 - E_{kr+k'r'+l''r''} + (ks + l's' + l''s'')\hbar\omega} \right) \right.
$$

$$
\left. \left. \times \left(|kr + k'r' + k''r''\rangle + \sum_{k'''>0} \cdots \right) \right] \right\}
\tag{8.37}
$$

for the modified "ground state" within the island. Given an excited quasimode n far from the interior of the island, the overlap $\langle n|\psi_0\rangle$ obtained from Equation 8.37 will in most cases be exponentially dominated by one or a few contributions. There is no systematic way to identify them *a priori*, although some guiding principle can be used in this respect [28].

Quite naturally, for instance, low-order resonances, with comparatively small r and s, will, in general, give larger contribution than high-order resonances with comparable winding numbers s/r but larger r and s, due to the strong differences in the sizes of the mean coupling matrix elements $V_{r:s}$ (see, e.g., Equation 8.22). In the same way, sequences of couplings involving small denominators, that is, energy differences like Equation 8.21 that are close to zero, and thus intermediate steps symmetrically located on each side of a resonance, will tend to give larger contributions. In the small $\hbar$ limit this will tend to favor multiresonance processes. Conversely, for intermediate values of $\hbar$ (in terms of the area of the regular region) the main contributions can be obtained from the lowest-order resonances. With few exceptions—especially concerning low-order resonances that are located close to the center of the island, thereby leading to relatively large energy denominators and small admixtures—this rule is generally observed for the semiclassical calculation of the eigenphase splittings we shall consider in Section 8.4.

8.3 COMBINATION WITH CHAOS-ASSISTED TUNNELING

We now discuss the implication of such nonlinear resonances on the tunneling process between the two symmetry-related regular islands under consideration. In the quantum system, these islands support (for not too large values of $\hbar$) locally quantized eigenstates or "quasimodes" with different node numbers n, which, due to the symmetry, have the same eigenvalues in both islands. In our case of a periodically driven system, these eigenvalues can be the eigenphases φ_n of the unitary time evolution (Floquet) operator $\hat{U}$ over one period τ of the driving, or, alternatively, the quasienergies E_n such that $\varphi_n = -E_n\tau/\hbar$ (modulo 2π).

The presence of a small (tunneling-induced) coupling between the islands lifts the degeneracy of the eigenvalues and yields the symmetric and antisymmetric linear combination of the quasimodes in the two islands as "true" eigenstates of the system. A nonvanishing splitting $\Delta\varphi_n \equiv |\varphi_n^+ - \varphi_n^-|$ consequently arises between the eigenphases $\varphi_n^\pm$ of the symmetric and the antisymmetric state, which is related to the splitting $\Delta E_n \equiv |E_n^+ - E_n^-|$ of the quasienergies $E_n^\pm$ through $\Delta\varphi_n = \tau\Delta E_n/\hbar$.

8.3.1 RESONANCE-ASSISTED TUNNELING IN NEAR-INTEGRABLE SYSTEMS

We start by considering a system in the nearly integrable regime. In that case, we can assume the presence of a (global) integrable Hamiltonian $H_0(p,q)$ that describes the dynamics in the entire phase space to a very good approximation.* The energy splittings for the corresponding quantum Hamiltonian $\hat{H}_0 \equiv H_0(\hat{p},\hat{q})$ can be semiclassically calculated via an analytic continuation of the invariant tori to the complex domain [8]. This generally yields the splittings

$$\Delta E_n^{(0)} = \frac{\hbar \Omega_n}{\pi} \exp(-\sigma_n/\hbar) \tag{8.38}$$

(up to a numerical factor of order one), where Ω_n is the classical oscillation frequency associated with the nth quantized torus and σ_n denotes the imaginary part of the action integral along the complex path that joins the two symmetry-related tori.

The main effect of nonlinear resonances in the nonintegrable system is, as was discussed in the previous subsections, to induce perturbative couplings between quasimodes of different excitation within the regular islands. For the nearly integrable systems this can already lead to a substantial enhancement of the splittings ΔE_n as compared to Equation 8.38 [27,28]. As can be derived within quantum perturbation theory, the presence of a prominent $r{:}s$ resonance modifies the splitting of the local "ground state" in the island (i.e., the state with vanishing node number $n = 0$) according to

$$\Delta\varphi_0 = \Delta\varphi_0^{(0)} + \sum_{k=1}^{k_c} |\mathcal{A}_{kr}^{(r:s)}|^2 \Delta\varphi_{kr}^{(0)} \tag{8.39}$$

(using $\Delta\varphi_n^{(0)} \gg \Delta\varphi_0^{(0)}$ for $n > 0$), where $\mathcal{A}_{kr}^{(r:s)} \equiv \langle kr|\psi_0\rangle$ denotes the admixture of the krth excited unperturbed component $|kr\rangle$ to the perturbed ground state $|\psi_0\rangle$ according to Equation 8.19 (possibly using Equation 8.36 instead of Equation 8.18). The maximal number k_c of coupled states is provided by the finite size of the island according to

$$k_c = \left[\frac{1}{r}\left(\frac{\text{area of the island}}{2\pi\hbar} - \frac{1}{2}\right)\right], \tag{8.40}$$

where the bracket stands for the integer part. The rapid decrease of the amplitudes $\mathcal{A}_{kr}^{(r:s)}$ with k is compensated by an exponential increase of the unperturbed splittings $\Delta\varphi_{kr}^{(0)}$, arising from the fact that the tunnel action σ_n in Equation 8.38 generally decreases with increasing n. The maximal contribution to the modified ground state splitting is generally provided by the state $|kr\rangle$ for which $I_{kr} + I_0 \simeq 2I_{r:s}$—that is, which in phase space is most closely located to the torus that lies symmetrically on the opposite side of the resonance chain. This contribution is particularly enhanced by a small energy denominator (see Equation 8.21) and typically dominates the sum in Equation 8.39.

As one goes further in the semiclassical $\hbar \to 0$ limit, a multiresonance process is usually the dominant one. Neglecting interference terms between different coupling pathways that connect the ground state with a given excited mode $|n\rangle$ (which is justified due to the fact that the amplitudes associated with those coupling pathways are, in general, much different from each other in size), we obtain from Equation 8.37 an expression of the form

$$\Delta\varphi_0 = \Delta\varphi_0^{(0)} + \sum_k |\mathcal{A}_{0,kr}^{(r:s)}|^2 \Delta\varphi_{kr}^{(0)} + \sum_k \sum_{k'} |\mathcal{A}_{0,kr}^{(r:s)}|^2 |\mathcal{A}_{kr,kr+k'r'}^{(r':s')}|^2 \Delta\varphi_{kr+k'r'}^{(0)} + \cdots \tag{8.41}$$

---

* Formally, this Hamiltonian is not identical with the unperturbed approximation $H_0(I)$ introduced in Section 8.2.1 as the definition of the latter is restricted to one well only. It is obvious, however, that $H_0(I)$ can be determined from $H_0(p,q)$, for example, by means of the Lie transformation method [39].

with the coupling amplitudes

$$A_{0,kr}^{(r:s)} = \prod_{l=1}^{k} \frac{\langle lr|\hat{H}_{\text{res}}^{(r:s)}|(l-1)r\rangle}{E_0 - E_{lr} + ls\hbar\omega} \tag{8.42}$$

$$A_{kr,kr+k'r'}^{(r':s')} = \prod_{l'=1}^{k'} \frac{\langle kr + l'r'|\hat{H}_{\text{res}}^{(r':s')}|kr + (l'-1)r'\rangle}{E_0 - E_{kr+l'r'} + (ks + l's')\hbar\omega} \tag{8.43}$$

$$A_{kr+k'r',kr+k'r'+k''r''}^{(r'':s'')} = \cdots$$

for the eigenphase splitting.

8.3.2 Coupling with Chaotic Sea

Turning now to the mixed regular–chaotic case, the integrable Hamiltonian $H_0(I)$ provides a good approximation only near the center of the regular island under consideration, and invariant tori exist only up to a maximum action variable I_c corresponding to the outermost boundary of the regular island in phase space. Beyond this outermost invariant torus, multiple overlapping resonances provide various couplings and pathways such that unperturbed states in this regime can be assumed to be strongly connected to each other. Under such circumstances, it is natural to divide the Hilbert space into two parts, integrable and chaotic, associated respectively with the phase-space regions within and outside the regular island.

For each symmetry class $\pm$ of the problem, let us introduce an effective Hamiltonian $\hat{H}_{\text{eff}}^{\pm}$ modeling the tunneling process. Let us furthermore denote $\hat{P}_{\text{reg}}$ and $\hat{P}_{\text{ch}}$ the (orthogonal) projectors onto the regular and chaotic Hilbert spaces. The diagonal blocks $\hat{H}_{\text{reg}}^{\pm} \equiv \hat{P}_{\text{reg}}\hat{H}_{\text{eff}}^{\pm}\hat{P}_{\text{reg}}$ and $H_{\text{ch}}^{\pm} \equiv \hat{P}_{\text{ch}}\hat{H}_{\text{eff}}^{\pm}\hat{P}_{\text{ch}}$ receive a natural interpretation: within $\hat{H}_{\text{reg}}^{\pm}$, on the one hand, the dynamics is exactly the same as in the nearly integrable regime above; $\hat{H}_{\text{ch}}^{\pm}$, on the other hand, is best modeled in a statistical manner by the introduction of random matrix ensembles. The only remaining delicate point is thus to connect the two, namely to model the off-diagonal block $\hat{P}_{\text{reg}}\hat{H}_{\text{eff}}^{\pm}\hat{P}_{\text{ch}}$. We stress that there is not yet a real consensus on the best way how to do this, although various approaches give good quantitative accuracy.

To state more clearly the problem, let us consider a regular state $|\bar{n}\rangle$ with quasienergy $E_{\bar{n}}^0$ close to the regular-chaos boundary. (Note that "close" here means that no resonance within the island can connect $|\bar{n}\rangle$ to a state $|n'\rangle$ within the island with $n' > \bar{n}$. This notion of "closeness" to the boundary is therefore $\hbar$-dependent.) The resonance assisted mechanism will connect any quasimode deep inside the island to such a state at the edge of the island. But to complete the description of the chaos-assisted tunneling process it is necessary to compute the variance $v_{\bar{n}}^2$ of the random matrix elements $v_{\bar{n}i}$ between $|\bar{n}\rangle$ and the eigenstates $|\psi_i^c\rangle$ of $\hat{H}_{\text{ch}}^{\pm}$ (the variance is independent of i if $\hat{H}_{\text{ch}}^{\pm}$ is modeled by the Gaussian orthogonal or unitary ensemble).

One possible approach to compute this quantity is the fictitious integrable system approach that was proposed by Bäcker et al. [22]. This method relies on the fact that, for the effective Hamiltonian $\hat{H}_{\text{eff}}$, the "direct" transition rate from a regular state $|n\rangle$ to the chaotic region is given using the Fermi's golden rule by

$$\Gamma_{n\rightarrow\text{chaos}}^d = \frac{2\pi}{\hbar} \frac{v_n^2}{\Delta_{\text{ch}}}, \tag{8.44}$$

(see, e.g., Section 5.2.2 of Bohigas et al. (1993) [11] for a discussion in the context of random matrix theory) where Δ_{ch} denotes the mean spacing between eigenenergies within the chaotic block. As a consequence, one obtains, in first order in $\tau v_n^2/\hbar\Delta_{\text{ch}}$,

$$||\hat{P}_{\text{ch}}\hat{U}|n\rangle||^2 = \tau\Gamma_{n\rightarrow\text{chaos}}^d = \frac{2\pi\tau}{\hbar} \frac{v_n^2}{\Delta_{\text{ch}}} \tag{8.45}$$

(see also Chapter 6 of this book). If one can explicitly construct a good integrable approximation $H_{reg} \equiv H_0(p,q)$ of the time-dependent dynamics (see Section 8.2.1), this allows one, by quantum or semiclassical diagonalization, to determine the unperturbed eigenstates $|n\rangle$ within the regular island, and to construct the projectors $\hat{P}_{reg}$ and $\hat{P}_{ch}$. The "direct" regular-to-chaotic tunneling matrix elements of the nth quantized state within the island is then evaluated by a simple application of the quantum time evolution operator $\hat{U}$ over one period of the driving.

This approach can be qualified as "seminumerical," as it requires to numerically perform the quantum evolution for one period of the map (although this can be done by analytical and semi-classical techniques in some cases [23], see also the contribution of Bäcker et al. in this book). It strongly relies on the quality of the integrable approximation $\hat{H}_{reg}$ of the Hamiltonian. If the latter was really diagonal (which, as a matter of principle, cannot be achieved by means of classical perturbation theory, due to the appearance of nonlinear resonances), Equation 8.45 would represent an exact result (apart from the first-order approximation in $\tau v_n^2/\hbar\Delta_{ch}$ which should not be a limitation in the tunneling regime). And indeed very good agreement between this prediction and numerically computed tunneling rates was found for quantum maps that were designed such as to yield a "clean" mixed regular–chaotic phase space, containing a regular island and a chaotic region which both do not exhibit appreciable substructures [22], as well as for the mushroom billiard [23]. In more generic situations, where nonlinear resonances are manifested within the regular island, this approach yields reliable predictions for the direct tunneling of regular states at the regular-chaos border, and its combination with the resonance assisted mechanism described in the previous sections leads to good quantitative predictions for the tunneling rates for the states deep in the regular island [35]. We shall illustrate this on the example of the kicked rotor system in Section 8.4.

8.3.3 INTEGRABLE SEMICLASSICAL MODELS FOR REGULAR-TO-CHAOTIC COUPLING

For now, however, we shall discuss other possible approaches of purely semiclassical nature (i.e., not involving any numerical evolution nor diagonalization of the quantum system) to the calculation of the coupling parameter $v_{\bar{n}}$ for a regular state $|\bar{n}\rangle$ at the edge of the regular-chaos boundary.

One way to obtain an order of magnitude of $v_{\bar{n}}^2$ is to consider the decay of the quasimode inside the regular island. For this purpose, let us assume that an integrable approximation $H_0(I)$, valid up to a maximum action I_c corresponding to the chaos boundary, has been obtained. Within the Birkhoff–Gustavson normal-form coordinates (P,Q) given by Equations 8.28 and 8.29, H_0 appears as a function of the harmonic-oscillator Hamiltonian $(P^2 + Q^2)/2$ and therefore, has the same eigenstates

$$\langle Q|\,\bar{n}\rangle = \frac{1}{\sqrt{2^{\bar{n}}\bar{n}!}}\frac{1}{(\pi\hbar)^{1/4}}\exp\left(-Q^2/2\hbar\right)\mathcal{H}_{\bar{n}}(Q/\hbar^{1/2}) \tag{8.46}$$

$$\simeq \frac{1}{2\sqrt{2\pi|P_{\bar{n}}(Q)|}}\exp\left(-\int_{Q_1}^{Q}|P_{\bar{n}}(Q)|\mathrm{d}Q\right)\qquad\text{for}\quad Q > Q_n, \tag{8.47}$$

where $\mathcal{H}_{\bar{n}}$ are the Hermite polynomials and $P_{\bar{n}}(Q) = \sqrt{2I_{\bar{n}} - Q^2}$. The last equation corresponds to the semiclassical asymptotics in the forbidden region on the right-hand side of the turning point $Q_1 = \sqrt{2I_{\bar{n}}}$ with $I_{\bar{n}} = \hbar(\bar{n} + 1/2)$. In Q representation, the regular island extends up to $Q_c = \sqrt{2I_c}$, at which point the state $|\bar{n}\rangle$ has decayed to $\psi_{\bar{n}}(Q_c) \simeq 1/(2\sqrt{2\pi|P_{\bar{n}}(Q_c)|})\,(\exp\left(-S(I_{\bar{n}},I_c)\right))$, where

$$S(I_{\bar{n}},I_c) = \int_{Q_1}^{Q_c}|P_{\bar{n}}(Q)|\mathrm{d}Q = \sqrt{I_c(I_c - I_{\bar{n}})} - I_n\ln\left((\sqrt{I_c - I_{\bar{n}}} + \sqrt{I_c})/\sqrt{I_{\bar{n}}}\right). \tag{8.48}$$

This expression is suspiciously simple (as it depends only on $I_{\bar{n}}$ and I_c, and on no other property of the system) and should not be taken too seriously as it is.

Indeed, it should be borne in mind that $v_{\bar{n}}^2$ is related not so much to the value of the wavefunction at the regular–chaos boundary than to the transition rate $\Gamma_{n\to\text{chaos}}^d$ through Equation 8.44, which we can equal to the current flux $J_{\bar{n}}$ through this boundary for the regular state $|\bar{n}\rangle$. Using $H_0(I)$ to compute this current leads to a zero result and it is, therefore, mandatory to use a better approximation of the nonintegrable Hamiltonian H to obtain a meaningful answer. What complicates the evaluation of the transition rate from the edge of the regular to the chaotic domain is therefore that one needs to find an approximation describing both the regular and chaotic dynamics—unless one actually uses there the exact quantum dynamics as was done in Bäcker et al. (2008) [22]. Since the regular-chaos border is typically the place where approximation schemes tend to be difficult to control, this will rely on some assumption to be made for the chaotic regions, two possible choices of which we shall describe now.

One scenario that has been considered amounts in some way to model the regular to chaos transition in the way depicted in Figure 8.4a, a kinetic-plus-potential Hamiltonian $p^2/2m + V(q)$ where the island itself correspond to a potential well and the edge of the regular region to the place where the potential decreases abruptly. The picture one has in mind in that case is that escaping from the edge of the regular island to the chaotic sea is akin to the standard textbook barrier tunneling [45]. Using Langer's connection formula [46] within that model, the semiclassical wavefunction for the quasimode $\bar{n}$ inside the potential well can be extended under the potential barrier and, beyond this, into the region where motion at energy $E_{\bar{n}}^0$ is again classically authorized. In the classically allowed region outside the well ($q > q_r'$) the semiclassical wavefunction can be written as

$$\psi_{\bar{n}}(q) \simeq \sqrt{\frac{\Omega_{\bar{n}}}{2\pi p_{\bar{n}}(q)}} \exp\left(\frac{i}{\hbar}\int_{q_r'}^q p_{\bar{n}}(q)\mathrm{d}q + i\pi/4\right)\exp\left(-\frac{S_t}{\hbar}\right), \tag{8.49}$$

with $\Omega_{\bar{n}}$ the angular frequency of the torus $E_{\bar{n}}^0$, $p_{\bar{n}}(q) = \sqrt{2m\left[E_{\bar{n}}^0 - V(q)\right]}$, and

$$S_t = \int_{q_r}^{q_r'} |p_{\bar{n}}(q)|dq \tag{8.50}$$

the tunneling action.* The current of probability leaving the well is then given by

$$J_{\bar{n}} \equiv \frac{\hbar}{m}\,\text{Im}[\psi_{\bar{n}}^*\nabla\psi_{\bar{n}}] = \frac{\Omega_{\bar{n}}}{2\pi}\exp\left(-2\frac{S_t}{\hbar}\right), \tag{8.51}$$

from which $v_{\bar{n}}^2$ is obtained through Equation 8.44 (identifying $J_{\bar{n}}$ with $\Gamma_{\bar{n}\to\text{chaos}}^d$).

Although Equation 8.51 is derived here for the particular case of a kinetic-plus-potential Hamiltonian, it applies more generally (up maybe to a factor of order one) to any system with a phase space portrait that is similar to the one of Figure 8.4b, where tori inside the island can be analytically continued in the complex plane to a manifold escaping to infinity. In that case, Equation 8.51 can be applied provided the tunneling action S_t is taken as the imaginary part of the action integral on a path joining the interior to the exterior of the island on this analytical continuation. As a last approximation, one may assume that the transition to the "open" part is extremely sharp once the separatrix is crossed. In the model of Figure 8.4a, this amounts to assume a very rapid decrease of the potential, in which case one may replace q_r by q_M in the tunneling action (Equation 8.50). In an actual calculation of the direct tunneling rate for a regular state at the edge of the chaos boundary

---

* Strictly speaking, outgoing (Siegert) boundary conditions [47] need to be employed in Equation 8.49 in order to properly describe the decay process from the well. Those outgoing boundary conditions involve, in addition, an exponential increase of the wavefunction's amplitude with increasing distance from the well, which is not taken into account in Equation 8.49 assuming that the tunneling rate from the well is comparatively weak.

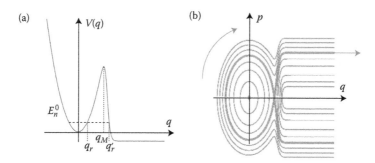

FIGURE 8.4 Modeling of the direct coupling between the edge of the regular region to the chaotic one by a potential barrier separating a potential well from an "open" region. (a) Sketch of the potential. (b) Corresponding phase space portrait.

(which we can reliably describe only coming from the interior of the island) this amounts to consider in the same way that the action $S(I_{\bar{n}}, I_c)$ (Equation 8.48) provides a good approximation to S_t. Under this hypothesis, one obtains for the coupling to the chaotic Hilbert space the prediction

$$v_{\bar{n}}^2 = \frac{\Delta_{\mathrm{ch}} \hbar \Omega_{\bar{n}}}{4\pi^2} \exp\left(-2\frac{S(I_{\bar{n}}, I_c)}{\hbar}\right) \tag{8.52}$$

(see Le Deunff and Mouchet (2010) [48] where this computation was proposed with the slightly different language of complex time trajectories).

This "potential-barrier" picture of the direct tunneling is in essence what is behind the approach of Podolskiy and Narimanov [19] (though their treatment of the problem is a bit more sophisticated). Its main virtue is that, beyond the quantized action $I_{\bar{n}} = \hbar(\bar{n} + 1/2)$, Equation 8.52 relies only on simple characteristics of the integrable regions: its area $2\pi I_c$ and the frequency $\Omega_{\bar{n}}$ of the torus $I_{\bar{n}}$. One needs to keep in mind, however, that Equation 8.52 implicitly assumes that the direct tunneling mechanism corresponds in some sense to the phase portrait of Figure 8.4b, which in general cannot be justified within any controlled approximation scheme.

8.3.4 Regular-to-Chaotic Coupling via a Nonlinear Resonance

Another possible, and presumably more realistic, approach to evaluate the direct coupling parameter $v_{\bar{n}}$ can be obtained assuming that the effective model (Equation 8.15) describes the vicinity of a resonance not only inside the regular region, but also within the chaotic sea in the near vicinity of the island. In this perspective, the model one has in mind for the chaotic region follows the spirit of Chirikov's overlapping criterion for the transition to chaos [39,49]: resonances still provide the couplings between quasimodes, but in the chaotic region these couplings become strong enough to completely mix the states. Near the regular-chaos edge, the transition between modes inside and outside the regular region is still dominated by one or several r:s resonances which might be within or possibly outside the regular island.

As an illustration, let us consider the simple case where a single nonlinear r:s resonance is responsible for all couplings, both within the island and from the island's edge to the chaotic region. Keeping in mind the discussion in Section 8.2.2 and in Brodier (2002) [28], we assume here that the couplings induced by the r:s resonance are dominantly described by the lowest non-vanishing Fourier component V_1 of the perturbation, that is, by the matrix elements $V_{r:s}^{(n+r)} \equiv \langle n + r | \hat{H}_{\mathrm{eff}} | n \rangle$, and set the phase ϕ_1 to zero without loss of generality.

The structure of the effective Hamiltonian that describes the coupling of the ground state E_0 to the chaotic sea is, in that case, given by

$$
H_{\text{eff}}^{\pm} =
\begin{pmatrix}
\tilde{E}_0 & V_{r:s}^{(r)} & & & \\
V_{r:s}^{(r)} & \ddots & & \ddots & \\
& \ddots & \tilde{E}_{k_c r} & V_{r:s}^{[(k_c+1)r]} & \\
& & V_{r:s}^{[(k_c+1)r]} & \boxed{\phantom{\begin{matrix} \\ \\ \\ \\ \\ \end{matrix}}} & \\
& & & \text{chaos}^{\pm} &
\end{pmatrix},
\tag{8.53}
$$

where $\tilde{E}_{kr} \equiv E_{kr} - \Omega_{r:s} I_{kr}$ are the eigenenergies of the unperturbed Hamiltonian $\mathcal{H}_0$ in the corotating frame, and the chaotic part (the square in the lower right corner) consists of a full subblock with equally strong couplings between all basis states with actions beyond the outermost invariant torus of the islands. In this example the last state within the island connected to the ground state is the quasi-mode $\bar{n} = k_c r$ and $v_{\bar{n}}^2 = (V_{r:s}^{[(k_c+1)r]})^2/N_{\text{ch}}$, with N_{ch} the number of states with a given parity in the chaotic Hilbert space. In a more general situation, many resonances may, as in Equation 8.37, be involved in connecting the ground stated to some $|\bar{n}\rangle$ at the edge of the island, but we would still write the variance $v_{\bar{n}}^2$ of the matrix elements providing the last coupling to the chaotic region as the ratio $(V_{r:s}^{[(k_c+1)r]})^2/N_{\text{ch}}$ for some $r:s$ resonance near the regular-chaos edge.

We stress here that it is necessary in this approach to have an explicit access to the number of states N_{ch} in the chaotic Hilbert space. This is obtained quite trivially for quantum maps, such as the kicked rotor we shall consider in Section 8.4, when $\hbar = 2\pi/N$ with integer N. In that case N is the total number of states in the full Hilbert space and N_{ch}/N represents the relative area of the chaotic region in phase space. For a two-dimensional conservative Hamiltonian, on the other hand, N and thus N_{ch} are related to the Thouless energy (see, e.g., Section 2.1.4 of Ullmo (2008) [50]), but this provides only an energy scale rather than a precise number, and a more detailed discussion is required to be quantitative. For the sake of clarity, we shall in the following limit our discussion to the simpler case that N is known.

8.3.5 THEORY OF CHAOS-ASSISTED TUNNELING

Let us consider now the effect of the chaotic block on the tunneling process. Eliminating intermediate states within the regular island leads for the effective Hamiltonian a matrix of the form

$$
H_{\text{eff}}^{\pm} =
\begin{pmatrix}
E_0 & V_{\text{eff}} & 0 & \cdots & 0 \\
V_{\text{eff}} & H_{11}^{\pm} & \cdots & \cdots & H_{1N_{\text{ch}}}^{\pm} \\
0 & \vdots & & & \vdots \\
\vdots & \vdots & & & \vdots \\
0 & H_{N_{\text{ch}}1}^{\pm} & \cdots & \cdots & H_{N_{\text{ch}}N_{\text{ch}}}^{\pm}
\end{pmatrix}
\tag{8.54}
$$

for each symmetry class. In the simplest case, Equation 8.53 where a single $r:s$ resonance needs to be considered, the effective coupling matrix element between the ground state and the chaos block $(H_{ij}^{\pm})$ is given by

$$
V_{\text{eff}} = V_{r:s}^{[(k_c+1)r]} \prod_{k=1}^{k_c} \frac{V_{r:s}^{(kr)}}{E_0 - E_{kr} + ks\hbar\omega},
\tag{8.55}
$$

where E_n are the unperturbed energies (Equation 8.20) of H_{eff} and $|k_c r\rangle$ represents the highest unperturbed state that is connected by the $r{:}s$ resonance to the ground state and located within the island (i.e., $I_{k_c r} < I_c < I_{(k_c+1)r}$). More generally, V_{eff} can be expressed in terms of the couplings associated with the various resonances that contribute to the transitions within the island and at the regular-chaos edge. The form of this expression (Equation 8.55) already provides an explanation for the appearance of plateau-like structures in the tunneling rates. Indeed, decreasing $\hbar$ leads to discontinous increments of the maximal number k_c of couplings through Equation 8.40 and hence to step-like reductions of the effective matrix element V_{eff}, while in between such steps V_{eff} varies smoothly through the action dependence of the coupling matrix elements $V_{r{:}s}^{(kr)}$, provided accidential near-degeneracies in the energy denominators do not occur.

In the simplest possible approximation, which follows the lines of Tomsovic and Ullmo (1994) [12] and Leyvraz and Ullmo (1996) [17], we neglect the effect of partial barriers in the chaotic part of the phase space [11] and assume that the chaos block $(H_{ij}^{\pm})$ is adequately modeled by a random Hermitian matrix from the Gaussian orthogonal ensemble (GOE). After a prediagonalization of $(H_{ij}^{\pm})$, yielding the eigenstates $\phi_j^{\pm}$ and eigenenergies $\mathcal{E}_j^{\pm}$, we can perturbatively express the shifts of the symmetric and antisymmetric ground state energies by

$$E_0^{\pm} = E_0 + \sum_{j=1}^{N_{\mathrm{ch}}} \frac{|v_{\mathrm{eff}\pm}^j|^2}{E_0 - \mathcal{E}_j^{\pm}}, \tag{8.56}$$

with $v_{\mathrm{eff}\pm}^j \equiv V_{\mathrm{eff}}\langle kr|\phi_j^{\pm}\rangle$. Performing the random matrix average for the eigenvectors, we obtain that $\langle\langle|\langle kr|\phi_j^{\pm}\rangle|^2\rangle\rangle \simeq 1/N_{\mathrm{ch}}$ for all $j = 1 \ldots N_{\mathrm{ch}}$, which simply expresses the fact that none of the basis states is distinguished within the chaotic block $(H_{ij}^{\pm})$. As a consequence, the variance of the $v_{\mathrm{eff}\pm}^j$'s is independent of j and equal to $v_{\mathrm{eff}}^2 = V_{\mathrm{eff}}^2/N_{\mathrm{ch}}$.

As was shown in Leyvraz and Ullmo (1996) [17], the random matrix average over the eigenvalues $\mathcal{E}_j^{\pm}$ gives rise to a Cauchy distribution for the shifts of the ground state energies, and consequently also for the splittings

$$\Delta E_0 = |E_0^+ - E_0^-| \tag{8.57}$$

between the symmetric and the antisymmetric ground state energy. For the latter, we specifically obtain the probability distribution

$$P(\Delta E_0) = \frac{2}{\pi} \frac{\overline{\Delta E_0}}{(\Delta E_0)^2 + (\overline{\Delta E_0})^2} \tag{8.58}$$

with

$$\overline{\Delta E_0} = \frac{2\pi v_{\mathrm{eff}}^2}{\Delta_{\mathrm{ch}}} \tag{8.59}$$

where Δ_{ch} denotes the mean level spacing in the chaos at energy E_0. This distribution is, strictly speaking, valid only for $\Delta E_0 \ll v_{\mathrm{eff}}$ and exhibits a cutoff at $\Delta E_0 \sim 2v_{\mathrm{eff}}$, which ensures that the statistical expectation value $\langle\Delta E_0\rangle = \int_0^{\infty} xP(x)\mathrm{d}x$ does not diverge.

Since tunneling rates and their parametric variations are typically studied on a logarithmic scale [i.e., $\log(\Delta E_0)$ rather than ΔE_0 is plotted vs. $1/\hbar$], the relevant quantity to be calculated from Equation 8.58 and compared to quantum data is not the mean value $\langle\Delta E_0\rangle$, but rather the average of the logarithm of ΔE_0. We therefore, define our "average" level splitting $\langle\Delta E_0\rangle_g$ as the *geometric* mean of ΔE_0 that is,

$$\langle\Delta E_0\rangle_g \equiv \exp[\langle\ln(\Delta E_0)\rangle] \tag{8.60}$$

and obtain as result the scale defined in Equation 8.59,

$$\langle\Delta E_0\rangle_g = \overline{\Delta E_0}. \tag{8.61}$$

This expression further simplifies for our specific case of periodically driven systems, where the time evolution operator $\hat{U}$ is modeled by the dynamics under the effective Hamiltonian (Equation 8.54) over one period τ. In this case, the chaotic eigenphases $\mathcal{E}_j^{\pm} \tau / \hbar$ are uniformly distributed in the interval $[0, 2\pi]$. We therefore obtain

$$\Delta_{ch} = \frac{2\pi\hbar}{N_{ch}\tau} \tag{8.62}$$

for the mean level spacing near E_0. This yields

$$\langle \Delta\varphi_0 \rangle_g \equiv \frac{\tau}{\hbar} \langle \Delta E_0 \rangle_g = \left(\frac{\tau V_{eff}}{\hbar} \right)^2 \tag{8.63}$$

for the geometric mean of the ground state's eigenphase splitting. Note that this final result does not depend on the number N_{ch} of chaotic states within the subblock $(H_{ij}^{\pm})$; as long as this number is sufficiently large to justify the validity of the Cauchy distribution (Equation 8.58) [17], the geometric mean of the eigenphase splitting is essentially given by the square of the effective coupling V_{eff} from the ground state to the chaos.

The distribution (Equation 8.58) also permits the calculation of the logarithmic variance of the eigenphase splitting: we obtain

$$\left\langle [\ln(\Delta\varphi_0) - \langle \ln(\Delta\varphi_0) \rangle]^2 \right\rangle = \frac{\pi^2}{4}. \tag{8.64}$$

This universal result predicts that the actual splittings may be enhanced or reduced compared to $\langle \Delta\varphi_0 \rangle_g$ by factors of the order of $\exp(\pi/2) \simeq 4.8$, independently of the values of $\hbar$ and external parameters. Indeed, as was discussed in Schlagheck et al. (2006) [32], short-range fluctuations of the splittings, arising at small variations of $\hbar$, are well characterized by the standard deviation that is associated with Equation 8.64.

It is interesting to note that the expression (Equation 8.63) for the (geometric) mean level spacing is quantitatively identical with the expression (Equation 8.45) for the mean escape rate from the regular island to the chaotic sea derived in Bäcker et al. (2008) [22] using Fermi's golden rule. This seems surprising as two different nonclassical processes, namely Rabi oscillations between equivalent islands and the decay from an island within an open system, underly these expressions. In one-dimensional single-barrier tunneling problems, these two processes would indeed give rise to substantially different rates; in Equation 8.1, to be more precise, the imaginary action integral in the exponent would have to be multiplied by two in order to obtain the corresponding expression for the decay rate (and the overall prefactor in front of the exponential function should be divided by two, which is not important here). The situation is a bit different, however, in our case of dynamical tunneling in mixed regular–chaotic systems. In such systems, level splittings between two equivalent regular islands involve *two* identical dynamical tunneling processes between the islands and the chaotic sea (namely one process for each island), while the decay into the chaotic sea involves only *one* such process, with, however, the square of the corresponding (exponentially suppressed) coupling coefficient. This explains from our point of view the equivalence of the expressions (Equations 8.63 and 8.45).

We finally remark that the generalization of the expression for the mean splittings to multi-resonance processes is straightforward and amounts to replacing the product of admixtures in Equation 8.55 by a product involving several resonances subsequently, in close analogy with Equation 8.41. In fact, the multi-resonance expression (Equation 8.41) can be directly used in this context replacing the "direct" splittings $\Delta\varphi_n^{(0)}$ by $(V_{r_f:s_f}^{[(k_c+1)r_f]} \tau / \hbar)^2$ where the $r_f:s_f$ resonance is the one that induces the final coupling step to the chaotic sea (provided $I_n < I_c < I_{n+r_f}$ holds for the corresponding action variables; otherwise, we would set $\Delta\varphi_n^{(0)} = 0$). This expression represents the basic formula that is used in the semiclassical calculations of the splittings in the kicked rotor model, to be discussed below.

8.3.6 ROLE OF PARTIAL BARRIERS IN CHAOTIC DOMAIN

In the previous section, we assumed a perfectly homogeneous structure of the Hamiltonian out-side the outermost invariant torus, which allowed us to make a simple random-matrix ansatz for the chaotic block. This assumption hardly ever corresponds to reality. As was shown in Bohigas et al. (1993) [11], Tomsovic and Ullmo (1994) [12], and Bohigas et al. (1990) [51] for the quartic oscillator, the chaotic part of the phase space is, in general, divided into several subregions which are weakly connected to each other through partial transport barriers for the classical flux (see, e.g., Figure 8 in Tomsovic and Ullmo (1994) [12]). This substructure of the chaotic phase space (which is generally not visible in a Poincaré surface of section) is particularly pronounced in the immediate vicinity of a regular island, where a dense hierarchical sequence of partial barriers formed by broken invariant tori and island chains is accumulating [52–54].

In the corresponding quantum system, such partial barriers may play the role of "true" tunneling barriers in the same spirit as invariant classical tori. This will be the case if the phase space area ΔW that is exchanged across such a partial barrier within one classical iteration is much smaller than Planck's constant $2\pi\hbar$ [55], while in the opposite limit $\Delta W \gg 2\pi\hbar$ the classical partial barrier appears completely transparent in the quantum system.* Consequently, the "sticky" hierarchical region around a regular island acts, for not extremely small values of $\hbar$, as a dynamical tunneling area and thereby extends the effective "quantum" size of the island in phase space. As a matter of fact, this leads to the formation of localized states (also called "beach" states in the literature [13]) which are supported by this sticky phase space region in the surrounding of the regular island [56] (see Figure 8.6b).

An immediate consequence of the presence of such partial barriers for resonance-assisted tunnel-ing is the fact that the critical action variable I_c defining the number k_c of resonance-assisted steps within the island according to Equation 8.55 should not be determined from the outermost invariant torus of the island, but rather from the outermost partial barrier that acts like an invariant torus in the quantum system. We find that this outermost quantum barrier is, for not extremely small values of $\hbar$, generally formed by the stable and unstable manifolds that emerge from the hyperbolic periodic points associated with a low-order nonlinear $r{:}s$ resonance. These manifolds are constructed until their first intersection points in between two adjacent periodic points, and iterated $r - 1$ times (or $r/2 - 1$ times in the case of period-doubling of the island chain due to discrete symmetries), such as to form a closed artificial boundary around the island in phase space.† As shown in Figure 8.5, one further iteration maps this boundary onto itself, except for a small piece that develops a loop-like deformation. The phase-space area enclosed between the original and the iterated boundary precisely defines the classical flux ΔW exchanged across this boundary within one iteration of the map [52,53].

The example in Figure 8.5 shows a boundary that arises from the inner stable and unstable manifolds (i.e., the ones that would, in a near-integrable system, form the inner separatrix structure) emerging from the unstable periodic points of a 4:1 resonance (which otherwise is not visible in the Poincaré section) in the kicked rotor system. Judging from the size of the flux area ΔW, this boundary should represent the relevant quantum chaos border for the tunneling processes that are discussed in the following section. We clearly see that it encloses a nonnegligible part of the chaotic classical phase space, which includes a prominent 10:3 resonance that, consequently, needs to be taken into account for the coupling process between the regular island and the chaotic sea. Thereby, we naturally arrive at *multi-step* coupling processes across a sequence of several resonances, which

---

* More precisely, the authors of Reference [55] claim that ΔW has to be compared with $\pi\hbar$ in order to find out whether or not a given partial barrier is transparent in the quantum system.

† This construction is also made in order to obtain the phase space areas $S_{r{:}s}^{\pm}$ that are enclosed by the outer and inner separatrix structures of an $r{:}s$ resonance, and that are needed in order to compute the mean action variable $I_{r{:}s}$ and the coupling strength $V_{r{:}s}$ of the resonance according to Equations 8.26 and 8.27.

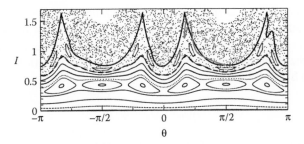

FIGURE 8.5 Classical phase space of the kicked rotor at $K = 3.4$ in approximate action-angle variables (I, θ). The thick solid line shows the location of the effective quantum boundary of the central island for the values of Planck's constant that are considered in Section 8.4. This effective boundary is constructed from segments of the stable and unstable manifolds that emerge from the hyperbolic periodic points of the 4:1 resonance at $\theta \simeq \pi/6$ and $5\pi/6$, respectively. Those segments were computed until the symmetry axis at $\theta = \pi/2$ and then iterated three times under the kicked rotor map, yielding the thick solid line. A further iteration of this boundary maps it onto itself, except for the piece between $\theta \simeq 0.15\pi$ and $\theta = 0.5\pi$ which is replaced by the lighter curve. The phase space area that is enclosed between the original (dark) and the iterated (light) boundary defines the classical flux that is exchanged across this boundary within one iteration of the map. The dashed line shows, in comparison, the actual classical chaos border defined by the outermost invariant torus of the island.

would have to be computed for a reliable prediction of the tunneling rates in the semiclassical regime.

8.4 APPLICATION TO KICKED ROTOR

8.4.1 TUNNELING IN KICKED ROTOR

To demonstrate the validity of our approach, we apply it to the "kicked rotor" model, which is described by the Hamiltonian

$$H(p,q,t) = p^2/2 - K \sum_{n=-\infty}^{\infty} \delta(t-n)\cos q. \tag{8.65}$$

The classical dynamics of this system is described by the "standard map" $(p,q) \mapsto (p',q')$ with

$$p' = p - K\sin q \tag{8.66}$$
$$q' = q + p', \tag{8.67}$$

which generates the stroboscopic Poincaré section at times immediately before the kick. The phase space of the kicked rotor is 2π periodic in position q and momentum p, and exhibits, for not too large perturbation strengths $K < 4$, a region of bounded regular motion centered around $(p,q) = (0,0)$.

The quantum dynamics of the kicked rotor is described by the associated time evolution operator

$$\hat{U} = \exp\left(-\frac{i}{\hbar}\frac{\hat{p}^2}{2}\right)\exp\left(-\frac{i}{\hbar}K\cos\hat{q}\right), \tag{8.68}$$

which contains two unitary operators that describe the effect of the kick and the propagation in between two kicks, respectively ($\hat{p}$ and $\hat{q}$ denote the position and momentum operators).

Because of the classical periodicity in both p and q, we can consider tunneling between the main regular island centered around $(p,q) = (0,0)$ and its counterparts that are shifted by integer multiples of 2π along the momentum axis or along the q-axis. To mimic a double-well configuration, we will restrict the boundary conditions for the eigenstates of $\hat{U}$ and consider tunneling between two islands centered around (i) $(p, q) = (0, 0)$ and $(2\pi, 0)$ or around (ii) $(p, q) = (0, 0)$ and $(0, 2\pi)$. The effective parity that allows to discriminate the eigenphases $\varphi_n^\pm$ of $\hat{U}$ manifests as $\tilde{\psi}_n^\pm(p + 2\pi) = \pm\tilde{\psi}_n^\pm(p)$ for the corresponding eigenstates in momentum representation in case (i) and as $\psi_n^\pm(q + 2\pi) = \pm\psi_n^\pm(q)$ for the eigenstates in position representation in case (ii). In both cases, tunneling will be characterized by the splitting

$$\Delta\varphi_n = |\varphi_n^+ - \varphi_n^-|. \tag{8.69}$$

Numerically, it can be convenient to deal with a finite Hilbert space of (even) size N, and this can be obtained provided the two-phase-space translation operators $\hat{T}_1 = \exp(2\pi i\hat{p}/\hbar)$ and $\hat{T}_2 = \exp(-2\pi i\hat{q}/\hbar)$ commute, which is the case if we choose $\hbar = 2\pi/N$ [57,58].

8.4.2 Numerical Computation of Eigenphase Splittings

Figure 8.6 shows the eigenphase splittings $\Delta\varphi_0$ [see Equation 8.69] in case (ii) (tunneling in position) for the local "ground state" ($n = 0$) in the central island of the kicked rotor, that is, for the state that is most strongly localized around the center of the island, at $K = 2.28$. While on average these splittings decrease exponentially with $1/\hbar$, significant fluctuations arise on top of that exponential decrease. In particular, large spikes are visible. As illustrated in Figure 8.6b, they can be related to the crossing of "excited states" within the island, which are coupled to the ground state by a classical resonance. Figure 8.6c shows the Husimi distribution of the relevant states involved, demonstrating that the coupling process is most effective when the states are symmetrically located on each side of the classical resonance. To illustrate that the influence of the resonances inside the regular islands is actually independent of the details of the chaotic regime (a major feature of the resonance-assisted and chaos-assisted tunneling schemes) we furthermore plot in Figure 8.7 a comparison between the splittings for the cases (i) and (ii) of tunneling in p and q direction, respectively. The (rough) matching between the dominant spikes of fluctuations in both cases confirms that tunneling outside one island is mainly isotropic in phase space.

From now on, we will consider case (i) only (tunneling in momentum) with $N = 2\pi/\hbar$ being an integer, corresponding to the number of Planck cells that fit into one Bloch cell. Figures 8.9 and 8.10 show the eigenphase splittings $\Delta\varphi_0$ of the island's ground state for $K = 2.6, 2.8, 3.0$ (Figure 8.9) as well as for $K = 3.2, 3.4, 3.6$ (Figure 8.10). As in Eltschka and Schlagheck (2005) [31] and Schlagheck et al. (2006) [32], these splittings were calculated with a diagonalization routine for complex matrices that is based on the GMP multiple precision library (http://gmplib.org) in order to obtain accurate eigenvalue differences below the ordinary machine precision limit.

8.4.3 Semiclassical Calculations

The role played by the nonlinear resonances in the tunneling mechanism is made explicit by comparing these numerically calculated splittings with semiclassical predictions based on the most relevant resonances that are encountered in phase space. In practice, we took those $r{:}s$ resonances into account that exhibit the smallest possible values of r and s for the winding numbers s/r under consideration. In all of the considered cases, the "quantum boundary" of the regular island, which determines the value of k_c through Equation 8.40, was defined by the partial barrier that results from the intersections of the inner stable and unstable manifolds associated with the hyperbolic periodic points of the 4:1 resonance (see also Figure 8.5). While this partial barrier lies rather close to the classical chaos border of the island for $K = 2.6$ (Figure 8.9), it encloses an appreciable part of the chaotic sea for $K = 3.6$ (Figure 8.10) including some relevant nonlinear resonances.

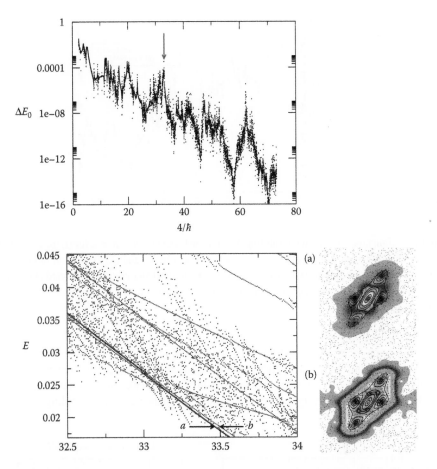

FIGURE 8.6 **(See color insert.)** The upper panel shows the quasienergy splittings in $\Delta E_0 = \hbar \varphi_0$ in the kicked rotor model for $K = 2.28$, corresponding to a 4:1 classical resonance. Here we are concerned with a tunneling in q. The lower panel shows the quasienergy spectrum $E_n^{\pm} = \hbar \varphi_n^{\pm}$ of $\hat{U}$ where only states with a significant overlap with a coherent state localized around $(p, q) = (0, 0)$ have been retained. The horizontal arrow (a) marks the central state doublet ($n = 0$, not resolved at that scale) and the arrow (b) indicates the third excited state localized in the island ($n = 4$). Their Husimi distribution superimposed with the Poincaré surface of section are shown in the panels on the right-hand side in order to illustrate the clear correspondence between the classical and the quantum resonance. As $1/\hbar$ increases, the crossing of the doublet by the resonant state provokes the large and wide spike indicated by the vertical arrow in the upper panel. Color version available online.

8.4.3.1 Pure Resonance-Assisted Tunneling

We stress that the semiclassical calculations shown in Figures 8.9 and 8.10 involve a few differences as compared to some of our previous publications [31–34]. To start with, (I) the action dependence of the coupling coefficients associated with the resonances has been included (see Equation 8.36). Furthermore, the unperturbed energy differences $E_n - E_{n+kr}$ of the quasimodes are not computed via the quadratic pendulum approximation (Equation 8.24). Instead, as illustrated in Figure 8.8, (II) a global parabolic fit to the action dependence of the frequency $\Omega \equiv \Omega(I)$ was applied on the basis of the classically computed values of the resonant actions $I_{r:s}$ and their frequencies $\Omega(I_{r:s}) = (s/r)\omega$, for a sequence of resonances with not too large r and s. These two modifications significantly improve the reproduction of individual peak structures in the tunneling rates.

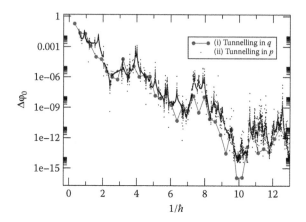

FIGURE 8.7 Comparison between the splitting $\Delta\varphi_0$ for the kicked rotor at $K = 2$ for (i) the case of tunneling between two islands centered at $(p,q) = (0,0)$ and at $(p,q) = (2\pi,0)$, and (ii) the case of tunneling between two islands centered at $(p,q) = (0,0)$ and at $(p,q) = (0,2\pi)$. In this latter case, we restricted ourselves to even integer values of $N = 2\pi/\hbar$ for which the splittings can be computed by diagonalizing finite $N \times N$ matrices beyond the double precision.

With these improvements (I) and (II), we generally find that the quantum splittings are quite well reproduced by our simple semiclassical theory based on nonlinear resonances. In particular, the location and height of prominent plateau structures and peaks in the tunneling rates can, in almost all cases, be quantitatively reproduced through resonance-assisted tunneling. The additional

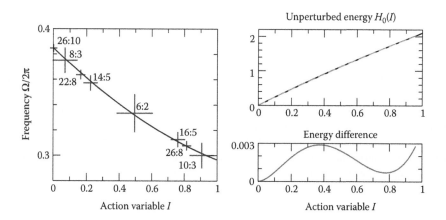

FIGURE 8.8 Unperturbed energies and oscillation frequencies within the regular island of the kicked rotor at $K = 3.5$. The left panel shows the action dependence of the oscillation frequencies as computed from a quadratic fit to individual nonlinear resonances (crosses), which results in the expression $\Omega(I) = 2.41886 - 0.790561I + 0.235191I^2$ (solid line). The upper right panel compares the unperturbed energies resulting from the integration of this quadratic expression (solid line) with the unperturbed energies used in Löck et al. (2010) [35] that were obtained by analyzing a dense set of quasiperiodic trajectories within the regular island (dashed line) [59].This latter approach yields $\Omega(I) = 2.41740 - 0.952917I + 1.00151I^2 - 1.00153I^3 + 0.368829I^4$, which essentially constitutes the definition of the fictitious integrable system used in Löck et al. (2010) [35]. Although the difference between these two approaches is rather small as shown in the lower right panel, it plays a significant role for the tunneling rates in the deep semiclassical limit (see Figure 8.12).

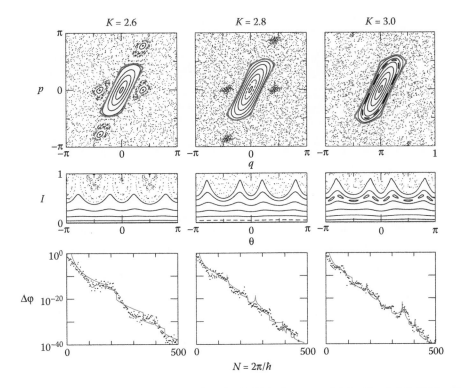

FIGURE 8.9 Quantum and semiclassical splittings in the kicked rotor model for $K = 2.6$ (left column) $K = 2.8$ (central column), and $K = 3$ (right column). The upper and middle panels show the classical phase space in the original phase-space variables p and q, with the thick curve marking the effective quantum boundary of the island, and in approximate action-angle variables I and θ. The lower panels display the quantum and semiclassical eigenphase splittings (dots and solid lines, respectively) of the ground state in the central regular island. For the semiclassical splittings, we used the 14:4 and 18:5 resonances for $K = 2.6$, the 10:3 and 14:4 resonances for $K = 2.8$, and the 10:3, 14:4, 16:5, and 22:7 resonances for $K = 3$. As pointed out in the text, the splittings were computed with a generalization of the multi-resonance Equation 8.41 to mixed systems, using (I) the corrected action dependence Equation 8.36 of the matrix elements and (II) unperturbed energies that were determined from a global parabolic fit of $\Omega(I)$, and (III) computing the coupling to the chaotic domain via the outermost nonlinear resonance, as explained at the end of Section 8.3.5. The discontinuous steps in the semiclassical splittings are actually induced by the discontinuity of the integer part in Equation 8.40 for the maximal number k_c of couplings within the island.

fluctuations of the splittings on a small scale of N, however, cannot be accounted for by our approach as they arise from the details of the eigenspectrum in the chaotic block of the Hamiltonian. Their average size, however, seems in good agreement with the universal prediction (Equation 8.64) for the variance of eigenphase splittings in chaos-assisted tunneling.

Note that there is a general tendency of the semiclassical theory to overestimate the exact splittings wherever the latter encounter local minima. We attribute those minima to the occurrence of destructive interferences between different pathways that connect the ground state to a given excited state $|n\rangle$. As pointed out in the discussion of Equation 8.41, such destructive interferences are not yet accounted for in our present implementation of resonance-assisted tunneling. Their inclusion would require to take into account the phases ϕ_k associated with individual resonances (see Equation 8.13), the discussion of which is beyond the scope of this article.

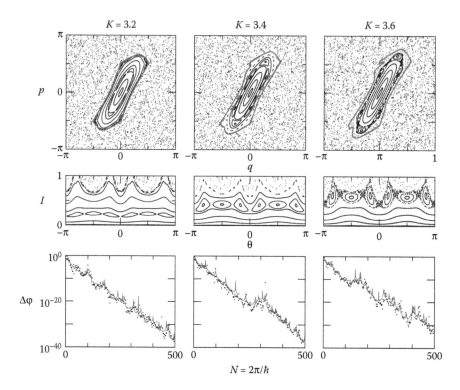

FIGURE 8.10 Same as Figure 8.9 for $K = 3.2$, $K = 3.4$, and $K = 3.6$. For the semiclassical splittings, we used the 6:2 and 10:3 resonances for $K = 3.2$, the 6:2, 10:3, and 14:5 resonances for $K = 3.4$, and the 6:2 and 8:3 resonances for $K = 3.6$.

8.4.3.2 Resonance-Assisted Tunneling at a Bifurcation

As a last comment, we note that the influence of a nonlinear resonance on tunneling processes in mixed systems may persist even if that resonance is not at all manifested in the classical phase space. This is precisely the case at the value of the perturbation parameter at which this resonance is bifurcating from the center of the island, that is, at which the central fixpoint of the island exhibits a rational winding number corresponding to the resonance under consideration. A prominent example in the kicked rotor model is found at $K = 2$ where the central fixpoint has the winding number 0.25. Indeed, $K = 2$ is exactly the critical value where two periodic orbits of period 4 coalesce in the center, both separately coming from the complex phase space (their action being strictly negative for $K < 2$) and then becoming real and distinct (one stable and one unstable) for $K > 2$.

Following the normal-form arguments in Section 8.2.3, the scaling of the "classical size" of the resonance with the perturbation parameter K is, in lowest order, provided by the effective pendulum matrix element $V_{r:s}(K) = \tilde{v}_1 [I_{r:s}(K)]^{rk/2}$ where $I_{r:s}(K)$ represents the dependence of the resonant action on K. However, as can be seen in Equation 8.36, the associated coupling matrix elements that affect tunneling only depend on the prefactor $\tilde{v}_1$, which ought to be a well-behaved function of K showing no singular behavior at the bifurcation point. Therefore, we have to conclude that these matrix elements remain finite even for $K \lesssim 2$.

This expectation is confirmed in Figure 8.11, which shows the tunneling rates in the kicked rotor for $K = 2$. For this value, the only major resonance that is manifested in the classical phase space is the 10:2 resonance whose island chain is located near the boundary of the main island. Taking into account this resonance alone (as well as combining it with other resonances of higher order, the result of which is not shown in Figure 8.11) apparently leads to a very strong underestimation of the

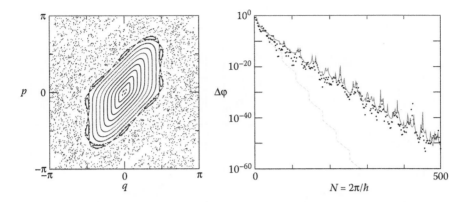

FIGURE 8.11 Resonance-assisted tunneling in the kicked rotor at a bifurcation. The left panel shows the classical phase space for $K = 2$, which contains a prominent 10:2 resonance close to the border of the island, and the right panel displays the corresponding quantum eigenphase splittings (dots). Semiclassical calculations of the splittings (solid and dashed lines, using the same levels of approximation as in Figures 8.9 and 8.10) were carried out using the 10:2 resonance only (dashed line) as well as a combination of the 4:1 resonance and the 10:2 resonance, the former emerging at the island's center right at $K = 2$. The parameter $\tilde{v}_1$ associated with that 4:1 resonance (see Equation 8.36) was determined from the classical phase space at $K = 2.001$.

quantum eigenphase splittings (in contrast to Eltschka and Schlagheck (2005) [31] where the action dependence of the coupling matrix elements was not properly incorporated). But once we take into account the 4:1 resonance and compute its corresponding classical parameters from the classical phase space at $K = 2.001$, we obtain a good reproduction of the quantum splittings.

8.4.3.3 Combination with Direct Regular-to-Chaotic Couplings

In the semiclassical predictions shown in Figures 8.9, 8.10, and 8.11, we have, following the discussion in Section 8.3.4, assumed that the resonance-assisted mechanism was providing the relevant couplings not only within the regular island, but also (III) from the edge of the regular island to the chaotic sea. As pointed out in Section 8.3.2, the regular-chaos boundary is, however, the place where approximation schemes are not controlled any longer. It is therefore, useful to compare the results obtained in this way with those derived from the more precise evaluation of the direct regular–chaotic couplings (IIIb) computed with Equation 8.45 through a numerical application of the quantum evolution operator $\hat{U}$.

Figure 8.12 shows the resulting comparison for the eigenphase splittings of the quantum kicked rotor at $K = 3.5$ [35] (see Figure 8.3 for the corresponding classical phase space). In addition to the quantum and semiclassical splittings obtained in the same way as in Figures 8.9 and 8.10 (upper left panel), two additional curves are shown. The green one corresponds to a fully semiclassical calculation for which the regular-to-chaotic couplings are evaluated by the resonance-assisted mechanism. In contrast to the curves in the upper left panel, and to the calculations in Figures 8.9 and 8.10, however, the action dependence $H_0(I)$ of the unperturbed energies within the island was not obtained by a fit to several relevant resonances as described above, but rather (IIb) by computing the action and the rotation number for a dense set of trajectories within the island, from which the energies are deduced via the relation $\Omega(I) = dH_0(I)/dI$ (see, e.g., Figure 11 of Bohigas et al. (1993) [11] and the associated text for a detailed discussion).* The curve in the lower left panel implements, in addition to these improved energies, an evaluation of the direct regular–chaotic coupling (IIIb)

---

* We are indebted to Steffen Löck, who carried out this calculation and sent us the resulting data.

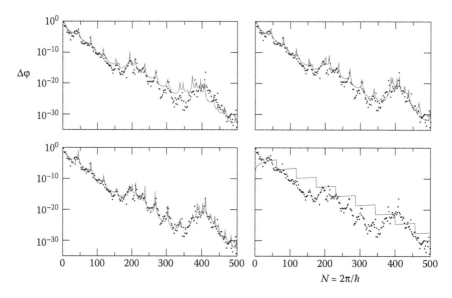

$N = 2\pi/\hbar$

FIGURE 8.12 Quantum and semiclassical splittings in the kicked rotor model for $K = 3.5$ (see Figure 8.3 for the classical phase space). As in Figures 8.9 and 8.10, the dots and solid lines represent, respectively, the quantum splittings and the semiclassical prediction. In the upper left panel, this prediction is based on our approach, where we take into account the 6:2, 8:3, 10:3, and 14:5 resonances. The upper right panel shows the same calculation, except that a more refined evaluation of the unperturbed energies (and their action dependence) is used (IIb) (see text and Figure 8.8). The lower left panel shows the prediction that is obtained with these improved energies and with an evaluation of the direct regular–chaotic coupling (IIIb) using the fictitious integrable system approach of Bäcker et al. [22] (which amounts to applying the quantum kicked rotor map onto the corresponding eigenstate of the integrable system, see Equation 8.45). Finally, the lower right panel shows the prediction that would be obtained with the 6:2 resonance according to the approach outlined in Eltschka and Schlagheck (2005) [31], that is, neglecting the action dependence of the coupling matrix elements, neglecting the occurence of partial barriers in the chaotic domain, and making a simple quadratic expansion of the unperturbed energies in the vicinity or the resonance.

using the approach of Bäcker et al. [22] (see Equation 8.45); it actually corresponds to the curve published in Figure 3 of Löck et al. (2010) [35].

What is observed in Figure 8.12 is that, although there is good agreement between the three theoretical curves and the numerical one for a large range of $N = 2\pi/\hbar$, some significant deviations arise in the range $300 < N < 450$. Remarkably, however, the most striking discrepancies that are encountered for the curve in the upper left panel (i.e., for the full semiclassical calculation *without* the improved energies) are essentially cured once the improved energies are implemented (curve in the upper right panel). Adding the seminumerical evaluation of the direct regular-to-chaotic coupling further improves the prediction, but to a significantly lesser degree.

Figure 8.8 compares $H_0(I)$ computed for the kicked rotor at $K = 3.5$ by the two methods under consideration, namely the fitting approach using the most relevant resonances (which is illustrated in the left panel of Figure 8.8) and the more refined approach based on a dense set of trajectories. Apparently, both approaches yield nearly identical energies, with relative differences being well below one percent (see the lower right panel of Figure 8.8). This underlines that small imprecisions in the prediction of the unperturbed eigenenergies of regular quasimodes may, under special conditions, lead to dramatic over- or underestimations of the tunneling rates in mixed systems. In the particular case considered here, it seems to be the transition across the 8:3 resonance, located rather close to the center of the island, which is not properly described using the more approximate energies. This gives rise to a horizontal shift of the predicted splittings towards smaller N, as is clearly seen in Figure 8.12. This issue obviously requires further investigations. It does, however, not seem to put into question the principal conclusion that the resonance-assisted coupling mechanism provides an accurate approach to evaluate regular-to-chaotic tunneling rates in a purely semiclassical manner.

8.5 CONCLUSION

In summary, we have provided a comprehensive description of the theory of resonance-assisted tunneling in mixed regular–chaotic systems. This description is partly based on previous publications of ours [27,28,31–33], but contains also some new aspects and significant improvements especially concerning the determination of the matrix elements associated with resonance-induced couplings. Moreover, partial barriers in the chaotic domain are now incorporated into the scheme of resonance-assisted tunneling, which generally gives rise to an effective enhancement of the size of the regular islands under consideration. In practice, we find that the most relevant partial barriers for tunneling are constituted by combinations of stable and unstable manifolds that are associated with hyperbolic periodic points of low-order nonlinear resonances within the chaotic domain. The application of this approach to the kicked rotor model yields rather good agreement between the "exact" eigenphase splittings of states that are localized in the center of the main regular island, and their semiclassical predictions based on nonlinear resonances.

The main message that we intend to communicate here is that resonance-assisted tunneling not only allows one to understand the origin of plateaus and peak structures in the tunneling rates, but it also provides a simple, readily implementable scheme to quantitatively predict the appearance of such structures on the basis of purely classical information. In practice, one needs for this purpose to identify the relevant resonances in the regular phase space region under consideration, to find their stable and unstable fixpoints in the Poincaré surface of section, to compute their stability indices and the areas enclosed by their separatrix structures, respectively by their stable and unstable manifolds, and finally to compute the flux enclosed by the turnstiles in order to determine effective "quantum chaos border" of the island. Even for a simple model like the kicked rotor, this programme requires much less numerical effort than a quantum calculation of the tunneling rates. As we demonstrated for the kicked rotor, it provides, on the other hand, a reproduction of the quantum tunneling rates which is extremely satisfactory from a quantitative point of view. This numerical accuracy requires, however, a careful evaluation of the various classical parameters entering into the semiclassical calculation of the splittings. This includes the action dependence of the resonance–Hamiltonian couplings, the effective size of the regular island, and the evaluation of the unperturbed energies within the island. It turns out in particular that in some circumstances, for example, at $K = 3.5$ in the range $300 < N < 450$, very small imprecision in the determination of the unperturbed eigenenergies may significantly affect the accuracy of the semiclassical predictions.

We expect that the framework of resonance-assisted tunneling can be generalized to systems with more than two effective degrees of freedom, although the identification of important resonances might become more involved in such systems and resonance-assisted (quantum) tunneling might compete there with (classical) Arnold diffusion in the deep semiclassical regime [30]. Another open problem which needs to be addressed in more detail concerns the role of nonlinear resonances in trajectory-based semiclassical approaches to tunneling, put forward by Shudo, Ikeda, and coworkers [36–38] (see also Chapter 7 in this book), which involve the complexified classical phase space and are more rigorous from a formal semiclassical point of view than our approach. We strongly believe that nonlinear resonances should leave their characteristic traces in the self-similar complex phase space structures that govern tunneling in this framework [36,37]. Such an insight should significantly contribute to rendering those approaches practicable for more complicated systems as well—and underline the semiclassical nature of resonance-assisted tunneling.

ACKNOWLEDGMENTS

We are grateful to Olivier Brodier, Dominique Delande, Christopher Eltschka, Kensuke Ikeda, Jérémy Le Deunff, Steffen Löck, Srihari Keshavamurthy, Roland Ketzmerick, Mirjam Schmid, Akira Shudo and Steve Tomsovic for helpful assistance and stimulating discussions. We further acknowledge financial support by the Deutsche Forschungsgemeinschaft (DFG) through the Forschergruppe FOR760 "Scattering systems with complex dynamics."

REFERENCES

1. M. J. Davis and E. J. Heller, Quantum dynamical tunneling in bound states, *J. Chem. Phys.* 75, 246, 1981.
2. W. K. Hensinger, H. Häffner, A. Browaeys, N. R. Heckenberg, K. Helmerson, C. McKenzie, G. J. Milburn, W. D. Phillips, S. L. Rolston, H. Rubinsztein-Dunlop, and B. Upcroft, Dynamical tunneling of ultracold atoms, *Nature* 412, 52, 2001.
3. D. A. Steck, W. H. Oskay, and M. G. Raizen, Observation of chaos-assisted tunneling between islands of stability, *Science* 293, 274, 2001.
4. A. Mouchet, C. Miniatura, R. Kaiser, B. Grémaud, and D. Delande, Chaos-assisted tunneling with cold atoms, *Phys. Rev. E* 64, 016221, 2001.
5. C. Dembowski, H.-D. Gräf, A. Heine, R. Hofferbert, H. Rehfeld, and A. Richter, First experimental evidence for chaos-assisted tunneling in a microwave annular billiard, *Phys. Rev. Lett.* 84, 867, 2000.
6. S. Creagh, *Tunneling in two dimensions*, in *Tunneling in Complex Systems*, S. Tomsovic (ed.), World Scientific, Singapore, pp. 1–65, 1998.
7. L. D. Landau and E. M. Lifshitz, *Quantum Mechanics: Non-Relativistic Theory*, Pergamon Oxford, 1958.
8. S. C. Creagh, Tunnelling in multidimensional systems, *J. Phys. A* 27, 4969, 1994.
9. W. A. Lin and L. E. Ballentine, Quantum tunneling and chaos in a driven anharmonic oscillator, *Phys. Rev. Lett.* 65, 2927, 1990.
10. O. Bohigas, D. Boosé, R. Egydio de Carvalho, and V. Marvulle, Quantum tunneling and chaotic dynamics, *Nucl. Phys. A* 560, 197, 1993.
11. O. Bohigas, S. Tomsovic, and D. Ullmo, Manifestation of classical phase space structures in quantum mechanics, *Phys. Rep.* 223, 43, 1993.
12. S. Tomsovic and D. Ullmo, Chaos-assisted tunneling, *Phys. Rev. E* 50, 145, 1994.
13. E. Doron and S. D. Frischat, Semiclassical description of tunneling in mixed systems: Case of the annular billiard, *Phys. Rev. Lett.* 75, 3661, 1995.
14. S. D. Frischat and E. Doron, Dynamical tunneling in mixed systems, *Phys. Rev. E* 57, 1421, 1998.
15. A. Ishikawa, A. Tanaka and A. Shudo, Recovery of chaotic tunneling due to destruction of dynamical localization by external noise, *Phys. Rev. E* 80, 046204, 2009.
16. F. Grossmann, T. Dittrich, P. Jung, and P. Hänggi, Coherent destruction of tunneling, *Phys. Rev. Lett.* 67, 516, 1991.
17. F. Leyvraz and D. Ullmo, The level splitting distribution in chaos-assisted tunneling, *J. Phys. A* 29, 2529, 1996.
18. A. Mouchet and D. Delande, Signatures of chaotic tunneling, *Phys. Rev. E* 67, 046216, 2003.
19. V. A. Podolskiy and E. E. Narimanov, Semiclassical description of chaos-assisted tunneling, *Phys. Rev. Lett.* 91, 263601, 2003.
20. V. A. Podolskiy and E. E. Narimanov, Chaos-assisted tunneling in dielectric microcavities, *Opt. Lett.* 30, 474, 2005.
21. A. Bäcker and R. Ketzmerick, private communication.
22. A. Bäcker, R. Ketzmerick, S. Löck, and L. Schilling, Regular-to-chaotic tunneling rates using a fictitious integrable system, *Phys. Rev. Lett.* 100, 104101, 2008.
23. A. Bäcker, R. Ketzmerick, S. Löck, M. Robnik, G. Vidmar, R. Höhmann, U. Kuhl, and H.-J. Stöckmann, Dynamical tunneling in mushroom billiards, *Phys. Rev. Lett.* 100, 174103, 2008.
24. A. M. Ozorio de Almeida, Tunneling and the semiclassical spectrum for an isolated classical resonance, *J. Phys. Chem.* 88, 6139, 1984.

25. T. Uzer, D. W. Noid, and R. A. Marcus, Uniform semiclassical theory of avoided crossings, *J. Chem. Phys.* 79, 4412, 1983.

26. L. Bonci, A. Farusi, P. Grigolini, and R. Roncaglia, Tunneling rate fluctuations induced by nonlinear resonances: A quantitative treatment based on semiclassical arguments, *Phys. Rev. E* 58, 5689, 1998.

27. O. Brodier, P. Schlagheck, and D. Ullmo, Resonance-assisted tunneling in near-integrable systems, *Phys. Rev. Lett.* 87, 064101, 2001.

28. O. Brodier, P. Schlagheck, and D. Ullmo, Resonance-assisted tunneling, *Ann. Phys.* 300, 88, 2002.

29. S. Keshavamurthy, On dynamical tunneling and classical resonances, *J. Chem. Phys.* 122, 114109, 2005.

30. S. Keshavamurthy, Resonance-assisted tunneling in three degrees of freedom without discrete symmetry, *Phys. Rev. E* 72, 045203(R), 2005.

31. C. Eltschka and P. Schlagheck, Resonance- and chaos-assisted tunneling in mixed regular-chaotic systems, *Phys. Rev. Lett.*, 94, 014101, 2005.

32. P. Schlagheck, C. Eltschka, and D. Ullmo, Resonance- and chaos-assisted tunneling, in *Progress in Ultrafast Intense Laser Science I*, K. Yamanouchi, S. L. Chin, P. Agostini, and G. Ferrante (eds.), Springer, Berlin, pp. 107–131, 2006.

33. A. Mouchet, C. Eltschka, and P. Schlagheck, Influence of classical resonances on chaotic tunnelling, *Phys. Rev. E* 74, 026211, 2006.

34. S. Wimberger, P. Schlagheck, C. Eltschka, and A. Buchleitner, Resonance-assisted decay of nondispersive wave packets, *Phys. Rev. Lett.* 97, 043001, 2006.

35. S. Löck, A. Bäcker, R. Ketzmerick, and P. Schlagheck, Regular-to-chaotic tunneling rates: From the quantum to the semiclassical regime, *Phys. Rev. Lett.* 104, 114101, 2010.

36. A. Shudo and K. S. Ikeda, Complex classical trajectories and chaotic tunneling, *Phys. Rev. Lett.* 74, 682, 1995.

37. A. Shudo and K. S. Ikeda, Stokes phenomenon in chaotic systems: pruning trees of complex paths with principle of exponential dominance, *Phys. Rev. Lett.* 76, 4151, 1996.

38. A. Shudo, Y. Ishii, and K. S. Ikeda, Julia set describes quantum tunnelling in the presence of chaos, *J. Phys. A* 35, L225, 2002.

39. A. J. Lichtenberg and M. A. Lieberman, *Regular and Stochastic Motion*, Springer-Verlag, New York, 1983.

40. S. Tomsovic, M. Grinberg, and D. Ullmo, Semiclassical trace formulas of near-integrable systems: Resonances, *Phys. Rev. Lett.* 75, 4346, 1995.

41. G. D. Birkhoff, *Dynamical Systems* Am. Math. Soc., New York, 1966.

42. F. G. Gustavson, On constructing formal integrals of a Hamiltonian system near an equilibrium point, *Astron. J.* 71, 670, 1966.

43. A. M. Ozorio de Almeida, *Hamiltonian systems*, Cambridge University Press, 1988.

44. P. Leboeuf and A. Mouchet, Normal forms and complex periodic orbits in semiclassical expansions of Hamiltonian systems, *Ann. Phys.* 275, 54, 1999.

45. E. Merzbacher, *Quantum mechanics*, 2nd ed. Wiley, New York, 1970.

46. R. E. Langer, On the connection formulas and the solutions of the wave equation, *Phys. Rev.* 51, 669, 1937.

47. A. J. F. Siegert, On the derivation of the dispersion formula for nuclear reactions, *Phys. Rev.* 56, 750, 1939.

48. J. Le Deunff and A. Mouchet, Instantons re-examined: Dynamical tunneling and resonant tunneling, *Phys. Rev. E* 81, 046205, 2010.

49. B. Chirikov, *Time-dependent quantum systems*, in *Chaos and Quantum Physics*, M. J. Giannoni, A. Voros, and J. Zinn-Justin (eds.), North-Holland, Amsterdam, 1991, p. 201.

50. D. Ullmo, Many-body physics and quantum chaos, *Rep. Prog. Phys.* 71, 026001, 2008.

51. O. Bohigas, S. Tomsovic, and D.Ullmo, Classical transport effects on chaotic levels, *Phys. Rev. Lett.* 65, 5, 1990.

52. R. S. MacKay, J. D. Meiss, and I. C. Percival, Stochasticity and transport in Hamiltonian systems, *Phys. Rev. Lett.* 52, 697, 1984.

53. R. S. MacKay, J. D. Meiss, and I. C. Percival, Transport in Hamiltonian systems, *Physica* 13D, 55, 1984.

54. J. D. Meiss and E. Ott, Markov-tree model of intrinsic transport in Hamiltonian systems, *Phys. Rev. Lett.* 55, 2742, 1985.

55. N. T. Maitra and E. J. Heller, Quantum transport through cantori, *Phys. Rev. E* 61, 3620, 2000.

56. R. Ketzmerick, L. Hufnagel, F. Steinbach, and M. Weiss, New class of eigenstates in generic Hamiltonian systems, *Phys. Rev. Lett.* 85, 1214, 2000.
57. F. Izrailev and D. Shepelyansky, Quantum resonance for a rotator in a nonlinear periodic field, *Sov. Phys. Dokl.* 24, 996, 1979.
58. S. Fishman, I. Guarneri, and L. Rebuzzini, Stable quantum resonances in atom optics, *Phys. Rev. Lett.* 89, 084101, 2002.

9 Dynamical Tunneling from the Edge of Vibrational State Space of Large Molecules

David M. Leitner

CONTENTS

9.1 QUANTUM ENERGY FLOW IN LARGE MOLECULES

The transport and localization of vibrational energy in molecules mediate a variety of molecular processes. A central motivation for the study of vibrational energy flow in molecules has long been its influence on spectroscopy [1–11], chemical reaction kinetics in gas and condensed phases [12–30], and the desire to control chemical reactions with lasers [31–37]. In large molecules, quantum mechanical effects can both enhance as well as impose severe limitations on energy flow. The former takes place by classically forbidden tunneling processes. The latter occurs when quantum mechanical coherence effects introduce localization in a system that exhibits transport in the classical limit. In many-dimensional systems such as molecules, both of these quantum properties play a key role in energy flow, and both are addressed in this chapter.

One example of classically forbidden transport exhibited in quantum mechanical systems is dynamical tunneling. Davis and Heller [38] argued that even when a classical system is confined in phase space to a torus identified by a specific set of action variables, the quantum system need not be so confined, and may tunnel among many nearly degenerate tori. Stuchebrukhov and Marcus, while studying [39,40] the puzzlingly narrow spectral lines corresponding to CH stretch excitation in t-butyl acetylene and several isomers that had been recently measured [5], explored the relation between dynamical tunneling and transitions among vibrational states via off-resonant states, terming the latter vibrational superexchange. They showed that vibrational superexchange and dynamical tunneling are essentially equivalent processes [39–41], where the former is reached from a quantum mechanical rather than semiclassical perspective. Resonant coupling between zero-order states that could be identified with a set of action variables can lead to energy transfer with a classical analog, whereas vibrational superexchange arises from virtual transitions, a quantum mechanical effect analogous to tunneling in phase space. In this chapter, we review our work on quantum energy flow in large molecules at relatively low energy [42,43], and focus on the regime of "large $\hbar$." We thus adopt the language of vibrational superexchange in our discussion of quantum ergodicity and energy flow in large molecules, recognizing throughout the equivalence to dynamical tunneling.

We address here the role of vibrational superexchange on energy flow and localization in large molecules, examining two aspects of this contribution. One is the role of vibrational superexchange in quantum localization, that is, the quantum ergodicity transition in molecules. Leitner and Wolynes [42,43] compared the contributions of vibrational superexchange and direct resonant coupling by high-order anharmonic coupling using local random matrix theory (LRMT) and we review here and illustrate these contributions to the quantum ergodicity transition of sizable organic molecules. We then examine how vibrational superexchange, or dynamical tunneling, facilitates energy transfer from vibrational states that lie at the edge of the vibrational state space.

The chapter is organized as follows. In Section 9.2 we review LRMT, developed to predict the location of the quantum ergodicity transition in a system of many coupled nonlinear oscillators. The theory centers on an ensemble of coupled, nonlinear oscillators (defined by Equations 9.1 and 9.2). We provide examples of application of LRMT to organic molecules using representative parameter values. The statistical analysis that we adopt addresses quantum localization among most vibrational states, what we refer to as the interior of the vibrational state space. Since spectroscopic experiments often interrogate vibrational states at the "edge," we turn our attention to the transfer of energy from the edge to the interior. Here both high-order anharmonic coupling in the potential energy and vibrational superexchange can potentially contribute, and we examine the relative contributions of each using representative coupling strengths and energies for organic molecules.

9.2 LRMT AND QUANTUM ERGODICITY TRANSITION

9.2.1 Quantum Localization in Interior of the Vibrational State Space

We aim to characterize properties that influence the extent and rate of vibrational energy flow in molecules with perhaps tens of vibrational modes. We do this within the framework of an ensemble theory for molecules, in this case a matrix ensemble where each member is a reasonable representation of the vibrational Hamiltonian of a molecule with many vibrational modes. We thereby address energy flow and localization from a random matrix perspective [44], an approach that has been very successful in describing statistical properties of spectra of highly excited molecules [45–47]. In perhaps the simplest application of random matrix theory, one identifies all conserved quantities, and constructs an ensemble of random matrices labeled by good quantum numbers. Statistical properties of eigenstates are then obtained with this ensemble and can be matched to those of a molecule whose dynamics are strongly chaotic. If a quantum number is partially conserved, a matrix ensemble that accounts for this may be constructed and useful statistical information about eigenstates of such systems may be found [48].

A similar but more complex situation pertains to molecular vibrations in general. At low energy, for instance energies corresponding to barriers to conformational change of a few kcal/mol, only a small number of vibrational modes of the molecule are excited, and anharmonic coupling among vibrational modes is generally small. The largest anharmonic terms are typically low-order terms. These properties suggest a reasonable representation for the Hamiltonian ensemble. A natural zero-order vibrational state space consists of a product space of states labeled by the occupation number of a collection of nonlinear oscillators [49]. These may, for example, be the normal modes of vibration and the diagonal anharmonic terms of the Hamiltonian. The zero-order states are coupled to each other by matrix elements that arise from low-order anharmonic coupling. In principle these matrix elements can be calculated for a given molecule, but many of the properties that we are interested in can be suitably captured by an ensemble of Hamiltonians, each one a reasonable representation of the one of interest. We thus take the elements of the vibrational Hamiltonian of a molecule as random subject to the constraint that states are coupled locally in the quantum number space. For this reason, we refer to the theory as LRMT. Several reviews of LRMT have already appeared [18,50,51], which have summarized the important steps toward arriving at prediction of the quantum ergodicity transition, energy flow in quantum number space above the transition, and applications to unimolecular rate theory. Here our focus is the role of vibrational superexchange on energy flow from edge states to the interior of the vibrational state space.

Our N-nonlinear oscillator Hamiltonian is defined by $H = H_0 + V$, where

$$H_0 = \sum_{\alpha=1}^{N} \varepsilon_\alpha(\hat{n}_\alpha), \tag{9.1}$$

$$V = \sum_m \prod_\alpha \Phi_m (b_\alpha^+)^{m_\alpha^+} (b_\alpha)^{m_\alpha}, \tag{9.2}$$

where $m = \{m_1^\pm, m_2^\pm, \ldots\}$. The zero-order Hamiltonian H_0 consists of a sum over the energies of the nonlinear oscillators, where each oscillator has frequency $\omega_\alpha(n_\alpha) = \hbar^{-1} \partial \varepsilon_\alpha / \partial n_\alpha$, and nonlinearity $\omega_\alpha'(n_\alpha) = \hbar^{-1} \partial \omega_\alpha / \partial n_\alpha$, and the number operator is defined by $\hat{n}_\alpha = b_\alpha^+ b_\alpha$. The nonlinearity is assumed to be sufficiently small so that $\hbar |\omega_\alpha'(n_\alpha)| \ll \omega_\alpha(n_\alpha)$, though finite nonlinearity is essential in removing correlations among matrix elements coupling states in the vibrational state space [49]. The set of zero-order energies, $\{\varepsilon_\alpha\}$, and coefficients of V, $\{\Phi_m\}$, are treated as random variables with suitable average and variance.

The vibrational Hamiltonian defined by Equations 9.1 and 9.2 includes direct resonant coupling terms of arbitrary order. In order for coupling of states in the matrix ensemble to be "local," we assume that the coefficients Φ_m on average decay exponentially. One form for the off-diagonal matrix elements that embodies this assumption has been proposed for modest-sized organic molecules and supported by the computational work of Gruebele and coworkers [52,53] and is used below. The larger low-order terms in V couple states close to one another in the vibrational quantum number space. Following Logan and Wolynes [49], we can thus picture the topology of the state space as an N-dimensional lattice where each lattice site is coupled locally to nearby sites by matrix elements arising from low-order anharmonic terms in the potential (Figure 9.1). The zero-order energy of a site in the vibrational state space is determined by the frequencies and nonlinearities of the vibrational modes, which may in principle be known. However, since one such site is coupled to a fairly large number of sites nearby in quantum number space, we assume the zero-order energies of all these sites are random from a distribution whose width is of the order of a vibrational frequency. We assume for now that the states do not lie at the edge of the lattice, where "surface" effects yield a different local density of states than found in the interior of the lattice.

These assumptions lead to a tight-binding picture in a many-dimensional vibrational quantum number space with random site energies. The problem of vibrational energy flow in molecules thus resembles the problem of single-particle transport on a many-dimensional disordered lattice, and theoretical approaches to address the condensed phase problem can be brought to bear on describing vibrational energy flow in molecules. Exploiting this connection, Logan and Wolynes found a transition for energy to flow globally on the energy shell that occurs at a critical value of the product of the anharmonic coupling and local density of states [49].

The problem of energy transport in the vibrational state space of the molecule centers on the self-energy for a particular state. The imaginary part is finite for extended states and proportional to the energy-transfer rate from that state, while it is infinitesimally small for a localized state. To calculate the real and imaginary parts of the self-energy, it is convenient to approximate the topology of the vibrational state space as a Cayley tree. This topology serves as an increasingly better approximation to the topology of the vibrational state space as the number of oscillators in the system increases,

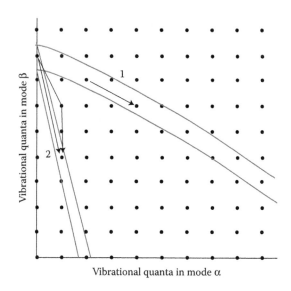

FIGURE 9.1 A two-dimensional slice through the vibrational quantum number space of a molecule, where coordinates corresponding to modes α and β are shown. Two possible energy shells are shown schematically, labeled 1 and 2, and indicated by pairs of gray lines. On energy shell 1 low-order (cubic) anharmonic coupling terms enable a transition between two states in the interior of the quantum number space, as indicated by the arrow. On energy shell 2 higher order (fifth or higher order) anharmonic coupling terms enable a transition between two states connected by an arrow. Alternatively, a transition between the two states can occur by vibrational superexchange, whereby one (or more generally a series of) off-resonant transition enables a transition between the same two states on energy shell 2. In the scenario illustrated by energy shell 2, mode α lies outside the main band of vibrational modes, mode β lies within the main band and states with excitation of mode α and few in other modes are edge states.

in line with our interest in coupled many-oscillator systems. The real and imaginary parts of the site self-energy are expressed by the Feenberg renormalized perturbation series, for which only the lowest order term is needed for a state space with a Cayley tree topology.

Details of the analysis, including incorporation of the higher-order terms accounting for vibrational superexchange, can be found in Leitner and Wolynes (1996) [42]. Here we highlight the main points, which are needed for discussion of dynamical tunneling from edge states in Section 9.3. The lowest order terms in the Feenberg renormalized perturbation series for the real, E_j, and imaginary, Δ_j, parts of the site self energy are given by

$$E_j(E) = \sum_{Q,k \neq j}^{K_Q} \frac{|V_{Q,jk}|^2 (E - \varepsilon_k - E_k)}{(E - \varepsilon_k - E_k)^2 + (\Delta_k(E) + \eta)^2}, \tag{9.3}$$

$$\Delta_j(E) = \sum_{Q,k \neq j}^{K_Q} \frac{|V_{Q,jk}|^2 (\Delta_k(E) + \eta)}{(E - \varepsilon_k - E_k)^2 + (\Delta_k(E) + \eta)^2}, \tag{9.4}$$

where we are interested in the limit $\eta \to 0$. Q represents the distance in quantum number space between two states; it corresponds to the total number of quanta lost and gained in all oscillators in making a transition from the state labeled by j to states labeled by k; for example, cubic coupling can give rise to at most an exchange of three quanta among all the oscillators, in which case $Q = 3$, whereas quartic coupling can give rise to at most an exchange of four quanta among the oscillators, in which case $Q = 4$. $|V_{Q,jk}|$ is a matrix element coupling state j to state k, which lies a distance Q away; and K_Q is the number of such states.

We incorporate higher-order terms in the Feenberg perturbation series so that the imaginary part of the site self-energy maintains the same form it has in Equation 9.4 and maintaining the simplicity of the Cayley tree topology. We do this by renormalizing the values of $\langle |V_Q| \rangle$ as follows [42]:

$$\Delta_j(E) = \sum_{M=2}^{\infty} \Delta_{j,M}(E), \tag{9.5}$$

$$\Delta_{j,M}(E) = \sum_{Q,k,l,\ldots,r}^{K_Q} V_{Q,jr}\ldots V_{Q,kj} \left[\sum_{\alpha=k}^{r} \prod_{\beta=k(\neq\alpha)}^{r} X_{Q,\beta} Y_{Q,\beta} \right], \tag{9.6}$$

where

$$X_{Q,k}(E) = \frac{\left(E - \varepsilon_{Q,k} - E_{Q,k}\right)}{\left(E - \varepsilon_{Q,k} - E_{Q,k}\right)^2 + \Delta_{Q,k}^2(E)}, \tag{9.7}$$

$$Y_{Q,k}(E) = \frac{\Delta_{Q,k}(E)}{\left(E - \varepsilon_{Q,k} - E_{Q,k}\right)^2 + \Delta_{Q,k}^2(E)}. \tag{9.8}$$

The real part of the site self-energy is calculated as

$$E_j(E) = \sum_{M=2}^{\infty} E_{j,M}(E), \tag{9.9}$$

$$E_{j,M}(E) = \sum_{Q,k,l,\ldots,r}^{K_Q} V_{Q,jr} X_{Q,r} \ldots X_{Q,k} V_{Q,kj}. \tag{9.10}$$

The real and imaginary parts of the site self-energy can be transformed into a form that mimics the $M = 2$ equations by a suitable mean field averaging over all intermediate states, as carried out originally by Logan and Wolynes for the problem of band-structure effects on Anderson localization in topologically disordered systems [54], and applied to the problem of the quantum ergodicity transition in many-oscillator systems by Leitner and Wolynes [42]. The result yields an equation for Δ_j, and therefore, the lifetime of the zero-order state labeled as j, containing an effective coupling, $\psi_{Q,jk}$, instead of the direct resonant coupling term, $V_{Q,jk}$. The effective coupling can be expressed as a sum of the following chain diagrams [42]:

$$\psi_{Q,jk} = \text{○—○} + \text{○⬤○} + \text{○⬤⬤○} + \ldots, \tag{9.11}$$

where each vertex (filled circle) corresponds to $K_q \langle X_q \rangle$ and each line segment between a filled and open circle to $\langle V_q \rangle$. Resonant coupling, $V_{Q,jk}$, between edge state j and interior state k is depicted by the first term. The index q corresponds to the distance in quantum number space between states that are intermediate between edge state j and interior states labeled by k. We then have

$$\Delta_j(E) = \sum_{Q,k \neq j}^{K_Q} \frac{|\psi_{Q,jk}|^2 (\Delta_k(E) + \eta)}{(E - \varepsilon_k - E_k)^2 + (\Delta_k(E) + \eta)^2}. \tag{9.12}$$

Since the matrix elements and zero-order energies that appear above are random, a statistical self-consistent procedure is used to obtain Δ_j. We assume that the couplings and distribution of zero-order states are statistically equivalent in going from tier to tier (Figure 9.2), and that there are no particular bottlenecks between the initial state and other states further along that disrupt such a statistical equivalence. We expect the former to hold for the vast majority of states where, near a given energy, the quanta are distributed among many modes of the main "band" of vibrations, that

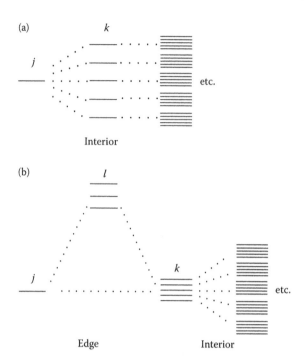

FIGURE 9.2 Tier structure diagram illustrating states coupled to each other by low-order resonances. In (a) a tier diagram is shown where initially excited state j is in the interior of the vibrational state space and coupled to states k; (b) illustrates a bottleneck between one state, j, at the edge of the vibrational state space, and states, k, of the interior. Transfer from state j to interior states can occur via the intermediate states l by dynamical tunneling.

is, states that lie in the "interior" of the vibrational state space. We refer to the other states, with energy distributed mainly in one or a few modes, particularly those that lie outside the main band of molecular vibrations, as "edge states." In this section we focus on the quantum localization and energy flow in the interior of the vibrational state space, and turn to energy transfer from the edge states in Section 9.3.

To establish energy flow in the interior of the vibrational state space we solve self-consistently for the probability distribution of Δ_j in the limit in $\eta \to 0$. In one approach to the self-consistent analysis, the most probable value for the imaginary part of the self-energy, Δ_{mp}, is found self-consistently [42,49]. In a second approach, the average inverse participation ratio is solved self-consistently [55]. In both cases, the eigenstates of H are assumed localized, so that the most probable value for the imaginary part of the self-energy must be infinitesimally small and the average value for the inverse participation ratio must be finite. A range of molecular parameters are then identified for which the most probable value of the self-energy is in fact infinitesimally small, or, similarly, where the average inverse participation ratio is finite. The result for both self-consistent approaches is nearly the same, differing only by a constant of order one. Solving for the most probable value of the imaginary part of the self-energy, we find that vibrational energy flow is unrestricted when [42,49]

$$T(E) \equiv \sqrt{\frac{2\pi}{3}} \sum_Q \langle |\psi_Q| \rangle \rho_Q \geq 1, \qquad (9.13)$$

while energy is localized in the vibrational state space at energy E when $T(E)$ is less than 1. Here, ρ_Q is the local density of states that lie a distance Q away in quantum number space, and $\langle |\psi_Q| \rangle$ is the average effective coupling matrix element to such states.

In large molecules, the quantum ergodicity transition depends significantly on high-order direct resonance coupling terms. Specifically, for N vibrational modes, the order of the anharmonic coupling terms contributing most to the transition, which we call Q_m, is [42]

$$Q_m \approx \frac{2N}{\sigma} - 1, \qquad (9.14)$$

where σ is a parameter that governs the size of higher-order anharmonic coupling terms, and is large when such terms are small. Typical values of σ are 3 to 10, as discussed in Section 9.2.2. We have found the role of vibrational superexchange in the quantum ergodicity threshold to be relatively minor, having evaluated ψ_Q for several specific examples [51]. All terms of ψ_Q beyond the first term, V_Q, in Equation 9.11 contain $\langle X_q \rangle$, which, due to the energy level distribution coupled to a given state of the interior of the vibrational state space, is typically too small for ψ_Q and V_Q to be very different when Q is near Q_m. The role of vibrational superexchange is found to be much more significant, indeed typically dominant, in energy transfer from the edge of the vibrational state space, as we discuss in Section 9.3, often due to a sparsity of states at small Q directly coupled to the edge state.

Above the transition, $T(E) > 1$, energy flows over all states of the energy shell. Schofield and Wolynes [56,57] have argued that energy flow in the vibrational state space both just above and well beyond the transition can be described by a random walk in quantum number space, a picture that has been supported by numerical calculations over a wide range of time scales [50,58–61]. The state-to-state energy transfer rate can be estimated by LRMT. Well above the transition we would expect the rate of quantum energy flow between the interior states of the vibrational state space to be given by $k(E) = \frac{2\pi}{\hbar} \sum_Q |\psi_Q|^2 \rho_Q(E)$. More generally, including the region near the transition, we find the most probable value of the imaginary part of the site self-energy to be given by [42]

$$\Delta_{mp}(E) = \pi \sqrt{1 - T^{-1}(E)} \sum_Q |\psi_Q|^2 \rho_Q(E), \quad T(E) > 1, \qquad (9.15)$$

which is related to the most probable energy transfer rate in the interior of the vibrational state space by $k(E) = 2\Delta_{mp}(E)/\hbar$. Equation 9.15 goes over to a golden rule-like expression that reveals the locality of energy flow through a crossover region just above the transition, which in practice we find to be quite narrow, particular when we account for higher-order resonances. While the transition itself is increasingly influenced by higher-order terms the larger the molecule, the influence of high-order anharmonic coupling on vibrational energy transfer rates is generally less pronounced [42].

9.2.2 EXAMPLE

In the following, we consider a simple, illustrative example using an approximate expression for T for a system of N coupled nonlinear oscillators. Our estimate for $T(E)$ for a molecule with N vibrational modes begins with approximations to the connectivity, K_Q, between a state in the vibrational state space to states a distance Q away; D_Q, the local density of states; and V_Q, the local coupling. We have found that if N is moderately large and Q is less than $\approx 2N^{1/2}$ we can approximate K_Q by [42]

$$K_Q \approx \frac{(2N)^Q}{Q!}. \qquad (9.16)$$

Below and near the transition we can approximate $D_Q(E)$ by ($\hbar = 1$) [42]

$$D_Q(E) \approx \left(\pi Q^{1/2} \omega_{\mathrm{rms}} \right)^{-1}, \qquad (9.17)$$

where ω_{rms} is the root mean square vibrational frequency. Finally, we use for V_Q the scaling relation adopted by us previously [42]

$$V_Q = \Phi_3 \sigma^{3-Q} M^{Q/2}, \qquad Q \geq 3, \tag{9.18}$$

where Φ_3 is a representative cubic anharmonic constant and σ quantifies how higher-order constants become smaller with the order [52,53]. The relevant parameters typically have ranges $0.1\ \text{cm}^{-1} < \Phi_3 < 10\ \text{cm}^{-1}$ and $3 < \sigma < 10$ for organic molecules [52,53].

Combining these, we have [42]

$$T(E) = \frac{2\pi}{3} \tilde{\phi}^2 \sigma^6 F^2(\kappa), \tag{9.19}$$

$$F(\kappa) = (2\pi)^{-1/2} \left[\sum_{Q=3} Q^{-1} \left(\frac{\kappa}{Q} \right)^Q e^Q \right], \tag{9.20}$$

$$\kappa = \frac{2NM^{1/2}}{\sigma}, \tag{9.21}$$

$$\tilde{\phi} = \Phi_3 \left(\pi \omega_{rms} \right)^{-1}, \tag{9.22}$$

where Stirling's approximation has been used in the estimate for K_Q. For simplicity, we take the average number of quanta per mode, M, to be $M \approx E/N\omega_{rms}$ at energy E. We also take $\langle X_q \rangle = 0$ for simplicity, which is reasonable for interior states. In practice, we have found $\langle X_q \rangle$ to be small for the interior of the vibrational state space when carrying out more detailed calculations by direct count of all states [62].

The main "band" of vibrational frequencies in a large organic molecule corresponds to frequencies between of order $10\ \text{cm}^{-1}$ up to about $1800\ \text{cm}^{-1}$. At higher frequency, between 2900 and $3500\ \text{cm}^{-1}$, there are stretches of CH, OH, or NH bonds. For estimating the location of the quantum ergodicity transition we have usually considered only coupling among modes in the main band of vibrational frequency, while considering changes in the high-frequency vibrations to occur at the edge. Of course, if we were to calculate $T(E)$ by direct count of all states using specific values of the anharmonic matrix elements we do not need to make this assumption. Here we adopt the approximate formulas above and in doing so we use for N the number of vibrational modes with frequency in the main band.

We consider examples of 12-, 18-, and 24-vibrational mode molecules. For simplicity, we take these modes to be randomly distributed between 0 and $1500\ \text{cm}^{-1}$. We take as a representative value for a cubic anharmonic constant, $\Phi_3 = 2\ \text{cm}^{-1}$. We allow the "decay rate constant" for higher-order anharmonicity, σ, to vary from 3 to 10, within the established limits for organic molecules [52,53]. We calculate $T(E)$ as a function of energy, E, with Equation 9.18 and determine the energy of the quantum ergodicity threshold, that is, the energy where $T = 1$. An example of this kind of "phase diagram" is presented in Figure 9.3. This figure provides an instructive schematic plot of how the quantum ergodicity transition varies with the number of vibrational modes in the lower band of vibrational frequencies and with the contribution of the higher-order anharmonic interactions, which become less important with larger σ. Of course, the value for the cubic constant is simply one representative value. For specific organic molecules, we generally use information from quantum chemical calculations of the low-order anharmonic constants and the vibrational modes themselves and calculate matrix elements and local densities of states by direct count [24,63]. This can yield quite different values for the transition energy than the schematic values plotted in Figure 9.3. For example, *trans*-stilbene has 60 modes in the main band, and we have calculated the energy of the quantum ergodicity threshold with Equations 9.5 and 9.6 to be around $1200\ \text{cm}^{-1}$ [17].

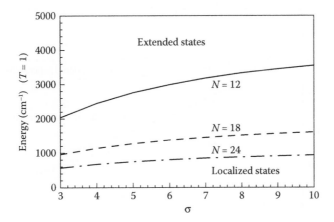

FIGURE 9.3 A "phase diagram" showing energies where vibrational states are localized or extended, calculated using LRMT for 12- (solid), 18- (dashed), and 24-mode (dot-dashed) molecules, where parameters representative of organic molecules have been chosen. The number of modes listed only counts those in the "main band" of up to about 1500 cm^{-1}, and does not include the CH stretches closer to 3000 cm^{-1}.

9.3 VIBRATIONAL SUPEREXCHANGE FROM EDGE OF VIBRATIONAL STATE SPACE

9.3.1 DYNAMICAL TUNNELING IN LARGE MOLECULES

The quantum ergodicity transition, located in the previous section, establishes the energy thresholds where vibrational energy can tunnel into the interior from the edge of the vibrational state space, thereby connecting states that are often probed spectroscopically with states of the interior of the vibrational state space. Once we have carried out the self-consistent analysis on the interior states and find the most probable value of the site self-energy, $\Delta_{mp}(E)$, to be finite, we can compute the transition rate to these interior states from the edge.

Let Q_{int} be the distance in quantum number space from the edge state to the first "tier" of the interior of the vibrational state space. We assume that the rate of energy transfer from states of this tier of interior states is determined by $\Delta_{mp}(E)$, whereas for tiers closer to the edge state this is not necessarily the case. We estimate $\Delta_j(E)$ for the edge state j by considering only the nearest interior states. In this case we have [42]

$$\Delta_j(E) = \pi K_{Q_{\text{int}}} \left\langle |\psi_{Q_{\text{int}}}|^2 \right\rangle D_{Q_{\text{int}}}(E), \tag{9.23}$$

where $K_{Q_{\text{int}}}$ is the number of states that lie a distance Q_{int} from the edge state and $D_{Q_{\text{int}}}$ is probability of finding a state at energy, E, a distance Q_{int} from the edge state.

9.3.2 EXAMPLE

We consider now illustrative examples where the interior states lie outside the reach of direct resonant interactions by low-order cubic and quartic anharmonic coupling terms. We consider cases where Q_{int} are 6 and 7, and several different values of $\langle X_q(E) \rangle$, defined below Equation 9.11, for the bridge states, specifically [30 cm^{-1}]$^{-1}$, [50 cm^{-1}]$^{-1}$, and [100 cm^{-1}]$^{-1}$. The former value is comparable to those encountered in the t-butyl acetylene calculations of Stuchebrukhov and Marcus [39], and the smaller values of $\langle X_q(E) \rangle$ illustrate cases where vibrational superexchange would have less of a chance to be favorable. Note that $\langle X_q(E) \rangle$ is obtained from the values of $\varepsilon_{q,l}$ for the intermediate levels, and one can often assume for simplicity that the real part, $E_j(E)$, of

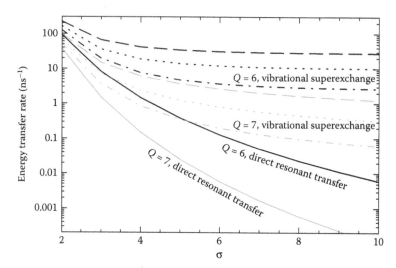

FIGURE 9.4 Plotted are rates of energy transfer from an edge state to interior states that lie $Q_{int} = 6$ (black) and 7 (gray) away in the vibrational quantum number space of an 18-mode organic molecule. Results for different average off-resonant energies of the bridge states are shown, specifically $\langle X_q(E) \rangle = [30\,cm^{-1}]^{-1}$ (dot-dashed), $[50\,cm^{-1}]^{-1}$ (dotted), and $[100\,cm^{-1}]^{-1}$ (dashed). These are compared with the rate of energy transfer when only direct resonant coupling is used in the calculation (solid). We observe that dynamical tunneling contributes more to the rate than does direct resonant coupling, typically by orders of magnitude, for all values of $\langle X_q(E) \rangle$ used and for all σ, which determines the strength of the higher-order anharmonic coupling terms.

the site-self energy is negligibly small compared to the difference between the energy, E, and the zero-order energy, $\varepsilon_{q,l}$, that is, $E - \varepsilon_{q,l} - E_l(E) \approx E - \varepsilon_{q,l}$.

We take the energy of the molecule to be 3000 cm^{-1}, as would roughly be the case if all the vibrational energy were in a CH stretch. We observe in Figure 9.3 that the vibrational states are extended at this energy for all values of σ at $N = 18$ or larger, and we take $N = 18$ in this example, noting that we find no qualitative difference with a larger choice of N. As noted, values of σ, which establish the magnitude of anharmonic coupling beyond cubic terms, typically range between 3 and 10 for the vibrations of organic molecules. However, they may be smaller for vibrations coupled to internal rotations, for example, methyl rotors [39,40], so that we allow a somewhat wider range of 2 to 10 in this example, recognizing that σ must be greater than 1 to maintain the locality of coupling that characterizes vibrational coupling in molecules. We then use Equations 9.16–9.18 and 9.23 to find $\Delta_j(E)$.

Results of this illustrative calculation are plotted in Figure 9.4, where we take the rate to be $2\Delta_j(E)/\hbar$. The results are plotted for $Q_{int} = 6$ and 7, and $\langle X_q(E) \rangle = [30\,cm^{-1}]^{-1}$, $[50\,cm^{-1}]^{-1}$, and $[100\,cm^{-1}]^{-1}$. We find the rates of energy transfer from the edge state to the interior states to vary from of order 0.1 to 100 ns^{-1}, depending on the value of σ. If only direct resonant coupling is accounted for, and no vibrational superexchange, the energy transfer rate is generally much smaller. For the parameters considered here we observe that vibrational superexchange is predominantly responsible for energy transfer from the edge state for all values of σ greater than 2 or 3, and even at 2–3 it contributes at least 50% to the rate, depending on $\langle X_q(E) \rangle$. For example, when $\sigma \approx 2$, direct resonance coupling only accounts for about 40% of the energy transfer rate when $\langle X_q(E) \rangle = 0.01$ cm and $Q_{int} = 6$ or 7. For values of σ even as small as 4 the estimate of the rate is about an order of magnitude too small, at larger σ very much smaller, when vibrational superexchange is neglected in the calculation of the energy transfer rate from the edge to the interior states of the vibrational state space. We thus observe, in line with an earlier analysis of a similar model [43], that vibrational superexchange facilitates energy transfer in large molecules, often far exceeding the contribution of direct resonant coupling. This conclusion is consistent with the role of dynamical tunneling in phase space transport found in a variety of semiclassical studies of low-dimensional systems, some of which are discussed in other chapters of this volume.

9.4 CONCLUDING REMARKS

Quantum mechanical effects mediate the extent and rate of vibrational energy flow in large molecules. On the one hand, energy flow can be limited to an extremely small region of the energy shell in the vibrational state space of a large molecule due to quantum coherent effects that give rise to localization. In the corresponding classical system, for example, widespread phase space transport occurs by Arnold diffusion [64], a process that cannot occur with sufficiently large $\hbar$ [65]. To determine where the quantum ergodicity transition lies in a quantum mechanical system of many coupled nonlinear oscillators, we have developed and adopted a statistical theory, called LRMT, summarized in Section 9.2. LRMT shows that it is the local density of states resonantly coupled to interior states of the vibrational state space that determines the energy of the quantum ergodicity transition, and that for larger molecules it is higher-order direct resonant coupling terms that contribute most to this.

Even when the vibrational states are extended and energy flows freely among the vast majority of states of the vibrational state space, there may be some states, often at the edge of vibrational state space, only weakly coupled to the interior. By edge states we refer to states where only one or a relatively small number of modes are excited at a given energy, such as states where modes lying outside the main band of vibrations are excited; for organic molecules this usually includes excitation of CH, OH, or NH stretches, with little or no vibrational excitation elsewhere. When edge states are detached from the interior due to a sparsity of states resonantly coupled by low-order anharmonic coupling, then quantum effects play a paramount role in energy transfer, as illustrated in Section 9.3.2 and Figure 9.4. Though we have considered overall low excitation in large molecules, and thus a regime of large $\hbar$, these effects are described semiclassically by dynamical tunneling, where a system tunnels between tori in phase space. As discussed here and other recent studies [66], dynamical tunneling plays a central role in intramolecular vibrational redistribution (IVR) and in processes that are mediated by IVR, such as spectroscopy. Future work may well expose its role in unimolecular reaction kinetics, which depends critically on IVR.

ACKNOWLEDGMENTS

The author has enjoyed many illuminating discussions with Srihari Keshavamurthy on energy flow in molecules, and is thankful to him for the invitation to contribute a chapter to this volume. Support from NSF grant CHE-0910669 is gratefully acknowledged.

REFERENCES

1. S. C. Farantos, R. Schinke, H. Guo, and M. Joyeux, Energy localization in molecules, bifurcation phenomena, and their spectroscopic signature: The global view. *Chem. Rev.* 109, 4248, 2009.
2. P. Manikandan, A. Semparithi, and S. Keshavamurthy, Decoding the dynamical information embedded in highly excited vibrational eigenstates: State space and phase space viewpoints. *J. Phys. Chem.* A 113, 1717, 2009.
3. E. J. Heller, Dynamical tunneling and molecular spectra. *J. Phys. Chem.* 99, 2625, 1995.
4. T. Uzer, Theories of intramolecular vibrational energy transfer. *Phys. Rep.* 199, 73, 1991.
5. K. K. Lehmann, G. Scoles, and B. H. Pate, Intramolecular dynamics from the eigenstate-resolved infrared spectra. *Annu. Rev. Phys. Chem.* 45, 241, 1994.
6. M. J. Davis, Analysis of highly excited vibrational eigenstates. *Int. Rev. Phys. Chem.* 14, 15, 1995.

7. G. A. Bethardy and D. S. Perry, Competing mechanisms for intramolecular vibrational redistribution in the v_{14} asymmetric methyl stretch band of trans-ethanol. *J. Chem. Phys.* 99, 9400, 1993.

8. A. Mehta, A. A. Stuchebrukhov, and R. A. Marcus, IVR in overtones of the acetylenic C-H stretch in propyne. *J. Phys. Chem.* 99, 2677, 1995.

9. Y.-S. Lin, S. G. Ramesh, J. M. Shorb, E. L. Sibert, and J. L. Skinner, Vibrational energy relaxation of the bend fundamental of dilute water in liquid chloroform and d-chloroform. *J. Phys. Chem. B* 112, 390, 2008.

10. E. L. Sibert III and M. Gruebele, Molecular vibrational energy flow and dilution factors in an anharmonic state space. *J. Chem. Phys.* 124, 024317, 2006.

11. P. M. Felker, W. R. Lambert, and A. H. Zewail, Dynamics of intramolecular vibrational-energy redistribution (IVR). IV. Excess energy dependence, t-stilbene. *J. Chem. Phys.* 82, 3003, 1985.

12. U. Lourderaj and W. L. Hase, Theoretical and computational studies of non-RRKM unimolecular dynamics. *J. Phys. Chem. A* 113, 2236, 2009.

13. S. Nordholm, Photoisomerization of stilbene—a theoretical study of deuteration shifts and limited internal vibrational redistribution. *Chem. Phys.* 137, 109, 1989.

14. S. Nordholm and A. Back, On the role of nonergodicity and slow IVR in unimolecular reaction rate theory—a review and a view. *Phys. Chem. Chem. Phys.* 3, 2289, 2001.

15. R. A. Kuharski, D. Chandler, J. A. Montgomery, F. Rabii, and S. J. Singer, Stochastic MD simulation of cyclohexane isomerization. *J. Phys. Chem.* 92, 3261, 1988.

16. D. M. Leitner, Influence of quantum energy flow and localization on molecular isomerization in gas and condensed phases. *Int. J. Quantum Chem.* 75, 523, 1999.

17. D. M. Leitner, B. Levine, J. Quenneville, T. J. Martinez, and P. G. Wolynes, Quantum energy flow and trans stilbene photoisomerization: An example of a non-RRKM reaction. *J. Phys. Chem.* 107, 10706, 2003.

18. D. M. Leitner, Heat transport in molecules and reaction kinetics: The role of quantum energy flow and localization. *Adv. Chem. Phys.* 130B, 205, 2005.

19. D. M. Leitner and P. G. Wolynes, Quantum energy flow during molecular isomerization. *Chem. Phys. Lett.* 280, 411, 1997.

20. J. C. Keske and B. H. Pate, Decoding the dynamical information embedded in highly mixed quantum states. *Ann. Rev. Phys. Chem.* 51, 323, 2000.

21. M. J. Davis, Bottlenecks to intramolecular energy transfer and the calculation of relaxation rates. *J. Chem. Phys.* 83, 1016, 1985.

22. M. J. Davis and S. K. Gray, Unimolecular reactions and phase space bottlenecks. *J. Chem. Phys.* 84, 5389, 1986.

23. D. M. Leitner and P. G. Wolynes, Quantum theory of enhanced unimolecular reaction rates below the ergodicity threshold. *Chem. Phys.* 329, 163, 2006.

24. D. M. Leitner and M. Gruebele, A quantum model of restricted vibrational energy flow on the way to the transition state in unimolecular reactions. *Mol. Phys.* 106, 433, 2008.

25. R. Hernandez, W. H. Miller, C. B. Moore, and W. F. Polik, A random matrix/transition state theory for the probability distribution of state-specific unimolecular decay rates: Generalization to include total angular momentum conservation and other dynamical symmetries. *J. Chem. Phys.* 99, 950, 1993.

26. N. DeLeone and B. J. Berne, Intramolecular rate process-isomerization dynamics in the transition to chaos. *J. Chem. Phys.* 75, 3495, 1981.

27. B. C. Dian, G. G. Brown, K. O. Douglass, F. S. Rees, J. E. Johns, P. Nair, R. D. Suenram, and B. H. Pate, Conformational isomerization kinetics of pent-1-en-4-yne with 3330 cm-1 of internal energy measured by dynamic rotational spectroscopy. *Proc. Natl. Acad. Sci. USA* 105, 12696, 2008.

28. A. Shojiguchi, C. B. Li, T. Komatsuzaki, and M. Toda, Fractional behavior in multi-dimensional Hamiltonian systems describing reactions. *Phys. Rev. E* 76, 056205, 2007.

29. C. B. Li, Y. Matsunaga, M. Toda, and T. Komatsuzaki, Phase space reaction network on a multisaddle energy landscape: HCN isomerization. *J. Chem. Phys.* 123, 184301, 2005.

30. M. Toda, Global aspects of chemical reactions in multidimensional phase space. *Adv. Chem. Phys.* 130A, 337, 2005.

31. D. Weidinger, M. F. Engel, and M. Gruebele, Freezing vibrational energy flow: A fitness function for interchangeable computational and experimental control. *J. Phys. Chem. A* 113, 4184, 2009.

32. M. Gruebele and P. G. Wolynes, Quantizing Ulam's control conjecture. *Phys. Rev. Lett.* 99, 060201, 2007.

33. B. C. Dian, A. Longarte, and T. S. Zwier, Conformational dynamics in a dipeptide after single-mode vibrational excitation. *Science* 296, 2369, 2002.
34. J. K. Agbo, D. M. Leitner, D. A. Evans, and D. J. Wales, Influence of vibrational energy flow on isomerization of flexible molecules: Incorporating non-RRKM kinetics in the simulation of dipeptide isomerization. *J. Chem. Phys.* 123, 124304, 2005.
35. F. F. Crim, Vibrationally mediated photodissociation—exploring excited-state surfaces and controlling decomposition pathways. *Ann. Rev. Phys. Chem.* 44, 397, 1993.
36. A. A. Stuchebrukhov, M. V. Kuzmin, V. N. Bagratashvili, and V. Lethokov, Threshold energy dependence of IVR in polyatomic molecules. *J. Chem. Phys.* 107, 429, 1986.
37. A. H. Zewail, Femtochemistry: Recent progress in studies of dynamics and control of reactions and their transition states. *J. Phys. Chem.* 100, 1996.
38. M. J. Davis and E. J. Heller, Quantum dynamical tunneling in large molecules. A plausible conjecture. *J. Phys. Chem.* 85, 307, 1981.
39. A. A. Stuchebrukhov and R. A. Marcus, Theoretical study of intramolecular vibrational relaxation of acetylenic CH vibration for v=1 and 2 in large polyatomic molecules $(CX_3)_3YCCH$, where X=H or D and Y=C or Si. *J. Chem. Phys.* 98, 6044, 1993.
40. A. A. Stuchebrukhov, A. Mehta, and R. A. Marcus, Vibrational superexchange mechanism of intramolecular vibrational relaxation in $(CH_3)_3CCCH$. *J. Phys. Chem.* 97, 12491, 1993.
41. A. A. Stuchebrukhov and R. A. Marcus, Perturbation theory approach to dynamical tunneling splitting of local mode vibrational states in ABA molecules. *J. Chem. Phys.* 98, 8443, 1993.
42. D. M. Leitner and P. G. Wolynes, Vibrational relaxation and energy localization in polyatomics: Effects of high-order resonances on flow rates and the quantum ergodicity transition. *J. Chem. Phys.* 105, 11226, 1996.
43. D. M. Leitner and P. G. Wolynes, Many-dimensional quantum energy flow at low energy. *Phys. Rev. Lett.* 76, 216, 1996.
44. C. E. Porter, *Statistical Theories of Spectra: Fluctuations*, Academic Press, New York, 1965.
45. W. F. Polik, D. R. Guyer, W. H. Miller, and C. B. Moore, Eigenstate-resolved unimolecular reaction dynamics—Ergodic character of S_0 formaldehyde at the dissociation threshold. *J. Chem. Phys.* 92, 3471, 1990) .
46. W. F. Polik, D. R. Guyer, and C. B. Moore, Stark level-crossing spectroscopy of S_0 formaldehyde eigenstates at the dissociation threshold. *J. Chem. Phys.* 92, 3453, 1990.
47. T. Zimmermann, H. Köppel, L. S. Cederbaum, G. Persch, and W. Demtröder, Confirmation of random-matrix fluctuations in molecular spectra. *Phys. Rev. Lett.* 61, 3, 1988.
48. D. M. Leitner, Real-symmetric random matrix ensembles of Hamiltonians with partial symmetry-breaking. *Phys. Rev. E* 48, 2536, 1993.
49. D. E. Logan and P. G. Wolynes, Quantum localization and energy flow in many-dimensional Fermi resonant systems. *J. Chem. Phys.* 93, 4994, 1990.
50. M. Gruebele, Molecular vibrational energy flow: A state space approach. *Adv. Chem. Phys.* 114, 193, 2000.
51. D. M. Leitner and P. G. Wolynes, in *ACH-Models in Chemistry, Special Issue, Symposium on Dynamical Information from Molecular Spectra*, A. Schubert (ed.), Library of the Hungarian Academy of Sciences, 1997.
52. R. Bigwood and M. Gruebele, A simple matrix model of intramolecular vibrational redistribution and its implications. *Chem. Phys. Lett.* 235, 604, 1995.
53. R. Bigwood, M. Gruebele, D. M. Leitner, and P. G. Wolynes, The vibrational energy flow transition in organic molecules: Theory meets experiment. *Proc. Natl. Acad. Sci. USA* 95, 5960, 1998.
54. D. E. Logan and P. G. Wolynes, Anderson localization in topologically disordered systems: The effects of band structure. *J. Chem. Phys.* 85, 937, 1986.
55. D. M. Leitner and P. G. Wolynes, Statistical properties of localized vibrational eigenstates. *Chem. Phys. Lett.* 258, 18, 1996.
56. S. Schofield and P. G. Wolynes, A scaling perspective on quantum energy flow in molecules. *J. Chem. Phys.* 98, 1123, 1993.
57. S. A. Schofield and P. G. Wolynes, in *Dynamics of Molecules and Chemical Reactions*, R. E. Wyatt and J. Z. H. Zhang (eds.) Marcel Dekker, New York, 1996.
58. M. Gruebele, Intramolecular vibrational dephasing obeys a power law at intermediate times. *Proc. Nat. Acad. Sci. USA* 95, 5965, 1998.

59. M. Gruebele and R. Bigwood, Molecular vibrational energy flow: Beyond the Golden Rule. *Int. Rev. Phys. Chem.* 17, 91, 1998.

60. A. Semparithi and S. Keshavamurthy, Intramolecular vibrational energy redistributions as diffusion in state space: Classical-quantum correspondence. *J. Chem. Phys.* 125, 141101, 2006.

61. S. Keshavamurthy, Scaling of the average survival probability for low dimensional systems. *Chem. Phys. Lett.* 300, 281, 1999.

62. D. M. Leitner and P. G. Wolynes, Vibrational mixing and energy flow in polyatomics: Quantitative prediction using local random matrix theory. *J. Phys. Chem.* A 101, 541, 1997.

63. J. K. Agbo, D. M. Leitner, E. M. Myshakin, and K. D. Jordan, Quantum energy flow and the kinetics of water shuttling between hydrogen bonding sites on trans-formanilide. *J. Chem. Phys.* 127, 064315, 2007.

64. A. J. Lichtenberg and M. A. Lieberman, *Regular and Stochastic Motion*, Springer, New York, 1983.

65. D. M. Leitner and P. G. Wolynes, Semiclassical quantization of the stochastic pump model of Arnold diffusion. *Phys. Rev. Lett.* 79, 55, 1997.

66. S. Keshavamurthy, Dynamical tunneling in molecules: Quantum routes to energy flow. *Int. Rev. Phys. Chem.* 26, 521, 2007.

10 Dynamical Tunneling and Control

Srihari Keshavamurthy

CONTENTS

10.1 INTRODUCTION

Tunneling is a supreme quantum effect. Every introductory text [1] on quantum mechanics gives the paradigm example of a particle tunneling through a one-dimensional potential barrier despite having a total energy less than the barrier height. Indeed, the reader typically works through a number of exercises, all involving one-dimensional potential barriers of one form or another modeling several key physical phenomena ranging from atom transfer reactions to the decay of α-particles [2]. However, one seldom encounters coupled multidimenisonal tunneling in such texts since an analytical solution of the Schrödinger equation in such cases is not possible. Interestingly, the richness and complexity of the tunneling phenomenon manifest themselves in full glory in the case of multidimensional systems [3]. Thus, for instance, the usual one-dimensional expectation of increasing tunneling splittings as one approaches the barrier top from below is not necessarily true as soon as one couples another bound degree of freedom to the tunneling coordinate. In the context of molecular reaction dynamics, multidimensional tunneling can result in strong mode-specificity and fluctuations in the reaction rates [4]. In fact, a proper description of tunneling of electrons and hydrogen atoms is absolutely essential [5,6] even in molecular systems as large as enzymes and proteins. Although one usually assumes tunneling effects to be significant in molecules involving light atom transfers, it is worth pointing out that neglecting the tunneling of even a heavy atom like carbon is the difference between a reaction occurring or not occurring. In particular, one can underestimate rates by nearly hundred orders of magnitude [7]. Interestingly, and perhaps paradoxically, several penetrating insights into the nature and mechanism of multidimensional barrier tunneling have been obtained from a phase-space perspective [8,9]. The contributions by Creagh, Shudo and Ikeda, and

Takahashi in the present volume provide a detailed account of the latest advances in the phase-space-based understanding of multidimensional barrier tunneling.

What happens if there are no coordinate space barriers? In other words, in situations wherein there are no static energetic barriers separating "reactants" from the "products" does one still have to be concerned about quantum tunneling? One such model potential is shown in Figure 10.1 which will be discussed in the next section. Here we have the notion of reactants and products in a very general sense. So, for instance, in the context of a conformational reaction they might correspond to the several near-degenerate conformations of a specific molecule. Naively one might expect that tunneling has no consequences in such cases. However, studies over last several decades [10–20] have revealed that things are not so straightforward. Despite the lack of static barriers, the dynamics of the system can generate barriers and quantum tunneling can occur through such dynamical barriers [21]. This, of course, immediately implies that dynamical tunneling is a very rich and subtle phenomenon since the nature and number of barriers can vary appreciably with changes in the nature of the dynamics over the timescales of interest. This would also seem to imply that deciphering the mechanism of dynamical tunneling is a hopeless task as opposed to the static potential barrier case wherein elegant approximations to the tunneling rate and splittings can be written down. However, recent studies have clearly established that even in the case of dynamical tunneling it is possible to obtain very accurate approximations to the splittings and rate. In particular, it is now clear that unambiguous identification of the local dynamical barriers is possible only by a detailed study of the structure of the underlying classical phase space. The general picture that has emerged is that dynamical tunneling connects two or more classically disconnected regions in the phase space. More importantly, and perhaps ironically, the dynamical tunneling splittings and rates are extremely sensitive to the various phase space structures like nonlinear resonances [22–29], chaos [16–19,30,31] and partial barriers [32,33]. It is crucial to note that although purely quantum approaches can be formulated for obtaining the tunneling splittings, any mechanistic understanding requires a detailed understanding of the phase-space topology. In this sense, the phenomenon of dynamical tunneling gets intimately linked to issues related to phase-space transport. Thus, one now has the concept of resonance-assisted tunneling (RAT) and chaos-assisted tunneling (CAT) and realistic systems typically involve both the mechanisms.

Since the appearance of the first book [21] on the topic of interest more than a decade ago, there have been several beautiful experimental studies [34–39] that have revealed various aspects of the phenomenon of dynamical tunneling. The most recent one by Chaudhury et al. realizes [40] the paradigmatic kicked top model using cold ^{133}Cs atoms and clearly demonstrate the dynamical tunneling occurring in the underlying phase space. Interestingly, good correspondence between the quantum dynamics and classical phase-space structures is found despite the system being in a deep quantum regime. As another example, I mention the experimental observation [41] by Fölling et al. of second-order co-tunneling of interacting ultracold rubidium atoms in a double-well trap. The similarities between this system and the studies on dynamical tunneling using molecular effective Hamiltonians is striking. In particular, the description of the cold atom study in terms of super-exchange (qualitative and quantitative) is reminiscent of the early work by Stuchebrukhov and Marcus [42] on understanding the role of dynamical tunneling in the phenomenon of intramolecular energy flow. Further details on the experimental realizations can be found in this volume in the articles by Steck and Raizen, and Hensinger. An earlier review [43] provides extensive references to the experimental manifestations of dynamical tunneling in molecular systems in terms of spectroscopic signatures. Undoubtedly, in the coming years, one can expect several other experimental studies which will lead to a deeper understanding of dynamical tunneling and raise many intriguing issues related to the subject of classical-quantum correspondence.

As remarked earlier, it seems ironic that a pure quantum effect like tunneling should bear the marks of the underlying classical phase-space structures. However, it is useful to to recall the statement by Heller that tunneling is only meaningful with classical dynamics as the baseline. Thus, insights into the nature of the classical dynamics translates into a deeper mechanistic insight into the corresponding quantum dynamics. Indeed, one way of thinking about classical-quantum correspondence is that classical mechanics is providing us with the best possible "basis" to describe the quantum evolution. The wide range of contributions in this volume are a testimony to the richness of the phenomenon of dynamical tunneling and the utility of such a classical-quantum correspondence perspective. In this chapter, I focus on the specific field of quantum control and show as to how dynamical tunneling can lead to useful insights into the control mechanism [44,45]. The hope is that more such studies will eventually result in control strategies which are firmly rooted in the intutive classical world, yet accounting for the classically forbidden pathways and mechanisms in a consistent fashion. This is a tall order, and some may even argue as an unnecessary effort in these days of fast computers and smart and efficient algorithms to solve fairly high-dimensional quantum dynamics. However, in this context, it is useful to remember the following, which was written by Born, Heisenberg, and Jordan nearly 80 years ago [46]:

> The starting point of our theoretical approach was the conviction that the difficulties that have been encountered at every step in quantum theory in the last few years could be surmounted only by establishing a mathematical system for the mechanics of atomic and electronic motions, which would have a unity and simplicity comparable with the system of classical mechanics ... further development of the theory, an important task will lie in the closer investigation of the nature of this correspondence and in the description of the manner in which symbolic quantum geometry goes over into visualizable classical geometry.

The above remark was made in an era when computers were nonexistent. Nevertheless, it is remarkably prescient since even in the present era one realizes the sheer difficulty in implementing an all-quantum dynamical study on even relatively small molecules [47]. In any case, it is not entirely unreasonable to argue that large-scale quantum dynamical studies will still require some form of an implicit classical-quantum correspondence approach to grasp the underlying mechanistic details. With the above remark in mind I start things off by revisiting the original paper [11] by Davis and Heller since, in my opinion, it is ideal from the pedagogical point of view.

10.2   DAVIS–HELLER SYSTEM REVISITED

Three decades ago, Davis and Heller in their pioneering study [11] gave a clear example of dynamical tunneling. A short recount of this work including the famous plots of the classical trajectories and the associated quantum eigenstates can be found in Heller's article in this volume. However, I revisit this model here in order to bring forth a couple of important points that seem to have been overlooked in subsequent works. First, the existence of another class of eigenstate pairs, called as circulating states [11], which can exert considerable influence on the usual tunneling doublets at higher energies. Second, a remark in the original paper [11] which can be considered as a harbinger for CAT. As shown below, there are features in the original model that are worth studying in some detail even after three decades since the original paper was published.

The Hamiltonian of choice is the two degrees of freedom (2DoF) Barbanis-like model [11]

$$H(s,u,p_s,p_u) = \frac{1}{2}\left(p_s^2 + p_u^2\right) + \frac{1}{2}\left(\omega_s^2 s^2 + \omega_u^2 u^2\right) + \lambda s u^2, \tag{10.1}$$

with the labels "s" and "u" denoting the symmetric and unsymmetric stretch modes respectively. The above Hamiltonian has also been studied [48] in great detail to uncover the correspondence between classical stability of the motion and quantum spectral features, wavefunctions, and energy transfer. The potential is symmetric with respect to $u \leftrightarrow -u$ as shown in Figure 10.1 but there is no potential barrier. Davis and Heller used the parameter values $\omega_s = 1.0, \omega_u = 1.1$, and $\lambda = -0.11$ for which the dissociation energy $E_{\text{dis}} \equiv \omega_s^2 \omega_u^4 / 8\lambda^2 = 15.125$. Note that the masses are taken to be unity and one is working in units such that $\hbar = 1$. The key observation by Davis and Heller was that despite

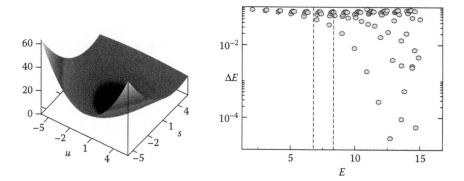

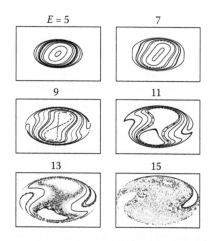

FIGURE 10.1 The left plot shows the two-dimensional potential used by Davis and Heller to illustrate the phenomenon of dynamical tunneling. Note that there are no barriers in the coordinate space. On the right the magnitude of the splitting between adjacent eigenstates are shown as a function of the eigenenergies. The first and second dotted vertical lines indicate the energies around which the symmetric (s) and unsymmetric (u) stretch modes become unstable. Sequences of doublets with very small splittings can be seen around the onset of the instabilities. On the other hand, the regularity of the sequences vanishes at higher energies.

the lack of any potential barriers several bound eigenstates came in symmetric–antisymmetric pairs $|\psi_1\rangle$ and $|\psi_2\rangle$ with energy splittings much smaller than the fundamental frequencies, i.e., $\Delta E \equiv |E_1 - E_2| \ll O(1)$. In Figure 10.1, the various splittings between adjacent eigenstates are shown and it is clear that several "tunneling" pairs appear above a certain threshold energy.

How can one understand the onset of such near degeneracies in the system? The crucial insight that Davis and Heller provided was that the appearance of such doublets is correlated with the large scale changes in the classical phase space. The nature of the phase space with increasing total energy is shown in Figure 10.2 using the (u, p_u) Poincaré surface of section. Such a surface of section, following standard methods, is constructed by recording the points, with momentum $p_s > 0$, of the intersection of a trajectory with the the $s = 0$ plane in the phase space. Such a procedure for

FIGURE 10.2 Evolution of the classical phase space with increasing energy shown as (u, p_u) Poincaré surface of sections. Note that large-scale chaos appears for $E \geq 11$. The formation of a separatrix and two classically disjoint regular regions can be seen at $E = 9$ due to the symmetric mode periodic orbit becoming unstable. The regular regions almost vanish near the dissociation energy. In the bottom row some of the chaotic orbits have been suppressed, for clarity of the figure, by showing them in gray.

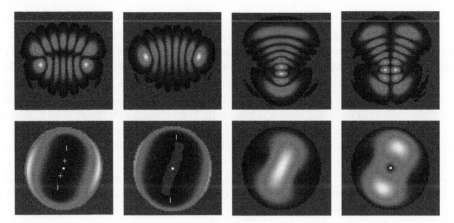

FIGURE 10.3 Four eigenstates close to an energy where the 1:1 resonance is just starting to appear. The upper panel shows the (u, s) coordinate space representations and the lower panels show the corresponding Husimi distributions in the (u, p_u) surface of section of the phase space. The first two states ($E \approx 8.4$) can be assigned approximate zeroth-order quantum numbers (n_s, n_u) whereas the last two states ($E \approx 9.0$) show perturbed nodal features. Clear difference in the phase-space nature of the eigenstates can be seen. See text for details.

several trajectories with specific total energy generates a typical surface of section as shown in Figure 10.2 and clearly indicates the global nature of the phase space. It is clear from the figure that the phase space for $E < 11$ is mostly regular while higher energy phase spaces exhibit mixed regular-chaotic dynamics. One of the most prominent change happens when the symmetric stretch periodic orbit ($u = 0, p_u = 0$) becomes unstable around $E \approx 6.8$. In Figure 10.2, the consequence of such a bifurcation can be clearly seen for $E = 9$ as the formation of two classically disjoint regular regions. In fact, the two regular regions signal a 1:1 resonance between the stretching modes. The crucial point to observe here is that classical trajectories initiated in one of the regular regions cannot evolve into the other regular region. With increasing energy, the classically disjoint regular regions move further apart and almost vanish near the dissociation energy. The result presented in Figure 10.1 in fact closely mirrors the topological changes, shown in Figure 10.2, in the phase space. Thus, in Figure 10.1 a sequence of eigenstates with very small splittings begins right around the energy at which the symmetric periodic orbit becomes unstable. The unsymmetric stretch, however, becomes unstable at a higher energy $E \approx 8.3$ and one can observe in Figure 10.1 another sequence that seemingly begins near this point. Note that at higher energies it is not easy to identify any more sequences, but small splittings are still observed. The important thing to note here is that not only do the splittings but the individual eigenstates also correlate with the changes in the phase space.

In Figure 10.3, we show a set of four eigenstates to illustrate an important point—as soon as the 1:1 resonance manifests itself in the phase space, the tunneling doublets start to form and an integrable normal form approximation is insufficient to account for the splittings. Note that the phase space is mostly regular at the energies of interest. A much more detailed analysis can be found in the chapter by Farrelly and Uzer [49]. We begin by noting* that the pair of eigenstates (counting from the zero-point) 35 and 36 are split by about $\Delta E \approx 6.41 \times 10^{-2}$, whereas the pair 37 and 38 are separated by about 2.09×10^{-2}. In both cases, the splitting is smaller than the fundamental frequencies,

---

* In this section, the eigenstate numbers that are reported come from solving the time-independent Schrödinger equation in the uncoupled harmonic oscillator basis $|n_s, n_u\rangle$ such that $n_s + n_u \leq 60$. This yields a total of 1891 basis states. Convergence of the bound state energies has been checked with increasing basis size. The states shown in this section are converged to sufficient accuracy.

which are of the order of unity. Note that the latter pair of states appears to be a part of the sequence in Figure 10.1 that starts right after the bifurcation of the symmetric stretch periodic orbit. Since the original mode frequencies are nonresonant, one can obtain a normal form approximation [49] to the original Hamiltonian and see if the obtained splittings can be explained satisfactorily. In other words, the observed ΔE are coming from the perturbation that couples both the modes and in such a case there is no reason for classifying them as tunneling doublets. However, from the coordinate space representations of the states shown in Figure 10.3 one observes that there is an important difference between the two pairs of states. The pair $(35, 36)$ seems to have a perturbed nodal structure and hence one can approximately assign the states using the zeroth-order quantum numbers (n_s, n_u). Inspecting the figure leads to the assignment $(1, 6)$ and $(0, 7)$ for states 35 and 36 respectively. If the above arguments are correct then the splitting $\Delta E_{35,36}$ should be obtainable from the normal form Hamiltonian. Since the theory of normal forms is explained in detail in several textbooks [50], I will provide a brief derivation below. Begin by using the unperturbed harmonic action-angle variables $(\mathbf{I}, \boldsymbol{\phi})$ defined via the canonical transformation (similar set for s mode):

$$u = \left(\frac{2I_u}{\omega_u}\right)^{1/2} \sin \phi_u, \tag{10.2}$$

$$p_u = (2\omega_u I_u)^{1/2} \cos \phi_u, \tag{10.3}$$

to express the original Hamiltonian as

$$H(\mathbf{I}, \boldsymbol{\phi}) = \omega_s I_s + \omega_u I_u + \tilde{\lambda} I_s^{1/2} I_u \left\{ \sin \phi_s - \frac{1}{2} \sin(\phi_s + 2\phi_u) - \frac{1}{2} \sin(\phi_s - 2\phi_u) \right\}$$
$$\equiv H_0(\mathbf{I}) + \tilde{\lambda} H_1(\mathbf{I}, \boldsymbol{\phi}), \tag{10.4}$$

where $\tilde{\lambda} \equiv \lambda\sqrt{2}/(\omega_u\sqrt{\omega_s})$. Since H_1 is purely oscillatory, the angle average

$$\bar{H}_1 \equiv \frac{1}{(2\pi)^2} \int d\boldsymbol{\phi} H_1(\mathbf{I}, \boldsymbol{\phi}) = 0 \tag{10.5}$$

and thus to $O(\tilde{\lambda})$ the normal form Hamiltonian can be identified with the zeroth-order $H_0(\mathbf{I})$ above. The first nontrivial correction arises at $O(\tilde{\lambda}^2)$ and can be obtained using the $O(\tilde{\lambda})$ generating function

$$W_1(\mathbf{I}, \boldsymbol{\phi}) = \sum_{\mathbf{k} \neq 0} \frac{i}{\mathbf{k} \cdot \boldsymbol{\omega}} H_{1,\mathbf{k}}(\mathbf{I}) e^{i\mathbf{k} \cdot \boldsymbol{\phi}}, \tag{10.6}$$

where $H_{1,\mathbf{k}}(\mathbf{I})$ are the coefficients in the Fourier expansion of the oscillatory part of $H_1(\mathbf{I}, \boldsymbol{\phi})$. One now obtains the $O(\tilde{\lambda}^2)$ correction to the Hamiltonian as

$$K_2 = \frac{1}{2}\overline{\{W_1, H_1\}}, \tag{10.7}$$

with the bar denoting angle averaging of the Poisson bracket involving W_1 and H_1. Performing the calculations the normal form at $O(\tilde{\lambda}^2)$ is obtained as

$$H_N^{(2)} = \omega_s I_s + \omega_u I_u + \tilde{\lambda}^2 K_2 \tag{10.8}$$

with

$$K_2 = \frac{1}{\omega_s \omega_u^2} \left[\left(\frac{2\omega_u}{\omega_s^2 - 4\omega_u^2}\right) I_s I_u - \left(\frac{3\omega_s^2 - 8\omega_u^2}{4\omega_s(\omega_s^2 - 4\omega_u^2)}\right) I_u^2 \right]. \tag{10.9}$$

The primitive Bohr–Sommerfeld quantization $I_j \to (n_j + 1/2)\hbar$ yields the quantum eigenvalues perturbatively to $O(\tilde{\lambda}^2)$. The procedure can be repeated to obtain the normal form Hamiltonian at higher orders. For instance, Farrelly and Uzer have [49] computed the normal form out to $O(\tilde{\lambda}^{12})$ and used Padé resummation techniques to improve in cases when the zeroth-order frequencies are near resonant. For our qualitative discussions, the $O(\tilde{\lambda}^2)$ normal form is sufficient.

Note that the normal form Hamiltonian above is integrable since it is ignorable in the angle variables. Indeed, using the normal form and the approximate assignments of the eigenstates shown in Figure 10.3 one finds $\Delta E_{35,36} \approx 7.4 \times 10^{-2}$ which is in fair agreement with the actual numerical value. Thus, a large part of the splitting $\Delta E_{35,36}$ can be explained classically. On the other hand, although the states 37 and 38 seem to have an identifiable nodal structure (cf. Figure 10.3), it is clear that they are significantly perturbed. Nevertheless, persisting with an approach based on counting the nodes, states 37 and 38 can be assigned as $(8,0)$ and $(7,1)$ respectively. Using the normal form one estimates $\Delta E_{37,38} \approx 5.12 \times 10^{-2}$ and this is about a factor of two larger than the numerically computed value. Thus, the splitting in this case is not accounted for solely by classical considerations. One might argue that a higher-order normal form might lead to better agreement, but the phase-space Husimi distributions of the eigenstates shown in Figure 10.3 suggests otherwise. The Husimi disributions, when compared to the classical phase spaces shown in Figure 10.2, clearly show that the pair $(37, 38)$ are localized in the newly created 1:1 resonance zone. Hence, this pair of states is directly influenced by the nonlinear resonance and the splitting between them cannot be accurately described by the normal form Hamiltonian. Indeed, as discussed by Farrelly and Uzer [49], in this instance one needs to consider a resonant Hamiltonian which explicitly takes the 1:1 resonance into account. This provides a clear link between dynamical tunneling, appearance of closely spaced doublets and creation of new, in this case a nonlinear resonance, phase-space structures. The choice of states in Figure 10.3 is different from what is usually shown as the standard example for dynamical tunneling pairs. However, in the discussion above the states were chosen intentionally with the purpose of illustrating the onset of near-degeneracy due to the formation of a nonlinear resonance. In a typical situation involving near-integrable phase spaces there are several such resonances ranging from low orders to fairly high orders. The importance of a specific resonance depends sensitively on the effective value of the Planck's constant [51,52]. Indeed, detailed studies [53–55] have shown that excellent agreement with numerically computed splittings can be obtained if proper care is taken to include the various resonances. Clear and striking examples in this context can be found in the contributions by Schlagheck et al. and Bäcker et al. in this volume.

To finish the discussion of the Davis–Heller system, I show an example which involves states forming tunneling pairs that are fairly complicated both in terms of their coordinate space representations as well as their phase-space Husimi distributions. The example involves three states 102, 103, and 104 around $E \approx 14.4$, which is rather close to the dissociation energy. In the original work [11] Davis and Heller noted that the splittings seem to increase by an order of magnitude in this high energy region. Notably, they commented that *perhaps the degree of irregularity between the regular regions plays some part.* In Figure 10.4 the variation of energy levels with the coupling parameter λ is shown. One can immediately see that the three states are right at the center of an avoided crossing. The coordinate space representations of the states shows extensive mixing for states 102 and 104 while the state 103 seems to be cleaner. The Husimi distributions for the respective states conveys the same message. Compared to the phase-space sections shown in Figure 10.2, it is clear that the Husimis for state 102 and 104 seem to be ignoring the classical regular-chaotic divison—a clear indication of the quantum nature of the mixing. Although it is not possible to strictly assign these states as chaotic, a closer inspection does show substantial Husimi contribution in the border between the regular and chaotic regions. Interestingly, the pairwise splitting between the states is nearly the same but localized linear combinations of any two states exhibits two-level dynamics. Thus, the situation here is not of the generic [18] CAT one wherein one of the states is chaotic and interacts with the other two regular states. Nevertheless, linear combinations of the three states reveal (not shown here) that they are mixed with each other. A look at the coordinate and phase-space representations of states 102 and 104 in Figure 10.4 reveals that a different kind of state is causing the three-way interaction. It turns out that in the Davis–Heller model there is another

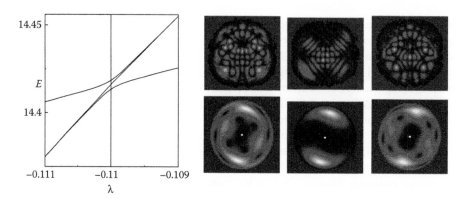

FIGURE 10.4 An avoided crossing in the high energy region ($E \approx 14.4$) of the Davis–Heller system involving three states. As in the previous figure, the (u, s) and phase-space Husimi representations are also shown. The first and the last states appear to be strongly mixed states. The middle state seems relatively cleaner. Does chaos play a role in this case? Detailed discussions in the text.

class of symmetry-related pairs that appear when the unsymmetric mode becomes unstable. In the original work they were refered to as "circulating" states which displayed much larger splitting then the so-called local mode pairs. The case shown in Figure 10.4 involves one of the circulating pairs interacting with the usual local mode doublet leading to the complicated three-way interaction. The subtle nature of this interaction is evident from the coordinate space representation of state 103 which exhibits a broken symmetry. Interestingly, and as far as I can tell, there has been very little understanding of such three-state interactions in the Davis–Heller system and further studies are needed to shed some light on the phase-space nature of the relevant eigenstates.

10.3 DYNAMICAL TUNNELING AND CONTROL: TWO EXAMPLES

In the previous section, the close connection between dynamical tunneling and phase space structures was introduced. The importance of a nonlinear resonance and a hint of the role played by the chaos (cf. Figure 10.4) is evident. Several other contributions in this volume discuss the importance of various phase-space structures using different models, both continuous Hamiltonians and discrete maps. In the molecular context, various mode–mode resonances play a critical role in the process of intramolecular vibrational redistribution (IVR) of energy [56]. The phenomenon of IVR is at the heart of chemical reaction dynamics and it is now well established that molecules at high levels of excitation display all the richness, complexity and subtelty that is expected from nonlinear dynamics of multidimensional systems [57]. In this context, dynamical tunneling is an important agent [43,58] of IVR and state mixing for a certain class of initial states (akin to the so-called NOON states [59]) which are typically prepared by the experiments. The importance of the anharmonic resonances to IVR and the regimes wherein dynamical tunneling, mediated by these resonances, is expected to be crucial is described in some detail in this volume by Leitner. The phase-space perspective on Leitner's viewpoint can be found in a recent review [43] (see also Heller's contribution in the present volume). In this regard it is interesting to note that there has been a renaissance of sorts in chemical dynamics with researchers critically examining the validity of the two pillars of reaction rate theory [60]—transition state theory (TST) and the Rice–Ramsperger–Kassel–Marcus (RRKM) theory. Since both theories have classical dynamics at their foundation, advances in our understanding of nonlinear dynamics and continuing efforts to characterize the phase-space structure of systems with three or more degrees of freedom are beginning to yield crucial mechanistic insights into the dynamics [61,62]. At the same time, rapid advances in experimental techniques and theoretical understanding of the reaction mechanisms has led researchers to focus on the issue of controlling the dynamics of molecules. What implications might dynamical tunneling have on our efforts to control the atomic and molecular dynamics? In the rest of this Chapter, I focus on this issue and use two seemingly simple and well studied systems as examples to highlight the role of dynamical tunneling in the context of coherent control. Both examples are in the context of periodically driven systems, and I refer the reader to the work of Flatté and Holthaus [63] for an exposition of the close quantum-classical correspondence in such systems.

10.3.1   DRIVEN QUARTIC DOUBLE WELL: CHAOS-ASSISTED TUNNELING

Historically, an early indication that dynamical tunneling could be sensitive to the chaos in the underlying phase space came from the study of strongly driven double-well potential by Lin and Ballentine [17]. The model Hamiltonian in this case can be written down as

$$H(x,p;t) = H_0(x,p) + \lambda_1 x \cos(\omega_F t), \tag{10.10}$$

with ω_F being the frequency of the monochromatic field (henceforth refered to as the driving field) and the unperturbed part

$$H_0(x,p) = \frac{1}{2M}p^2 + Bx^4 - Dx^2 \tag{10.11}$$

is the Hamiltonian corresponding to a double-well potential with two symmetric minima at $x = \pm(D/2B)^{1/2}$ and a maximum at $x = 0$. Following the original work [17], the parameters of the unperturbed system (assuming atomic units) are taken to be $M = 1$, $B = 0.5$, and $D = 10$ for which the potential has a barrier height $V_B = 50$ and supports about eight tunneling doublets. As is well known, in the absence of the driving field, a wavepacket prepared in the left well can coherently tunnel into the right well with the time scale for tunneling being inversely proportional to the tunnel splitting. This unperturbed scenario, however, is significantly altered in the presence of a strong driving field. In the presence of a strong field the phase space of the system exhibits large-scale chaos coexisting with two symmetry-related regular regions. Lin and Ballentine observed that a coherent state localized in one of the regular region tunnels to the other symmetry-related regular region on timescales which are orders of magnitude smaller than in the unperturbed case. It was suspected that the extensive chaos in the system might be assisting the tunneling process.

In order to illustrate the tunneling process, Figure 10.5a shows the stroboscopic surface of section for the case of strong driving with $\lambda_1 = 10$. Note the extensive chaos and the two regular islands (left and right) in the phase space. A coherent state $|z\rangle \equiv |x_0, p_0\rangle$ is placed in the center of the left island and time evolved using the Floquet approach which is ideally suited for time-periodic driven systems. In this instance, one is interested in the time at which the coherent state localized on the left tunnels over to the regular region on the right. In order to obtain this information it is necessary to compute the survival probability of the initial coherent state. Briefly, Floquet states $\{|\chi_n\rangle\}$ are eigenstates of the Hermitian operator $H - i\hbar\partial/\partial t$ and form a complete orthonormal basis. An arbitrary time-evolved state $|z(t)\rangle$ can be expressed as

$$|z(t)\rangle = \sum_n A_n e^{-iE_n t}|\chi_n(t)\rangle, \tag{10.12}$$

with E_n being the quasienergy associated with the Floquet state $|\chi_n\rangle$. The expansion coefficients A_n are independent of time and given by

$$A_n = \langle\chi_n(0)|z(0)\rangle, \tag{10.13}$$

yielding the expansion

$$|z(t)\rangle = \sum_n e^{-iE_n t}|\chi_n(t)\rangle\langle\chi_n(0)|z(0)\rangle. \tag{10.14}$$

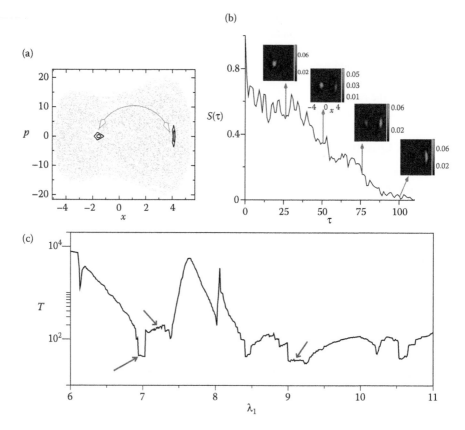

FIGURE 10.5 (a) Phase space for the driven double well system with field strength $\lambda_1 = 10$. A coherent state localized in the left regular island tunnels (indicated by a green arrow) over to the right regular island. The chaotic regions (gray) have been suppressed for clarity. (b) Monitoring the survival probability of the initial state to determine the timescale of tunneling ($\sim 100\tau$ in this case). Snapshots of the evolving Husimi distributions are also shown at specific intervals, with green (gray) indicating maxima of the distributions. (c) The decay time determined as in (b) for a variety of field strengths with the initial state localized in the left island. Note the fluctuations in the decay time over several orders of magnitude despite very similar nature of the phase spaces over the entire range of the driving field strength. The arrows highlight some of the plateau regions which are crucial for bichromatic control.

Measuring time in units of field period (T_f) and owing to the periodicity of the Floquet states, $|\chi_n(t)\rangle = |\chi_n(t + T_f)\rangle$, the above equation simplifies to

$$|z(\tau)\rangle = \sum_n e^{-iE_n\tau} |\chi_n(0)\rangle\langle\chi_n(0)|z(0)\rangle \equiv \hat{U}(\tau)|z(0)\rangle, \qquad (10.15)$$

with $\tau \equiv kT_f$ and integer k.

The time evolution operator

$$\hat{U}(\tau) = \sum_n e^{-iE_n\tau} |\chi_n(0)\rangle\langle\chi_n(0)| \qquad (10.16)$$

is determined by successive application of the one-period time evolution operator $U(T_f, 0)$, i.e.,

$$\hat{U}(kT_f, 0) = [\hat{U}(T_f, 0)]^k. \qquad (10.17)$$

This allows us to express the survival probability of the initial coherent state in terms of Floquet states as

$$
\begin{aligned}
S(\tau) &\equiv |\langle z(0)|z(\tau)\rangle|^2 = \left| \sum_n e^{-iE_n\tau} \langle z(0)|\chi_n(0)\rangle \langle \chi_n(0)|z(0)\rangle \right|^2 \\
&= \sum_{m,n} p_{zn} p_{zm} e^{-i(E_n - E_m)\tau},
\end{aligned}
\tag{10.18}
$$

where the overlap intensities are denoted by $p_{zn} \equiv |\langle z(0)|\chi_n(0)\rangle|^2$. In order to determine the "lifetime" (T) of the initial coherent state, we monitor the time at which $S(\tau)$ goes to zero (minimum) at the first instance. In other words, it is the time when the coherent state leaves its initial position (left regular island) for the first time. In Figure 10.5b the $S(\tau)$ for the initial state of interest is shown along with the snapshots of Husimi distribution at specific times. Clearly, the initial state tunnels in about $100 T_f$, which suggests that chaos assists the dynamical tunneling process. However, the issue is subtle and highlighted in Figure 10.5c which shows the decay time plot for a range of driving field strengths for an intial state localized in the left regular island. It is important to note that the gross features of the phase space are quite similar over the entire range. However, Figure 10.5c exhibits strong fluctuations over several orders of magnitude and this implies that a direct association of the decay time with the extent of chaos in the phase space is not entirely correct.

The above discussion and results summarized in Figure 10.5 bring up the following key question. What is the mechanism by which the initial state $|z\rangle$ decays out of the regular region? In turn, this is precisely the question that modern theories of dynamical tunneling strive to answer. According to the theory of RAT [24,25,51], the mechanism is possibily one wherein $|z\rangle$ couples to the chaotic sea via one or several nonlinear resonances provided certain conditions are satisfied. Specifically, the local structure of the phase space surrounding the regular region is expected to play a critical role. The theoretical underpinnings of RAT along with several illuminating examples can be found in the contribution by Schlagheck et al. and here we suggest a simple numerical example which points to the importance of the local phase-space structure around $|z\rangle$.

Preliminary evidence for the role of field-matter nonlinear resonances in controlling the decay of $|z\rangle$ is given in Figure 10.6, which shows the effect of changing the driving field frequency ω_F on the local phase-space structures and the decay times. From Figure 10.6 it is apparent that detuning ω_F by ± 0.05 leads to a significant change in the local phase-space structure and the decay time of the initially localized state. In Figure 10.6b, corresponding to the field frequency $\omega_F = 6.02$, a prominent 6:1 field-matter resonance is observed. It is plausible that the the decay time is only a few hundred field periods in this case due to assistance from the nonlinear resonance. However, the decay time increases for $\omega_F = 6.07$ and becomes even larger by an order of magnitude for $\omega_F = 6.12$, due to absence of the 6:1 resonance. This indicates that decay dynamics is highly sensitive to the changes in local phase space structure of the left regular island. Therefore, taking into account the relatively large-order resonances, since $\hbar = 1$, is unavoidable in order to understand the decay time plot in Figure 10.5 and for smaller values of the effective Planck constant one expects a more complicated behavior. Note, however, that the very high-order island chain visible in the last case in Figure 10.6 is unable to assist the decay. This is where we believe that an extensive $\hbar$-scaling study will help in gaining a deeper understanding of the decay mechanism. Such an extensive calculation can be found, for example, in the recent work [53] by Mouchet, Eltschka, and Schlagheck on the driven pendulum. There is sufficient evidence [53] in the driven pendulum system for a mechanism in which nonlinear resonances play a central role in coupling initial states localized in regular phase-space regions to the chaotic sea. One might be able to provide a clear qualitative and quantitative explanation for the results in Figure 10.5c based on the recent advances.

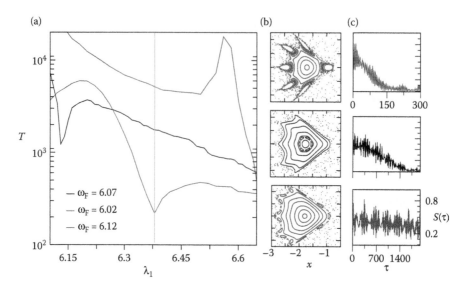

FIGURE 10.6 Decay time plot for $\omega_F = \omega_F^{(0)} \equiv 6.07$ (black) and for $\omega_F^{(0)} + 0.05$ (blue) and $\omega_F^{(0)} - 0.05$ (red) as a function of λ_1. The local phase structure in the vicinity of the initial coherent state and the survival probability for fixed $\lambda_1 = 6.4$ (indicated by green line in (a)) corresponding to red shifted, fundamental and blue shifted field frequency is shown in panels (b) and (c) respectively. The 6:1 field-matter resonance in case of $\omega_F^{(0)} - 0.05$ is clearly visible and correlates with the very short decay time as compared to the other two cases. Note the different time axis scale in the first survival probability plot.

10.3.1.1 CAT Spoils Bichromatic Control

This brings us to the second important issue—what is the role of chaos in this dynamical tunneling process? An earlier work [64] by Utermann, Dittrich, and Hänggi on the driven double-well system showed that there is indeed a strong correlation between the splittings of the Floquet states and the overlaps of their Husimi distributions with the chaotic regions in the phase space. However, a different perspective yields clear insights into the role of chaos with potential implications for coherent control. In order to highlight this perspective I start with a simple question: *is it possible to control the decay of the localized initial state with an appropriate choice of a control field?* It is important to note that the Lin–Ballentine system parameters imply that one is dealing with a multilevel control scenario and there has been a lot of activity over the last few years to formulate control schemes involving multiple levels in both atomic and molecular systems. The driven double-well system has been a particular favorite in this regard, more so in recent times due to increased focus on the physics of trapped Bose–Einstein condensates [65]. More specifically, several studies have explored the possibility of controlling various atomic and molecular phenomenon using bichromatic fields with the relative phase between the fields providing an additional control parameter. In the present context, for example, Sangouard et al. exploited the physics of adiabatic passage to show that an appropriate combination of $(\omega_F, 2\omega_F)$ field leads to suppression of tunneling in the driven double-well model [66]. The choice of relative phase between the two fields allowed them to localize the initial state in one or the other well. However, the parameter regimes in the work by Sangouard et al. correspond to the underlying phase space being near integrable and hence a minimal role of the chaotic sea.

More relevant to the mixed regular-chaotic phase-space case presented in Figure 10.5 is an earlier work [67] by Farrelly and Milligan wherein it was demonstrated that one can suppress the tunneling dynamics in a driven double-well system using a $(\omega_F, 2\omega_F)$ bichromatic field. In other words, the original Hamiltonian of Equation 10.10 is modified as follows:

$$H(x,p;t) = H_0(x,p) + \lambda_1 x \cos(\omega_F t) + \lambda_2 x \cos(2\omega_F t + \phi), \qquad (10.19)$$

with $H_0(x,p)$ being the same as in Equation 10.11 and the additional $2\omega_F$-field is taken to be the control field. Moreover, modulating the turn-on time of the control field can trap the wavepacket in the left or right well of the double-well potential. Hence, it was argued [67] that the tunneling dynamics in a driven double well can be controlled at will for specific choices of the control field parameters (λ_2,ϕ). Note that in the presence of control field, i.e., $\lambda_2 \neq 0$ with the relative phase $\phi = 0$ the Hamiltonian in Equation 10.19 transforms under symmetry operations as

$$H\left(-x,-p;t+\frac{\pi}{\omega_F}\right) = H_0(x,p) - x[\lambda_1 \cos(\omega_F t + \pi) + \lambda_2 \cos(2\omega_F t + 2\pi)]$$
$$= H_0(x,p) - x[-\lambda_1 \cos(\omega_F t) + \lambda_2 \cos(2\omega_F t)]$$
$$\neq H(x,p;t). \qquad (10.20)$$

Similarly, except at $\phi = \pi/2$, the discrete symmetry of the Hamiltonian is broken under the influence of the additional $2\omega_F$-field. Farrelly and Milligan thus argued [67] that the control field with strength smaller than the driving field will lead to localization due to the breaking of the generalized symmetry of the Hamiltonian and the Floquet states. The impact of a small symmetry breaking control field with $(\lambda_2,\phi) = (0.7,0)$ can be clearly seen in Figure 10.7 in terms of the changes in the classical phase space structures. However, there is a subtlety which is not obvious upon inspecting the phase spaces shown in Figure 10.7 for two different but close values of the driving field strength. In case of $\lambda_1 = 9.2$ corresponding to Figure 10.7a, the control field is unable to suppress the decay of the initial state $|z\rangle$ localized in the left regular region. On the other hand, Figure 10.7b corresponds to $\lambda_1 = 10$ and computations show that the control field is able to suppress the decay of $|z\rangle$ to an appreciable extent. Thus, although in both cases the control field breaks the symmetry and the resulting phase spaces show very similar structures, the extent of control exerted by the $2\omega_F$-field is drastically different.

The discussions above and the results summarized in Figures 10.6 and Figure 10.7 clearly indicate that the decay of $|z\rangle$ out of the left regular region can be very different and deserves to be understood in greater detail. In fact, computations show that such cases of complete lack of control are present for other values of λ_1 as well. This is confirmed by inspecting Figure 10.8 which shows the control landscape for the bichromatically driven double well in the specific case of $\phi = 0$. Other

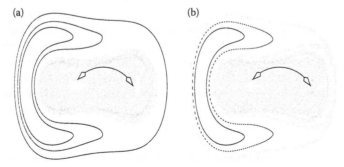

FIGURE 10.7 Stroboscopic surface of sections for the bichromatically driven double-well system with the control field strength fixed at $\lambda_2 = 0.7$ and primary field strength (a) $\lambda_1 = 9.2$ and (b) $\lambda_1 = 10.0$. As in the earlier plots, the chaotic regions have been suppressed (in gray) for clarity. Note the breaking of the symmetry in both cases. However, significant suppression of the decay of the state localized in the left regular island happens in case (b) only.

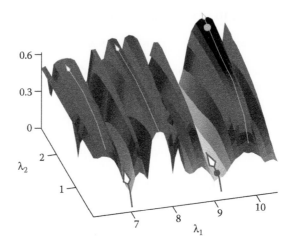

FIGURE 10.8 Time-smoothed survival probability representation of the control landscape for $\phi = 0$ as a function of the field parameters (λ_1, λ_2). The initial state in every case is localized in the left regular region in the classical phase space. Notice the convoluted form of the landscape with the regions of low probability indicating little to no control (red thick arrows). The green lines (thin arrows) indicate regions where a high degree of control can be achieved.

choices for ϕ also show similarly convoluted landscapes. There are several ways of presenting a control landscape and in Figure 10.8 the time-smoothed survival probability (cf. Equation 10.18)

$$\langle S \rangle = \frac{1}{\tau} \int_0^\tau d\tau' S(\tau') \tag{10.21}$$

associated with the initial state $|z\rangle$ is used to map the landscape as a function of the field strengths (λ_1, λ_2). Note that the choice of $\langle S \rangle$ to represent the landscape is made for convenience; the decay time is a better choice which requires considerable effort but the gross qualitative features of the control landscape do not change upon using $\langle S \rangle$. Large (small) values of $\langle S \rangle$ indicate that the decay dynamics is suppressed (enhanced). It is clear from Figure 10.8 that the landscape comprises of regions of control interspersed with regions exhibiting lack of control. Such highly convoluted features of control landscape are a consequence of the simple bichromatic choice for the control field and the nonlinear nature of the corresponding classical dynamics. From a control point of view, there are regions on the landscape for which a monotonic increase of λ_2 leads to increasing control. Interestingly, the lone example of control illustrated in Farrelly and Milligan's paper [67] happens to be located on one of the prominent hills on the control landscape (shown as a green dot in Figure 10.8). A striking feature that can be seen in Figure 10.8 is the deep valley around $\lambda_1 = 9.2$ which signals an almost complete lack of control even for significantly large strengths of the $2\omega_F$-field. The valleys in Figure 10.8 correspond precisely to the plateaus seen in the decay time plot shown in Figure 10.6 for $\lambda_2 = 0$ (red arrows). It is crucial to note that this "wall of no control" is robust even upon varying the relative phase ϕ between the driving and the control field.

Insights into the lack of control for driving field strength $\lambda_1 = 9.2$ (and other values as well which are not discussed here) can be obtained by studying the variation of the Floquet quasienergies with λ_2, the control field strength. In Figure 10.9a the results of such a computation are shown and it is immediately clear that even in the presence of $2\omega_F$-control field six states contribute to the decay of $|z\rangle$ over the entire range of λ_2. This is further confirmed in Figure 10.9b where the plot of the overlap intensities shows multiple Floquet states participating nearly equally in the decay dynamics of the initial state. The final clue comes from inspecting the Husimi distributions shown

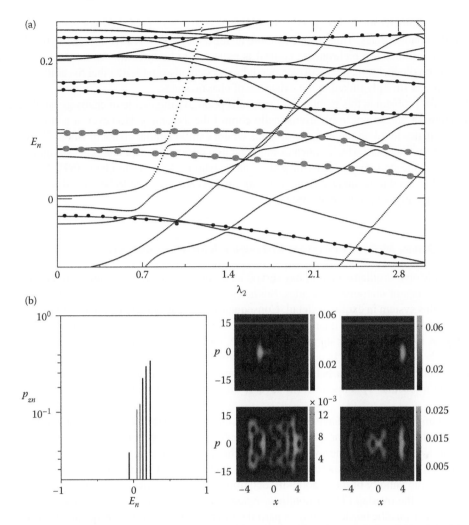

FIGURE 10.9 (a) Variation of the Floquet quasienergies with λ_2. The primary driving field strength is fixed at $\lambda_1 = 9.2$. Six states that have appreciable overlap with $|z\rangle$ are highlighted by circles. (b) Overlap intensity p_{zn} for $\lambda_1 = 9.2$ and $\lambda_2 = 2.1$ indicates multilevel interactions involving the states shown in (a). Husimi distribution function of the Floquet states regulating the decay of $|z\rangle$ are also shown. Notice that the nature of delocalized states (gray) does not change much with λ_2.

in Figure 10.9, highlighting the phase space delocalized nature of some of the participating states. Despite the symmetry of the tunneling doublets being broken due to the bichromatic field, two or more of the participating states are extensively delocalized in the chaotic reigons of the phase space. Moreover, the participation of the chaotic Floquet states persists even for larger values of λ_2. Hence, using the symmetry breaking property of the $2\omega_F$-field for control purposes is not very effective when chaotic states are participating in the dynamics. *Therefore, the lack of control, signaled by plateaus in Figure 10.6 and the valleys in Figure 10.8, is due to the dominant participation by chaotic states, i.e., chaos-assisted tunneling.* The plateaus arise due to the fact that the coupling between the localized states and the delocalized states vary very little with increasing control field strength λ_2—something that is evident from the Floquet level motions shown in Figure 10.9 and established earlier by Tomsovic and Ullmo in their seminal work on CAT in coupled quartic oscillators [19]. It is important to note that for $\lambda_2 = 0$ the chaotic states, as opposed to the regular states, do not have a definite parity. Consequently, the presence of the $2\omega_F$-field does not have a major influence on the chaotic states. Thus, if one or more chaotic states are already influencing the dynamics of $|z\rangle$ at $\lambda_2 = 0$ then the bichromatic control is expected to be difficult. An earlier study [68] by Latka et al. on the bichromatically driven pendulum system also suggested that the ability to control the dynamics is strongly linked to the existence of chaotic states.

The model problem in this section and the results point to a direct role of chaos-assisted pathways in the failure of an attempt to bichromatically control the dynamics. However, it is not yet clear if control strategies involving more general fields would exhibit similar characteristics. There is some evidence in the literature which indicates that quantum optimal control landscapes might be highly convoluted if the underlying classical phase space exhibits large-scale chaos [69]. Nevertheless, further studies need to be done and the resulting insights are expected to be crucial in any effort to control the dynamics of multilevel systems.

10.3.2 DRIVEN MORSE OSCILLATOR: RESONANCE-ASSISTED TUNNELING

The driven Morse oscillator system has served as a paradigm model for understanding the dissociation dynamics of diatomic molecules. Studies spanning nearly three decades have explored the physics of this system in exquisite detail. Consequently, a great deal is known about the mechanism of dissociation both from the quantum and classical perspectives. Indeed, the focus of researchers nowadays is to control, either suppress or enhance, the dissociation dynamics and various suggestions have been put forward. In addition, one hopes that the ability to control a single vibrational mode dynamics can lead to a better understanding of the complications that arise in the case of polyatomic molecular systems wherein several vibrational modes are coupled at the energies of interest.

Several important insights have originated from classical-quantum correspondence studies which have established that molecular dissociation, in analogy to multiphoton ionization of atoms, occurs due to the system gaining energy by diffusing through the chaotic regions of the phase space. For example, an important experimental study by Dietrich and Corkum has shown [70], amongst other things, the validity of the chaotic-dissociation mechanism. Thus, the formation of the chaotic regions due to the overlap [71] of nonlinear resonances (field-matter), hierarchical structures [72] near the regular-chaotic borders acting as partial barriers, and their effects on quantum transport [73] have been studied in a series of elegant papers [74–76]. In the context of this chapter an interesting question is as follows. Since a detailed mechanistic understanding of the role of various phase-space structures of the driven Morse system is known, is it possible to design local phase space barriers to effect control over the dissociation dynamics? In particular, the central question here is whether the local phase space barriers are also able to suppress the quantum-dissociation dynamics. There is an obvious connection between the above question and the theme of this volume—quantum mechanics can "shortcircuit" the classical phase-space barriers due to the phenomenon of dynamical tunneling. Thus, such phase-space barriers might be very effective in controlling the classical-dissociation dynamics but might fail completely when it comes to controlling the quantum-dissociation dynamics. There is a catch here, however, since there is also the possibility that the cantori barrier in the classical phase space may be even more restrictive in the quantum case. Thus, creation or existence of a phase-space barrier invariably leads to subtle competition between classical transport and quantum dynamical tunneling through the barrier. As expected, the delicate balance between the classical and quantum mechanism is determined by the effective Planck constant of the system of interest. An earlier detailed review [32] by Radons, Geisel, and Rubner is highly recommended for a nice introduction to the subject of classical-quantum correspondence perspective on phase-space transport through Kolmogoroff–Arnold–Moser (KAM) and cantori barriers. In the driven Morse oscillator case, Brown and Wyatt showed [74] that the cantori barriers do leave their imprint on the quantum dissociation dynamics and act as even stronger barriers as compared to the classical system. Maitra and Heller in their study [33] on transport through cantori in the whisker map have clearly highlighted the classical versus quantum competition.

From the above discussion it is apparent that any approach to control the dissociation dynamics by recreating local phase space barriers will face the subtle classical–quantum competition. In fact, it is tempting to think that every quantum control algorithm works by creating local phase space dynamical barriers and the efficiency of the control is decided by the classical–quantum competition. However, at this point of time there is very little work towards making such a connection and the above statement is, at best, a conjecture. For the purpose of this Chapter, I turn to the driven Morse oscillator system to provide an example for the importance of RAT in controlling the dissociation dynamics.

The model system is inspired from the early work [74] by Wyatt and Brown (see also the work by Breuer and Holthaus [77]) and the Hamiltonian can be written as

$$H(x, p; t) = H_0(x, p) - \mu(x)\lambda_1 \cos(\omega_F t) \tag{10.22}$$

in the dipole approximation. The zeroth-order Hamiltonian

$$H_0(x, p) = \frac{1}{2M}p^2 + D_0[1 - e^{-\alpha(x-x_e)}]^2 \tag{10.23}$$

represents the Morse oscillator modeling the anharmonic vibrations of a diatomic molecule. In the above, $\mu(x)$ is the dipole moment function, λ_1 is the strength of the laser field, ω_F is the driving field frequency, and

$$M = \frac{m_1 m_2}{m_1 + m_2}, \tag{10.24}$$

is the reduced mass of the diatomic molecule with m_1 and m_2 being the two atomic masses. In Equation 10.23, D_0 is the dissociation energy, α is the range of the potential and x_e is the equilibrium bond length of the molecule.

Rather than attempting to provide a general account as to how RAT might interfere with the process of control, I feel that it is best to illustrate with a realistic molecular example. Hopefully, the generality of the arguments will become apparent later on. For the present purpose I choose the diatomic molecule hydrogen fluoride (HF) as the specific example. Any diatomic molecule could have been chosen but HF is studied here due to the fact that Brown and Wyatt have already discussed the role of cantori barriers to the dissociation dynamics in some detail. The Morse oscillator parameters for HF are $D_0 = 0.225, \alpha = 1.174$, $x_e = 1.7329$, and $M = 1744.59$. These parameters correspond to ground electronic state of the HF molecule supporting $N_B = 24$ bound states. Note that atomic units are used for both the molecular and field parameters with time being measured in units of the field period $\tau_F = 2\pi/\omega_F$. The only difference between the present work and that of Brown and Wyatt has to do with the field-matter coupling. Brown and Wyatt use the dipole function

$$\mu(x) = Axe^{-Bx^4}, \tag{10.25}$$

with $A = 0.4541$ and $B = 0.0064$, obtained from ab initio data on HF. Here a linear approximation for $\mu(x)$,

$$\mu(x) \approx \mu(x_e) + \left(\frac{\partial \mu}{\partial x}\right)_{x_e} (x - x_e)$$

$$\equiv \mu(x_e) + d_1(x - x_e), \tag{10.26}$$

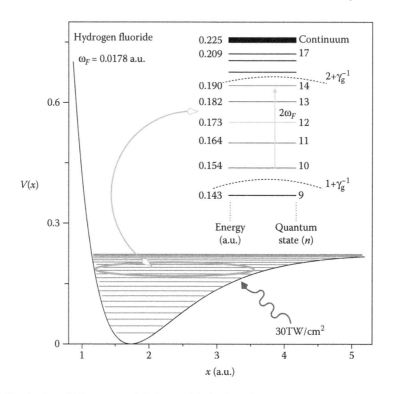

FIGURE 10.10 A plot of Morse potential along with the bound states supported by the potential well. A schematic view of energy level diagram for Morse eigenstates under study in this chapter is shown as inset. The laser field of intensity $\lambda_1 = 0.0287$ and frequency $\omega_F = 0.0178$ connects state $n = 10$ and $n = 14$ via a two-photon resonance transition. The location of cantori in presence of the field with $\omega_F = 0.0178$ are indicated as dotted curves in the figure.

is employed with $d_1 \approx 0.33$ in case of HF. There are quantitative differences in the dissociation probabilities due to the linearization approximation but the main qualitative features remain intact despite the linearization approximation.

In Figure 10.10 the Morse potential for HF is shown along with a summary of the key features such as the energy region of interest, the quantum state whose dissociation is to be controlled, and the classical phase-space structures that might play an important role in the dissociation dynamics. The driving field parameters are chosen as $(\omega_F, \lambda_1) = (0.0178, 0.0287)$, same as in the earlier work [74], and the field strength corresponds to about 30 TW/cm$^2 \equiv 30 \times 10^{12}$ W/cm^2. As shown in Figure 10.10 (inset), the focus is on understanding and controlling the dissociation dynamics of the $n = 10$ excited Morse oscillator eigenstate. There are several reasons for such a choice and I mention two of the most important reasons. First, the earlier study [74] has established that $n = 10$ of HF happens to be in an energy regime wherein two specific cantori barriers in the classical phase space affect the dissociation dynamics. Moreover, for driving frequency $\omega_F = 0.0178$, two of the zeroth-order eigenstates $n = 10$ and $n = 14$ have unperturbed energies such that $E_{14} - E_{10} \approx 2\hbar\omega_F$ and hence corresponds to a two-photon resonant situation. Second, for a field strength of 30 TW/cm^2, the dissociation probability of the ground vibrational state is negligible. Far stronger field strengths are required to dissociate the $n = 0$ state and ionization process starts to compete with the dissociation at such high intensities. Thus, in order to to illustrate the role of phase space barriers in the dissociation dynamics without such additional complications, the specific initial state $n = 10$ is chosen. Incidentally, such a scenario is quite feasible since a suitably chirped laser field can populate the $n = 10$ state very efficiently from the initial ground state $n = 0$ and one imagines coming in with the monochromatic laser to dissociate the molecule.

10.3.2.1 Nature of Classical Phase Space

The monochromatically driven Morse system studied here has a dimensionality such that one can conveniently visualize the phase space in the original Cartesian (x, p) variables. However, since the focus is on suppressing dissociation by creating robust KAM tori in the phase space, action-angle variables (J, θ) which are canonically conjugate to (x, p) are convenient and a natural representation to work with. The action-angle variables (J, θ) of the unperturbed Morse oscillator, appropriate for the bound regions, are given by [78]

$$J = \sqrt{\frac{2MD_0}{\alpha^2}} \left(1 - \sqrt{1 - E}\right) \tag{10.27a}$$

$$\theta = -sgn(p) \cos^{-1} \left[\frac{1 - E}{\sqrt{E}} e^{\alpha(x - x_e)} - \frac{1}{\sqrt{E}}\right]. \tag{10.27b}$$

In the above equations, $E = H_0/D_0 < 1$ denotes the dimensionless bound state energy, and $sgn(p) = 1$ for $p \geq 0$, $sgn(p) = -1$ for $p < 0$. In terms of the action-angle variables it is possible to express the cartesian (x, p) as follows:

$$x = x_e + \frac{1}{\alpha} \ln \left[\frac{1 + \sqrt{E_0(J)} \cos \theta}{(1 - E_0(J))}\right], \tag{10.28a}$$

$$p = \frac{-(2MD_0)^{1/2} [E_0(J)(1 - E_0(J))]^{1/2} \sin \theta}{1 + \sqrt{E_0(J)} \cos \theta}, \tag{10.28b}$$

where $E_0(J) = H_0(J)/D_0$. Substituting for (x, p) in terms of (J, θ), the unperturbed Morse oscillator Hamiltonian in Equation 10.23 is transformed into

$$H_0(J) = \omega_0 \left(J - \frac{\omega_0}{4D_0} J^2\right), \tag{10.29}$$

where $\omega_0 = \sqrt{2\alpha^2 D_0/M}$ is the harmonic frequency at the minimum. The zeroth-order nonlinear frequency is easily obtained as

$$\Omega_0(J) \equiv \frac{\partial H_0}{\partial J} = \omega_0 \left(1 - \frac{\omega_0}{2D_0} J\right), \tag{10.30}$$

and with increasing excitation—i.e., increasing action (quanta) J, the nonlinear frequency $\Omega_0(J)$ decreases monotonically and eventually vanishes, signaling the onset of unbound dynamics leading to dissociation.

The driven system can now be expressed in terms of the variables (J, θ) as

$$H(J, \theta; t) = H_0(J) - \frac{\varepsilon}{\alpha} \ln \left[\frac{1 + \sqrt{E_0(J)} \cos \theta}{(1 - E_0(J))}\right] \cos(\omega_F t). \tag{10.31}$$

In addition, since x is a periodic function of θ, one has the Fourier expansion

$$x = 2 \left[V_0(J) + \sum_{n=1}^{\infty} V_n(J) \cos(n\theta)\right]. \tag{10.32}$$

As a consequence, the driven Hamiltonian in Equation 10.31 can be written as

$$H(J,\theta;t) = H_0(J) - \varepsilon v(J,\theta;t), \tag{10.33}$$

where the matter-field interaction term is denoted as

$$v(J,\theta;t) = 2\left[V_0(J) + \sum_{n=1}^{\infty} V_n(J)\cos(n\theta)\right]\cos(\omega_F t).$$

The Fourier coefficients $V_0(J)$ and $V_n(J)$ are known analytically and given by

$$V_0(J) = \frac{1}{2\alpha}\ln\left[\frac{D_0 + \sqrt{D_0^2 - D_0 E_0(J)}}{2(D_0 - E_0(J))}\right], \tag{10.34a}$$

$$V_n(J) = \frac{(-1)^{n+1}}{\alpha n}\left[\frac{\sqrt{D_0 E_0(J)}}{D_0 + \sqrt{D_0^2 - D_0 E_0(J)}}\right]^n. \tag{10.34b}$$

The stroboscopic surface of section in the (J,θ) variables is shown in Figure 10.11 and is a typical mixed regular-chaotic phase space. A few important points are worth noting at this stage. First, the initial state of interest is located close to a cantorus with $\omega_F : \Omega_0(J) = 1 + \gamma_g^{-1}$ with $\gamma_g \equiv (1 + \sqrt{5})/2$ being the golden ratio. The importance of this cantorus to the resulting dissociation dynamics of the $n = 10$ state was the central focus of the work by Brown and Wyatt [74]. In particular, the extensive stickiness around this region can be clearly seen and hence one expects nontrivial influence on the classical dissociation dynamics as well. Second, a prominent $\omega_F : \Omega_0(J) = 2:1$ nonlinear resonance is also observed in the phase space and represents the classical analog of the quantum 2-photon resonance condition. Interestingly, the area of this resonance is about $\hbar$ and, therefore, can support

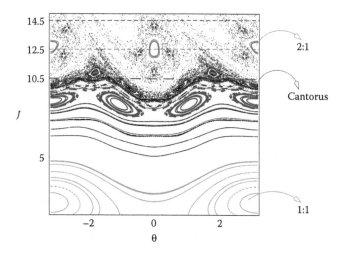

FIGURE 10.11 The phase space, as a stroboscopic surface of section, for the driven Morse system with laser field of intensity $\lambda_1 = 0.0287$ and frequency $\omega_F = 0.0178$ corresponding to a on-resonance situation. The 2-photon resonance is clearly seen in the phase space and marked as 2:1 resonance in dark green. The initial Morse eigenstate $n = 10$ (thick dashed line) is situated rather close to a cantorus (indicated), $\omega_F/\Omega_0(J) = 1 + \gamma_g^{-1}$ with $\gamma_g \equiv (1 + \sqrt{5})/2$ being the golden ratio. The $n = 10$ state is connected to the $n = 14$ Morse eigenstate via the 2:1 resonance. Note that the states $n = 10, 12$, and 14 are symmetrically located about the resonance with $n = 12$ being localized in the resonance.

one quantum state. It turns out that the Husimi density of the Morse state $n = 12$ is localized inside the 2:1 resonance island. In addition, the states $n = 10$ and $n = 14$ are nearly symmetrically located about the 2:1 resonance. A way to see this is to use secular pertubation theory on the driven Morse Hamiltonian. One can show that in the vicinity of the 2:1 resonance (cf. Figure 10.11) an effective pendulum Hamiltonian

$$H_{eff}(J,\phi) \simeq \frac{1}{2\tilde{m}_{2:1}}(\Delta J)^2 + 2\tilde{V}_{2:1}(J_{2:1})\cos(2\phi), \tag{10.35}$$

is obtained with $\Delta J = J - J_{2:1}$, $\tilde{m}_{2:1} = 2D_0/\omega_0^2$ and $V_{2:1}(J) = \varepsilon V_1(J)/2$. The resonant action

$$J_{2:1} = \frac{2D_0}{\omega_0}\left[1 - \frac{\omega_F}{2\omega_0}\right] \approx 12.6, \tag{10.36}$$

for the parameters used in this work. Using action values $J = 10.5$ (quantum state $n = 10$) and $J' = 14.5$ (quantum state $n' = 14$), the energy difference is calculated as

$$\begin{aligned} |E_J - E_{J'}| &= \left|\frac{1}{2\tilde{m}_{2:1}}(J - J')(J + J' - 2J_{2:1})\right| \\ &\approx 3.2 \times 10^{-4} \ll E_J. \end{aligned} \tag{10.37}$$

In other words, the states $n = 10$ and $n = 14$ are nearly symmetrical with respect to the state $n = 12$, which is localized in the 2:1 resonance. Therefore, the nonzero coupling $V_{2:1}$ will efficiently connect the states $n = 10$ and $n = 14$. Moreover, for the given parameters, using the definition of $V_{r:s}(J)$ in terms of Fourier coefficient V_1, the strength of the resonance is estimated as $V_{2:1}(J_{2:1}) \approx 0.01464$, clearly a fairly strong resonance. Consequently, the situation in Figure 10.11 is a perfect example where RAT can play a crucial role in the dissociation dynamics. Indeed, quantum computations (not shown here) show that there is a Rabi-type cycling of the probabilities between the three Morse states. I now turn to the issue of *selectively* controlling the dissociation dynamics of the initial Morse eigenstate $n = 10$ by creating local barriers in the phase space shown in Figure 10.11, bearing in mind the possibility of quantum dynamical tunneling interfering with the control process.

10.3.2.2 Creating a Local Phase-Space Barrier

If the classical mechanism of chaotic diffusion leading to dissociation holds in the quantum domain as well then a simple way of controlling the dissociation is to create a local phase-space barrier between the state of interest and the chaotic region. In a recent work [79], Huang, Chandre, and Uzer provided the theory for recreating local phase-space barriers for time-dependent systems and showed that such barriers indeed suppress the ionization of a driven atomic system. However, Huang et al. were only concerned with the classical ionization process. Thus, potential complications due to dynamical tunneling were not addressed in their study. The driven Morse system studied here presents an ideal system to understand the interplay of quantum and classical dissociation mechanisms. In what follows, I provide a brief introduction to the methodology with an explicit expression for the classical control field needed to recreate an invariant KAM barrier, preferably an invariant torus with sufficiently irrational frequency Ω_r.

To start with, the nonautonomous Hamiltonian is mapped into an autonomous one by considering $(t(\mathrm{mod}\,2\pi), E)$ as an additional angle-action pair. Denoting the action and angle variables by $\mathbf{A} \equiv (J, E)$ and $\boldsymbol{\theta} \equiv (\theta, t)$, the original driven system Hamiltonian (see Equation 10.33) can be expressed as

$$H(\mathbf{A}, \boldsymbol{\theta}) = H_0(\mathbf{A}) - \varepsilon V(\mathbf{A}, \boldsymbol{\theta}), \tag{10.38}$$

with $V(\mathbf{A}, \boldsymbol{\theta}) \equiv v(J, \theta; t)$. Note that for a fixed driving field strength λ_1 and the value of d_1 corresponding to a diatomic molecule, $\varepsilon \equiv \lambda_1 d_1$ is also fixed. Moreover, for physically meaningful values of d_1 for most diatoms and typical field strengths far below the ionization threshold one always has $\varepsilon \ll 1$. In the absence of the driving field ($\varepsilon = 0$), the zeroth-order Hamiltonian is integrable and the phase space is foliated with invariant tori labeled by the action $\mathbf{A}$ corresponding to the frequency $\boldsymbol{\omega} \equiv \partial H_0/\partial \mathbf{A} = (\Omega_0, \omega_F)$. However, in the presence of the driving field ($\varepsilon \neq 0$) the field-matter interaction renders the system nonintegrable with a mixed regular-chaotic phase space. More specifically, for field strengths near or above a critical value ε_c one generally observes a large scale destruction of the field-free invariant tori leading to significant chaos and hence the onset of dissociation. The critical value ε_c itself is clearly dependent on the specific molecule and the initial state of interest. The aim of the local control method is to rebuild a nonresonant torus $\mathbf{A}_0 = (J_0, 0)$, $\mathbf{k} \cdot \boldsymbol{\omega} \neq 0$ with integer $\mathbf{k}$, which has been destroyed due to the interaction with the field. Assuming that the destruction of $\mathbf{A}_0$ is responsible for the significant dissociation observed for some initial state of interest, the hope is that locally recreating the $\mathbf{A}_0$ will suppress the dissociation, i.e., $\mathbf{A}_0$ acts as a local barrier to dissociation. Ideally, one would like to recreate the local barrier by using a second field (appropriately called as the control field) which is much weaker and distinct from the primary driving field.

Following Huang et al. [79] such a control field $f(\boldsymbol{\theta})$ can be analytically derived and has the form

$$f(\boldsymbol{\theta}) = -H(\mathbf{A}_0 - \partial_\theta \Gamma b(\boldsymbol{\theta}), \boldsymbol{\theta}), \tag{10.39}$$

where $b(\boldsymbol{\theta}) \equiv H(\mathbf{A}_0, \boldsymbol{\theta}) = \sum_\mathbf{k} b_\mathbf{k} e^{i\mathbf{k} \cdot \boldsymbol{\theta}}$ and Γ being a linear operator defined by

$$\Gamma b(\boldsymbol{\theta}) \equiv \sum_{\mathbf{k} \cdot \boldsymbol{\omega} \neq 0} \frac{b_\mathbf{k}}{i\mathbf{k} \cdot \boldsymbol{\omega}} e^{i\mathbf{k} \cdot \boldsymbol{\theta}}. \tag{10.40}$$

The classical control Hamiltonian can now be written down as

$$\begin{aligned} H_c(\mathbf{A}, \boldsymbol{\theta}) &= H(\mathbf{A}, \boldsymbol{\theta}) + f(\boldsymbol{\theta}) \\ &\equiv H_0(\mathbf{A}) - \varepsilon V(\mathbf{A}, \boldsymbol{\theta}) + f(\boldsymbol{\theta}). \end{aligned} \tag{10.41}$$

In case of the driven Morse system the control field can be obtained analytically and to leading order is given by

$$H_c(J, \theta; t) \approx H_0(J) - \varepsilon v(J, \theta; t) + \varepsilon^2 g_a(\theta, t), \tag{10.42}$$

where

$$g_a(\theta, t) = \frac{\omega_0^2}{4D_0}(\partial_\theta \Gamma b)^2 - 2V_{01}(\partial_\theta \Gamma b)\cos(\omega_F t) - (\partial_\theta \Gamma b)\zeta(J, \theta; t), \tag{10.43}$$

and it can be shown that

$$V_{01} = \frac{\omega_0^2}{8\alpha\Omega_r D_0}\left(\frac{2\omega_0 + \Omega_r}{\omega_0 + \Omega_r}\right),$$

$$V_{n1} = (-1)^{n+1}\left(\frac{\omega_0^3}{2\alpha D_0}\right)\frac{(\omega_0 - \Omega_r)^{\frac{n}{2}-1}}{(\omega_0 + \Omega_r)^{\frac{n}{2}+1}}, \tag{10.44}$$

$$\zeta(J, \theta; t) = \sum_{n=1}^{\infty} V_{n1}(J_r)[(\cos(n\theta + \omega_F t) + \cos(n\theta - \omega_F t)].$$

I skip the somewhat tedious derivation of the result above and refer to the original literature [79] as well as a recent thesis [45] for details. Note that in the above Ω_r is the frequency of the invariant torus that is to be recreated corresponding to the unperturbed action

$$J_r = (\omega_0 - \Omega_r)\frac{2D_0}{\omega_0^2}, \tag{10.45}$$

and to $O(\varepsilon)$ is located at $J(\theta) = J_r - \varepsilon\partial_\theta \Gamma b$, assuming the validity of the perturbative treatment.

The leading order control field in Equation 10.43 is typically weaker than the driving field and has been shown [45] to be quite effective in off-resonant cases in suppressing the classical dissociation. However, in order to study the effect of the control field on the quantum dissociation probabilities, it is necessary to make some simplifications. One of the main reasons for employing the simplified control fields via the procedure given below has to do with the fact that the classical action-angle variables do not have a direct quantum counterpart [80]. Essentially, the dominant Fourier modes F_{k_1,k_2} of Equation 10.43 are identified and one performs the mapping

$$F_{k_1,k_2}\cos(k_1\theta + k_2\omega_F t) \rightarrow \lambda_2(k_1,k_2)\cos(k_2\omega_F t), \tag{10.46}$$

yielding the simplified control Hamiltonian

$$H_c = H(J,\theta;t) + \mu(x)\lambda_2(k_1,k_2)\cos(k_2\omega_F t). \tag{10.47}$$

If more than one dominant Fourier modes are present then they will appear as additional terms in Equation 10.47. Note that the above simplified form is equivalent to assuming that the control field is polychromatic in nature, which need not be true in general. Nevertheless, a qualitative understanding of the role of the various Fourier modes towards local phase space control is still obtained. More importantly, and as shown next, in the on-resonant case of interest here, the simplified control field already suggests the central role played by RAT.

10.3.2.3   RAT Spoils Local Phase-Space Control

In Figure 10.12 a summary of the efforts to control the dissociation of the initial $n = 10$ state is shown. Specifically, Figure 10.12a and b show the phase spaces where two different KAM barriers, $\omega_F/\Omega_r = 1 + \gamma_g^{-1}$ and $\omega_F/\Omega_r = \sqrt{3}$ respectively are recreated. The control Hamiltonian in both instances, in cartesian variables, has the following form:

$$H_c = H_0(x,p) - \lambda_1\mu(x)\cos(\omega_F t) + \lambda_2\mu(x)\cos(2\omega_F t), \tag{10.48}$$

with $\lambda_2 \approx 0.01 \equiv 3$ TW/cm^{-2}. Thus, as desired, the control field strengths are an order of magnitude smaller than the driving field strength. The simplified control Hamiltonian reflects the dominance of the $F_{3,-2}$ Fourier mode of the leading order control field in Equation 10.43 and obtained using the method outlined before. Also, note that the control field comes with a relative phase $\phi = \pi$ with respect to the driving field. As expected, from the line of thinking presented in the previous section, Figure 10.12c and d show that the KAM barriers indeed suppress the classical dissociation significantly. *However, the quantum dissociation in both cases increases slightly!* Clearly, the recreated KAM barriers are ineffective and suggests that the quantum dissociation mechanism is somehow bypassing the KAM barriers.

A clue to the surprising quantum results comes from comparing the phase spaces in Figures 10.11 and 10.12 which show the uncontrolled and controlled cases respectively. Although, the KAM barriers seem to have reduced the extent of stochasticity, the 2:1 resonance is intact and appears to occupy slightly larger area in the phase space. Thus, this certainly indicates that a significant amount of the quantum dissociation is occurring due to the RAT mechanism involving the three Morse states $n = 10, 12$, and 14 as discussed before. In particular, the $n = 12$ state must still be actively providing a route to couple the initial state to the chaotic region via RAT. How can one test the veracity of such an explanation? One way is to scale the Planck constant down from $\hbar = 1$ and monitor the quantum dissociation process. Reduced $\hbar$ implies that the 2:1 island can support several states as well as the fact that other higher order resonances now become relevant to the RAT mechanism. Such a study is not presented here but one would anticipate that the dissociation mechanism can be understood based on the theory of RAT that already exists (e.g., see contributions by Schlagheck et al. and Bäcker et al.). Another way is to directly interfere locally with the resonance and see if the

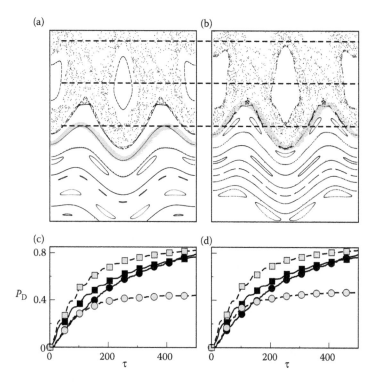

FIGURE 10.12 Phase space for the driven Morse system in the presence of the simplified classical control fields (cf. Equation 10.47 with $\lambda_2 \approx 0.01$) designed to recreate specific KAM barriers (a) $\omega_F/\Omega_r = 1 + \gamma_g^{-1}$ and (b) $\omega_F/\Omega_r = \sqrt{3}$. Note that in both cases the dominant $F_{3,-2}$ Fourier amplitude of the leading order control field of Equation 10.43 have been utilized and the desired KAM barriers are clearly seen (thick gray). The effect of the barriers seen in (a) and (b) on the classical (gray circles) and quantum (gray squares) dissociation probabilities are shown in (c) and (d) respectively. The uncontrolled results are indicated by the corresponding black symbols. Interestingly, the classical dissociation is suppressed but the quantum dissociation is slightly enhanced.

quantum dissociation is actually suppressed. Locally interfering with a specific phase-space structure, keeping the gross features unchanged, is not necessarily a straightforward approach. However, the tools in the previous section allow for such local interference and I present the results below.

The control field in Equation 10.43 corresponding to the case of Figure 10.12b, i.e., recreating the $\omega_F/\Omega_r = \sqrt{3}$ KAM barrier, turns out to be well approximated by

$$H_c(J,\theta,t) \approx H(J,\theta,t) + \sum_{n=3,4} F_{n,-2}\cos(n\theta - 2\omega_F t), \qquad (10.49)$$

The above Hamiltonian comes about due to the fact [45] that two Fourier modes $F_{3,-2}$ and $F_{4,-2}$ are significant in this case. Note that the specific KAM barrier of interest has a frequency between that of the $1 + \gamma_g^{-1}$ cantorus (around which the initial state is localized, cf. Figure 10.11) and the 2:1 nonlinear resonance. From a perturbative viewpoint, creation of this KAM barrier using Equation 10.49 is not expected to be easy due to the proximity to the 2:1 resonance and the fact that the $F_{4,-2}$ Fourier component is nothing but the 2:1 resonance. Nevertheless, Figure 10.13a shows that the specific KAM barrier is restored and, compared to Figure 10.11, the controlled phase space does exhibit reduced amount of chaos. Consistently, Figure 10.13c shows that the classical dissociation is suppressed by nearly a factor of two as in the case shown in Figure 10.12b wherein only the $F_{3,-2}$ component of the control Hamiltonian was retained. As mentioned earlier, in order to

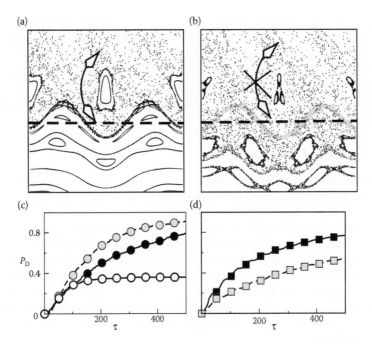

FIGURE 10.13 Phase space for the driven Morse system in the presence of the classical control fields designed to recreate the $\omega_F/\Omega_r = \sqrt{3}$ KAM barrier. In (a) two dominant Fourier modes $F_{3,-2}$ and $F_{4,-2}$ of the leading order control term in Equation 10.43 are retained. In (b) the simplified control field as in Equation 10.47 is used with an effective field strength estimated using $F_{3,-2}$ and $F_{4,-2}$. Notice that the desired KAM barrier is recreated in (a) but not in (b) (thick gray line indicates the expected location). In (c) the classical dissociation probabilities of the uncontrolled (black circles), control using field (a) (open circles), and control using field (b) (gray circles) are shown. The quantum results are shown in (d) for the uncontrolled (black squares) and control using field (b) (gray squares) are shown. See text for discussion.

calculate the quantum dissociation probability the control Hamiltonian in Equation 10.49 needs to be mapped into a from as in Equation 10.47. Since two Fourier modes need to be taken into account, the effective control field strength is given by

$$\lambda_2 = \frac{F_{3,-2}}{V_3(J_r)} + \frac{F_{4,-2}}{V_4(J_r)}. \tag{10.50}$$

Such a procedure yields $\lambda_2 \approx -0.016$ and thus the control field, still less intense than the primary field, comes with a relative phase of zero. Interestingly, as shown in Figure 10.13b, the resulting simplified control Hamiltonian fails to create the desired barrier. Moreover, the phase space also exhibits increased stochasticity and as a consequence the classical dissociation is enhanced (cf. Figure 10.13c). However, Figure 10.13b reveals an interesting feature—the 2:1 resonance is severly perturbed. This perturbation is a consequence of including the $F_{4,-2}$ Fourier component into the effective control Hamiltonian. *The key result, however, is shown in Figure 10.13d where one observes that the quantum dissociation probability is reduced significantly.* It seems like the quantum dynamics feels the barrier when there is none! The surprising and counterintuitive results summarized in Figures 10.12 and 10.13 can be rationalized by a single phenomenon—dynamical (resonance assisted in this case) tunneling. The main clue comes from the observation that quantum suppression happens as soon as the 2:1 resonance is perturbed.

Although the results here are shown with a specific example, the phenomenon is general. Indeed computations (not published) for different sets of parameters have supported the viewpoint expressed above. Interestingly, the work of Huang et al. focused on classical suppression of ionization by creating local phase-space barriers in case of the driven one-dimensional hydrogen atom [79]. Around the same time Brodier et al. highlighted [81] the importance of the RAT mechanism in order to obtain accurate decay lifetimes of localized wavepackets in the same system. Work is currently underway to see if the conclusions made in this section hold in the driven atomic system as well, i.e., whether the attempt to suppress the ionization by creating local phase-space barriers is foiled by the phenomenon of RAT.

10.3.3 Summary and Future Outlook

The two examples discussed in this work illustrate a key point—*Dynamical tunneling plays a nontrivial role in the process of quantum control.* In the first example of the driven quartic double well it is clear that bichromatic control fails in regimes where CAT is important. In the second example of a driven Morse oscillator it is apparent that efforts to control by building phase-space KAM barriers fail when RAT is possible. In both instances the competition between classical and quantum mechanisms is brought to the forefront. Although the two examples shown here represent the failure of specific control schemes due to dynamical tunneling, one must not take this to be a general conclusion. It is quite possible that some other control schemes might owe their efficiency to the phenomenon of dynamical tunneling itself. Further classical-quantum correspondence studies on control with more general driving fields are required in order to confirm (or refute) the conclusions presented in this chapter.

At the same time the two examples presented here are certainly not the the last word; establishing the role of dynamical tunneling in quantum coherent control requires one to step into the murky world of three or more degrees-of-freedom systems [82]. I mention two model systems, currently being studied in our group, in order to stress upon some of the key issues that might crop up in such high-dimensional systems. For example, two coupled Morse oscillators which are driven by a monochromatic field already present a number of challenges both from the technical as well as conceptual viewpoints. The technical challenge arises due to the fact that dimensionality constraints do not allow one to visualize the global phase space structures as easily as done in this chapter. One approach is to use the method of local frequency analysis [83] to construct the Arnold web, i.e., the network of nonlinear resonances that regulate the multidimensional phase space transport. In a previous work [84], involving a time-independent Hamiltonian system, the utility of such an approach and the validity of the RAT mechanism has been established. However, "lifting" quantum dynamics onto the Arnold web is an intriguing possibility which is still an open issue. On the conceptual side there are several issues with multidimensional systems. I mention a few of them here. First, even at the classical level one has the possibility of transport like Arnold diffusion [50,71] which is genuinely a three or more DoF effect and has no counterpart in systems with less than three DoFs. Note that Arnold diffusion is typically a very long time process and is notoriously difficult to observe in realistic physical systems [85]. Moreover, arguments can be made for the irrelevance of Arnold diffusion (or some similar process) in quantum systems due to the finiteness of the Planck constant. Second, an interesting competition occurs in systems such as the driven coupled Morse oscillators. Even in the absence of the field the dynamics is nonintegrable and one can be in a regime where the modes are exchanging energy but none of the modes gain enough energy to dissociate. On the other hand, in the absence of mode–mode coupling, a weak enough field can excite the system without leading to dissociation. However, in the presence of such a weak field and the mode–mode coupling one can have significant dissociation of a specific vibrational mode. Clearly, there is nontrivial competition between transport due to mode–mode resonances, field-mode resonances, and the chaotic regions [45]. Selective control of such driven coupled systems is an active research area [86] today and the lessons learnt from the two examples suggest that dynamical tunneling in one form or another can play a central role. In the context of studying the potential competition between Arnold diffusion and dynamical tunneling, I should mention the driven coupled quartic oscillator system with the Hamiltonian

$$H(x,y,p_x,p_y;t) = \frac{1}{2}(p_x^2 + p_y^2) + \frac{1}{4}(x^4 + y^4) - \mu x^2 y^2 - x f_0(\cos\Omega_1 t + \cos\Omega_2 t). \qquad (10.51)$$

In the absence of the field $f_0 = 0$ the Hamiltonian reduces to the case originally studied by Tomsovic, Bohigas, and Ullmo wherein the existence of CAT was established in exquisite detail [18,19]. At the same time, with $f_0 \neq 0$ the system represents one of the few examples for which the phenomenon of Arnold diffusion has been investigated over a number of years. Nearly a decade ago, Demikhovskii, Izrailev, and Malyshev studied [87] a variant of the above Hamiltonian to uncover the fingerprint of Arnold diffusion on the quantum eigenstates and dynamics. An interesting question, amongst many others, is this: *Will the fluctuations in the chaos-assisted tunneling splittings for $f_0 = 0$, observed for varying $\hbar$, survive in the presence of the field?*

A related topic which I have not touched upon in this chapter has to do with the control of IVR using weak external fields. Based on the insights gained from studies done until now on field-free IVR (see also the contribution by Leitner in this volume), it is natural to expect that dynamical tunneling could play spoilsport for certain class of initial states that are prepared experimentally. The issue, however, is far more subtle and more studies in this direction might shed light on the mechanism by which quantum optimal control methods work. For instance, recent proposals on quantum control by Takami and Fujisaki [88] and the so-called "quantized Ulam conjecture" by Gruebele and Wolynes [89] take advantage (implicitly) of the system having a completely chaotic phase space. Dynamical tunneling is not an issue in such cases. However, in more generic instances of systems with mixed regular-chaotic phase space, a quantitative and qualitative understanding of dynamical tunneling becomes imperative. Given the level of detail at which one is now capable of studying the clasically forbidden processes, as reflected by the varied contributions in this volume, I expect exciting progress in this direction.

ACKNOWLEDGMENTS

I think that this is an appropriate forum to acknowledge the genesis of the current and past research of mine on dynamical tunneling. In this context I am grateful to Greg Ezra, whose first suggestion for my postdoc work came in the form of a list of some of the key papers on dynamical tunneling. We never got around to work on dynamical tunneling *per se* but those references came in handy nearly a decade later. It is also a real pleasure to thank Peter Schlagheck for several inspiring discussions on tunneling in general and dynamical tunneling in particular.

REFERENCES

1. L. D. Landau and E. M. Lifshitz, *Course of Theoretical Physics, Vol. 3: Quantum mechanics*, Butterworth Heinemann, Oxford, 1998; For molecular examples and applications, see W. H. Flygare, *Molecular Structure and Dynamics*, Prentice-Hall, New Jersey, 1978.

2. See, E. Merzbacher, The early history of quantum tunneling. *Physics Today*, 44, August 2002, for a short and interesting account of the history of tunneling.

3. See, for example, V. A. Benderskii, D. E. Makarov, and C. A. Wight, Chemical dynamics at low temperatures. *Adv. Chem. Phys.* 88, Wiley-Interscience, NY, 1994.

4. W. H. Miller, Tunneling and state specificity in unimolecular reactions. *Chem. Rev.* 87, 19, 1987.

5. A. Kohen and J. P. Klinman, Hydrogen tunneling in biology. *Chem. and Biol.* 6, R191, 1999.

6. I. A. Balabin and J. N. Onuchic, Dynamically controlled protein tunneling paths in photosynthetic reaction centers. *Science* 290, 114, 2000.

7. P. S. Zuev, R. S. Sheridan, T. V. Albu, D. G. Truhlar, D. A. Hrovat, and W. T. Borden, Carbon tunneling from a single quantum state. *Science* 299, 867, 2003.

8. G. C. Smith and S. C. Creagh, Tunnelling in near-integrable systems. *J. Phys. A: Math. Gen.* 39, 8283, 2006; S. C. Creagh, Tunneling in multidimensional systems. *J. Phys. A: Math. Gen.* 27, 4969, 1994;

S. C. Creagh and N. D. Whelan, Statistics of chaotic tunneling. *Phys. Rev. Lett.* 84, 4084, 2000; S. C. Creagh and N. D. Whelan, Complex periodic orbits and tunneling in chaotic potentials. *Phys. Rev. Lett.* 77, 4975, 1996; S. C. Creagh and N. D. Whelan, A matrix element for chaotic tunnelling rates and scarring intensities. *Ann. Phys. (NY)* 272, 196, 1999.

9. A. Shudo and K. S. Ikeda, Complex classical trajectories and chaotic tunneling. *Phys. Rev. Lett.* 74, 682, 1995; A. Shudo and K. S. Ikeda, Stokes phenomenon in chaotic systems: Pruning trees of complex paths with principle of exponential dominance. *Phys. Rev. Lett.* 76, 4151, 1996; A. Shudo and K. S. Ikeda, Chaotic tunneling: A remarkable manifestation of complex classical dynamics in non-integrable quantum phenomena. *Physica* 115D, 234, 1998; T. Onishi, A. Shudo, K. S. Ikeda, and K. Takahashi, Tunneling mechanism due to chaos in a complex phase space. *Phys. Rev. E* 64, 025201, 2001; A. Shudo, Y. Ishii, and K. S. Ikeda, Julia set describes quantum tunnelling in the presence of chaos. *J. Phys. A: Math. Gen.* 35, L225, 2002; K. Takahashi and K. S. Ikeda, Complex-classical mechanism of the tunnelling process in strongly coupled 1.5-dimensional barrier systems. *J. Phys. A: Math. Gen.* 36, 7953, 2003; K. Takahashi and K. S. Ikeda, An intrinsic multi-dimensional mechanism of barrier tunneling. *Europhys. Lett.* 71, 193, 2005.

10. R. T. Lawton and M. S. Child, Local mode vibrations of water. *Mol. Phys.* 37, 1799, 1979.

11. M. J. Davis and E. J. Heller, Quantum dynamical tunneling in bound states. *J. Chem. Phys.* 75, 246, 1981.

12. W. G. Harter and C. W. Patterson, Rotational energy surfaces and high-J eigenvalue structure of poly-atomic molecules. *J. Chem. Phys.* 80, 4241, 1984.

13. A. M. Ozorio de Almeida, Tunneling and semiclassical spectrum for an isolated classical resonance. *J. Phys. Chem.* 88, 6139, 1984.

14. E. L. Sibert III, W. P. Reinhardt, and J. T. Hynes, Classical dynamics of energy transfer between bonds in ABA triatomics. *J. Chem. Phys.* 77, 3583, 1982; E. L. Sibert III, J. T. Hynes, and W. P. Reinhardt, Quantum mechanics of local mode ABA triatomic molecules. *J. Chem. Phys.* 77, 3595, 1982.

15. K. Stefanski and E. Pollak, An analysis of normal and local mode-dynamics based on periodic orbits.1.Symmetric ABA triatomic molecules. *J. Chem. Phys.* 87, 1079, 1987.

16. R. Grobe and F. Haake, Dissipative death of quantum coherences in a spin system. *Z. Phys. B* 68, 503, 1987.

17. W. A. Lin and L. E. Ballentine, Quantum tunneling and chaos in a driven anharmonic oscillator. *Phys. Rev. Lett.* 65, 2927, 1990; W. A. Lin and L. E. Ballentine, Quantum tunneling and regular and irregular quantum dynamics of a driven double-well oscillator. *Phys. Rev. A* 45, 3637, 1992; W. A. Lin and L. E. Ballentine, Dynamics quasidegeneracies and quantum tunneling—reply. *Phys. Rev. Lett.* 67, 159, 1991.

18. O. Bohigas, S. Tomsovic, and D. Ullmo, Manifestations of classical phase-space structures in quantum-mechanics. *Phys. Rep.* 223, 43, 1993.

19. S. Tomsovic and D. Ullmo, Chaos-assisted tunneling. *Phys. Rev. E* 50, 145, 1994.

20. E. Doron and S. D. Frischat, Semiclassical description of tunneling in mixed systems—case of the annu-lar billiards. *Phys. Rev. Lett.* 75, 3661, 1995; S. D. Frischat and E. Doron, Dynamical tunneling in mixed systems. *Phys. Rev. E* 57, 1421, 1998.

21. S. Tomsovic, *Tunneling in Complex Systems*, World Scientific, NJ, 1998.

22. O. Brodier, P. Schlagheck, and D. Ullmo, Resonance-assisted tunneling in near-integrable systems. *Phys. Rev. Lett.* 87, 064101, 2001.

23. O. Brodier, P. Schlagheck, and D. Ullmo, Resonance-assisted tunneling. *Ann. Phys. (NY)* 300, 88, 2002.

24. P. Schlagheck, C. Eltschka, and D. Ullmo Resonance- and chaos-assisted tunneling. In *Progress in Ultra-fast Intense Laser Science I*, eds. K. Yamanouchi, S. L. Chin, P. Agostini, and G. Ferrante, Springer, Berlin, 2006, pp. 107–131.

25. C. Eltschka and P. Schlagheck, Resonance- and chaos-assisted tunneling in mixed regular-chaotic sys-tems. *Phys. Rev. Lett.* 94, 014101, 2005.

26. L. Bonci, A. Farusi, P. Grigolini, and R. Roncaglia, Tunneling rate fluctuations induced by nonlin-ear resonances: A quantitative treatment based on semiclassical arguments. *Phys. Rev. E* 58, 5689, 1998.

27. Y. Ashkenazy, L. Bonci, J. Levitan, and R. Roncaglia, Classical nonlinearity and quantum decay: The effect of classical phase-space structures. *Phys. Rev. E* 64, 056215, 2001.

28. R. Roncaglia, L. Bonci, F. M. Izrailev, B. J. West, and P. Grigolini, Tunneling versus chaos in the kicked Harper model. *Phys. Rev. Lett.* 73, 802, 1994.

29. M. Sheinman, S. Fishman, I. Guarneri, and L. Rebuzzini, Decay of quantum accelerator modes. *Phys. Rev.* A 73, 052110, 2006; M. Sheinman, *Decay of Quantum Accelarator Modes*, Master's thesis, Chapter 3, Technion, 2005.

30. J. Ortigoso, Anomalous asymmetry splittings in a molecule with internal rotation in the presence of classical chaos. *Phys. Rev.* A 54, R2521, 1996.

31. V. Averbukh, N. Moiseyev, B. Mirbach, and H. J. Korsch, Z. Dynamical tunneling through a chaotic region—a continuously driven rigid rotor. *Phys. D* 35, 247, 1995.

32. G. Radons, T. Geisel, and J. Rubner, Classical chaos versus quantum dynamics: KAM tori and cantori as dynamical barriers. *Adv. Chem. Phys.* LXXIII, 891, 1989; T. Geisel, G. Radons, and J. Rubner, Kolmogorov–Arnold–Moser barriers in the quantum dynamics of chaotic systems. *Phys. Rev. Lett.* 57, 2883, 1986.

33. N. T. Maitra and E. J. Heller, Quantum transport through cantori. *Phys. Rev.* E 61, 3620, 2000.

34. J. U. Nöckel and A. D. Stone, Ray and wave chaos in asymmetric resonant optical cavities. *Nature* (London) 385, 45, 1997.

35. R. Hofferbert, H. Alt, C. Dembowski, H.-D. Gräf, H. L. Harney, A. Heine, H. Rehfeld, and A. Richter, Experimental investigations of chaos-assisted tunneling in a microwave annular billiard. *Phys. Rev.* E 71, 046201, 2005; C. Dembowski, H.-D. Gräf, A. Heine, R. Hofferbert, H. Rehfeld, and A. Richter, First experimental evidence for chaos-assisted tunneling in a microwave annular billiard. *Phys. Rev. Lett.* 84, 867, 2000.

36. W. K. Hensinger, H. Häffner, A. Browaeys, N. R. Heckenberg, K. Helmerson, C. McKenzie, G. J. Milburn et al., Dynamical tunnelling of ultracold atoms. *Nature* (London) 412, 52, 2001.

37. D. A. Steck, W. H. Oskay, and M. G. Raizen, Observation of chaos-assisted tunneling between islands of stability. *Science* 293, 274, 2001.

38. J. P. Bird, R. Akis, D. K. Ferry, A. P. S. de Moura, Y. C. Lai, and K. M. Indlekofer, Interference and interactions in open quantum dots. *Rep. Prog. Phys.* 66, 583, 2003.

39. S. Shinohara, T. Harayama, T. Fukushima, M. Hentschel, T. Sasaki, and E. E. Narimanov, Chaos-assisted directional light emission from microcavity lasers. *Phys. Rev. Lett.* 104, 163902, 2010.

40. S. Chaudhury, A. Smith, B. E. Anderson, S. Ghose, and P. S. Jessen, Quantum signatures of chaos in a kicked top. *Nature* 461, 768, 2009.

41. S. Fölling, S. Trotzky, P. Cheinet, M. Feld, R. Saers, A. Widera, T. Müller, and I. Bloch, Direct observation of second-order atom tunnelling. *Nature* 448, 1029, 2007.

42. A. A. Stuchebrukhov and R. A. Marcus, Theoretical study of intramolecular vibrational energy relaxation of acteylinic CH vibration for $v = 1$ and 2 in large polyatomic molecules $(CX_3)_3YCCH$, where X=H or D and Y=C or Si. *J. Chem. Phys.* 98, 6044, 1993.

43. S. Keshavamurthy, Dynamical tunnelling in molecules: Quantum routes to energy flow. *Int. Rev. Phys. Chem.* 26, 521, 2007.

44. Most of the current article is based on and adapted from two recent publications: A. Sethi and S. Keshavamurthy, Bichromatically driven double well: Parametric perspective of the strong field control landscape reveals the influence of chaotic states. *J. Chem. Phys.* 128, 164117, 2008; A. Sethi and S. Keshavamurthy, Local phase space control and interplay of classical and quantum effects in dissociation of a driven Morse oscillator. *Phys. Rev.* A 79, 033416, 2009.

45. A. Sethi, *Mechanistic insights into the control of driven quantum systems*, Ph.D. thesis, IIT Kanpur, 2010.

46. M. Born, W. Heisenberg, and P. Jordan, Zur Quantenmechanik II. *Z. Phys.* 35, 557, 1926. Translated version in, B. L. Van der Waerden, *Sources of Quantum Mechanics*, North-Holland, Amsterdam, 1967.

47. For a recent detailed tutorial, see L. Lodi and J. Tennyson, Theoretical methods for small-molecule rovibrational spectroscopy. *J. Phys. B: At. Mol. Opt. Phys.* 43, 133001, 2010.

48. E. J. Heller, E. B. Stechel, and M. J. Davis, Molecular spectra, Fermi resonances, and classical motion. *J. Chem. Phys.* 73, 4720, 1980.

49. D. Farrelly and T. Uzer, Semiclassical quantization of slightly nonresonant systems—avoided crossings, dynamic tunneling, and molecular spectra. *J. Chem. Phys.* 85, 308, 1986.

50. A. J. Lichtenberg and M. A. Lieberman, *Regular and Chaotic Dynamics*, Springer-Verlag, NY, 1992.

51. P. Schlagheck, *Tunneling in the Presence of Chaos and Interactions*, Habilitationsschrift, Universität Regensburg, 2006.

52. S. Löck, A. Bäcker, R. Ketzmerick, and P. Schlagheck, Regular-to-chaotic tunneling rates: From the quantum to the semiclassical regime. *Phys. Rev. Lett.* 104, 114101, 2010.

53. A. Mouchet, C. Eltschka, and P. Schlagheck, Influence of classical resonances on chaotic tunneling. *Phys. Rev. E* 74, 026211, 2006.

54. S. Keshavamurthy, Dynamical tunneling in molecules: Role of the classical resonances and chaos. *J. Chem. Phys.* 119, 161, 2003.

55. S. Keshavamurthy, On dynamical tunneling and classical resonances. *J. Chem. Phys.* 122, 114109, 2005.

56. See for example, D. J. Nesbitt and R. W. Field, Vibrational energy flow in highly excited molecules: Role of intramolecular vibrational redistribution. *J. Phys. Chem.* 100, 12735, 1996; M. Gruebele and P. G. Wolynes, Vibrational energy flow and chemical reactions. *Acc. Chem. Res.* 37, 261, 2004; D. M. Leitner, Heat transport in molecules and reaction kinetics: The role of quantum energy flow and localization. *Adv. Chem. Phys.* 130B, 205, 2005.

57. S. C. Farantos, R. Schinke, H. Guo, and M. Joyeux, Energy localization in molecules, bifurcation phenomena, and their spectroscopic signatures: The global view. *Chem. Rev.* 109, 4248, 2009; M. E. Kellman and V. Tyng, The dance of molecules: New dynamical perspectives on highly excited molecular vibrations. *Acc. Chem. Res.* 40, 243, 2007; G. S. Ezra, Classical-quantum correspondence and the analysis of highly excited states: Periodic orbits, rational tori, and beyond. *Adv. Clas. Traj. Meth.* 3, 35, 1998.

58. E. J. Heller, Dynamical tunneling and molecular spectra. *J. Phys. Chem.* 99, 2625, 1995.

59. H. Lee, P. Kok, and J. P. Dowling, A quantum Rosetta stone for interferometry. *J. Mod. Opt.* 49, 2325, 2002.

60. U. Lourderaj and W. L. Hase, Theoretical and computational studies of non-RRKM unimolecular dynamics. *J. Phys. Chem. A* 113, 2236, 2009; U. Lourderaj, K. Park, and W. L. Hase, Classical trajectory simulations of post-transition state dynamics. *Int. Rev. Phys. Chem.* 27, 1, 2008.

61. H. Waalkens, R. Schubert, and S. Wiggins, Wigner's dynamical transition state theory in phase space: Classical and quantum, nonlinearity. *Nonlinearity* 21, R1, 2008; G. S. Ezra, H. Waalkens, and S. Wiggins, Microcanonical rates, gap times, and phase space dividing surfaces. *J. Chem. Phys.* 130, 164118, 2009.

62. C. Jaffe, S. Kawai, J. Palacian, P, Yanguas, and T. Uzer, A new look at the transition state: Wigner's dynamical perpective revealed. *Adv. Chem. Phys.* 130A, 171, 2005; R. Paskauskas, C. Chandre, and T. Uzer, Bottlenecks to vibrational energy flow in carbonyl sulfide: Structures and mechanisms. *J. Chem. Phys.* 130, 164105, 2009.

63. M. E. Flatté and M. Holthaus, Classical and quantum dynamics of a periodically driven particle in a triangular well. *Ann. Phys. (NY)* 245, 113, 1996.

64. R. Utermann, T. Dittrich, and P. Hänggi, Tunneling and the onset of chaos in a driven bistable system. *Phys. Rev. E* 49, 273, 1994.

65. O. Morsch and M. Oberthaler, Dynamics of Bose-Einstein condensates in optical lattices. *Rev. Mod. Phys.* 78, 179, 2006; R. Gati and M. K. Oberthaler, A bosonic Josephson junction. *J. Phys. B—At. Mol. Opt. Phys.* 40, R61, 2007.

66. N. Sangouard, S. Guérin, M. Aminat-Talab, and H. R. Jauslin, Control of localization and suppression of tunneling by adiabatic passage. *Phys. Rev. Lett.* 93, 223602, 2004.

67. D. Farrelly and J. A. Milligan, 2-frequency control and suppression of tunneling in the driven double well. *Phys. Rev. E* 47, R2225, 1993.

68. M. Latka, P. Grigolini, and B. J. West, Chaos-induced avoided level-crossing and tunneling. *Phys. Rev. A* 50, 1071, 1994.

69. C. D. Schweiters and H. Rabitz, Optimal control of nonlinear classical systems with applications to unimolecular dissociation reactions and chaotic potentials. *Phys. Rev. A* 44, 5224, 1991.

70. P. Dietrich and P. B. Corkum, Ionization and dissociation of diatomic molecules in intense infrared laser fields. *J. Chem. Phys.* 97, 3187, 1992.

71. B. V. Chirikov, Universal instability of many-dimensional oscillator systems. *Phys. Rep.* 52, 263, 1979.

72. R. S. Mackay, J. D. Meiss, and I. C. Percival, Stochasticity and transport in Hamiltonian systems. *Phys. Rev. Lett.* 52, 697, 1984; D. Bensimon and L. E. Kadanoff, Extended chaos and disappearance of KAM trajectories. *Physica D* 13, 82, 1984.

73. G. Radons and R. E. Prange, Wave-functions at the critical Kolmogorov–Arnold–Moser surface. *Phys. Rev. Lett.* 61, 1691, 1988.

74. R. C. Brown and R. E. Wyatt, Quantum mechanical manifestation of cantori—wave packet localization in stochastic regions. *Phys. Rev. Lett.* 57, 1, 1986; R. C. Brown and R. E. Wyatt, Barriers to chaotic classical motion and quantum mechanical localization in multiphoton dissociation. *J. Phys. Chem.* 90, 3590, 1986.

75. Y. Gu and J. M. Yuan, Classical dynamics and resonance structures in laser-induced dissociation of a Morse oscillator. *Phys. Rev. A* 36, 3788, 1987.
76. R. Graham and M. Höhnerbach, Quantum effects on the multiphoton dissociation of a diatomic molecule. *Phys. Rev. A* 43, 3966, 1991.
77. H. P. Breuer and M. Holthaus, Adiabatic control of molecular excitation and tunneling by short laser pulses. *J. Phys. Chem.* 97, 12634, 1993.
78. C. C. Rankin and W. H. Miller, Classical S-matrix for linear reactive collisions of H + Cl_2. *J. Chem. Phys.* 55, 3150, 1971.
79. S. Huang, C. Chandre, and T. Uzer, Reducing multiphoton ionization in a linearly polarized microwave field by local control. *Phys. Rev. A* 74, 053408, 2006.
80. See for example, P. Carruthers and M. M. Nieto, *Rev. Mod. Phys.* 40, 411, 1968.
81. S. Wimberger, P. Schlagheck, C. Eltschka, and A. Buchleitner, Resonance-assisted decay of nondispersive wavepackets. *Phys. Rev. Lett.* 97, 043001, 2006.
82. See, for example, the various contributions in *Geometric Structures of Phase Space in Multidimensional Chaos*, eds. M. Toda, T. Komatsuzaki, T. Konishi, R. S. Berry, and S. A. Rice, *Adv. Chem. Phys.* 130, Wiley-Interscience, 2005.
83. L. V. Vela-Arevalo and S. Wiggins, Time-frequency analysis of classical trajectories of polyatomic molecules. *Int. J. Bifur. Chaos.* 11, 1359, 2001; C. Chandre, S. Wiggins, and T. Uzer, Time-frequency analysis of chaotic systems. *Physica D* 181, 171, 2003; C. C. Martens, M. J. Davis, and G. S. Ezra, Local frequency analysis of chaotic motion in multidimensional systems—energy transport and bottlenecks in planar OCS. *Chem. Phys. Lett.* 142, 519, 1987; J. Laskar, Frequency analysis for multidimensional systems—global dynamics and diffusion. *Physica D* 67, 257, 1993.
84. S. Keshavamurthy, Resonance-assisted tunneling in three degrees of freedom without discrete symmetry. *Phys. Rev. E* 72, 045203R, 2005.
85. G. Haller, Diffusion at intersecting resonances in Hamiltonian systems. *Phys. Lett.* A 200, 34, 1995; P. Lochak in *Hamiltonian Systems with Three or More Degrees of Freedom*, Ed. C. Simó, NATO ASI, Kluwer, Dordrecht, 1999, p. 168; M. Guzzo, E. L. C. Froeschlé, First numerical evidence of Arnold diffusion in quasi-integrable systems. *Disc. Cont. Dyn. Sys.* B 5, 687, 2005.
86. H. Fielding, M. Shapiro, and T. Baumert, Coherent control. *J. Phys B: At. Mol. Opt. Phys.* 41, 070201, 2008; J. Gong and P. Brumer, Quantum chaos meets coherent control. *Ann. Rev. Phys. Chem.* 56, 1, 2005; M. Shapiro and P. Brumer, Coherent control of molecular dynamics. *Rep. Prog. Phys.* 66, 859, 2003; K. Ohmori, Wavepacket and coherent control dynamics. *Ann. Rev. Phys. Chem.* 60, 487, 2009; R. Chakrabarti and H. A. Rabitz, Quantum control landscapes. *Int. Rev. Phys. Chem.* 26, 671, 2007.
87. V. Ya. Demikhovskii, F. M. Izrailev, and A. I. Malyshev, Manifestation of Arnold diffusion in quantum systems. *Phys. Rev. Lett.* 88, 154101, 2002; V. Ya. Demikhovskii, F. M. Izrailev, and A. I. Malyshev, Quantum Arnold diffusion in a simple nonlinear system. *Phys. Rev. E* 66, 036211, 2002.
88. T. Takami and H. Fujisaki, Analytic approach for controlling quantum states in complex systems. *Phys. Rev. E* 75, 036219, 2007.
89. M. Gruebele and P. G. Wolynes, Quantizing Ulam's control conjecture. *Phys. Rev. Lett.* 99, 060201, 2007.

11 Tunneling of Ultracold Atoms in Time-Independent Potentials

Ennio Arimondo and Sandro Wimberger

CONTENTS

11.1 INTRODUCTION

Tunneling as a quantum mechanical effect takes place in a classically forbidden region between two regions of classically allowed motion. While the term "dynamical tunneling" typically refers to tunneling of quantum states across dynamical barriers in classical phase space [1], the original problem simply intended tunneling across a potential barrier. Both types of tunneling are addressed in this chapter, with major focus on situations in which external forces make the studied systems intrinsically time-dependent and allow for a dynamical control of tunneling through potential barriers or across band gaps which are dynamically explored by the system.

A standard example of tunneling across static barriers is the motion in a double-well potential. The two potential wells are separated by a potential barrier which is impenetrable for a low-energy classical particle. The quantum mechanical solution shows that the wave packet initially localized in one of the wells performs oscillations between the two classically allowed region. Tunneling takes place between two levels nearly degenerate in energy, and in most cases the investigated tunneling takes place between the lowest energy states—for instance of a double well. However, in a potential configuration as the asymmetric double well shown in Figure 11.1a, an energy matching between a ground state on one side and an excited state on the other side leads to a tunneling between those states resonantly enhanced by the energy matching. In the resonantly enhanced tunneling (RET) the probability for the quantum tunneling of a particle between two potential wells is increased when the energies of the initial and final states of the process coincide. In the one-dimensional double potential barrier of Figure 11.1b, the narrow central potential well has weakly quantized (or quasistationary) bound states, of which the energies are denoted by E_1 and E_2 in Figure 11.1. If the energy E of electrons incident on the barrier coincides with these energies, the electrons may tunnel through both barriers without any attenuation. The transmission coefficient reaches unity at the electron energy $E = E_1$ or $E = E_2$. It is interesting that while the transmission coefficient of a potential barrier is always lower than one, two barriers in a row can be completely transparent for certain energies of the incident particle.

In the early 1970s, Tsu, Esaki, and Chang computed the two terminal current–voltage characteristics of a finite superlattice, and predicted that RET to be observed not only in the transmission coefficient but also in the current-voltage characteristic [2,3]. Resonant tunneling also occurs in potential profiles with more than two barriers. Technical advances led to the observation of negative differential conductance at terahertz frequencies and triggered a considerable research effort to study tunneling through multibarrier structures. Owing to the fundamental nature of this effect and the practical interest [4], in the last few years much progress has been made in constructing solid-state systems such as superlattices [5–7], quantum wells [8], and waveguide arrays [9] which enable the controlled observation and application of RET. The potential profiles required for resonant tunneling and realized in semiconductor system using heterojunctions allowed the manufacture of resonant-tunneling diodes. These devices have important applications such as in high-frequency signal generation and multivalued data storage, as reviewed in Mizuta and Tanoue (1995) [10].

In the last decade, the experimental techniques used in atom and quantum optics have made it possible to control the external and internal degrees of freedoms of ultracold atoms with a very high degree of precision. Thus, ultracold bosons or fermions loaded into the periodic optical potential created by interfering laser beams (double-well, lattices and superlattices) are optimal realizations of quantum mechanical processes and phenomena proposed and studied in other contexts of solid-state physics. Ultracold atoms and Bose–Einstein condensates (BEC), for instance, have been used

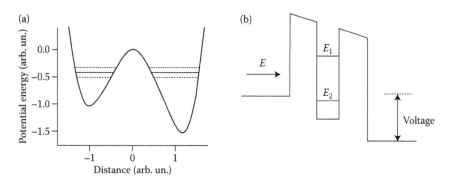

FIGURE 11.1 (a) Schematic representation of the energy levels within an asymmetric double well. The unperturbed energies within the left and right wells are indicated by the continuous lines. Because of the resonant tunneling between the ground state in the left well and the first excited one in the right well, the asymmetric and antisymmetric states are energy indicated by the dashed lines. (b) Schematic band diagram of a resonant-tunneling diode structure under a voltage bias between the incoming (left) and outgoing (right) regions.

to simulate phenomena such as Bloch oscillations in tilted periodic potentials [11–15] and to study quantum phase transitions driven by atom–atom interactions [16].

RET-like effects have been observed in a number of experiments till date. In Teo et al. (2002) [17], resonant tunneling was observed for cold atoms trapped by an optical lattice when an applied magnetic field produced a Zeeman splitting of the energy levels. Resonant tunneling has been observed in a Mott insulator within an optical lattice, where a finite amount of energy given by the on-site interaction energy is required to create a particle–hole excitation [18]. Tunneling of the atoms is therefore suppressed. If the lattice potential is tilted by application of a potential gradient, RET is allowed whenever the energy difference between neighboring lattice sites due to the potential gradient matches the on-site interaction energy. This RET control in a Mott insulator allowed Fölling et al. to observe a second-order coherence, that is, a two-atom RET [19].

Most of the quantum transport phenomena investigated with Bose–Einstein condensates within periodic optical lattices focused on the atomic motion in the ground state band of the periodic lattice. Only a few experiments examined the quantum transport associated with interband transitions "vertical" in the energy space. Interband transitions were induced by additional electromagnetic fields, as in the case of the spectroscopy of Wannier–Stark levels [20], or by quantum tunneling between the bands. Tunneling between otherwise uncoupled energy bands occurs when the bands are coupled by an additional force, which can be a static Stark force (tilting the otherwise periodic lattice) [14], or also by strong atom–atom interactions as observed for fermions in Köhl et al. (2005) [21] and discussed for bosons in Lee et al. (2007) [22]. The quantum tunneling between the ground and the first excited band is particularly pronounced in the presence of degeneracies of the single-well energy levels within the optical lattice leading to RET. In Sias et al. (2007) [23] and Zenesini et al. (2008) [24] such a type of RET was investigated for a Bose–Einstein condensate in a one-dimensional optical lattice, which allows for a high level of control on the potential depth and the lattice tilt. Those experimental investigations concentrated on the regime of parameters for which the tilting force—at RET conditions equal to the energy difference between neighboring wells—dominated the dynamics of the condensate. The RET tunneling of the ground band and the first two excited energy bands were measured in a wide range of experimental conditions. In addition the RET process is modified by the atom–atom interactions, bringing new physics to the quantum tunneling.

This chapter is organized as follows. Section 11.2 sets the stage discussing optical lattices and giving the necessary background. While Section 11.3 reports on RET in closed two- and three-well systems, Section 11.4 focuses on our main subject, the control of tunneling by RET in open quantum systems. Section 11.4 reports on our experimental data in the linear tunneling regime—that is, in the absence of atom–atom interactions, as well as on interaction induced effects. In Section 11.5, a model for many-body tunneling is introduced before we summarize the recent advances concerning RET in Section 11.6.

11.2 OPTICAL LATTICES

The investigations of tunneling for cold/ultracold atoms (Bose–Einstein condensates or Fermi degenerate gases) are based on the use of optical lattices [14,25]. For a 1D optical lattice a standing wave is created by the interference of two linearly polarized traveling waves counter-propagating along the x-axis with frequency ω_L and wave-vector λ_L. The amplitude of the generated electric field is $\mathcal{E}(r,t) = 2\mathcal{E}_0 \sin(\omega_L t)\sin(2\pi x/\lambda_L)$. When the laser detuning from the atomic transition is large enough to neglect the excited state spontaneous emission decay, the atom experiences a periodically varying conservative potential

$$V_{ol}(x) = V_0 \sin^2\left(\frac{\pi x}{d_L}\right),\tag{11.1}$$

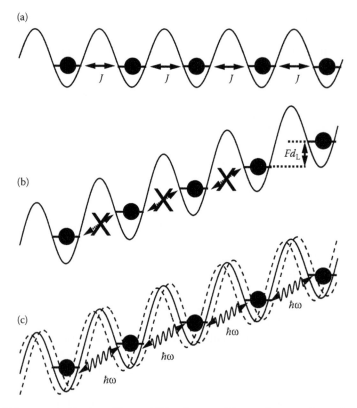

FIGURE 11.2 (a) In an optical lattice without additional external forces, the ground-state levels are resonantly coupled, leading to a tunneling energy J. (b) When a linear potential is applied, for example, by applying a force F, the levels are shifted out of resonance and tunneling is suppressed (Wannier–Stark localization). (c) If an additional potential energy oscillating at an appropriate frequency ω is applied, the levels can again be coupled through photons of energy $\hbar\omega$ and tunneling is partially restored.

schematically represented in Figure 11.2a. The amplitude V_0 depends on the laser detuning from the atomic transition and on the square of the $\mathcal{E}_0$ electric field amplitude [26]. The periodic potential has a spacing $d_L = \lambda_L/2$. This potential derives from the quantum mechanical interaction between atom and optical lattice photons. Therefore, the lattice quantities are linked to the recoil momentum $p_{rec} = 2\pi\hbar/\lambda_L$ acquired by an atom after the absorption or the emission of one photon. V_0 will be expressed in units of E_{rec}, the recoil energy acquired by an atom having mass M following one photon exchange

$$E_{rec} = \frac{h^2}{2M\lambda_L^2}. \tag{11.2}$$

Neglecting the atom–atom interactions in a Bose–Einstein condensate, our 1D system is described by the following Hamiltonian

$$H = -\frac{\hbar^2}{2M}\frac{d^2}{dx^2} + V_0 \sin^2\left(\frac{\pi x}{d_L}\right). \tag{11.3}$$

For this periodic potential, the associated single-particle eigenstates in the lowest band are Bloch plane waves with quasimomentum q. The energies $E_n(q)$ of the Bloch waves for the lowest bands $n = 1, 2, 3$ are plotted in Figure 11.3 versus quasimomentum. Ultracold atoms are loaded into the

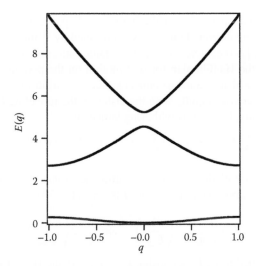

FIGURE 11.3 Plot of the energies for the energy bands $E_n(q)$ versus quasimomentum q for an optical lattice with optical depth $V_0 = 4E_{rec}$.

ground-state band having a minimum gap Δ at the edge of the Brillouin zone. The atomic evolution within that band or the excitation to a higher band is typically investigated.

If a force F is applied to the atom, as schematized in Figure 11.2b, the following Hamiltonian describes the atomic evolution neglecting for a moment atom–atom interactions in a Bose–Einstein condensate

$$H = -\frac{\hbar^2}{2M}\frac{d^2}{dx^2} + V_0 \sin^2\left(\frac{\pi x}{d_L}\right) + Fx. \tag{11.4}$$

This Hamiltonian defines the well-known Wannier–Stark problem for the electrons moving within a crystal lattice in the presence of an external electric field [27–29]. For small Stark forces F, one can picture the evolution of a momentum eigenstate induced by Equation 11.4 as an oscillatory motion in the ground energy band of the periodic lattice with Bloch period T_B [14,28,29], where

$$T_B = \frac{2\pi\hbar}{F d_L}. \tag{11.5}$$

At stronger applied forces, a wave packet prepared in the ground band has a significant probability to tunnel at the band edge to the first excited band. This process of the quantum tunnel across an energy gap at an avoided crossing of the system's energy levels is described by the Landau–Zener tunneling [30,31]. For a single tunneling event, the Landau–Zener tunneling probability is [29]

$$P_{LZ} = e^{-\frac{\pi^2}{8F_0}\left(\frac{\Delta}{E_{rec}}\right)^2}, \tag{11.6}$$

where we introduced the F_0 dimensionless force

$$F_0 = \frac{F d_L}{E_{rec}}. \tag{11.7}$$

In the presence of a sequence of Landau–Zener tunneling events, the Landau–Zener rate Γ_{LZ} to the excited band is obtained by multiplying P_{LZ} with the Bloch frequency $\nu_B = 1/T_B$ [28]. By introducing the recoil frequency $\nu_{rec} = E_{rec}/h$, Γ_{LZ} may be written

$$\Gamma_{LZ} = \nu_{rec} F_0 e^{-\frac{\pi^2}{8F_0}\left(\frac{\Delta}{E_{rec}}\right)^2}. \tag{11.8}$$

For the optical lattice periodic potential, an alternative single-particle basis useful for describing the tunneling of particles among discrete lattice sites is provided by Wannier functions [16,27–29,32]. The jth Wannier function $|j\rangle$ is centered around the j lattice site, and the functions are orthonormal. In a given energy band, the Hamiltonian for free motion on the periodic lattice is determined by hopping matrix elements, which in general connect lattice sites arbitrarily spaced. However, because the hopping amplitude decreases rapidly with the distance, the tunneling Hamiltonian may include only the J tunneling hopping between neighboring lattice sites

$$H = \sum_j E_j |j\rangle\langle j| - J \sum_j (|j\rangle\langle j+1| + |j+1\rangle\langle j|), \tag{11.9}$$

where E_j defines the energy of the jth site. For ultracold atoms in an optical lattice with depth $V_0 \gg E_{\text{rec}}$, the nearest-neighbor tunneling energy J is given by [33]

$$J = \frac{4}{\sqrt{\pi}} E_{\text{rec}} \left(\frac{V_0}{E_{\text{rec}}}\right)^{3/4} \exp\left(-2\sqrt{\frac{V_0}{E_{\text{rec}}}}\right). \tag{11.10}$$

In the presence of an applied force F, supposing $E_j \equiv E_0 = 0$, the Hamiltonian becomes

$$H = F d_{\text{L}} \sum_j j |j\rangle\langle j| - J \sum_j (|j\rangle\langle j+1| + |j+1\rangle\langle j|). \tag{11.11}$$

However, this Hamiltonian may be used to describe the atomic evolution in the ground band only when the Landau–Zener tunneling to the excited band can be neglected. Figure 11.4 reports for a given value of the dimensionless force F_0, the V_0 optical depth where the hopping constant J is 10 times larger than Γ_{LZ}.

The simulation of the temporal evolution of the Bose–Einstein condensate wavefunction is based either on the Gross–Pitaevskii equation based on a global mean-field description or on a many-body approach where the atomic number of the lattices sites is quantized [16,34,35]. Apart from the theoretical results reported in Section 11.5, we will concentrate here on the mean-field approach applied to describe experimental configurations and results reviewed in detail in Section 11.4. For a realistic description of those experiments, the Gross–Pitaevskii equation was used to simulate the temporal evolution of the condensate wave function $\psi(\vec{r},t)$ subjected to the optical lattice and to a confining harmonic potential, for instance with cylindrical symmetry

$$i\hbar\frac{\partial}{\partial t}\psi(\vec{r},t) = \left[-\frac{\hbar^2}{2M}\nabla^2 + \frac{1}{2}M\left(\omega_x^2 x^2 + \omega_r^2 \rho^2\right) + V_0 \sin^2\left(\frac{\pi x}{d_{\text{L}}}\right) + Fx + g|\psi(\vec{r},t)|^2\right]\psi(\vec{r},t). \tag{11.12}$$

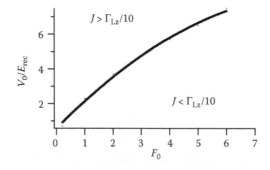

FIGURE 11.4 Plots of the line in the space of the optical lattice depth V_0, in E_{rec} units, and the dimensionless force F_0 dividing the upper (lower) region where the interwell tunneling is ten times larger (smaller) than the Landau–Zener tunneling to the upper band.

The frequencies ω_x and ω_r characterize the longitudinal and transverse harmonic confinement. The atom–atom interactions are modeled by the nonlinear term in Equation 11.12, with the nonlinear coupling constant given by $g = 4\pi\hbar^2 a_s/M$, where a_s is the s-wave scattering length [34,35]. Morsch and Oberthaler (2006) and Cristiani et al. (2002) [14,25] introduced the $\tilde{g}$ dimensionless nonlinearity parameter

$$\tilde{g} = \frac{gn_0}{8E_{\text{rec}}}, \tag{11.13}$$

computed from the peak density n_0 of the condensate initial state, to describe the nonlinear coupling relevant for optical lattice experiments. In the Thomas–Fermi regime of the condensate [34,35], for given ω_x and ω_r the density n_0, and therefore, $\tilde{g}$, is proportional to $N^{2/5}$ where N is the number of atoms in the condensate.

11.3 RESONANT TUNNELING IN CLOSED SYSTEMS

11.3.1 Two Levels

Quantum tunneling of a two-level system takes place in the double-well potential. The quantum mechanical solution shows that the wave packet initially localized in one of the wells performs oscillations between the two classically allowed regions. The period of these oscillations is related to the inverse of the energy difference between the symmetric and antisymmetric quantum states of the double-well system, that is, to the energy corresponding to the tunneling splitting. That energy is equal to the interaction Hamiltonian between the eigenstates of the two wells. In an asymmetric double well as that shown in Figure 11.1a, an energy matching between a ground state on one side and an excited state on the other side leads to a RET between those states. The dashed lines in Figure 11.1a denote the eigenenergies for the symmetric and antisymmetric quantum superposition of the wavefunctions in left and right wells. The tunneling evolution is described by the following Hamiltonian:

$$H = \sum_{j=1,2} E_j |j><j| - J(|1><2| + |1><2|) + U \sum_{j=1,2} n_j(n_j - 1), \tag{11.14}$$

where $|1>$ and $|2>$ denote the wavefunctions of the resonant states in the left and right wells, $\Delta = E_1 - E_2$ is the energy difference between the two wells, and J is the tunneling energy, U is the interatomic interaction energy and n_j is the atom number in the left or right well. For the following analysis U represents a shift in energy of the left or right well. By treating at first the $U = 0$ case, the atomic wavefunction may be expanded as a superposition of the $|1,2\rangle$ states

$$|\Psi(t)>= \sum_{j=1,2} C_j(t)|j\rangle, \tag{11.15}$$

the atomic evolution is characterized by Rabi oscillations between the two wells. For instance by supposing as initial condition $C_1(0) = 1$ and $C_2(0) = 0$, the occupation probabilities of the left well at time t are given by

$$|C_2(t)|^2 = \frac{J^2}{\Delta^2 + J^2} \sin^2 \sqrt{J^2 + \frac{\Delta^2}{4}}, \tag{11.16}$$

$$|C_1(t)|^2 = 1 - |C_2(t)|^2. \tag{11.17}$$

Therefore for the $\Delta = 0$ resonance condition of RET, a complete oscillation between the two wells at frequency $2J/\hbar$ takes place. The atomic interaction term U shifting the $E_{i=1,2}$ energies of the two wells may be included into the above equations for the occupation probabilities as a contribution to the Δ energy difference. Therefore, the presence of the U interatomic energy modifies the RET condition.

Periodic double-well structures may be created in properly chosen optical lattice or superlattice geometries. For cold atoms theoretical and experimental investigations were performed by Teo et al. (2002) [17], Castin et al. (1994) [36], Dutta et al. (1999) [37], and Haycock et al. (2000) [38]. For cold atoms the coherence length of the atomic wavefunction is comparable to the extent of each double well, so that the long range periodicity of the optical lattice plays a minor role on the tunneling properties. Therefore, those investigations will be mentioned here. Those studies examined the new features appearing when the double-well potential depends on the internal atomic structure, for instance on the two electron spin states. This case was theoretically analyzed by Castin et al. (1994) [36] within the context of two dimensional Sisyphus cooling. Resonant tunneling between the adjacent potential wells of the periodic potential for the two internal states, not present in a 1D geometry, contributes with quantum processes to the cooling phenomena in optical lattices. Dutta et al. (1999) [37] studied periodic well-to-well tunneling of ^{87}Rb atoms on adiabatic potential surfaces of a 1D optical lattice. Atoms that tunnel between neighboring wells of the lattice are an excellent tool for a careful study of topological potentials associated to the optical lattice. RET-like effects have been observed in a number of experiments to date. In Teo et al. (2002) [17], resonant tunneling was observed for cold atoms trapped by an optical lattice when an applied magnetic field produced a Zeeman splitting of the energy levels. At certain values of the applied magnetic field, the states in the up-shifting and down-shifting energy levels were tuned into resonance with one another. This led to RET drastically altering the quantum dynamics of the system and producing a modulation of the magnetization and lifetime of the atoms trapped by the optical lattice. Hacock et al. (2000) [38] observed the quantum coherent dynamics of atomic spinor wave packets in the double-well potentials. With appropriate initial conditions the atomic system performed Rabi oscillations between the left and right localized states of the ground doublet, with the atomic wavepacket corresponding to a coherent superposition of these mesoscopically distinct quantum states.

For ultracold atoms, Rabi oscillations in double-well geometries have been investigated and measured by Fölling et al. (2007) [19] and Kierig et al. (2008) [39]. A highly parallel structure of double wells is created using optical lattice or optical superlattice configurations. In the superlattice configuration of [19] the periodic potentials created by two laser standing waves at wavelength λ_L and $\lambda_L/2$ are applied to create a large set of individual wells. By changing the intensity of the standing wave lasers at the two wavelengths and their relative spatial phase, any configuration of symmetric or asymmetric double wells is created. In that experiment the double-well investigation was performed with ultracold atoms in a Mott-insulator configuration having single atom occupation of the wells [16]. The modification of the optical lattice potential from a periodic structure of single wells to a periodic structure of double wells, by adiabatically raising an energy bump within each single well, allowed to produce the asymmetric loading of each double well.

Figure 11.5 summarizes experimental results obtained in Fölling et al. (2007) [19] for the RET features in symmetric and asymmetric double wells. The tunneling of the ultracold atoms was measured as a function of the energy bias Δ between the wells. The left upper inset schematizes the case of single atom tunneling. The right lower one schematizes the tunneling of one atom in the presence of an energy shift produced by the atomic interaction (U term in Equation 11.14). A conditional resonant tunneling resonance occurs, where a single atom can tunnel only in the presence of a second atom and the interaction energy U is matched by the bias. For these two cases the measured atomic Rabi-type dynamical evolution between the two wells is shown in the right upper inset. Because the presence of an atom in the left well shifts by U the level energies, a bias $\Delta = -U$ is applied in order to compensate the shift. Thus, a resonant tunneling condition is verified and the gray data denote the periodic occupation of the left well and right well, located at positions -1 and 0, respectively. In the absence of an atom in the left well and without application of the bias, the tunneling is not resonant and the Rabi oscillations take place with a reduced amplitude and at a higher frequency, in agreement with the description of Equations 11.16 and 11.17. The left lower inset schematizes the case of a correlated atomic pair tunneling, as produced in a second-order tunneling process. The central part of that figure reports the amplitude of the Rabi oscillations versus the Δ bias for the different

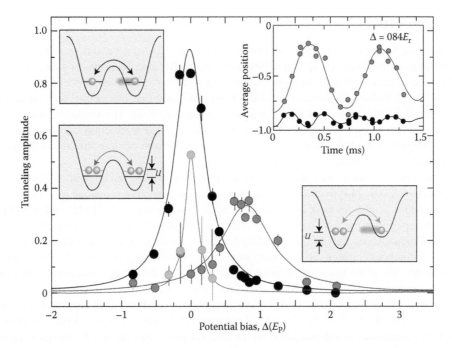

FIGURE 11.5 Tunneling configuration and experimental results for the resonant tunneling of single and double atoms in a superlattice. The periodic double-well potential for ultracold rubidium atoms was realized by superimposing two periodic potentials with periodicities of $\lambda_L = 765.0$ nm (long lattice) and $\lambda_L/2 = 382.5$ nm (short lattice), and controllable intensities and relative phase. The depth was $V_0 = 12E_{rec}$ for the short optical lattice, $V_0 = 9.5E_{rec}$ for the long lattice. The upper right, lower left and lower right insets describe the resonant tunneling configurations for one or two atoms per well. The upper left inset describes the oscillating motion of the atoms between the the two wells for the conditional resonant tunneling resonance where a single atom can tunnel only in the presence of a second atom and the interaction energy U is matched by an applied bias. In the central part the amplitude of the tunneling Rabi oscillations, and the Lorentzian fit, are shown as a function of the bias energy Δ for each of the tunneling configurations represented in the insets, black dots and Lorentzian centered at $\Delta = 0$ for upper left one, light gray dots and Lorentzian centered at $\Delta = 0$ for lower left one, and gray dots and Lorentzian centered at $\Delta = 0.78E_{rec}$ for lower right one. (From S. Fölling et al. *Nature* 448, 1029, 2007. With permission by MacMillan.)

tunneling configurations, and their fits by the Lorentzian line shapes predicted by Equation 11.16. The tunneling amplitude versus the potential bias is measured for the case of single atoms (black data points) and initially doubly occupied lattice sites (gray line and light gray data points). The gray data points and the Lorentzian fitted to the data point with center at $\Delta = 0.78(2)E_{rec}$ correspond to the conditional resonant tunneling resonance. The correlated pair tunneling (light gray circles) and the Lorentzian fit are resonant for zero bias because energies of both left and right wells are modified by the interaction energy U.

While the previous description applies to single particle tunneling, quantum tunneling of macroscopic N-body atomic systems introduces qualitatively new aspects to the quantum evolution of ultracold atoms, as investigated in Dounas-Frazer et al. (2007) [40] for Bose–Einstein condensate in a tilted multilevel double-well potential. For a double well without tilt as experimentally investigated by Albiez et al. (2005) [41], the so-called self trapping regimes is realized where the bosonic nonlinear interaction term of the equation, Equation 11.12 modifies the level energies and inhibits the resonant tunneling between the wells. Khomeriki et al. (2006) [42] demonstrated that for a double-well structure by a pulse-wise change of the intermediate barrier height, it is possible to switch between the tunneling regime and the self-trapped one.

11.3.2 THREE LEVELS

The idea of controlling the tunneling rate between two states has led several researchers to consider the effect of external forces on the tunneling oscillations. Because the tunneling rate is related to the difference in the energies of the quantum states, a number of complicated scenarios arise when one of the states undergoes interaction with a third state, and that interaction may be controlled by an external parameter, for instance a magnetic or electric field. The tunneling wavepacket is described as a linear combination of the three initial states. Their interaction can drastically affect the eigenenergies of the Hamiltonian and it would be possible to explore different regimes, from strong suppression to enhancement of tunneling.

This three-level control was theoretically investigated in Averbukh et al. (2002) [43] and Hensinger et al. (2004) [44] in connection to the dynamical tunneling produced by time-dependent potentials and for conditions as in an experiment by Raizen's group in 2001 [45] and at NIST [46]. The tunneling period in the time-dependent systems is related to the differences between quasienergies of the Floquet states, just as the tunneling period in the time-independent case has to do with the energy differences between the stationary states. The experimental and theoretical investigations considered the case of the tunneling doublet interacting with a third state associated with a chaotic region. The underlying classical phase space of the systems had a mixed regular-chaotic structure, giving the scenario of chaos-assisted [47] or, more generally speaking, of dynamical tunneling [1].

We present here the basic of the three-level tunneling in the case of time independent potentials. Figure 11.6 schematizes the dependence on an external parameter for the E_j energies for the $|j\rangle$ states, with $j = 1 \ldots 3$, in the absence of interactions between them. We will discuss the modifications to those energies produced by atomic interactions between states, supposing the presence of the interactions U_{12} between states $|1\rangle$ and $|2\rangle$, and U_{23} between states $|2\rangle$ and $|3\rangle$, and supposing no interaction between states $|1\rangle$ and $|3\rangle$. Notice that these interactions modify the E_j energies in the regions close to the energy crossings, boxes 1, 2, and 3 in the Figure 11.6, and that the tunneling

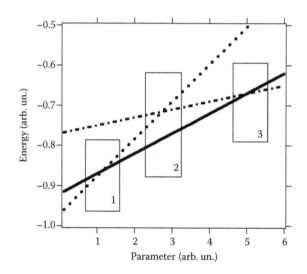

FIGURE 11.6   Unperturbed energies E_j, with $j = (1.3)$ (in arbitrary units) of three states experiencing crossings and anticrossings as a function of a parameter (also in arbitrary units). Continuous lines corresponds to state $|1\rangle$, the dotted one to state $|2\rangle$ and the dot-dashed to state $|3\rangle$. The boxes marked with 1 and 3 denote regions where the tunneling is dominated by two-state interactions. The box marked 2 denotes a region where the three-state interaction may modify the tunneling rate between state $|2\rangle$ and $|3\rangle$. In region 3 without direct interaction between states $|1\rangle$ and $|3\rangle$ a locking of tunneling, corresponding to a level crossing with $E_1 = E_3$, takes place.

frequency is determined by the splitting of the perturbed energies. In the box with number 1 the $E_1 - E_2$ energy separation, that is, the tunneling, is dominated by the interaction between states $|1\rangle$ and $|2\rangle$. In the box denoted as 2, a three-state interaction takes place and the amplitude of the interaction between states $|1\rangle$ and $|2\rangle$ may be used to enhance or suppress the tunneling frequency between the states $|2\rangle$ and $|3\rangle$. Within the region denoted as 3, in the absence of a direct interaction between the $|1\rangle$ and $|3\rangle$ states a $E_1 = E_3$ crossing point exists. This crossing produces an absence of tunneling, this configuration being indicated as locking of the wavefunction in the initial state of preparation [43].

11.4 TUNNELING IN OPEN SYSTEMS

11.4.1 OPTICAL LATTICE WITHOUT/WITH TILT

An optical lattice is composed of an infinite number of neighboring wells uniformly distributed along one direction and spacing $d_L = \lambda/2$ between the minima, where λ is the wavelength of the standing wave laser required to create the periodic potential for the atoms [14]. This configuration corresponds to Figure 11.2a. The tunneling in this system has strong similarities to the double well discussed above, when the presence of physical boundaries, as in the physical realizations, plays no role.

For a more general treatment we consider the case where an applied external force F produces an energy difference Fd_L between neighboring wells, see Figure 11.2b. The atomic evolution may be studied by considering the localized Wannier wavefunction $|i>$ and the perturbations originating from the atomic occupation in neighboring sites [32]. This approximation is valid when the overlap of atomic wavefunctions introduces corrections to the localized atom picture, but they are not large enough to render the single site description irrelevant. The H Wannier–Stark Hamiltonian determining the atomic evolution in the absence of the interatomic interactions U is given by

$$H = -J\sum_j (|j><j+1| + |j+1><j|) + Fd_L\sum_j j|j><j|. \tag{11.18}$$

In analogy to Equation 11.15 the generic atomic wave function can be written as a superposition of the $|j>$ localized wavefunctions where the sum extends over all lattice sites. The temporal evolution for the C_i coefficients under the Hamiltonian H is given by

$$i\hbar\frac{dC_j}{dt} = jFd_LC_j - J\left(C_{j+1} + C_{j-1}\right), \tag{11.19}$$

and in the following the ground state energy E_0 will be supposed to be equal to zero. The solution of these coupled equations with $t = 0$ initial condition of atomic occupation of the $i = 0$ site—that is, $C_j(t = 0) = \delta_{j=0}$, leads to [48]

$$|C_j(t)|^2 = \mathcal{J}_j^2\left[\frac{2JT_R}{\hbar}\sin\left(\frac{\pi t}{T_R}\right)\right], \tag{11.20}$$

having introduced the Bessel functions $\mathcal{J}_j$ of jth order. The argument of the Bessel functions in Equation 11.20 is an oscillatory function of time. T_R represents the recurrence time for the evolution of the atomic wavefunction. For the present case of the resonant tunneling modified by the presence of a force F, $T_R = T_B$ whence the recurrence time coincides with the Bloch period T_B defined in Equation 11.5 and is inversely proportional to the applied external force. The temporal recurrence of the atomic wavefunction is shown in Figure 11.7 for different times expressed in units of T_R. Notice that the parameter $2JT_R/\hbar$ of the Bessel function determines the range of lattice sites occupied by the periodic wavefunction expansion. The corresponding atomic mean-square displacement is

$$\frac{\sqrt{<m^2(t)>}}{d_L} = \frac{2\sqrt{2}JT_R}{\pi\hbar}\left|\sin\left(\frac{\pi t}{T_R}\right)\right|. \tag{11.21}$$

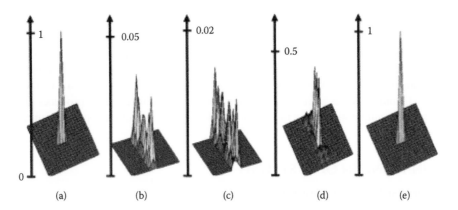

     (a)   (b)   (c)   (d)   (e)

FIGURE 11.7 Temporal recurrence of the occupation probability $|C_n|^2$ versus the n position of the lattice site at different interactions times. From (a) to (e) interaction times are 0, 0.2, 0.5, 0.95, 1 measured in units of T_R. The occupation probabilities are connected by lines. Notice the reduced vertical scale at the intermediate times. The plots are obtained for the parameter $2JT_R/\hbar = 28$.

In the limit of $Fd_L \gg J$ the mean-square displacement is largely decreased because of the suppression of the resonant tunneling, as schematized in Figure 11.2b. This suppression and the related Wannier–Stark localization of the wavefunction have been intensively discussed in the solid-state physics theoretical literature [28,49]. Korsch and coworkers [50,51] have considered the case of an atomic distribution not initially concentrated on a single site, and instead described by a Gaussian distribution with root mean-square σ_0. For that case the temporal evolution of the mean-square displacement is given by

$$\frac{<m^2(t)>}{d_L^2} = \left(\frac{\sigma_0}{d_L}\right)^2 + 8\left(\frac{JT_R}{\pi\hbar}\right)^2 \sin^2\left(\frac{\pi t}{T_R}\right)\left[1 - e^{-d_L^2/2\sigma_0^2}\cos\left(\frac{2\pi t}{T_R}\right) - 2e^{-d_L^2/8\sigma_0^2}\sin^2\left(\frac{\pi t}{T_R}\right)\right].$$

$$(11.22)$$

In the absence of external force, taking the limit of $F \to 0$, we recover the result of a diffusion process for the atomic wavefunction

$$|C_j(t)|^2 = \mathcal{J}_j^2\left[\frac{2Jt}{\hbar}\right],$$

$$(11.23)$$

$$\frac{\sqrt{<m^2>}}{d_L} = \frac{\sqrt{2}Jt}{\hbar}.$$

$$(11.24)$$

11.4.2 Photon-Assisted Tunneling

The above analysis can be applied also to the photon-assisted tunneling occurring when the ground states of adjacent potential wells tuned out of resonance by the Fd_L static potential are coupled by photons at frequency ω as schematized in Figure 11.2c. When the photon energy bridges the gap created by the static potential, tunneling is (partly) restored. The resonant tunneling is restored by a photon-assisted process when the energy provided by n photons matches the separation energy Fd_L between neighboring wells. The energy resonance condition for the frequency ω_R is given by

$$n\hbar\omega_R = Fd_L,$$

$$(11.25)$$

with the integer n denoting the order of the photon-assisted resonance. This resonance may be expressed as $\omega_R = 2\pi\nu_B/n$ in terms of the Bloch frequency. The frequency detuning from the resonance is $\Delta\omega = \omega - \omega_R$.

In solid-state systems, the photons are typically in the microwave frequency range and the static potential is provided by an electric bias field applied to the structure. Photon-assisted tunneling has been observed in superconducting diodes [52], semiconductor superlattices [53,54], and quantum dots [55,56].

For the photon-assisted tunneling of cold and ultracold atoms, a theoretical analysis was performed by Eckardt et al. (2005) [57] and by Kolovsky and Korsch (2009) [58], with experiments performed by Sias et al. (2008) [59], Ivanov et al. (2008) [60], Alberti et al. (2009) [61], and Haller et al. (2010) [62]. In these experiments a periodic time-dependent potential was applied to the cold atoms through a periodic spatial oscillation of the optical lattice minima/maxima, to be referred to as shaking in the following. In the lattice reference frame such a backward and forward motion of the periodic potential at frequency $\omega \approx \omega_R$ along one direction is equivalent to a periodic force $F_\omega \cos(\omega t)$ applied to the atoms. Thus, using the localized Wannier wavefunction introduced above for a deep lattice the atomic evolution is determined by the following Hamiltonian:

$$H_{\text{shaking}} = -J \sum_j (|j><j+1| + |j+1><j|) + [Fd_L + K\cos(\omega t)] \sum_j j|j><j|, \qquad (11.26)$$

once again not including the U interaction term. Here $K = F_\omega d_L$, denoted as shaking amplitude, is the shaking energy difference between neighboring sites of the linear chain associated to the shaking. The theory of Eckardt et al. (2005) [57] predicts that when the driving takes place at the frequency $\omega_R \gg J/E_{\text{rec}}$ and the resonance condition of Equation 11.25 is satisfied, the shaking leads to an effective tunneling rate

$$J_{\text{eff}}(K, \omega_R) = J \mathcal{J}_n \left(\frac{K}{\hbar \omega_R} \right). \qquad (11.27)$$

Therefore, a modification of the tunneling rate is obtained when the ratio of the rescaled shaking amplitude $K = F_\omega d_L$ and the shaking frequency times $\hbar$ is varied. In the experimental realization [59] the shaking frequency was fixed and the shaking amplitude was scanned to verify the relation of Equation 11.27.

The previous analysis for the evolution of the atomic wavefunction under resonant tunneling can be applied also to the photon-assisted tunneling by using the approximation of a resonant dynamics introduced by Thommen et al. (2002) [63] or equivalently by restricting our attention to the resonant Floquet states [64]. In the presence of a driving at frequency ω and taking into account the static energy difference Fd_L between neighboring wells, we write for the atomic wavefunction

$$|\Psi(t)> = \sum_{j,m} \tilde{C}_{j,m} e^{-i(jFd_L + m\hbar\omega)t/\hbar} |j>, \qquad (11.28)$$

where the j index labels the well and the m index the component in the Floquet spectrum. For ω close to the nth order resonance condition we may restrict the terms to the resonant ones in two sums of the above expansion

$$|\Psi(t)> = \sum_j e^{-ij\Delta\omega t} \tilde{C}_j^n |j>, \qquad (11.29)$$

where we have simplified the notation introducing the resonant coefficients $\tilde{C}_j^n$.

The temporal evolution of the $\tilde{C}_j^n$ is described by an equation similar to Equation 11.19 where J_{eff} determines the tunneling energy of the nth order resonance. Therefore, for the photon-assisted tunneling, the occupation of the jth lattice site and the mean-square displacement of the atoms are the analogs to those derived previously

$$|C_j^n(t)|^2 = \mathcal{J}_j^2 \left[\frac{2J_{\text{eff}} T_R}{\hbar} \sin \left(\frac{\pi t}{T_R} \right) \right], \qquad (11.30)$$

$$\frac{\sqrt{<m^2>}}{d_L} = \frac{2\sqrt{2} J_{\text{eff}} T_R}{\pi \hbar} \left| \sin \left(\frac{\pi t}{T_R} \right) \right|, \qquad (11.31)$$

with T_R the recurrence time for this process given by

$$T_R = 2\pi/\Delta\omega. \tag{11.32}$$

This recurrence process was named as super-Bloch oscillations in Kolovsky and Korsch (2009) [58] and Haller et al. (2010) [62]. For the resonant case $\Delta\omega = 0$, the mean-square displacement is given by Equation 11.24 and the occupation probabilities are given by Equation 11.23. Notice that for both Wannier–Stark localization and photon-assisted tunneling, the mean-square displacement and the occupation probabilities have the same functional dependence if we introduce a unifying parameter for the detuning from the resonant tunneling. This parameter is Fd_L for the case of an applied external force and $\hbar\Delta\omega$ for the case of the photon-assisted tunneling. Thus, the data of Figure 11.7 applies also to the occupation probabilities in the photon-assisted tunneling.

A few recent experiments on optical lattices have verified or made use of the theoretical predictions of this Section. In the following the experiments will be characterized by the depth V_0 of the optical lattice expressed in units E_{rec}, and the photon-assisted frequency detuning $\Delta\omega_0$.

The linear time dependence of atomic mean-square displacement predicted by Equation 11.24 in the conditions of $F = 0$ was applied by Lignier et al. (2007) [65] to measure the J tunneling energy and to verify that the experimental procedure reproduced the J dependence on the lattice depth V_0 predicted by Equation 11.10. The photon-assisted tunneling experiments [59,60] made use of that linear dependence to measure the effective tunneling rate. In these experiments, the linear dependence was tested for a total time larger than 10,000 tunneling times. Notice that in all these experimental observations the initial distribution of the atomic wavefunction was not concentrated on a single well as in our theoretical analysis and instead covered several wells. Nevertheless, a Gaussian convolution of the initial wavefunction spread and of the linearly expanding mean-square displacement represented a good fit of the experimental observations, even at earlier times where the initial width is comparable to the tunneling spread.

The Wannier–Stark localization of the atomic cloud in the presence of an applied force F was examined by Sias et al. (2008) [59] as a reduction of the mean-square displacement increasing the force amplitude at a given interrogation time. Figure 11.8a reports the temporal dependence of $\sqrt{< m^2 >}/d_L$ as predicted by Equation 11.21, at different values of the parameter Fd_L/J scanned in that experiment within the interval $(0,1)$. In order to provide a unified description the time is measured in units of T_B. It appears that $\sqrt{< m^2 >}$ is periodic in time with period T_B while the

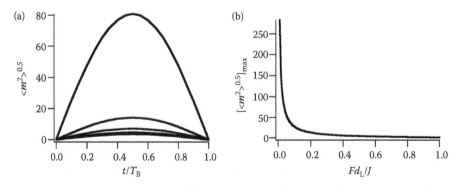

FIGURE 11.8   (a) Mean-square displacement versus time for different values of the unified RET energy mismatch, Fd_L/J for the Wannier–Stark localization and $\hbar\Delta\omega/J$ for the photon assisted tunneling. Results for values 0.2, 0.4, 0.6, 0.8, and 1.0 of the detuning parameter, with the displacement maximum decreasing at higher values. The time dependence of $\sqrt{< m^2 >}$ is periodic in time with period T_B. In (b) the maximum of the mean-square displacement is plotted versus Fd_L/J. The mean-square displacements are measured in units of the d_L lattice spacing.

amplitude of the oscillation decreases with the force until the Wannier–Stark localization regime is reached where the atomic motion is blocked. Figure 11.8b shows the amplitude of the oscillation predicted by Equation 11.24 versus the Fd_L/J parameter. By comparing this dependence to the Lorentzian one occurring for a two-level system of the previous Section, it appears that for an infinite systems of wells the oscillation amplitude decreases more rapidly increasing Fd_L/J. For different values of the applied force, the maximum of the oscillation occurs at a different value of t. Therefore, the experiment of Sias et al. (2008) [59] that measured the oscillation amplitude at a given interaction time, obtained results similar to those of Figure 11.8b, not precisely fitted by the inverse law as sketched in Figure 11.8.

For the photon-assisted tunneling the functional dependence on time of the wavefunction spreading on the lattice and the mean-square displacement was measured in Alberti et al. (2009) [61] for a total time equivalent up to seven recurrence times in the case of a drive detuned by $\Delta\omega/2\pi = \pm 5$ Hz and up to one recurrence time for the $\Delta\omega/2\pi = \pm 0.260$ Hz detuning. The measured sinusoidal evolutions are in reasonable agreement with the sinusoidal function predicted by our model and represented in Figures 11.7 and 11.8a. Our model does not take into account the initial atomic distribution over several optical lattice sites, and in Alberti et al. (2009) [61], because the atomic de Broglie wavelength was shorter than the lattice period, the coherence degree among adjacent Wannier–Stark eigenstates was negligible. The quantum–mechanical evolution of the atomic wavefunction under the tunneling Hamiltonian described by our analysis is limited by the presence of decoherence processes, and in Alberti et al. (2009) [61] a decoherence time of 28 s was measured. It would be interesting to investigate theoretically the role of a decoherence process on the tunneling evolution.

For the photon-assisted tunneling the mean-square amplitude dependence on the detuning $\Delta\omega$ is given by Equation 11.21 with $T_R = 2\pi/\Delta\omega$. That functional dependence predicts that the full width of the resonance line-shape $\Delta\omega_{FW}$, defined by the first zeros of the sin function, is determined by the experimental interrogation time T

$$\Delta\omega_{FW} = \frac{\pi}{T}. \tag{11.33}$$

For interrogation times between 0.5 and 2 s of the experimental investigations line widths in the few Hertz range were measured. In the investigation of Ivanov et al. (2008) [60] where the external force was gravity, the measurement of the resonance frequency for the photon-assisted tunneling with the accuracy reached by the above interrogation time allowed those authors to measure the gravity acceleration with ppm resolution. This shows that sensitive RET effects have a great potential for applications, for example, for precision measurements.

The recurrence process of super-Bloch oscillations was recently investigated by Haller et al. (2010) [62] for V_0/E_{rec} values in the 3–7 range, and $\Delta\omega/2\pi$ in the $0.1 = 2$ Hz range. The recurrence oscillations were measured up to 2.5 s.

11.4.3 RET in Optical Lattices with Tilt

In spite of the fundamental RET nature and of its practical interest, for a long time the experimental observation was restricted to the motion of electrons in superlattice structures [6]. In 2007 Sias et al. [23] observed resonant tunneling using Bose–Einstein condensates in accelerated optical lattice potentials. The nearly perfect control over the parameters of this system allowed the authors to prepare the condensates with arbitrary initial conditions and also to study the effects of nonlinearity and a loss of coherence. Such observation can be generalized to studying noise and thermal effects in resonant tunneling and underlines the usefulness of Bose–Einstein condensates in optical lattices as model systems for the solid state.

A schematic representation of RET is shown in Figure 11.9. In a tilted periodic potential, atoms can escape by tunneling to the continuum via higher-lying levels. The tilt of the potential is proportional to the applied force F acting on the atoms, and the tunneling rate Γ_{LZ} can be calcu-

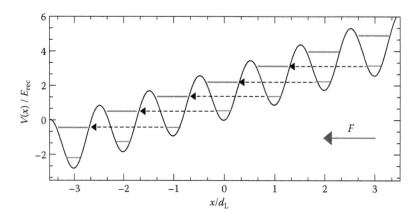

FIGURE 11.9 Schematic of the RET process between second nearest-neighboring wells, that is, for $\Delta i = 2$. The tunneling of atoms is resonantly enhanced when the energy difference between lattice wells matches the separation between the energy levels in different potential wells.

lated using the Landau–Zener formula of Equation 11.8. The actual rates can dramatically deviate from Equation 11.8 when two Wannier–Stark levels in different potentials wells are strongly coupled owing to the accidental degeneracy of Figure 11.9 where the tilt-induced energy difference between wells i and $i + \Delta i$ matches the separation between two quantized energy levels, as pointed out for cold atoms by Bharucha et al. (1997) [66]. Indeed, the tunneling probability can be enhanced by a large factor over the Landau–Zener prediction (see theoretical and experimental results of Figure 11.10).

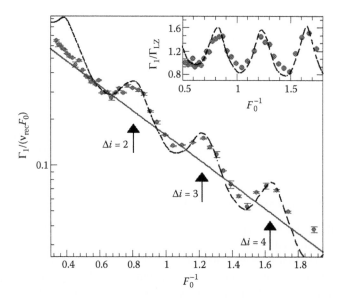

FIGURE 11.10 Resonant tunneling in the linear regime. Shown here is the tunneling rate from the lowest energy bands of the lattice as a function of the normalized inverse force F_0^{-1} for $V_0 = 2.5\,E_{\text{rec}}$ lattice depth. The straight line represents the prediction of the Landau–Zener theory. Inset: Deviation from the Landau–Zener prediction of Equation 11.6. (Adapted from C. Sias et al. *Phys. Rev. Lett.* 98, 120403, 2007. Copyright 2007 of American Physical Society.)

By imposing an energy resonance between the Wannier–Stark levels in different wells of an optical lattice shifted by the potential of the external force, one finds that the energy degeneracies occur at the values F at which $Fd_L\Delta i$ (Δi integer) is close to the mean band gap between two coupled bands of the $F = 0$ problem [7,28]. The actual peak positions are slightly shifted with respect to this simplified estimate, because the Wannier–Stark levels in the potential wells are only approximately defined by the averaged band gap of the $F = 0$ problem, a consequence of field-induced level shifts [28].

11.4.3.1 Linear Regime and Decay Rates

Although the finite and positive scattering length of ^{87}Rb atoms means that the linear Hamiltonian of Equation 11.4 is never exactly realized in experiments, the approximation of a noninteracting BEC is valid if the condensate density is maintained low. In that case, the interaction energy can be made much smaller than all the other energy scales of the system (recoil energy, bandwidth, gap width) and hence it is negligible for the present analysis of RET in a condensate.

Figure 11.10 shows the results of Sias et al. (2007) [23] for experimental investigations with low-density condensates and the nonlinearity parameter $\tilde{g}$ less than $\approx 1 \times 10^{-2}$, defined as the limit of the linear regime. The tunneling rate Γ_1 out of the first band is shown as a function of F_0^{-1}. Super-imposed on the overall exponential dependence of Γ_1/F_0 on F_0^{-1}, one clearly sees the resonant tunneling peaks corresponding to the various resonances $\Delta i = 1,2,3,4$. Which of the resonances were visible in the experiment depended on the choice of lattice parameters and the finite experimental resolution. The limit $n = 3$ for the highest band explored in Sias et al. (2007) [23] was given by the maximum lattice depth achievable.

By measuring the positions of the $\Delta i = 1,2,3$ tunneling resonances for different values of the lattice depth V_0, it appeared that the resonances were shifted according to the variation of the energy levels. For deep enough lattices, the resonance positions may be derived from a numerical simulation but can also be approximately calculated by making a harmonic approximation in the lattice wells, which predicts a separation of the two lowest energy levels ($n = 1$ and $n = 2$) of

$$\Delta E_{2-1} = 2E_{\text{rec}}\sqrt{\frac{V_0}{E_{\text{rec}}}}. \tag{11.34}$$

By imposing the resonance condition $\Delta E_{2-1} = F^{\text{res}}d_L\Delta i$, the calculated F^{res} resonance position results in good approximation with that predicted in Glück et al. (2002) [28] and Wimberger et al. (2005) [67].

11.4.3.2 Avoided Crossings

The accessibility of higher energy levels allowed an experimental measurement of the tunneling rates around RET conditions of two strongly coupled bands. The dependence of those rates on the system parameters was phrased into the frame of level crossing for states experiencing a loss rate. The modification of the level tunneling rate by the presence of a degeneracy may be described by a simple model of a two-level Hamiltonian with an energy separation ε described by an energy cross-ing splitting $\varepsilon = 0$ and with a single level characterized by a decay rate [68,69]. Real and imaginary parts of the Hamiltonian eigenvalues are different for $\varepsilon \neq 0$, and two different scenarios take place with crossings or anticrossings of the real and imaginary part of the Hamiltonian eigenvalues. In one case, denoted as type-I crossing, the imaginary parts of the eigenvalues cross while the real parts anticross. In the second case, denoted as type-II crossing, the eigenvalues anticross while the real parts cross. The numerical simulations of Zenesini et al. (2008) [24] pointed out that the large majority of the RET explored experimentally correspond to type-II crossings. As a consequence if a resonance takes place between the energy of the lower state and that of the decaying upper level, the tunneling rate of the lower state increases significantly. In addition the upper state experiences

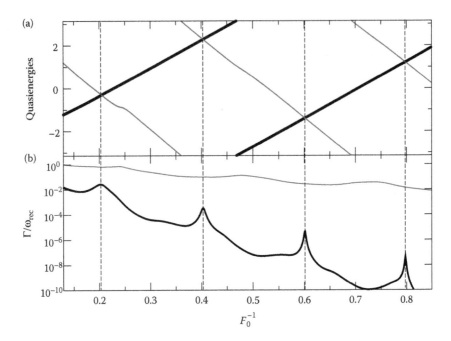

FIGURE 11.11 In (a) real parts of the eigenenergies and in (b) tunneling rate Γ in ω_{rec} units for a lattice depth of $V_0/E_{rec} = 10$ and the Hamiltonian from Equation 11.4. The eigenenergies and the tunneling rates are associated with two Wannier–Stark ladders or, equivalently, with two energy bands: ground state (thick black lines) and first excited state (thin gray lines). The maxima of the ground-state tunneling rates corresponds to $\Delta i = 1, 2, 3$, and 4. (Reproduced from A. Zenesini et al. *NJP* 10, 0530388, 2008. With permission. Copyright Institute of Physics.)

a resonantly stabilized tunneling (RST) with a decrease of its tunneling rate. Figure 11.11a shows theoretical predictions for type-II crossing and anticrossings for the real parts of the eigenenergies associated with a RET configuration investigated experimentally as a function of the experimental control parameter, the Stark force determined by the F_0 dimensionless parameter of Equation 11.7. The associated Wannier–Stark states tunneling rates are shown in Figure 11.11b as a function of F_0. The strong modulations on top of the global exponential decrease arise from RET processes originated by the energy crossings. The resonance eigenstates and eigenenergies for the noninteracting atoms described by Equation 11.4 were obtained in Zenesini et al. (2008) [24] by diagonalizing an open version of the Hamiltonian [28,70–73].

Experimental data on anticrossings in the tunneling rates are in Figure 11.12 taken from Sias et al. (2007) [23]. Although a direct observation of the discussed anticrossing scenario in two different levels for the same set of parameters was not possible, the experimental investigation compared the ground and excited state tunneling rates Γ_1 and Γ_2 with the theoretical predictions for two different parameter sets, as shown in Figure 11.12. This figure nicely reveals the anticrossing of the corresponding tunneling rates of strongly coupled levels as a function of the control parameter F_0 around RET conditions.

11.4.3.3　Nonlinearity

This section discusses how the experimental investigation of RET in tilted optical lattices are modified by the atom–atom interactions in the Bose–Einstein condensate. We focus on a parameter regime where the Stark force essentially dominates the dynamics of the condensate. Here the

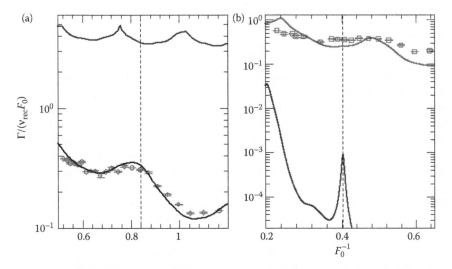

FIGURE 11.12 Anticrossing scenario of the RET rates. (a) Theoretical plot of $\Gamma_{1,2}$ for $V_0 = 2.5 E_{\text{rec}}$ with experimental points for Γ_1. (b) Theoretical plot of $\Gamma_{1,2}$ for $V_0 = 10 E_{\text{rec}}$ with experimental points for Γ_2. (Adapted from C. Sias et al. *Phys. Rev. Lett.* 98, 120403, 2007.)

quantum tunneling between the energy bands is significant and most easily detected experimentally. The critical field values for which such excitations are relevant can be estimated by comparing, for instance, the potential energy difference between neighboring wells, $F d_{\text{L}}$, with the coupling parameters of the many-body Bose–Hubbard model, that is, the hopping constant J and interaction constant U [14].

Our analysis will exclude the regime of $F_0 \leq J/E_{\text{rec}} \approx U/E_{\text{rec}}$ where a quantum chaotic system is realized [74–78]. The origin of quantum chaos, that is, of the strongly force-dependent and non-perturbative mixing of energy levels can be understood as a consequence of the interaction-induced lifting of the degeneracy of the multiparticle Wannier–Stark levels in the crossover regime from Bloch to Wannier spectra, making nearby levels strongly interact, for comparable magnitudes of hopping matrix elements and Stark shifts.

For the regime of $F_0 \gg J/E_{\text{rec}}$, the effect of weak atomic interactions is just a perturbative shifting and a small splitting of many-body energy levels [71,77]. In order to access the tunneling rates measured in the experiment of Sias et al. [23], we determine the temporal evolution of the survival probability $P_{\text{sur}}(t)$ for the condensate to remain in the energy band, in which it has been prepared initially. As proposed in Wimberger et al. (2005) [67] and applied in the experimental investigation, such a survival probability is best measured in momentum space, since, experimentally, the most easily measurable quantity is the momentum distribution of the condensate obtained from a free expansion after the evolution inside the lattice. Such probability decays exponentially

$$P_{\text{sur}}(t) = P_{\text{sur}}(t = 0) \exp(-\Gamma t). \tag{11.35}$$

In the absence of interatomic interactions in the Gross–Pitaevskii equation, for example, for nonlinearity parameter $g = 0$ in Equation 11.12, the individual tunneling events occurring when the condensate crosses the band edge are independent. Hence $P_{\text{sur}}(t)$ globally, that is, fitted over many Bloch periods, has a purely exponential form, apart from the $t \to 0$ limit [79]. When the nonlinear interaction term is present, the condensate density decays with time too. As a consequence, the rates Γ are at best defined locally in time, and in the presence of RET a sharp nonexponential decay may occur, as discussed in Schlagheck and Wimberger (2007) [72] and Carr et al. (2005) [80]. Nevertheless, for short evolution times and the weak nonlinear coupling strengths $\tilde{g}$ experimentally accessible ($\tilde{g}$ defined in Equation 11.13, the global decay of the condensate is well fitted by an exponential law [23,81])

$$P_{\text{sur}}(t) = P_{\text{sur}}(t = 0)\exp\left(-\Gamma_n t\right),\tag{11.36}$$

with rates Γ_n for the band $n = 1$ (ground band), 2 (first excited band), 3 (second excited band), in which the atoms are initially prepared.

We start our study of the tunneling rate in presence of a nonlinearity by discussing the position of RET peaks. These peaks, whose positions for the single-particle evolution are studied in the previous part of this Section 11.4.3, are affected by the nonlinear interaction term appearing in the Gross–Pitaevskii Equation 11.12 for BEC. The RET resonances originate from an exact matching of energy levels in neighboring potential wells, and hence they are very sensitive to slight perturbations. A shift of the RET peaks in energy or in the position of the Stark force, predicted in Wimberger et al. (2006) [71] for large value of the $\tilde{g}$ parameter, is negligible for the experimental investigated nonlinearities $\tilde{g} < 0.06$, the resonance shift corresponding to the extremely small $\Delta F_0 < 5 \times 10^{-4}$ value [71].

The $\tilde{g} \gtrsim 1 \times 10^{-2}$ regime was entered by carrying out the acceleration experiments in radially tighter traps (radial frequency $\gtrsim 100\,\text{Hz}$) and hence at larger condensate densities. Figure 11.13a shows the $\Delta i = 2$ and $\Delta i = 3$ resonance peaks of the ground-state band ($n = 1$) for increasing values of $\tilde{g}$, starting from the linear case and going up to $\tilde{g} \approx 3 \times 10^{-2}$. As the nonlinearity increases, two effects occur. First, the overall (off-resonant) level of Γ_1 increases linearly with $\tilde{g}$. This is in agreement with earlier experiments on nonlinear Landau–Zener tunneling [82,83] and can be modeled by a condensate evolution taking place within a nonlinearity-dependent effective potential $V_{\text{eff}} = V_0/(1 + 4\tilde{g})$ [84]. Second, with increasing nonlinearity, the contrast of the RET peak is decreased and the peak eventually vanishes, as evident from the different on-resonance and off-resonance dependence of the tunneling rate as a function of the atom number N (and hence the nonlinearity) (cf. Figure 11.13b).

The critical value of $\tilde{g}$ for which the nonlinearity affects the resonance peak is estimated by comparing the width of the RET peaks of a band n (which essentially is determined by the tunneling width Γ_{n+1} of the band into which the atoms tunnel) with the energy scale of the nonlinearity. In the experimental investigation of Sias et al. (2007) [23] atomic nonlinearities corresponding to this order-of-magnitude argument were reached. For the parameters of Figures 11.10 and 11.13a and the RET peak with $\Delta i = 2$, the typical width Γ_2 of the decaying state to which the atoms tunneling energy is of the order of $0.2\ldots0.5 \times E_{\text{rec}}$. Since $\tilde{g}$ reflects the nonlinearity expressed in units of $8 \times E_{\text{rec}}$, this means that substantial deviations from the linear behavior are expected when $\tilde{g} \gtrsim 0.025\ldots0.06$. The experimental observations confirmed that this threshold is a good estimate for the onset of the destruction of the RET peak, observed to occur around $\tilde{g} = 0.02$ in Figure 11.13a.

The role of nonlinearity on the time evolution of an Wannier–Stark state localized in a single site of the optical lattice was also studied by Krimer et al. (2009) [85]. They predict that the nonlinearity strength leads to different regimes, where the nonlinearity induced shift in the energy of the lattice may enhance or inhibit RET.

11.5 MANY-BODY TUNNELING

In state-of-the-art experiments the interatomic interactions can be tuned by the transversal confinement and by Feshbach resonances [16], resulting in strong interaction-induced correlations. A good starting point for the discussion of true many-body effects is to use a lattice model, as introduced above for a single particle (cf. Equation 11.11) and widely used in the context of strongly correlated ultracold quantum gases [16]. Such a lattice description has the great advantage that the number of degrees of freedom automatically is bounded as compared to field theoretical approaches (see, e.g., Kühner and Monien (1998) [86] and Duine

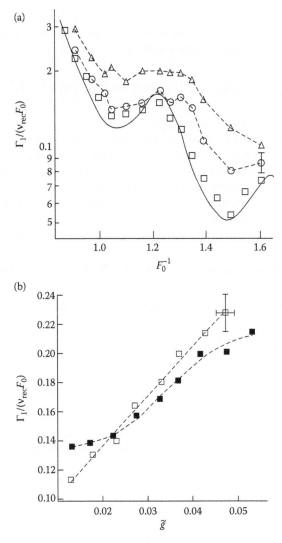

FIGURE 11.13 Resonant tunneling in the nonlinear regime. (a) The tunneling rates for $\Delta i = 3$ from the lowest energy band of the optical lattice as a function of the normalized inverse force F_0^{-1} for a lattice depth $V_0/E_{rec} = 2.5$ and different values of the nonlinearity parameter, $\tilde{g} \approx 0.01, 0.022, 0.033$ from bottom to top. The continuous line is the theoretical prediction in the linear regime. The dashed lines connect the data obtained at large $\tilde{g}$ values. As the nonlinearity increases, the overall tunneling rate increases and the resonance peak becomes less pronounced. (b) Dependence of the tunneling rate on the nonlinear parameter $\tilde{g}$ at the position $F_0^{-1} = 1.21$ (solid symbols) of the RET spectrum peak and at $F_0^{-1} = 1.03$ (open symbols) a the RET spectrum local minimum, for $V_0/E_{rec} = 3.0$. (Adapted from C. Sias et al. *Phys. Rev. Lett.* 98, 120403, 2007. Copyright 2007 of American Physical Society.)

and Stoof (2003) [87] and references therein), and one can use it for practical numerical simulations.

Using a single-band model, the regime of strong correlations in the Wannier–Stark system was addressed in Buchleitner and Kolovsky (2003) [74], Tomadin et al. (2007) [76], Tomadin et al. (2008) [77], Buonsante and Wimberger (2008) [78], and Kolovsky and Buchleitner (2003) [88], revealing the sensitive dependence of the system's dynamics on the Stark force F. The single-band

Bose–Hubbard system of Buchleitner and Kolovsky (2003) [74] and Kolovsky and Buchleitner (2003) [88] is defined by the following Hamiltonian with the creation $a_l^\dagger$, annihilation a_l, and number operators n_l^a for the first band of a lattice with sites $l = 1 \ldots L$:

$$H_{1B} = \sum_{l=1}^{L} \left[F_0 E_{rec} l n_l^a - \frac{J_a}{2} \left(a_{l+1}^\dagger a_l + \text{h.c.} \right) + \frac{U_a}{2} n_l^a (n_l^a - 1) + \varepsilon_a n_l^a \right], \quad (11.37)$$

where the last term describes the on-site energy.

In order to describe interband tunneling and phenomena related to those discussed in the previous Section 11.4, such a model has to be extended to include at least the equivalent of two single-particle energy bands (as plotted in Figure 11.3). In the presence of strong interatomic interactions parameterized by U terms, the single-band model of Equation 11.37 should be extended to allow for interband transitions, as for example, realized at $F_0 = 0$ in experiments with fermionic interacting atoms [21]. Doing so, Tomadin et al. [77] and Plöte et al. [89] arrived at the following full model Hamiltonian for a closed two-band system schematically sketched in Figure 11.14:

$$H(t) = \varepsilon_a \sum_{l=1}^{L} n_l^a + \varepsilon_b \sum_{l=1}^{L} n_l^b \quad \text{onsite energy} + F_0 D E_{rec} \sum_{l=1}^{L} (b_l^\dagger a_l + \text{h.c.}) \quad \text{force coupling}$$

$$- \frac{1}{2} J_a \sum_{l=1} (e^{i2\pi t/T_B} a_{l+1}^\dagger a_l + \text{h.c.}) + \frac{1}{2} J_b \sum_{l} (e^{i2\pi t/T_B} b_{l+1}^\dagger b_l + \text{h.c.}) \quad \text{hopping in the bands}$$

$$+ \frac{1}{2} U_a \sum_{l=1}^{L} n_l^a (n_l^a - 1) + \frac{1}{2} U_b \sum_{l=1}^{L} n_l^b (n_l^b - 1) \quad \text{onsite interaction}$$

$$+ 2 U_x \sum_{l=1}^{L} n_l^a n_l^b + \frac{1}{2} U_x \sum_{l=1}^{L} (b_l^\dagger b_l^\dagger a_l a_l + \text{h.c.}) \quad \text{interband interaction}, \quad (11.38)$$

where the b index and the $b_l, b_l^\dagger$ creation/annihilation operators are associated to the terms of the second band. D is the "dipole" matrix element between the ground and excited single-particle states in a single lattice site (measured in $2\pi/d_L$ length units, cf. the appendix A of Tomadin et al. (2008) [77] for a detailed explanation of how parameters are computed from the physical model).

Within this full two-band system, *two* dominating mechanisms promote to the second band particles starting from the ground band. The first one is a single-particle coupling arising from the force term

$$H_1 = F_0 D E_{rec} \sum_{l=1}^{L} \left(b_l^\dagger a_l + \text{h.c.} \right), \quad (11.39)$$

where the dipole matrix element D depends only on the lattice depth V_0 (measured in recoil energies according to the definition above, cf. Equation 11.2). The second one is a many-body effect,

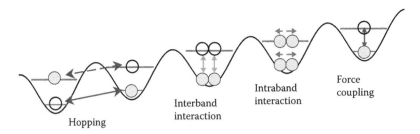

FIGURE 11.14 Sketch of most of the terms of the Hamiltonian Equation 11.38. This model can be used to *fully* describe RET, since it contains excited levels in each potential well, in contrast to the effective model of Section 11.5.1.

describing cotunneling of two particles from the first band into the second band

$$H_2 = \frac{U_x}{2} \sum_{l=1}^{L} \left(b_l^\dagger b_l^\dagger a_l a_l + \text{h.c.} \right). \tag{11.40}$$

In Equation 11.38 the tilting terms arising from the Stark force F_0 have been transformed into a phase factor $e^{\pm i 2\pi t/T_B}$ for the hopping terms by changing into the accelerated frame of Kolovsky and Buchleitner (2003) [88]. This transformation nicely shows that the present problem is intrinsically time-dependent. Since $H(t) = H(t + T_B)$ is periodic with the Bloch period T_B, a Floquet analysis can be used to derive the eigenbasis of the one-period evolution operator generated by $H(t)$. This trick allows also the application of periodic boundary conditions, which is reasonable in order to model large experimental systems, typically extending over a large number of lattice sites. The Hamiltonian of Equation 11.38 contains hopping terms linking nearest-neighboring wells in both bands (J_a and J_b), and terms couplings different bands at a fixed lattice site l either by the force presence ($F_0 D$) or by interactions (U_x). Other terms can, in principle, be included, yet they turn out to be exponentially suppressed for sufficiently deep lattices which are well described by Bose–Hubbard like models [16].

Because of its complex form and the large number of participating many-particle states, the above Hamiltonian is hard to interprete and to treat even numerically, for reasonable numbers of atoms N and lattice sites L. Two approximate treatments will be presented in the following. Section 11.5.1 uses an effective one band model which nevertheless takes the coupling terms between the bands of Equation 11.38 into account. While this model is valid for small interband couplings, Section 11.5.2 presents analytical and numerical results for the full model Equation 11.38, which on the other hand is valid for arbitrary interband couplings but is perturbative in the atom–atom interaction terms U_a, U_b, and U_x.

11.5.1 OPEN ONE-BAND MODEL

Instead of using a numerically hardly tractable complete many-band model, we introduce here a perturbative decay of the many-particle modes in the ground band to a second energy band. This novel approach when applied to the Landau–Zener-like tunneling between the first and the second band [23,25,66,67,82,83] predicts the expected tunneling rates and their statistical distributions.

To justify this perturbative approach, it is crucial to realize that the terms of Equations 11.39 and 11.40 must be small compared with the band gap $\Delta E \equiv \varepsilon_b - \varepsilon_a$ and indeed $F_0 D, U_x \ll \Delta E$ for the parameters of Figure 11.15. As exercised in detail by Tomadin et al. [76,77], from these two coupling terms by using Fermi's golden rule one can compute analytically the corresponding tunneling rates $\Gamma_1(s)$ and $\Gamma_2(s)$ for each basis state labeled by s. Those rates allow the computation of the total width $\Gamma(s) = \Gamma_1(s) + \Gamma_2(s)$ defined by the two analyzed coupling processes for each basis state $|s\rangle$ of the single-band problem given in Equation 11.37. The $\Gamma(s)$ are inserted as complex potentials in the diagonal of the single-band Hamiltonian matrix. Along with the statistics of the level spacings defined by the real parts of its eigenspectrum $\text{Re}\{E_j\}$ studied in Buchleitner and Kolovsky (2003) [74], Tomadin et al. (2007) [76], Tomadin et al. (2008) [77], Buonsante and Wimberger (2008) [78], and Kolovsky and Buchleitner (2003) [88], the statistical distributions of the tunneling rates $\Gamma_j = -2\text{Im}\{E_j\}$ may be analyzed, as done in Figure 11.15. For the regime where the motion of the atoms is localized along the lattice [28] that distribution is in good agreement with the expected log-normal distribution of tunneling rates (or of the similarly behaving conductance) [90]. In that regime the Stark force dominates and the system shows nearly perfect single-particle Bloch oscillations [74], the distributions agreeing with those predicted from the localization theory [90,91]. On the other hand, when the Stark force is comparable with J_a and U_a and all modes of our Bose–Hubbard model are strongly coupled, the rate distribution of Figure 11.15b follows the expected power-law for open quantum chaotic systems in the diffusive regime [91]. This regime shows strong signatures of quantum chaos [74,76–78,88], which manifest also in the rate distributions [76,77].

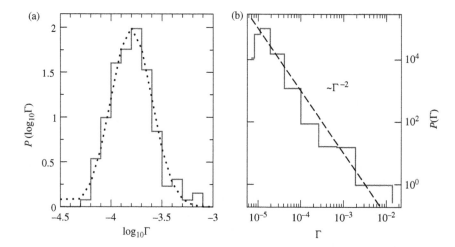

FIGURE 11.15 Rate distributions for the spectrum of an open one-band Bose–Hubbard model in (a) for $F_0 \simeq 0.47, J_a/E_{\mathrm{rec}} = 0.22, U_a/E_{\mathrm{rec}} = 0.2, U_x/E_{\mathrm{rec}} \simeq 0.1$ (for system size $(N,L) = (7,6)$) and in (b) for $F_0 \simeq 0.17, J_a/E_{\mathrm{rec}} = 0.22, U_a/E_{\mathrm{rec}} = 0.2, U_x/E_{\mathrm{rec}} \simeq 0.1$ $((N,L) = (9,8))$. In the regime where the Stark force dominates a log-normal distribution fits well the data (dotted in (a)), whilst a power-law $P(\Gamma) \propto \Gamma^{-x}$ distribution is found with $x \approx 2$ in the strongly coupled case (dashed line in (b)).

11.5.2 CLOSED TWO-BAND MODEL

Since the model introduced in the previous Section 11.5.1 cannot account for resonant tunneling between a ground level of one well and an excited level of another well, a different model which applies also for strong transitions between the bands was investigated by Plötz et al. (2010) [89]. This model is based on the full Hamiltonian of Equation 11.38 sketched schematically in Figure 11.14.

When the Stark force is tuned to the value where RET occurs for the single particle problem (cf. Section 11.4.3), the strong coupling of the atoms prepared in the ground band into the excited band plays an important role. Since the model is closed, that is, higher bands are neglected, there is no asymptotic tunneling as in the experimental situation described in Section 11.4.3. As a consequence, we observe an oscillation of the probability of occupying the lower and upper band, respectively, which is particularly pronounced at RET conditions. For a single particle in our lattice model, such RET oscillations can be understood easily, since in Floquet space (remembering that our Hamiltonian of Equation 11.38 is periodically time-dependent) the problem reduces to an effective two-state model of resonantly coupled states [89,92]. In this effective description, the evolution corresponds to the two-level Rabi problem of quantum optics of Plötz et al. (2010) [93]. For nonvanishing atom–atom interaction, the situation complicates, of course, and we expect a degradation of those single-particle Rabi oscillations. This is illustrated in Figure 11.16. The period of the single-particle interband oscillation is given by the following formula derived in Plötz et al. (2010) [89]:

$$\frac{t_{\mathrm{osc}}}{T_{\mathrm{Bloch}}} \approx \frac{1}{\left| 2D J_{\Delta i} \left(\frac{J_b - J_a}{F_0} \right) \right|}, \tag{11.41}$$

where Δi is the resonance order introduced in Section 11.4.3 and $J_{\Delta i}$ the Bessel function of the same order.

For a Stark force F_0 not satisfying the RET conditions, the coupling to the upper band is strongly suppressed, and almost negligible at least for small particle–particle interband interactions U_x. On the other hand if U_x dominates, strong interband coupling is possible even for small forces F_0. The

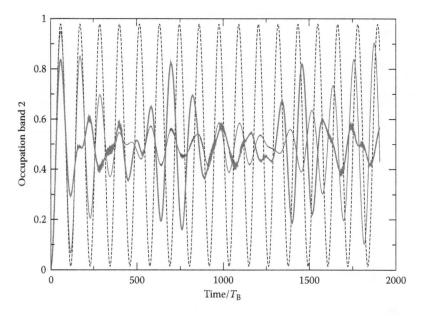

FIGURE 11.16 Population in the upper band as a function of time for the rescaling parameter $\alpha = 0$ (black dotted line), 0.2 (faint gray line) and 0.5 (gray thick line) in a closed two-band model. Clearly visible are the interaction induced collapses and revivals of the RET oscillations between the bands. Other parameters are $F_0 = 1.87$ (dominating energy scale!) and $J_a = 0.1, J_b = 0.77, U_a = 0.023, U_b = 0.014, U_x = 0.01, \varepsilon_b - \varepsilon_a = 3.38$ (all in recoil energy units) and $D = -0.16$ in length units, and $(N, L) = (5, 6)$.

latter strongly correlated regime of two energy bands is extremely hard to deal with, especially if one is searching for analytical predictions for the interband dynamics. The results shown in Figure 11.16 are just a small step in this direction. In the limit of small atom–atom interactions, the observed collapse and revival times can be determined analytically in good approximation. We quantify small interactions by artificially rescaling the parameters U_a, U_b, U_x, which would be obtained by a given scattering length and a given depth of the optical lattice potential [16], by a constant factor $0 < \alpha < 1$. From the results of Figure 11.16, α was chosen to be zero (black dotted line), 0.2 (faint gray line) and 0.5 (gray thick line). The analogy with the Rabi oscillation problem even carries over to those values of interaction strength, since we observe a collapse and later on a revival of the periodic oscillation of the population. Collapse and revival timescale inversely proportional with the strength factor α, as shown in Figure 11.17, where the revival time is well approximated by the formula derived in Plötz et al. (2010) [89]

$$\frac{t_{\text{revival}}}{T_{\text{Bloch}}} \approx \frac{2F_0}{\alpha U_x J_0^2 \left(\frac{J_a}{F_0}\right) J_0^2 \left(\frac{J_b}{F_0}\right)}, \tag{11.42}$$

with the zeroth-order Bessel function J_0. This formula arises from a perturbative calculation of the effect of atom–atom interactions for small $\alpha U_{a,b,x} \ll F_0$ starting from the single-particle solution, which itself is known within the effective two-state model, and assuming a delocalized initial state along the lattice. From Equation 11.42 the collapse time was estimated in Meystre and Sargent (2007) [93] as $t_{\text{collapse}} \approx t_{\text{revival}}/(\pi\sigma_s)$, with the effective number σ_s of additionally coupled many-particle states as compared to the single-particle two-state model. This collapse is analogous to that of the Rabi oscillations in the presence of atomic interactions, or to the collapse arising whenever the phase evolution of each s basis state is nonlinear in the particle number. Notice that the collapse

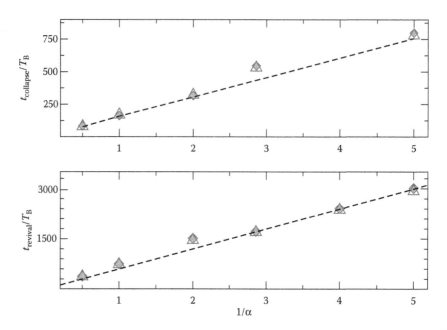

FIGURE 11.17 Collapse and revival times extracted from data (symbols for two different system parameter sets) as shown in Figure 11.16 versus the inverse of the atom–atom interaction rescaling factor α. As expected for a two-state Rabi problem perturbed by a coupling to additional states, both times scale inverse proportionally to α. The dashed lines should guide the eye.

and revival phenomena of Figure 11.16 stem from a degradation (arising from interactions) of single particle *interband* oscillations (with original period given by Equation 11.41 which just depends on the force F_0). So, even if there are analogies to the collapses and revivals observed in BEC [18, 94–96], their origins are different. In the BEC investigations the collapse-revival oscillations were produced by the interaction within a single-band (in Will et al. (2009) [96] by atomic interactions depending on higher power of the well occupation number). Therefore, those oscillations would *not* at all occur when the lower band nonlinear interaction(s) is (are) suppressed, equivalent to $U_a = 0$ in the model here discussed.

The above steps may be expanded in different directions within the realm of true many-body dynamics and tunneling, with great perspectives for many-body induced RET effects. Remaining questions are, for instance, the study of the strongly correlated regime of strong particle and strong interband interactions simultaneously, and the enlargement of our closed two band model in order to allow for a realistic description of experiments similar to the ones reported in Section 11.4.3 now carried over into the realm of strong many-body interactions.

11.6 CONCLUSIONS AND PERSPECTIVES ON RET

This chapter has presented and discussed the RET investigations performed with cold and ultracold atoms. Owing to the reached high level of control on the atom initial preparation and on the realization of potentials with arbitrary shapes, the atomic physics community has reproduced and analyzed basic quantum mechanics phenomena well established, and with important applications, within the solid-state physics community. An important feature associated to the investigations on the atoms, compared to those on electrons in a solid, is the absence of decoherence phenomena. Therefore, quantum interference phenomena may play an enormous role on the tunneling temporal evolution of the cold atoms. For the ultracold atoms an additional characteristic is the presence of interatomic interactions, that modify the position of the energy levels and therefore greatly influence the RET. In more complex configurations the atomic interactions lead to a very complex Hamiltonian whose action on the atoms requires large computational efforts or analyses based on perturbation approaches.

Our analysis was restricted to potentials which are either not explicitly time-dependent or lead to a temporal evolution of the atomic wavefunction corresponding to an adiabatic evolution of the atomic system. Tunneling processes produced by a nonadiabatic atomic evolution are described in other chapters of this volume.

Macroscopic quantum tunneling is an important direction of research well investigated by the solid-state physics community. Up to now no clear evidence of that tunneling was reported by the BEC community even if configurations for the occurrence of macroscopic quantum tunneling in Bose–Einstein condensates have been proposed by different authors. Ueda and Leggett [97,98] examined the instability of a collective mode in a BEC with attractive interaction induced by macroscopic tunneling. Thus, a collective variable, the spatial width of BEC is analyzed as a tunneling variable. Carr et al. [80] studied BEC in a potential of finite depth, harmonic for small radii and decaying as a Gaussian for large radii, which supports both bound and quasibound states. The atomic nonlinearity transforming bound states into quasibound ones, leads to macroscopic quantum tunneling. The experimental observation of such macroscopic tunneling would enlarge the quantum simulation configurations explored with ultracold atoms.

ACKNOWLEDGMENTS

Ennio Arimondo thanks the IFRAF, Paris (France), for the financial support at the École Normale Supérieure where this work was initiated. We also gratefully acknowledge the support from the Project NAMEQUAM of the Future and Emerging Technologies (FET) programme within the Seventh Framework Programme for Research of the European Commission (FET-Open grant number: 225187), the PRIN Project of the MIUR of Italy, and the Excellence Initiative by the German Research Foundation (DFG) through the Heidelberg Graduate School of Fundamental Physics (Grant No. GSC 129/1), the Frontier Innovation Funds and the Global Networks Mobility Measures of the University of Heidelberg. Sandro Wimberger is especially grateful to the Heidelberg Academy of Sciences and Humanities for the Academy Award 2010 and to the Hengstberger Foundation for support by the Klaus-Georg and Sigrid Hengstberger Prize 2009. This review was stimulated by the tunneling experiments performed in Pisa by D. Ciampini, H. Lignier, O. Morsch, C. Sias, and A. Zenesini, and we thank all of them for the continuous valuable discussions. Finally, we would like to thank our theory collaborators, G. Tayebirad, N. Lörch, A. Tomadin, P. Schlagheck, A. Kolovsky, P. Plötz, D. Witthaut, J. Madroñero, and R. Mannella for their help in pushing forward this work.

REFERENCES

1. M. J. Davis and E. J. Heller, Quantum dynamical tunneling in bound states. *J. Chem. Phys.* 75, 246, 1981.
2. R. Tsu and L. Esaki, Tunneling in a finite superlattice. *Appl. Phys. Lett.* 22, 562, 1973.
3. L. L. Chang, L. Esaki, and R. Tsu, Resonant tunneling in semiconductor double barriers. *Appl. Phys. Lett.* 24, 593, 1974.
4. L. L. Chang, E. E. Mendez, and C. Tejedor, eds., *Resonant Tunneling in Semiconductors*, Plenum Press, Amsterdam, 1991.
5. L. Esaki, A bird's-eye view on the evolution of semiconductor superlattices and quantum wells. *IEEE J. Quant. Electr.* QE-22, 1611, 1986.
6. K. Leo, *High-Field Transport in Semiconductor Superlattices*, Springer-Verlag, Berlin, 2003.
7. S. Glutsch, Nonresonant and resonant Zener tunneling. *Phys. Rev. B* 69, 235317, 2004.

8. M. Wagner and H. Mizuta, Complex-energy analysis of intrinsic lifetimes of resonances in biased multiple quantum wells. *Phys. Rev. B* 48, 14393, 1993.

9. B. Rosam, K. Leo, M. Glück, F. Keck, H. J. Korsch, F. Zimmer, and K. Köhler, Lifetime of Wannier-Stark states in semiconductor superlattices under strong Zener tunneling to above-barrier bands. *Phys. Rev. B* 68, 125301, 2003.

10. H. Mizuta and T. Tanoue, *The Physics and Applications of Resonant Tunnelling Diodes*, Cambridge University Press, Cambridge, UK, 1995.

11. M. Ben Dahan, E. Peik, J. Reichel, Y. Castin, and C. Salomon, Oscillations of atoms in an optical potential. *Phys. Rev. Lett.* 76, 4508, 1996.

12. M. Raizen, C. Salomon, and Q. Niu, New light on quantum transport. *Phys. Today* 50, 30, 1997.

13. G. Roati, E. de Mirandes, F. Ferlaino, H. Ott, G. Modugno, and M. Inguscio, Atom interferometry with trapped Fermi gases. *Phys. Rev. Lett.* 92, 230402, 2004.

14. O. Morsch and M. Oberthaler, Dynamics of Bose–Einstein condensates in optical lattices. *Rev. Mod. Phys.* 78, 179, 2006.

15. M. Gustavsson, E. Haller, M. J. Mark, J. G. Danzl, G. Rojas-Kopeinig, and H.-C. Nägerl, Control of interaction-induced dephasing of Bloch oscillations. *Phys. Rev. Lett.* 100, 080404, 2008.

16. I. Bloch, J. Dalibard, and W. Zwerger, Many-body physics with ultracold gases. *Rev. Mod. Phys.* 80, 885, 2008.

17. B. K. Teo, J. R. Guest, and G. Raithel, Tunneling resonances and coherence in an optical lattice. *Phys. Rev. Lett.* 88, 173001, 2002.

18. M. Greiner, O. Mandel, T. Esslinger, T. W. Hänsch, and I. Bloch, Quantum phase transition from a superfluid to a Mott insulator in a gas of ultracold atoms. *Nature* 415, 39, 2002.

19. S. Fölling, S. Trotzky, P. Cheinet, M. Feld, R. Saers, A. Widera, T. Müller, and I. Bloch, Direct observation of second-order atom tunnelling. *Nature* 448, 1029, 2007.

20. S. R. Wilkinson, C. F. Bharucha, K. W. Madison, Q. Niu, and M. G. Raizen, Observation of atomic Wannier-Stark ladders in an accelerating optical potential. *Phys. Rev. Lett.* 76, 4512, 1996.

21. M. Köhl, H. Moritz, T. Stöferle, K. Günter, and T. Esslinger, Fermionic atoms in a three dimensional optical lattice: Observing Fermi surfaces, dynamics, and interactions. *Phys. Rev. Lett.* 94, 080403, 2005.

22. C. Lee, E. A. Ostrovskaya, and Y. Kivshar, Nonlinearity-assisted quantum tunnelling in a matter-wave interferometer. *J. Phys. B: At. Mol. Opt. Phys.* 40, 4235, 2007.

23. C. Sias, A. Zenesini, H. Lignier, S. Wimberger, D. Ciampini, O. Morsch, and E. Arimondo, Resonantly enhanced tunneling of Bose-Einstein condensates in periodic potentials. *Phys. Rev. Lett.* 98, 120403, 2007.

24. A. Zenesini, C. Sias, H. Lignier, Y. Singh, D. Ciampini, O. Morsch, R. Mannella, E. Arimondo, A. Tomadin, and S. Wimberger, Resonant tunneling of Bose-Einstein condensates in optical lattices. *NJP* 10, 0530388, 2008.

25. M. Cristiani, O. Morsch, J. H. Müller, D. Ciampini, and E. Arimondo, Experimental properties of Bose-Einstein condensates in one-dimensional optical lattices: Bloch oscillations, Landau-Zener tunneling, and mean-field effects. *Phys. Rev. A* 65, 063612, 2002.

26. R. Grimm, M. Weidemüller, and Y. B. Ovchinnikov, Optical dipole traps for neutral atoms. *Adv. At. Mol. Opt. Phys.* 42, 95, 2000.

27. G. Nenciu, Dynamics of band electrons in electric and magnetic fields: Rigorous justification of the effective Hamiltonians. *Rev. Mod. Phys.* 63, 91, 1991.

28. M. Glück, A. R. Kolovsky, and H. J. Korsch, Wannier-Stark resonances in optical and semiconductor superlattices. *Phys. Rep.* 366, 103, 2002.

29. M. Holthaus, Bloch oscillations and Zener breakdown in an optical lattice. *J. Opt. B: Quant. Semicl. Opt.* 2, 589, 2000.

30. L. Landau, On the theory of transfer of energy at collisions II. *Phys. Z. Sowjetunion* 2, 46, 1932.

31. C. Zener, Non-adiabatic crossing of energy levels. *Proc. R. Soc. London, Ser.* A 137, 696, 1932.

32. N. Aschcroft and N. Mermin, *Solid State Physics*, Saunders College, Philadelphia, 1976.

33. W. Zwerger, Mott–Hubbard transition of cold atoms in optical lattices. *J. Opt. B: Quantum Semicl. Opt.* 5, S9, 2003.

34. C. J. Pethick and H. Smith, *Bose–Einstein Condensation in Dilute Gases*, Cambridge University Press, Cambridge, UK, 2002.

35. L. Pitaevskii and S. Stringari, *Bose–Einstein Condensation*, Oxford University Press, Oxford, UK, 2003.

36. Y. Castin, K. Berg-Sörensen, J. Dalibard, and K. Mölmer, Two-dimensional Sisyphus cooling. *Phys. Rev. A* 50, 5092, 1994.
37. S. K. Dutta, B. K. Teo, and G. Raithel, Tunneling dynamics and gauge potentials in optical lattices. *Phys. Rev. Lett.* 83, 1934, 1999.
38. D. L. Haycock, P. M. Alsing, I. H. Deutsch, J. Grondalski, and P. S. Jessen, Mesoscopic quantum coherence in an optical lattice. *Phys. Rev. Lett.* 85, 3365, 2000.
39. E. Kierig, U. Schnorrberger, A. Schietinger, J. Tomkovic, and M. K. Oberthaler, Single-particle tunneling in strongly driven double-well potentials. *Phys. Rev. Lett.* 100, 190405, 2008.
40. D. R. Dounas-Frazer, A. M. Hermundstad, and L. D. Carr, Ultracold bosons in a tilted multilevel double-well potential. *Phys. Rev. Lett.* 99, 200402, 2007.
41. M. Albiez, R. Gati, J. Fölling, S. Hunsmann, M. Cristiani, and M. K. Oberthaler, Direct observation of tunneling and nonlinear self-trapping in a single bosonic Josephson junction. *Phys. Rev. Lett.* 95, 010402, 2005.
42. R. Khomeriki, S. Ruffo, and S. Wimberger, Driven collective quantum tunneling of ultracold atoms in engineered optical lattices. *Europhys. Lett.* 77, 40005, 2006.
43. V. Averbukh, S. Osovski, and N. Moiseyev, Controlled tunneling of cold atoms: From full suppression to strong enhancement. *Phys. Rev. Lett.* 89, 253201, 2002.
44. W. K. Hensinger, A. Mouchet, P. S. Julienne, D. Delande, N. R. Heckenberg, and H. Rubinsztein-Dunlop, Analysis of dynamical tunneling experiments with a Bose–Einstein condensate. *Phys. Rev. A* 70, 013408, 2004.
45. D. A. Steck, W. H. Oskay, and M. G. Raizen, Observation of chaos-assisted tunneling between islands of stability. *Science* 293, 274, 2001.
46. W. Hensinger, H. Haffner, A. Browaeys, N. Heckenberg, K. Helmerson, C. McKenzie, G. Milburn, W. Phillips, S. Rolston, H. Rubinsztein-Dunlop, and B. Upcroft, Dynamical tunnelling of ultracold atoms. *Nature* 412, 52, 2001.
47. O. Bohigas, S. Tomsovic, and D. Ullmo, Manifestations of classical phase space structures in quantum mechanics. *Phys. Rep.* 223, 43, 1993.
48. D. H. Dunlap and V. M. Kenkre, Dynamic localization of a charged particle moving under the influence of an electric field. *Phys. Rev. B* 34, 3625, 1986.
49. F. Rossi, Coherent phenomena in semiconductors. *Semicon. Sci. Technol.* 13, 147, 1998.
50. H. J. Korsch and S. Mossmann, An algebraic solution of driven single band tight binding dynamics. *Phys. Lett. A* 317, 54, 2003.
51. A. Klumpp, D. Witthaut, and H. J. Korsch, Quantum transport and localization in biased periodic structures under bi- and polychromatic driving. *J. Phys. A: Math. Theor.* 40, 2299, 2007.
52. P. K. Tien and J. P. Gordon, Multiphoton process observed in the interaction of microwave fields with the tunneling between superconductor films. *Phys. Rev.* 129, 647, 1963.
53. B. J. Keay, S. J. Allen, J. Galán, J. P. Kaminski, K. L. Campman, A. C. Gossard, U. Bhattacharya, and M. J. W. Rodwell, Photon-assisted electric field domains and multiphoton-assisted tunneling in semiconductor superlattices. *Phys. Rev. Lett.* 75, 4098, 1995.
54. B. J. Keay, S. Zeuner, S. J. Allen, K. D. Maranowski, A. C. Gossard, U. Bhattacharya, and M. J. W. Rodwell, Dynamic localization, absolute negative conductance, and stimulated, multiphoton emission in sequential resonant tunneling semiconductor superlattices. *Phys. Rev. Lett.* 75, 4102, 1995.
55. L. P. Kouwenhoven, S. Jauhar, J. Orenstein, P. L. McEuen, Y. Nagamune, J. Motohisa, and H. Sakaki, Observation of photon-assisted tunneling through a quantum dot. *Phys. Rev. Lett.* 73, 3443, 1994.
56. T. H. Oosterkamp, L. P. Kouwenhoven, A. E. A. Koolen, N. C. van der Vaart, and C. J. P. M. Harmans, Photon sidebands of the ground state and first excited state of a quantum dot. *Phys. Rev. Lett.* 78, 1536, 1997.
57. A. Eckardt, C. Weiss, and M. Holthaus, Superfluid-insulator transition in a periodically driven optical lattice. *Phys. Rev. Lett.* 95, 260404, 2005.
58. A. R. Kolovsky and H. J. Korsch, Dynamics of interacting atoms in driven tilted optical lattices in Bose-Einstein condensates in optical lattices. *J. Sib. Fed. Un.: Math, Phys*, 3, 211, 2010.
59. C. Sias, H. Lignier, Y. P. Singh, A. Zenesini, D. Ciampini, O. Morsch, and E. Arimondo, Observation of photon-assisted tunneling in optical lattices. *Phys. Rev. Lett.* 100, 040404, 2008.
60. V. V. Ivanov, A. Alberti, M. Schioppo, G. Ferrari, M. Artoni, M. L. Chiofalo, and G. M. Tino, Coherent delocalization of atomic wave packets in driven lattice potentials. *Phys. Rev. Lett.* 100, 043602, 2008.

61. A. Alberti, V. Ivanov, G. Tino, and G. Ferrari, Engineering the quantum transport of atomic wavefunctions over macroscopic distances. *Nat. Phys.* 5, 547, 2009.

62. E. Haller, R. Hart, M. J. Mark, J. Danzl, L. Reichsöllner, and H. Nägerl, Inducing transport in a dissipation-free lattice with super Bloch oscillations. *Phys. Rev. Lett.* 104, 200403, 2010.

63. Q. Thommen, J. C. Garreau, and V. Zehnle, Theoretical analysis of quantum dynamics in one-dimensional lattices: Wannier–Stark description. *Phys. Rev. A* 65, 053406, 2002.

64. A. Eckardt and M. Holthaus, Dressed matter waves. *J. Phys: Conf. Ser.* 99, 012007, 2008.

65. H. Lignier, C. Sias, D. Ciampini, Y. Singh, A. Zenesini, O. Morsch, and E. Arimondo, Dynamical control of matter-wave tunneling in periodic potentials. *Phys. Rev. Lett.* 99, 220403, 2007.

66. C. F. Bharucha, K. W. Madison, P. R. Morrow, S. R. Wilkinson, B. Sundaram, and M. G. Raizen, Observation of atomic tunneling from an accelerating optical potential. *Phys. Rev. A* 55, R857, 1997.

67. S. Wimberger, R. Mannella, O. Morsch, E. Arimondo, A. R. Kolovsky, and A. Buchleitner, Nonlinearity-induced destruction of resonant tunneling in the Wannier–Stark problem. *Phys. Rev. A* 72, 063610, 2005.

68. J. E. Avron, The lifetime of Wannier ladder states. *Ann. Phys.* 143, 33, 1982.

69. F. Keck, H. J. Korsch, and S. Mossmann, Unfolding a diabolic point: A generalized crossing scenario. *J. Phys. A: Math. Gen.* 36, 2125, 2003.

70. M. Glück, A. R. Kolovsky, and H. J. Korsch, Lifetime of Wannier–Stark states. *Phys. Rev. Lett.* 83, 891, 1999.

71. S. Wimberger, P. Schlagheck, and R. Mannella, Tunnelling rates for the nonlinear Wannier–Stark problem. *J. Phys. B: At. Mol. Opt. Phys.* 39, 729, 2006.

72. P. Schlagheck and S. Wimberger, Nonexponential decay of Bose–Einstein condensates: A numerical study based on the complex scaling method. *Appl. Phys. B: Lasers Opt.* 86, 385, 2007.

73. D. Witthaut, E. M. Graefe, S. Wimberger, and H. J. Korsch, Bose–Einstein condensates in accelerated double-periodic optical lattices: Coupling and crossing of resonances. *Phys. Rev. A* 75, 013617, 2007.

74. A. Buchleitner and A. R. Kolovsky, Interaction-induced decoherence of atomic Bloch oscillations. *Phys. Rev. Lett.* 91, 253002, 2003.

75. Q. Thommen, J. C. Garreau, and V. Zehnlé, Classical chaos with Bose–Einstein condensates in tilted optical lattices. *Phys. Rev. Lett.* 91, 210405, 2003.

76. A. Tomadin, R. Mannella, and S. Wimberger, Many-body interband tunneling as a witness of complex dynamics in the Bose–Hubbard mode. *Phys. Rev. Lett.* 98, 130402, 2007.

77. A. Tomadin, R. Mannella, and S. Wimberger, Many-body Landau–Zener tunneling in the Bose–Hubbard mode. *Phys. Rev. A* 77, 013606, 2008.

78. P. Buonsante and S. Wimberger, Engineering many-body quantum dynamics by disorder. *Phys. Rev. A* 77, 041606(R), 2008.

79. S. R. Wilkinson, C. F. Bharucha, M. C. Fischer, K. W. Madison, P. R. Morrow, Q. Niu, B. Sundaram, and M. G. Raizen, Observation of atomic Wannier-Stark ladders in an accelerating optical potential. *Nature* 387, 575, 1997.

80. L. D. Carr, M. J. Holland, and B. A. Malomed, Macroscopic quantum tunnelling of Bose–Einstein condensates in a finite potential well. *J. Phys. B: At. Mol. Opt. Phys.* 38, 3217, 2005.

81. S. Wimberger, D. Ciampini, O. Morsch, R. Mannella, and E. Arimondo, Engineered quantum tunnelling in extended periodic potentials. *J. Phys.: Conf. Ser.* 67, 012060, 2007.

82. O. Morsch, J. H. Müller, M. Cristiani, D. Ciampini, and E. Arimondo, Bloch oscillations and mean-field effects of Bose–Einstein condensates in 1D optical lattices. *Phys. Rev. Lett.* 87, 140402, 2001.

83. M. Jona-Lasinio, O. Morsch, M. Cristiani, N. Malossi, J. H. Müller, E. Courtade, M. Anderlini, and E. Arimondo, Asymmetric Landau–Zener tunneling in a periodic potential. *Phys. Rev. Lett.* 91, 230406, 2003; ibidem (93), 119903(E), 2004.

84. D.-I. Choi and Q. Niu, Bose–Einstein condensates in an optical lattice. *Phys. Rev. Lett.* 82, 2022, 1999.

85. D. O. Krimer, R. Khomeriki, and S. Flach, Delocalization and spreading in a nonlinear Stark ladder. *Phys. Rev. E* 80, 036201, 2009.

86. T. D. Kühner and H. Monien, Phases of the one-dimensional Bose–Hubbard model. *Phys. Rev. B* 58, R14741, 1998.

87. R. A. Duine and H. T. C. Stoof, Many-body aspects of coherent atom-molecule oscillations. *Phys. Rev. Lett.* 91, 150405, 2003.

88. A. R. Kolovsky and A. Buchleitner, Floquet-Bloch operator for the Bose–Hubbard model with static field. *Phys. Rev. E* 68, 056213, 2003.
89. P. Plötz, J. Madroñero, and S. Wimberger, Collapse and revival in inter-band oscillations of a two-band Bose–Hubbard mode. *J. Phys. B: At. Mol. Opt. Phys.* 43, 081001, 2010.
90. C. W. J. Beenakker, Random-matrix theory of quantum transport. *Rev. Mod. Phys.* 69, 731, 1997.
91. T. Kottos, Statistics of resonances and delay times in random media: Beyond random matrix theory. *J. Phys. A: Math. General* 38, 10761, 2005.
92. Y. Nakamura, Y. A. Pashkin, and J. S. Tsai, Rabi oscillations in a Josephson-junction charge two-level system. *Phys. Rev. Lett.* 87, 246601, 2001.
93. P. Meystre and M. I. Sargent, *Elements of Quantum Optics*, Springer-Verlag, Heidelberg, 2007.
94. M. Anderlini, J. Sebby-Strabley, J. Kruse, J. V. Porto, and W. Phillips, Controlled atom dynamics in a double-well optical lattice. *J. Phys. B: At. Mol. and Opt. Phys.* 39, S199, 2006.
95. J. Sebby-Strabley, B. L. Brown, M. Anderlini, P. J. Lee, W. D. Phillips, J. V. Porto, and P. R. Johnson, Preparing and probing atomic number states with an atom interferometer. *Phys. Rev. Lett.* 98, 200405, 2007.
96. S. Will, T. Best, U. Schneider, L. Hackermüller, D.-S. Lühmann, and I. Bloch, Multi-orbital quantum phase diffusion. *Nature* 465, 197, 2010.
97. M. Ueda and A. J. Leggett, Macroscopic quantum tunneling of a Bose–Einstein condensate with attractive interaction. *Phys. Rev. Lett.* 80, 1576, 1998.
98. M. Ueda and A. J. Leggett, Ueda and Leggett reply. *Phys. Rev. Lett.* 81, 1343, 1998.

12 Dynamic Localization in Optical Lattices

Stephan Arlinghaus, Matthias Langemeyer, and Martin Holthaus

CONTENTS

12.1 INTRODUCTION

The concept of dynamic localization goes back to an observation reported by Dunlap and Kenkre in 1986: The wave packet of a single particle moving on a single-band tight-binding lattice endowed with only nearest-neighbor couplings remains perpetually localized when driven by a spatially homogeneous ac force, provided the amplitude and the frequency of that force obey a certain condition [1]. When trying to overcome the limitations of the model, it is comparatively straightforward to deal with an arbitrary form of the dispersion relation—thus abandoning the nearest-neighbor approximation—and with arbitrary time-periodic forces, thus doing away with the restriction to purely sinusoidal driving [2]. But in any real lattice system an external time-periodic force will induce interband transitions, and it is by no means obvious whether dynamic localization can survive when these come into play.

In this chapter we consider ultracold atoms in driven optical lattices, which provide particularly attractive, experimentally well accessible examples of quantum particles in spatially periodic structures exposed to time-periodic forcing [3–5]. Such systems are much cleaner, and more easy to control, than electrons in crystal lattices under the influence of ac electric fields, for which the original idea had been developed [1]. With the help of results obtained by numerical calculations we illustrate that such ultracold atoms in kHz-driven optical lattices exhibit dynamic localization in almost its purest form if the parameters are chosen judiciously, despite the potentially devastating presence of interband transitions.

When viewing dynamic localization as resulting from a band collapse [6,7], far-reaching further possibilities emerge. Namely, the actual strength of deviations from exact spatial periodicity, be they isolated [8], random [9], or quasiperiodic [10,11], is measured relative to the effective bandwidth. Thus, when the band in question almost collapses in response to time-periodic driving, the effects of even slight deviations from exact lattice periodicity are strongly enhanced. This allows one, in particular, to coherently control the "metal-insulator"-like incommensurability transition occurring in sufficiently deep quasiperiodic optical lattices [10–12]. While the very transition has already been observed with Bose–Einstein condensates in bichromatic optical potentials [13], its coherent control by means of time-periodic forcing still awaits its experimental verification.

12.2 THE BASIC IDEA

The one-dimensional tight-binding system described by the Hamiltonian

$$H_0 = -J \sum_\ell \Big(|\ell + 1\rangle\langle\ell| + |\ell\rangle\langle\ell + 1| \Big), \tag{12.1}$$

where $|\ell\rangle$ denotes a Wannier state localized at the ℓth lattice site, and J is the hopping matrix element connecting neighboring sites, is about the simplest model for the formation of Bloch bands. Assuming that the unspecified number of sites is so large that finite-size effects may be neglected, its energy eigenstates are Bloch waves

$$|\varphi_k\rangle = \sum_\ell |\ell\rangle \exp(i\ell ka) \tag{12.2}$$

labeled by a wave number k; the lattice period is given by a. The corresponding energy dispersion relation reads

$$E(k) = -2J \cos(ka); \tag{12.3}$$

here we assume $J > 0$, so that its minimum is located at $k = 0 \bmod 2\pi/a$. Now we let an external time-dependent, spatially homogeneous force $F(t)$ act on the system, such that the total Hamiltonian becomes

$$H(t) = H_0 + H_1(t) \tag{12.4}$$

with

$$H_1(t) = -F(t) \sum_\ell |\ell\rangle a\ell\langle\ell|. \tag{12.5}$$

It is easy to verify that the wave functions

$$|\psi_k(t)\rangle = \exp\left(-\frac{i}{\hbar} \int_0^t d\tau\, E\big(q_k(\tau)\big) \right) \sum_\ell |\ell\rangle \exp\big(i\ell q_k(t)a\big) \tag{12.6}$$

then are solutions to the time-dependent Schrödinger equation, provided the time-dependent wave numbers $q_k(t)$ introduced here obey the "semiclassical" relation

$$\hbar \dot{q}_k(t) = F(t). \tag{12.7}$$

We demand that $q_k(t)$ be equal to k at time $t = 0$, and therefore set

$$q_k(t) = k + \frac{1}{\hbar} \int_0^t d\tau\, F(\tau). \tag{12.8}$$

These wave functions (Equation 12.6), originally considered by Houston in the context of crystal electrons exposed to a uniform electric field superimposed on a periodic lattice potential [14], are known as "accelerated Bloch waves," or Houston states.

In the particular case of a monochromatic force with angular frequency ω and amplitude F_1, given by

$$F(t) = F_1 \cos(\omega t), \tag{12.9}$$

one has

$$q_k(t) = k + \frac{F_1}{\hbar\omega} \sin(\omega t), \tag{12.10}$$

so that $q_k(t)$ naturally acquires the temporal period $T = 2\pi/\omega$ of the driving force. Then also $E(q_k(t))$ is T-periodic, but the Houston state (Equation 12.6) is not, because the integral appearing in the exponential prefactor acquires a contribution which grows linearly with time. In order to extract that contribution, we calculate the one-cycle average

$$
\begin{aligned}
\varepsilon(k) &\equiv \frac{1}{T} \int_0^T d\tau\, E(q_k(\tau)) \\
&= -2J_{\text{eff}} \cos(ka),
\end{aligned}
\tag{12.11}
$$

thus obtaining an effective hopping matrix element given by

$$
J_{\text{eff}} = J J_0\left(\frac{F_1 a}{\hbar\omega}\right),
\tag{12.12}
$$

with $J_0(z)$ denoting the Bessel function of zero order. We then write

$$
\exp\left(-\frac{\mathrm{i}}{\hbar} \int_0^t d\tau\, E(q_k(\tau))\right) = \exp\left(-\frac{\mathrm{i}}{\hbar} \int_0^t d\tau\, \Big[E(q_k(\tau)) - \varepsilon(k)\Big]\right) \exp\left(-\mathrm{i}\varepsilon(k)t/\hbar\right),
\tag{12.13}
$$

so that the first exponential on the right-hand side now is T-periodic by construction. Hence, for the T-periodic force (Equation 12.9) the Houston states (Equation 12.6) can be cast into a form

$$
|\psi_k(t)\rangle = |u_k(t)\rangle \exp\left(-\mathrm{i}\varepsilon(k)t/\hbar\right)
\tag{12.14}
$$

with T-periodic functions $|u_k(t)\rangle$,

$$
|u_k(t)\rangle = |u_k(t+T)\rangle.
\tag{12.15}
$$

This leads to a remarkable conclusion. Any wave packet governed by the full Hamiltonian (Equation 12.4) with periodic forcing (Equation 12.9) can be expanded with respect to these states (Equation 12.14) with coefficients that are constant in time, because the time-dependence already is fully incorporated into the states themselves. After each cycle T the T-periodic functions $|u_k(t)\rangle$ are restored, so that the time evolution of the wave packet, when viewed stroboscopically at intervals T, is determined by the different "speed" of rotation of the complex phase factors $\exp(-\mathrm{i}\varepsilon(k)t/\hbar)$ of the packet's individual components. But if all quantities $\varepsilon(k)$ are equal, which according to Equations 12.11 and 12.12 occurs when the scaled driving amplitude

$$
K_0 \equiv \frac{F_1 a}{\hbar\omega}
\tag{12.16}
$$

equals a zero of the Bessel function J_0, all phase factors evolve at the same speed, so that the wave packet reproduces itself exactly after each period: There is some T-periodic wiggling, but no long-term motion. This, in short, is dynamic localization [1].

The above argument appears so special, and the decisive step (Equation 12.13) so swift, that it is not easy to see how to transfer this finding to more realistic situations: How can one incorporate deviations from exact lattice periodicity into this reasoning? How to proceed when several bands are coupled by interband transitions? The answer to these questions is provided by the Floquet picture, which does not directly take recourse to the spatial lattice periodicity, but rather builds on the temporal periodicity of the Hamiltonian: when $H(t) = H(t + T)$, there exists a complete set of solutions to the time-dependent Schrödinger equation of the particular form

$$
|\psi_n(t)\rangle = |u_n(t)\rangle \exp(-\mathrm{i}\varepsilon_n t/\hbar),
\tag{12.17}
$$

where the functions $|u_n(t)\rangle = |u_n(t+T)\rangle$ inherit the T-periodicity of the underlying Hamiltonian. These states are known as Floquet states; the quantities ε_n are dubbed as quasienergies [15–18]. Obviously, the Houston states (Equation 12.6) with time-periodic forcing are particular examples of such Floquet states; from now on we employ an abstract state label n instead of the wave number k in order to also admit settings without lattice periodicity. In the case of the Houston–Floquet states, the determination of their quasienergies (Equation 12.11) essentially was a by-product of the solution of an initial value problem. The general case, however, has to proceed along a more sophisticated route: Floquet states and quasienergies are determined by solving the eigenvalue problem

$$\left(H(t) - i\hbar\frac{\partial}{\partial t} \right) |u_n(t)\rangle\rangle = \varepsilon_n |u_n(t)\rangle\rangle, \tag{12.18}$$

posed in an *extended Hilbert space* of T-periodic functions; in that space time plays the role of a *coordinate*. Therefore, if $\langle u_1(t)|u_2(t)\rangle$ is the scalar product of two T-periodic functions in the usual physical Hilbert space, their scalar product in the extended space reads [18]

$$\langle\langle u_1|u_2\rangle\rangle \equiv \frac{1}{T} \int_0^T dt \, \langle u_1(t)|u_2(t)\rangle. \tag{12.19}$$

Hence, we write $|u_n(t)\rangle$ for a Floquet eigenfunction when viewed in the physical Hilbert space, and $|u_n(t)\rangle\rangle$ when that same function is regarded as an element of the extended space.

A most important consequence of this formalism stems from the fact that when $|u_n(t)\rangle\rangle$ is a solution to the problem (Equation 12.18) with eigenvalue ε_n, then $|u_n(t)\exp(im\omega t)\rangle\rangle$ is a further solution with eigenvalue $\varepsilon_n + m\hbar\omega$, where we have set $\omega = 2\pi/T$, and m is any (positive, zero, or negative) integer, in order to comply with the required T-periodic boundary condition. For $m \neq 0$ these two solutions are orthogonal with respect to the scalar product (Equation 12.19). But when going back to the physical Hilbert space, one has

$$|u_n(t)\exp(im\omega t)\rangle \exp\left(-i(\varepsilon_n + m\hbar\omega)/\hbar \right) = |u_n(t)\rangle \exp(-i\varepsilon_n t/\hbar), \tag{12.20}$$

so that the two different solutions represent *the same* Floquet state (Equation 12.17). We conclude that a physical Floquet state does not simply correspond to an individual solution to the eigenvalue problem (Equation 12.18), but rather to a whole class of such solutions labeled by the state index n, whereas the "photon" index m distinguishes different representatives of such a class. Likewise, a quasienergy should not be regarded as a single eigenvalue, but rather as a set $\{\varepsilon_n + m\hbar\omega \mid m = 0, \pm1, \pm2, \dots\}$ associated with one particular state n, while m ranges through all integers. Therefore, each "quasienergy Brillouin zone" of width $\hbar\omega$ contains one quasienergy representative of each state.

The time evolution of any wave function can then be written as a Floquet-state expansion,

$$|\psi(t)\rangle = \sum_n c_n |u_n(t)\rangle \exp(-i\varepsilon_n t/\hbar), \tag{12.21}$$

where the coefficients c_n remain constant in time. This is one of the main benefits offered by the Floquet picture, and allows one to draw many parallels to the evolution of systems governed by a time-independent Hamiltonian.

Equipped with this set of tools, it is now clear how to investigate the possible occurrence of dynamic localization in realistic lattice structures: one has to solve the eigenvalue problem (Equation 12.18) for the Hamiltonian with the respective full lattice potential, and to enquire whether the resulting quasienergy bands collapse at least approximately, that is, acquire negligible widths for certain parameters. If so, any wave packet prepared in a quasienergy band at a collapse point will suffer from "prohibited dephasing," as in the archetypal model specified by Equations 12.1,

12.5, and 12.9; and thus remain dynamically localized. Interband transitions then are automatically included, with multiphoton-like resonances manifesting themselves through quasienergy-curve anticrossings [19].

In the following section we will carry through this program for ultracold atoms in driven one-dimensional optical lattices.

12.3 DOES IT WORK?

A one-dimensional optical lattice is created by two counterpropagating laser beams with wave number k_L, suitably detuned from a dipole-allowed transition of the atomic species moving in this standing light wave. By means of the ac Stark effect, the spatially periodic electric field experienced by the atoms then translates into a cosine potential

$$V_{\text{lat}}(x) = \frac{V_0}{2} \cos(2k_L x) \tag{12.22}$$

for their translational motion along the lattice, with a depth V_0 that is proportional to the laser intensity [20,21]. The characteristic energy scale then is given by the single-photon recoil energy

$$E_{\text{rec}} = \frac{\hbar^2 k_L^2}{2M}, \tag{12.23}$$

where M denotes the atomic mass. To give a numerical example, when working with ^{87}Rb in a lattice generated by laser radiation with wavelength $\lambda = 2\pi/k_L = 842$ nm [4,5] one has $E_{\text{rec}} = 1.34 \times 10^{-11}$ eV. Thus, typical lattice depths of 5–10 recoil energies are on the order of 10^{-10} eV—which means that one encounters many phenomena with ultracold atoms in optical lattices which are known from traditional solid-state physics, but scaled down in energy by no less than 10 orders of magnitude.

This also tells us what "ultracold" means. Taking an ensemble of atoms with a temperature T_{ens} such that $k_B T_{\text{ens}}$ is roughly equal to E_{rec}, say, where k_B is Boltzmann's constant, the de Broglie wavelength of these atoms, given by

$$\lambda_{\text{deBroglie}} = \frac{h}{\sqrt{2\pi M k_B T_{\text{ens}}}} \approx \frac{2}{\sqrt{\pi}} \frac{\lambda}{2}, \tag{12.24}$$

is barely longer than the lattice constant $a = \lambda/2$. But in order to experience quantum mechanical lattice effects, the particles have to be able to "feel" the periodic structure, so that $\lambda_{\text{deBroglie}}$ should cover *at least* a few lattice constants—which means that being this cold is not cold enough: we even require $k_B T_{\text{ens}} \ll E_{\text{rec}}$.

With hardly any thermal excitation energy left the atoms occupy only the lowest Bloch band of their optical lattice, so that the single-particle Hamiltonian with the lattice potential (Equation 12.22) translates directly into the single-band tight-binding model (Equation 12.1) when working in a basis of Wannier functions pertaining to that lowest band, and neglecting all couplings other than those between nearest neighbors, denoted as J. The accuracy of this approximation increases with increasing lattice depth [4,12]: for $V_0/E_{\text{rec}} = 5$ the magnitude of the ratio of the neglected matrix element connecting next-to-nearest neighbors to J still reaches about 5%, but it decreases to about 1% when $V_0/E_{\text{rec}} = 10$. Moreover, when expressing the exact band structure of a cosine lattice in terms of characteristic values of the Mathieu equation, and noting that the width W of the cosine energy band (Equation 12.3) is $4J$, one finds the approximation [21]

$$J/E_{\text{rec}} \sim \frac{4}{\sqrt{\pi}} \left(\frac{V_0}{E_{\text{rec}}} \right)^{3/4} \exp\left(-2\sqrt{\frac{V_0}{E_{\text{rec}}}} \right) \qquad \text{for } V_0/E_{\text{rec}} \gg 1. \tag{12.25}$$

The requisite still missing now is the time-periodic force corresponding to the model (Equation 12.5). This can be effectuated either by introducing a small oscillating frequency difference between the two lattice-generating laser beams, as detailed later, or by retro-reflecting one such beam off an oscillating mirror back into itself [3–5]. In a frame of reference co-moving with the oscillating lattice, one then obtains the single-particle Hamiltonian

$$H(t) = \frac{p^2}{2M} + V_{\text{lat}}(x) - F_1 x \cos(\omega t + \phi), \tag{12.26}$$

where p is the atomic center-of-mass momentum in the lattice direction, the driving force is parametrized in accordance with Equation 12.9, and we have also admitted an arbitrary phase ϕ.

In all our model calculations we consider a lattice with depth $V_0/E_{\text{rec}} = 5.7$, implying that the width of the lowest Bloch band is $W/E_{\text{rec}} = 0.22$, whereas the gap between this lowest band and the first excited one figures as $\Delta/E_{\text{rec}} = 2.76$. Even for such a comparatively shallow lattice, which is routinely being realized in current experiments [5], the dispersion of the lowest band already is reasonably well described by the tight-binding cosine approximation (Equation 12.3), setting $J = W/4$. In order to obtain dynamic localization, the driving frequency should then be chosen such that the quantum $\hbar\omega$ is significantly smaller than the gap Δ, so that, perturbatively speaking, interband transitions require higher order multiphoton-like processes, which would be suppressed as long as the driving amplitude F_1 is not too strong [19]. On the other hand, it is reasonable to demand that $\hbar\omega$ be larger than the bandwidth, so that the band fits into a single quasienergy Brillouin zone. A good choice of the driving frequency should therefore adhere to the chain $4J = W < \hbar\omega < \Delta$; we take $\hbar\omega/E_{\text{rec}} = 0.5$ in all numerical scenarios depicted below. For ^{87}Rb atoms in a lattice with $\lambda = 842$ nm this choice fixes the frequency at $\omega/(2\pi) = 1.62$ kHz.

Figure 12.1 shows one Brillouin zone of quasienergies for these parameters versus the scaled driving amplitude K_0, as defined by Equation 12.16. Observe that the first quasimomentum Brillouin zone ranges from $-\hbar\pi/a = -\hbar k_L$ to $+\hbar\pi/a = +\hbar k_L$; the homogeneous force does not mix states

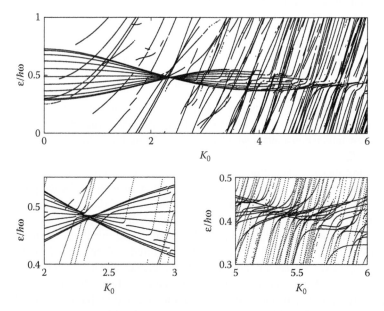

FIGURE 12.1   Above: One Brillouin zone of quasienergies for the optical lattice (Equation 12.22) with depth $V_0/E_{\text{rec}} = 5.7$, driven with scaled frequency $\hbar\omega/E_{\text{rec}} = 0.5$, versus the scaled driving amplitude K_0. The lower left panel testifies that the first band collapse is almost perfect, whereas the second one, enlarged in the lower right panel, is already thwarted by multiphoton-like resonances.

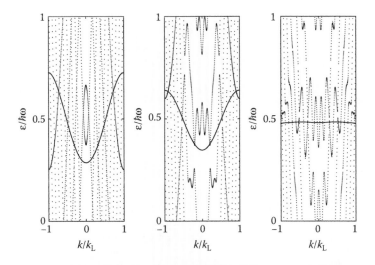

FIGURE 12.2 "Lowest" quasienergy band for the optical lattice (Equation 12.22) with depth $V_0/E_{rec} = 5.7$, driven with scaled frequency $\hbar\omega/E_{rec} = 0.5$, and scaled amplitudes $K_0 = 0$ (left), 1.18 (middle), and 2.35 (right). Additional curves result from higher bands.

with different wave numbers [10]. Hence, we combine quasienergies for states with $k = (i/10)k_L$ in this plot, with $i = 0, 1, 2, \ldots, 10$. In this way, the comparison of the ideal quasienergy band (Equation 12.11) with the one appearing in the actual optical lattice is greatly facilitated. Evidently, the first band collapse is almost perfect, although it is slightly shifted from $K_0 = 2.405$, the first zero of J_0, to $K_0 \approx 2.35$. In contrast, the second collapse, expected at $K_0 = 5.520$, already is significantly affected by a host of anticrossings, indicating multiphoton-like resonances. Thus, with $V_0/E_{rec} = 5.7$ and $\hbar\omega/E_{rec} = 0.5$ we may expect almost perfect dynamic localization at the first collapse point, whereas there will be strong disturbances of the ideal dynamics at the second one.

In Figure 12.2 we depict the lowest quasienergy band for $K_0 = 0$, where it coincides with the original energy band; $K_0 = 1.18$, where its width is reduced by a factor of $J_0(1.18) = 0.681$; and at the first collapse point, $K_0 = 2.35$. Ideally, a collapsed quasienergy band is completely flat, so that dynamic localization is associated with an infinite effective mass of the driven Bloch particle. Here we still observe some residual dispersion, probably resulting from both next-to-nearest neighbor couplings and couplings to higher bands, but the degree of band flattening achieved by the driving force is nonetheless impressive.

The ultimate demonstration of dynamic localization requires, of course, the inspection of wave-packet dynamics. To this end, we first compute the Bloch states $\langle x|\varphi_{1,k}\rangle$ of the lowest energy band of the lattice (Equation 12.22), and use them to design an initial wave packet

$$\langle x|\psi(t=0)\rangle = \int_{-k_L}^{k_L} dk\, g_1(k,t=0)\,\langle x|\varphi_{1,k}\rangle \tag{12.27}$$

with a Gaussian k-space distribution

$$g_1(k,t=0) = \frac{1}{\sqrt{2k_L\sqrt{\pi}\Delta k}}\exp\left(-\frac{(k-k_c)^2}{2(\Delta k)^2}\right) \tag{12.28}$$

centered around some predetermined wave number k_c, with width Δk. The corresponding probability density $|\langle x|\psi(t=0)\rangle|^2$ is concentrated in the wells of the lattice potential, equipped with a Gaussian envelope that varies the more slowly with x the narrower its distribution (Equation 12.28), that is, the smaller Δk. We then take this packet (Equation 12.27) as initial condition, and compute the

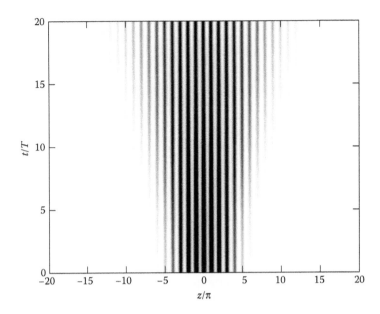

FIGURE 12.3 Spreading of the Bloch wave packet (Equation 12.27) with initial k-space width $\Delta k/k_L = 0.1$, and initial momentum $k_c/k_L = 0$, in the unforced optical lattice. In this and the following figures, density is encoded in shades of gray.

wave function $\langle x|\psi(t)\rangle$ for $t > 0$ by solving the time-dependent Schrödinger equation numerically, fixing the phase ϕ in the Hamiltonian (Equation 12.26) at the value $\phi = \pi/2$. This means that the force $F(t) = F_1 \cos(\omega t + \phi)$ is instantaneously switched on at $t = 0$.

Figure 12.3 shows a density plot of the wave packet when it evolves in the undriven lattice, that is, for $K_0 = 0$; the density is encoded in shades of gray. In this and the following figures, spatial extensions are measured in terms of the dimensionless coordinate $z = k_L x$, so that a distance $\Delta z/\pi = 1$ corresponds to one lattice period; moreover, the time scale is set by the period $T = 2\pi/\omega$. With $k_c/k_L = 0$ the initial packet carries no net momentum; its width is chosen as $\Delta k/k_L = 0.1$. As expected, the width of the packet then grows in the course of time by well-to-well tunneling.

In Figure 12.4 we depict the density of the wave packet that evolves from the same initial condition when the driving amplitude is tuned to the first band collapse at $K_0 = 2.35$. Here we observe dynamic localization at its very best: the spreading has stopped, the packet is "frozen."

It is then also of interest to monitor the evolution at the supposed second collapse, at $K_0 = 5.52$; this is done in Figure 12.5. While the "regular spreading" that has been prominent in Figure 12.3 indeed seems to have stopped, small probability wavelets leak out of the initial packet almost immediately, spreading rapidly over the lattice. This is an effect of the multiphoton-like resonances previously spotted in Figure 12.1, which assist parts of the wave function in getting to higher bands, allowing them to escape on a short time scale.

As long as interband transitions remain negligible, the resulting single-band dynamics can often be regarded as "semiclassical" [22]: Namely, if an initial packet is strongly centered in k-space around some arbitrary wave number $k_c \equiv k_c(0)$, this center wave number evolves in time according to Bloch's famous "acceleration theorem"

$$\hbar \dot{k}_c(t) = F(t), \tag{12.29}$$

similar to the evolution (Equation 12.7) of the index of a single Houston state. The model Hamiltonian (Equation 12.26) specifies $F(t) = F_1 \cos(\omega t + \phi)$, so that in this case

$$k_c(t) = k_c(0) + \frac{F_1}{\hbar\omega}\left(\sin(\omega t + \phi) - \sin(\phi)\right). \tag{12.30}$$

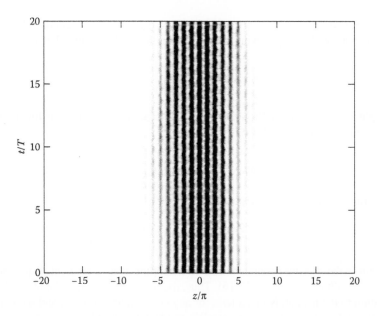

FIGURE 12.4 Evolution of the same initial wave packet as in Figure 12.3 at the first band collapse ($K_0 = 2.35$): Here one encounters almost perfect dynamic localization; wave-packet spreading is disabled because the quasienergy band is dispersionless.

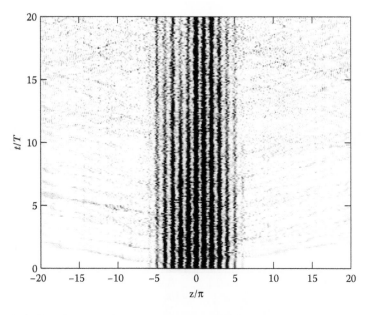

FIGURE 12.5 Evolution of the same inital wave packet as in Figure 12.3 at the supposed second band collapse ($K_0 = 5.52$): Here the multiphoton-like resonances visible in Figure 12.1 lead to a marked degradation of the localization.

The packet's group velocity then is given by the derivative of the dispersion relation $E(k)$ of the band it lives in, evaluated at this moving center wave number (Equation 12.30):

$$v_{\text{group}}(t) = \frac{1}{\hbar} \frac{dE}{dk}\bigg|_{k_c(t)}. \tag{12.31}$$

Taking the tight-binding relation (Equation 12.3) as a good approximation for the actual lowest energy band of our model, this yields

$$v_{\text{group}}(t) = \frac{2Ja}{\hbar} \sin\left(k_c(t)a\right). \tag{12.32}$$

Upon time averaging, one is therefore left with

$$\bar{v}_{\text{group}} = \frac{2J_{\text{eff}}a}{\hbar} \sin\left(k_c(0)a - K_0 \sin(\phi)\right), \tag{12.33}$$

where J_{eff} again is the driving-dependent effective hopping matrix element (Equation 12.12), and K_0 is the scaled amplitude (Equation 12.16). Thus, the initial phase ϕ may be utilized for imparting some momentum to the packet. Nonetheless, for any combination of $k_c(0)$ and ϕ the average group velocity vanishes when $J_{\text{eff}} = 0$, as corresponding to ideal dynamic localization.

This semiclassical behavior is illustrated by a further set of figures. In Figure 12.6 we plot the evolution of the exact k-space density that originates from the initial condition (Equation 12.28). Again we set $\Delta k/k_L = 0.1$, meaning that the distribution is sufficiently narrow to ensure the validity of Equation 12.31; moreover, $k_c/k_L = 0$ and $K_0 = 1.2$. Since $\phi = \pi/2$, the distribution then oscillates around $\bar{k} = -F_1/(\hbar\omega)$, or $\bar{k}/k_L = -K_0/\pi$, following precisely the k-space trajectory predicted by Equation 12.30.

A nonzero average momentum of the packet can likewise be achieved by selecting some suitable value of k_c/k_L. Figure 12.7 shows an example with $k_c/k_L = 0.8$, while $K_0 = 0.4$ and $\Delta k/k_L = 0.1$.

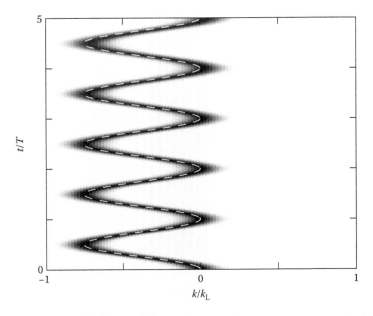

FIGURE 12.6 Evolution of the k-space distribution initially given by Equation 12.28 with width $\Delta k/k_L = 0.1$, for $K_0 = 1.2$ and $k_c/k_L = 0$. The white-dashed line indicates the "classical" solution (Equation 12.30).

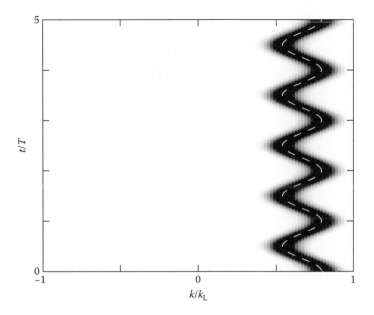

FIGURE 12.7 Evolution of the k-space distribution initially given by Equation 12.28 with width $\Delta k/k_{\mathrm{L}} = 0.1$, but now for $K_0 = 0.4$ and $k_{\mathrm{c}}/k_{\mathrm{L}} = 0.8$, so that the white-dashed "classical" solution (Equation 12.30) starts from a nonzero value.

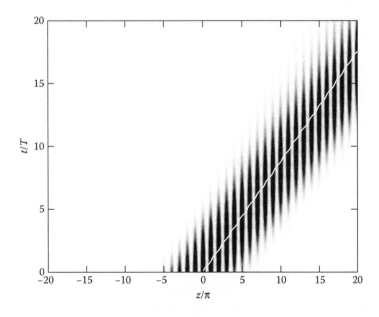

FIGURE 12.8 Evolution of the Bloch wave packet (Equation 12.27) with initial k-space width $\Delta k/k_{\mathrm{L}} = 0.1$ and initial momentum $k_{\mathrm{c}}/k_{\mathrm{L}} = 0.8$, driven with scaled amplitude $K_0 = 0.4$, as corresponding to the k-space distribution depicted in Figure 12.7. The white line marks the trajectory obtained by integrating the oscillating group velocity (Equation 12.32).

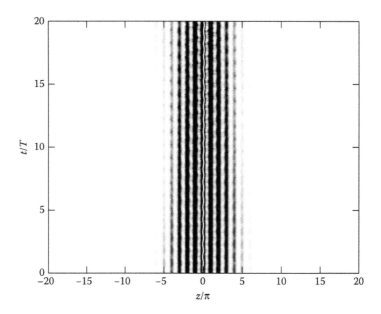

FIGURE 12.9 Evolution of the Bloch wave packet (Equation 12.27) with initial k-space width $\Delta k/k_L = 0.1$ and initial momentum $k_c/k_L = 0.8$, now driven with scaled amplitude $K_0 = 2.35$: Despite the nonzero average momentum, the average group velocity vanishes because of the band collapse. The white line in the center is obtained as described in Figure 12.8.

This obviously corresponds to a wave function $\langle x | \psi(t) \rangle$ which moves into the positive x-direction all the time; the density of this wave function is displayed in Figure 12.8. Here the white line indicates the classical trajectory that results from integrating the group velocity (Equation 12.32); indeed, this trajectory describes the motion of the packet's center quite well. When adjusting the driving amplitude to the first collapse, as in Figure 12.9, the average motion stops despite the nonzero average momentum, as it should; when increasing K_0 to still higher values, so that J_{eff} becomes negative, the packet's direction of motion can even be reversed.

While the semiclassical approach to dynamic localization may be helpful, insofar as it appeals to our intuition, its explanation in terms of "prohibited dephasing" resulting from a quasienergy band collapse is *much* more powerful: this view immediately reveals that not only does the average motion of a wave packet come to a complete standstill, but so does its spreading; moreover, prohibited dephasing applies to *any* initial condition, regardless whether its envelope varies sufficiently slowly to justify the semiclassical approximation. As an extreme example of "nonclassical" motion, we consider in Figure 12.10 the undriven evolution of a wave function that coincides with a single Wannier function of the optical lattice [12] at $t = 0$, and therefore certainly does not possess a slowly varying envelope then, giving rise to a fairly complex spreading pattern which differs substantially from the semiclassical one previously visualized in Figure 12.3. Nonetheless, when driven with the amplitude $K_0 = 2.35$ marking the first quasienergy band collapse, one observes another occurrence of dynamic localization, as witnessed by Figure 12.11; the difference between the two evolution patterns depicted in Figures 12.10 and 12.11 could hardly be more striking.

In actual laboratory experiments it is advantageous to work with a phase-coherent atomic Bose–Einstein condensate, rather than with individual atoms; if the density of the condensate is sufficiently low, or if the interatomic s-wave scattering length is tuned close to zero by means of a Feshbach resonance [13], the condensate is practically ideal, so that one effectively can perform a measurement on an ensemble of identically prepared noninteracting atoms in a single shot.

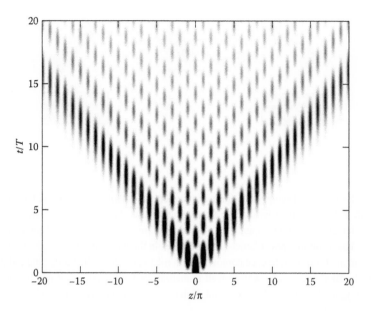

FIGURE 12.10 Evolution of the wave function that originates from an initial single Wannier state in the undriven lattice. This state does not possess a slowly varying envelope, and thus does not conform to semiclassical dynamics.

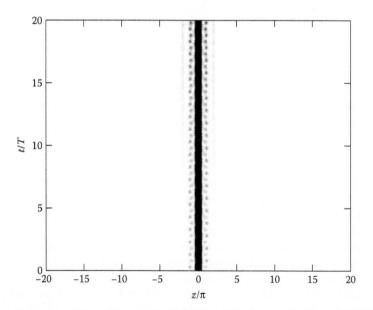

FIGURE 12.11 Evolution of the wave function that originates from an initial single Wannier state when driven with scaled amplitude $K_0 = 2.35$. The semiclassical approximation is not applicable here, but dynamic localization works nonetheless.

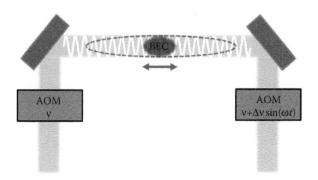

FIGURE 12.12 Experimental setup for *in-situ* measurement of dynamic localization of a Bose–Einstein condensate (BEC) in a driven optical lattice: the frequencies of two laser beams are shifted with the help of acousto-optic modulators (AOMs) by ν and by $\nu + \Delta\nu \sin(\omega t)$, respectively, before being directed against each other by mirrors. The resulting optical lattice then oscillates in the laboratory frame, giving rise to an oscillating inertial force in the frame of reference comoving with the lattice. After the initial longitudinal confinement is switched off, the BEC expands by well-to-well tunneling; its final width (indicated by the dashed line) is recorded by imaging its shadow cast by a resonant flash onto a CCD chip. (Courtesy of O. Morsch.)

Figure 12.12 shows a possible experimental setup [3,4]: The optical lattice is formed by two laser beams of wavelength λ, which are directed against each other with the help of mirrors. Each beam passes through an acousto-optic modulator which shifts its frequency by ν and by $\nu + \Delta\nu(t)$, respectively. Because of the frequency difference $\Delta\nu(t)$ thus introduced between the counterpropagating beams, the condensate experiences the potential

$$V_{\text{lab}}(x,t) = \frac{V_0}{2} \cos\left(2k_{\text{L}} \left[x + \frac{\lambda}{2} \int_0^t d\tau \, \Delta\nu(\tau) \right] \right) \tag{12.34}$$

in the laboratory frame, which means that the lattice position shifts in time according to the prescribed protocol $\Delta\nu(t)$. In a frame of reference comoving with the lattice, this shift translates into the inertial force

$$F(t) = M \frac{\lambda}{2} \frac{d\Delta\nu(t)}{dt}. \tag{12.35}$$

Therefore, choosing $\Delta\nu(t) = \Delta\nu_{\text{max}} \sin(\omega t + \phi)$ leads to the desired Hamiltonian (Equation 12.26) in the comoving frame, with the driving amplitude

$$F_1 = M\omega \frac{\lambda}{2} \Delta\nu_{\text{max}}. \tag{12.36}$$

Now a Bose–Einstein condensate initially trapped in the center of the oscillating lattice is allowed to expand freely in the lattice direction by well-to-well tunneling after switching off the longitudinal confinement, while maintaining a weak transversal confinement in order to keep the condensate in the lattice. After a variable expansion time, the *in situ* width of the condensate is determined by a resonant flash, the shadow cast by which is imaged onto a CCD chip [3]. The measured expansion rate then is to a good approximation proportional to $|J_{\text{eff}}|$, that is, to the absolute value of the effective hopping matrix element (Equation 12.12); in principle, even the sign of J_{eff} can be deduced from additional time-of-flight measurements [3]. In Figure 12.13 we display data for the ratio J_{eff}/J acquired in this manner by the Pisa group with a condensate of ^{87}Rb atoms in a lattice of depth $V_0/E_{\text{rec}} = 6.0$, driven with frequency $\omega/(2\pi) = 4.0$ kHz, after expansion times of 150 ms. Evidently, these data match the expected Bessel function $J_0(K_0)$ quite well even up to the second zero. Note that here one has $\hbar\omega/E_{\text{rec}} = 1.24$, so that the frequency employed in these measurements is significantly higher than in our model calculations. This means that the inequality $4J < \hbar\omega$

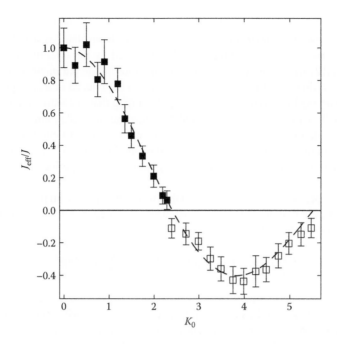

FIGURE 12.13 Experimental results for the ratio J_{eff}/J of the effective hopping matrix element (Equation 12.12) to the bare one as a function of the scaled driving amplitude (Equation 12.16), obtained with a Bose–Einstein condensate of ^{87}Rb atoms in an optical lattice with a depth of $V_0 = 6 E_{\mathrm{rec}}$ ($\lambda = 842$ nm), driven with frequency $\omega/(2\pi) = 4.0$ kHz. The dashed line corresponds to the expected Bessel function $J_0(K_0)$. (Courtesy of O. Morsch.)

is satisfied in a stronger manner, while $\hbar\omega$ still remains reasonably small compared to the band gap. As a consequence, even the second band collapse can be quite well developed. In any case, this figure strikingly demonstrates that the concept of dynamic localization by now has crossed, in the context of mesoscopic matter waves, the threshold from an idealized theoretical concept to a well-controllable laboratory reality.

12.4 WHAT IS IT GOOD FOR?

Up to this point we have considered no more than a possible realization of dynamic localization which comes fairly close to the theoretical ideal [1]. Apart from its observation with dilute Bose–Einstein condensates in time-periodically shifted optical lattices [3,4], this type of quantum wave propagation has meanwhile also been made visible by means of an optical analog based on sinusoidally curved lithium–niobate waveguide arrays [23]. This is certainly interesting, but it is not what one would call "deep"; the "prohibited dephasing"-view clearly reveals that the only physics entering here is summarized by stating that an initial state is "frozen" in time if the phase factors of all of its spectral components evolve at the same speed. Yet, the accompanying band collapse furnishes a strong hint that there may be more in stock. Namely, when the ideal dynamics is somehow perturbed it is the bandwidth which sets the scale with respect to which the strength of such a perturbation has to be gauged. A prominent example is provided by the repulsive interaction between ultracold atoms in an optical lattice; the strength of this interaction is expressed in terms of a parameter U which quantifies the repulsion energy of one pair of atoms occupying the same lattice site [21]. Accordingly, the characteristic dimensionless parameter then is the ratio U/J; here $J = W/4$ is taken instead of the bandwidth W. Indeed, it is this ratio U/J which decides which quantum phase a gas of ultracold, repulsively interacting atoms in an optical lattice adopts. For $U/J \ll 1$ the system is superfluid, but becomes a Mott insulator when this ratio exceeds a critical value [21]. Hence, when recalling that J is replaced by the effective hopping matrix element (Equation 12.12) when the system is driven with appropriate parameters, it is only natural to predict that this superfluid-to-Mott insulator transition can be induced in a time-periodically shifted optical lattice by varying the driving force [24,25]. Assuming that one starts in the superfluid phase, J_{eff} can then virtually be made arbitrarily small by adjusting the scaled amplitude K_0 to a zero of J_0, resulting in a value of U/J_{eff} so large that the system is forced to enter the Mott regime. The experimental confirmation of this scenario, achieved by the Pisa group [5], probably constitutes the first known example of coherent control exerted by means of time-periodic forcing on a quantum phase transition.

There are other types of perturbations, associated with deviations from perfect translational symmetry, which affect even noninteracting ultracold atoms in optical lattices. Most notably, the system governed by the tight-binding Hamiltonian

$$H_{\text{AA}} = -J \sum_{\ell} \Big(|\ell + 1\rangle\langle\ell| + |\ell\rangle\langle\ell + 1| \Big) + V \sum_{\ell} \cos(2\pi g\ell + \delta)|\ell\rangle\langle\ell|, \qquad (12.37)$$

differing from its antecedent (Equation 12.1) through additional on-site energies which oscillate along the lattice with amplitude V, shows a quite peculiar behavior when the number g is irrational, so that this system becomes *quasiperiodic* [26–28]: As long as $|V/J| < 2$, so that the on-site perturbations are relatively weak, all of its energy eigenstates still remain extended over the entire lattice in a Bloch-like manner, whereas they are all exponentially localized, with one common localization length, when $|V/J| > 2$. Thus, there is a metal-insulator-like, incommensurability-induced transition at $|V/J| = 2$, originally studied by Harper (1955) [26] in the context of conduction electrons in a magnetic field, and later by Aubry and André (1980) [27]; this transition can be realized approximately with ultracold atoms in a *bichromatic* optical lattice described by the potential

$$V_{\text{bic}}(x) = \frac{V_0}{2} \cos(2k_{\text{L}}x) + V_1 \cos(2gk_{\text{L}}x + \delta). \qquad (12.38)$$

The guiding idea here is to employ a primary lattice with depth V_0 for setting up the hosting tight-binding system (Equation 12.1), as before, and then to invoke a secondary lattice with much smaller depth $2V_1$ for achieving the required modulation of the local energies at the sites of the host [10,11]. When the primary lattice is comparatively shallow, possessing a depth of only a few recoil energies, the transition occurs stepwise upon increasing V_1 [12], featuring pronounced mobility edges resulting mainly from the next-to-nearest neighbor couplings between the host's sites which are present in the full bichromatic potential (Equation 12.38), but do not occur in the Aubry–André model (Equation 12.37). When $V_0/E_{\text{rec}} \gg 1$, so that the primary lattice is so deep that these additional couplings may be safely neglected, the transition occurring in the actual bichromatic lattice (Equation 12.38) is fairly sharp. The parameter J then again is given approximately by Equation 12.25; moreover, one has

$$V/E_{\text{rec}} \sim \frac{V_1}{E_{\text{rec}}} \exp\left(-\frac{g^2}{\sqrt{V_0/E_{\text{rec}}}} \right) \qquad \text{for } V_0/E_{\text{rec}} \gg 1 \qquad (12.39)$$

with reasonably chosen g on the order of unity. Therefore, the equation $|V/J| = 2$ marking the metal-insulator-like transition in the ideal Aubry–André model now translates into the estimate [12]

$$\frac{V_1^c}{E_{\text{rec}}} \sim \frac{8}{\sqrt{\pi}} \left(\frac{V_0}{E_{\text{rec}}} \right)^{3/4} \exp\left(-2\sqrt{\frac{V_0}{E_{\text{rec}}}} + \frac{g^2}{\sqrt{V_0/E_{\text{rec}}}} \right) \qquad (12.40)$$

for the critical strength V_1^c of the secondary optical lattice, given a sufficient depth of the primary one. Indeed, this transition has been observed with a Bose–Einstein condensate consisting of ^{39}K atoms, using a magnetically tunable Fesbach resonance for rendering these atoms practically noninteracting [13].

When ultracold atoms in such a bichromatic lattice (Equation 12.38) are subjected to time-periodic forcing, one obtains an additional knob which can be turned to induce the transition: Because J is replaced by the effective hopping strength (Equation 12.12) when the system is suitably driven, one can cross the critical border $|V/J_{\mathrm{eff}}| = 2$ by varying the parameters of the driving force; the critical parameters then are linked approximately by the relation

$$|J_0(K_0)| \sim \frac{\sqrt{\pi}}{8} \frac{V_1}{E_{\mathrm{rec}}} \left(\frac{V_0}{E_{\mathrm{rec}}}\right)^{-3/4} \exp\left(+2\sqrt{\frac{V_0}{E_{\mathrm{rec}}}} - \frac{g^2}{\sqrt{V_0/E_{\mathrm{rec}}}}\right). \tag{12.41}$$

Hence, it is feasible to coherently control the metal-insulator-like transition exhibited by noninteracting ultracold atoms in properly designed bichromatic optical potentials through time-periodic forcing [10,11]. In order to substantiate this prediction, we now display the results of further numerical wave-packet calculations. In all of these we employ a primary lattice with depth $V_0/E_{\mathrm{rec}} = 5.7$, as in our preceding studies, and fix the incommensurability parameter at the golden mean $g = (\sqrt{5} - 1)/2$ up to numerical accuracy. With this choice, the above estimate (Equation 12.40) yields $V_1^c/E_{\mathrm{rec}} \approx 0.165$ for the critical strength of the secondary lattice. The driving frequency is given by $\hbar\omega/E_{\mathrm{rec}} = 0.5$ throughout.

Figure 12.14 visualizes the evolution of a wave function that originates from the same Gaussian initial state as already employed in Figure 12.3. Here the driving force is still absent, and the depth of the secondary lattice is $V_1/E_{\mathrm{rec}} = 0.10$, placing the system in its metallic phase; accordingly, the wave function readily explores the entire lattice. In contrast, when $V_1/E_{\mathrm{rec}} = 0.25$ and the drive is still switched off, the wave function remains localized as shown in Figure 12.15; this indicates that we are encountering the insulating phase now. But the wave function also remains localized when the secondary lattice is tuned back to $V_1/E_{\mathrm{rec}} = 0.10$ and the driving force acts with scaled amplitude $K_0 = 1.7$, as depicted in Figure 12.16; the relation (Equation 12.41) predicts the transition from the

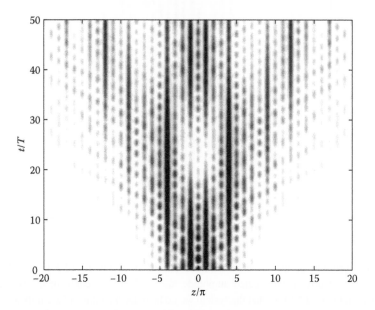

FIGURE 12.14 Evolution of the same initial wave packet as in Figure 12.3 in an undriven bichromatic optical lattice (Equation 12.38). Here the strength of the secondary potential is $V_1/E_{\mathrm{rec}} = 0.10$, so that the system is in its mobile "metallic" phase, allowing the wave function to spread.

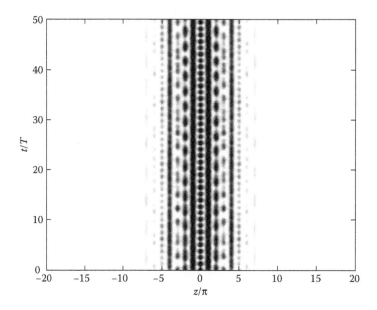

FIGURE 12.15 Evolution of the same initial wave packet as in Figure 12.3 in an undriven bichromatic optical lattice (Equation 12.38). Here the strength of the secondary potential is $V_1/E_{rec} = 0.25$, so that the system is in its "insulating" phase, keeping the wave function localized.

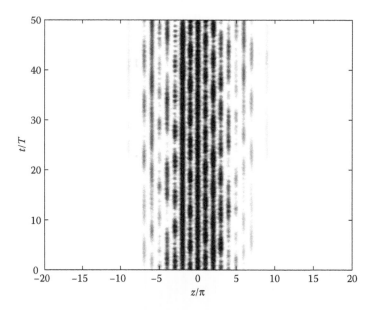

FIGURE 12.16 Evolution of the same initial wave packet as in Figure 12.3 in a bichromatic optical lattice (Equation 12.38) driven with scaled amplitude $K_0 = 1.7$. The strength of the secondary lattice is $V_1/E_{rec} = 0.10$, as in Figure 12.14, so that the system would be in its "metallic" phase if there were no forcing.

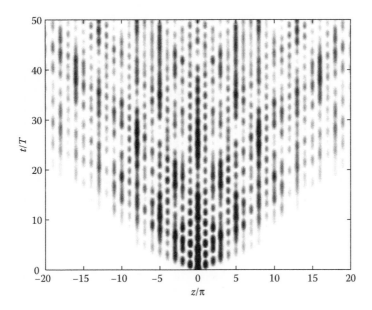

FIGURE 12.17 Evolution of the wave function originating from a single Wannier state of the primary lattice in an undriven bichromatic optical lattice (Equation 12.38). Here the strength of the secondary potential is $V_1/E_{rec} = 0.10$, so that the system is in its "metallic" phase.

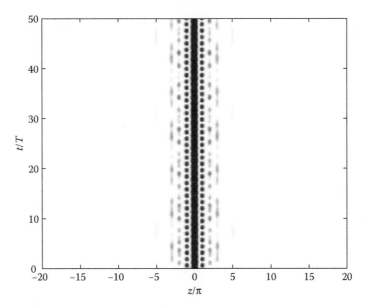

FIGURE 12.18 Evolution of the wave function originating from a single Wannier state of the primary lattice in an undriven bichromatic optical lattice (Equation 12.38). Here the strength of the secondary potential is $V_1/E_{rec} = 0.25$, so that the system is in its "insulating" phase.

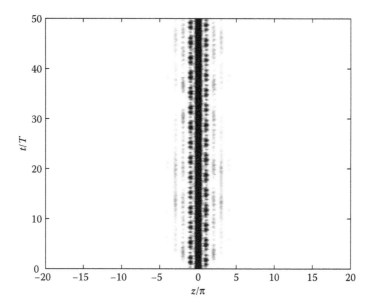

FIGURE 12.19 Evolution of the wave function originating from a single Wannier state of the primary lattice in a bichromatic optical lattice (Equation 12.38) driven with scaled amplitude $K_0 = 1.7$. The strength of the secondary lattice is $V_1/E_{rec} = 0.10$, as in Figure 12.17, so that the system would be in its "metallic" phase if there were no forcing.

metallic to the insulating phase to have occurred already at about $K_0 \approx 1.3$. It should be noted that there is a pronounced difference from the ideal dynamic localization reviewed in the preceding section: there the wave packet remains localized only when K_0 is exactly equal to a zero of J_0. In contrast, here one switches from the metallic into the insulating phase already when $|J_0(K_0)|$ becomes sufficiently small.

Finally, we show a corresponding sequence of results for wave functions which evolve from an initial Wannier state of the primary lattice. In Figure 12.17 we again consider an undriven bichromatic lattice with $V_1/E_{rec} = 0.10$, so that the mobile metallic phase enables uninhibited spreading; in Figure 12.18, where $V_1/E_{rec} = 0.25$, the system's insulating character then keeps the wave function strongly localized. But that same high degree of localization may also be obtained when again resetting the strength of the secondary lattice to $V_1/E_{rec} = 0.10$, and switching on the driving force with scaled amplitude $K_0 = 1.7$, as done in Figure 12.19.

These figures vividly illustrate the main message: in the presence of time-periodic forcing it is the width of the underlying quasienergy band which determines the effective strength of deviations from perfect spatial periodicity. In an ideal lattice without such deviations one encounters "only" dynamic localization, but in lattices with isolated, quasiperiodic, or random perturbations the strengths of these can be adjusted at will by suitably selecting the parameters of the drive. With regard to experimental tests, the enormous flexibility offered by ultracold atoms in optical potentials makes such systems far superior to electrons in AC-driven crystal lattices.

When the concept of controlling the incommensurability-induced metal-insulator transition exhibited by the Aubry–André model (Equation 12.37) by means of time-periodic forcing was conceived [10,11] the experimental investigation of ultracold atoms in optical lattices was still in its infancies. But now that this transition has been unambiguously observed with a noninteracting Bose–Einstein condensate [13], the demonstration of its coherent control has come into immediate reach. Besides the already established coherent control of the interaction-induced superfluid-to-Mott insulator transition [5], this demonstration would constitute a further milestone achievement in the on-going effort to explore the newly emerging prospects provided by dressed matter waves.

ACKNOWLEDGMENTS

We thank O. Morsch and E. Arimondo for continuing discussions of their experiments [3–5], and O. Morsch in particular for providing Figures 12.12 and 12.13. This work was supported by the Deutsche Forschungsgemeinschaft under Grant No. HO 1771/6.

REFERENCES

1. D. H. Dunlap and V. M. Kenkre, Dynamic localization of a charged particle moving under the influence of an electric field, *Phys. Rev. B* 34, 3625–3633, 1986.
2. M. M. Dignam and C. M. de Sterke, Conditions for dynamic localization in generalized AC electric fields, *Phys. Rev. Lett.* 88, 046806, 2002.
3. H. Lignier, C. Sias, D. Ciampini, Y. Singh, A. Zenesini, O. Morsch, and E. Arimondo, Dynamical control of matter-wave tunneling in periodic potentials, *Phys. Rev. Lett.* 99, 220403, 2007.
4. A. Eckardt, M. Holthaus, H. Lignier, A. Zenesini, D. Ciampini, O. Morsch, and E. Arimondo, Exploring dynamic localization with a Bose–Einstein condensate, *Phys. Rev. A* 79, 013611, 2009.
5. A. Zenesini, H. Lignier, D. Ciampini, O. Morsch, and E. Arimondo, Coherent control of dressed matter waves, *Phys. Rev. Lett.* 102, 100403, 2009.
6. M. Holthaus, Collapse of minibands in far-infrared irradiated superlattices, *Phys. Rev. Lett.* 69, 351–354, 1992.
7. M. Holthaus and D. Hone, Quantum wells and superlattices in strong time-dependent fields, *Phys. Rev. B* 47, 6499–6508, 1993.
8. D. W. Hone and M. Holthaus, Locally disordered lattices in strong AC electric fields, *Phys. Rev. B* 48, 15123–15131, 1993.
9. M. Holthaus, G.H. Ristow, and D.W. Hone, AC-field-controlled Anderson localization in disordered semiconductor superlattices, *Phys. Rev. Lett.* 75, 3914–3917, 1995.
10. K. Drese and M. Holthaus, Ultracold atoms in modulated standing light waves, *Chem. Phys.* 217 (Special issue: *Dynamics of Driven Quantum Systems*), 201–219, 1997.
11. K. Drese and M. Holthaus, Exploring a metal-insulator transition with ultracold atoms in standing light waves?, *Phys. Rev. Lett.* 78, 2932–2935, 1997.
12. D. J. Boers, B. Goedeke, D. Hinrichs, and M. Holthaus, Mobility edges in bichromatic optical lattices, *Phys. Rev. A* 75, 063404, 2007.
13. G. Roati, C. D'Errico, L. Fallani, M. Fattori, C. Fort, M. Zaccanti, G. Modugno, M. Modugno, and M. Inguscio, Anderson localization of a non-interacting Bose–Einstein condensate, *Nature* 453, 895–898, 2008.
14. W. V. Houston, Acceleration of electrons in a crystal lattice, *Phys. Rev.* 57, 184–186, 1940.
15. J. H. Shirley, Solution of the Schrödinger equation with a Hamiltonian periodic in time, *Phys. Rev.* 138, B979–B987, 1965.
16. Ya. B. Zel'dovich, The quasienergy of a quantum-mechanical system subjected to a periodic action, *Zh. Eksp. Theor. Fiz.* 51, 1492–1495, 1966 [*Sov. Phys. JETP* 24, 1006–1008, 1967].
17. V. I. Ritus, Shift and splitting of atomic energy levels by the field of an electromagnetic wave, *Zh. Eksp. Theor. Fiz.* 51, 1544–1549, 1966 [*Sov. Phys. JETP* 24, 1041–1044, 1967].
18. H. Sambe, Steady states and quasienergies of a quantum-mechanical system in an oscillating field, *Phys. Rev. A* 7, 2203–2213, 1973.
19. S. Arlinghaus and M. Holthaus, Driven optical lattices as strong-field simulators, *Phys. Rev. A* 81, 063612, 2010.
20. O. Morsch and M. Oberthaler, Dynamics of Bose–Einstein condensates in optical lattices, *Rev. Mod. Phys.* 78, 179–215, 2006.
21. I. Bloch, J. Dalibard, and W. Zwerger, Many-body physics with ultracold gases, *Rev. Mod. Phys.* 80, 885–964, 2008.
22. N. W. Ashcroft and N. D. Mermin, *Solid State Physics*, Harcourt College Publishers, Fort Worth, 1976 (see Chapter 12 therein).
23. S. Longhi, M. Marangoni, M. Lobino, R. Ramponi, P. Laporta, E. Cianci, and V. Foglietti, Observation of dynamic localization in periodically curved waveguide arrays, *Phys. Rev. Lett.* 96, 243901, 2006.

24. A. Eckardt, C. Weiss, and M. Holthaus, Superfluid-insulator transition in a periodically driven optical lattice, *Phys. Rev. Lett.* 95, 260404, 2005.

25. C. E. Creffield and T. S. Monteiro, Tuning the Mott transition in a Bose–Einstein condensate by multiple photon absorption, *Phys. Rev. Lett.* 96, 210403, 2006.

26. P. G. Harper, Single band motion of conduction electrons in a uniform magnetic field, *Proc. Phys. Soc. A* 68, 874–878, 1955.

27. S. Aubry and G. André, Analyticity breaking and Anderson localization in incommensurate lattices, *Ann. Israel Phys. Soc.* 3, 133–140, 1980.

28. J. B. Sokoloff, Unusual band structure, wave functions and electrical conductance in crystals with incommensurate periodic potentials, *Phys. Rep.* 126, 189–244, 1985.

13 Control of Photonic Tunneling in Coupled Optical Waveguides

Stefano Longhi

CONTENTS

13.1 INTRODUCTION

Tunneling is a universal phenomenon to wave physics which arises when an evanescent field transfers energy through a barrier to a region where a propagating wave is allowed. Tunneling phenomena have been predicted and observed in countless quantum and classical physical systems. In particular, optical (or photonic) tunneling usually refers to light propagation across forbidden photonic barriers, such as in photonic band gap structures or undersized waveguides, as well as to evanescent field coupling among waveguides or resonators or to light passage via evanescent coupling in frustrated total internal reflection (FTIR). The first observation of FTIR and of photonic tunneling date back to three centuries ago, when Newton reported on light transmission through a gap between prisms. Nowadays, FTIR at plane interfaces is generally viewed as an optical analog of quantum mechanical tunneling across a one-dimensional potential barrier [1]. Optical tunneling across photonic barriers, realized by for example, FTIR or photonic band gaps, has attracted a great deal of attention in the past two decades as an experimentally accessible system with which to measure tunneling times [2–7]. FTIR has been also exploited to study wave packet reshaping [8] and to provide an optical realization of curious quantum–mechanical effects such as quantum evaporation [9]. Similarly, photonic tunneling in spherical dielectric media has been proposed as an optical analog of nuclear decay arising form particle escape from a three-dimensional potential well [10]. In deformed dielectric microcavities, chaos-assisted tunneling phenomena have been investigated as well [11–14]. A different optical system that provides an experimentally accessible laboratory tool to visualize and control photonic tunneling, which is at the focus of this chapter, is represented by evanescently coupled optical waveguides or fibers [15]. An optical waveguide or fiber may be viewed as a potential well which traps light in the transverse plane, still allowing wave propagation along the waveguide or fiber optical axis z [16]. Evanescent field coupling among adjacent optical wells then leads to photonic tunneling and to light energy transfer along the propagation direction, a phenomenon which is at the basis of many integrated photonic devices such as directional couplers and splitters. These optical structures can be viewed as the optical analog of coupled quantum wells, in which the usually fast temporal evolution of quantum mechanical tunneling is replaced by spatial light propagation along the waveguide axis z. The quantum–optical analogy basically stems from the formal analogy between the temporal Schrödinger equation for a nonrelativistic quantum particle and paraxial propagation of light in weakly guiding dielectric media [17]. For a z-invariant structure, like in a straight waveguide, in the scalar and paraxial approximations light evolution along the z axis is described by a two-dimensional Schrödinger-like equation, in which the refractive index profile of the guide determines the shape of the potential well, the refractive index of the bulk medium is analogous to the particle mass, the paraxial propagation distance z is equivalent to time, and the Planck constant is played by the reduced wavelength of photons [17]. As compared to quantum mechanical tunneling, the study of photonic tunneling in coupled waveguide structures offer a few advantages, such as the possibility of a direct visualization in space of typical ultrafast phenomena in time, the possibility to explore coherent dynamical regimes of difficult access in quantum systems, and the ability to mimic coherent control of quantum mechanical tunneling by simple geometric bending or twisting of the guiding photonic structures. The dynamical process of tunneling can be precisely mapped and visualized by means of fluorescence imaging [18–20] or tunneling optical microscopy [21,22] techniques. In the optical analog, dynamical control of quantum mechanical tunneling by external driving forces [23], leading, for example, to such phenomena as coherent destruction of tunneling [24] or field-induced barrier transparency [25], can be realized by the introduction of a geometric bending [26,27] or twisting [28,29] of waveguide axis (for a recent review see, for instance, Longhi (2009) [30]). Owing to the equivalence between geometry and inertial forces, fictitious forces for photons are in this way introduced by topological means [30]. Coherent control of photonic tunneling in a double-well potential, originally proposed for quantum particles [24], has been in this way proposed [31] and experimentally visualized for the first time in a sinusoidally curved optical directional coupler [18]. Similarly, the optical analog of field-induced barrier transparency has been proposed for FTIR on an undulating glass–air–glass interface [32]. Other dynamical techniques of quantum tunneling control, such as coherent quantum transport in space of the wave function among tunneling-coupled quantum wells by adiabatic passage [33,34] or control of quantum mechanical decay for macroscopic quantum tunneling [35–37], can be similarly implemented for photonic tunneling in suitably engineered coupled waveguides [19,22,38–45]. Finally, photonic tunneling enables to visualize some basic dynamical aspects embodied in tunneling processes, such as level crossing and Landau–Zener dynamics [20,46] encountered in different areas of physics.

In this Chapter, a review on dynamical tunneling of light waves in photonic structures is presented, with a special focus to the recent theoretical and experimental advances in the visualization and control of photonic tunneling in evanescently coupled optical waveguides. The topics covered by this Chapter include: coherent destruction of tunneling in periodically curved directional couplers, the optical analog of field-induced barrier transparency, coherent tunneling of light by adiabatic passage in coupled waveguides, control of photonic tunneling in waveguide arrays and the optical analog of the quantum Zeno effect, and the photonic analog of Landau–Zener tunneling.

13.2 PHOTONIC TUNNELING IN COUPLED OPTICAL WAVEGUIDES: BASIC EQUATIONS

The problem of tunneling of monochromatic light waves in coupled optical waveguides, as described by Maxwell's equations, can be cast under certain conditions in a form familiar to quantum mechanical tunneling. The similarity is basically grounded on the formal equivalence between the scalar and paraxial optical wave equation describing *spatial* propagation of a monochromatic light beam in weakly curved (or twisted) guiding dielectric structures and the *temporal* Schrödinger equation for a quantum particle (e.g., an electron) in potential wells driven by an external electromagnetic field. On the basis of such a formal equivalence, the fast temporal evolution of the electronic wave function in a quantum mechanical system is similar to the spatial light propagation along the guiding structure, which can be precisely mapped and visualized by means of fluorescence imaging [18–20] or tunneling optical microscopy [21,22] techniques. To highlight the optical–quantum analogy, let us consider propagation of monochromatic light waves at wavelength (in vacuum) λ in a weakly guiding dielectric structure, whose axis is allowed to be weakly curved along the paraxial propagation direction z (Figure 13.1a). In the scalar and paraxial approximations, light propagation is described by the following equation for the slowly varying electric field amplitude ψ:

$$i\hbar \frac{\partial \psi}{\partial z} = -\frac{\hbar^2}{2n_s} \frac{\partial^2 \psi}{\partial x^2} + V(x - x_0(z))\psi, \tag{13.1}$$

where z is the paraxial propagation distance, $\hbar \equiv \lambda/(2\pi)$ is the reduced wavelength, $V(x) = [n_s^2 - n^2(x)]/(2n_s) \simeq n_s - n(x)$, $n(x)$ is the refractive index profile of the guiding structure, n_s is the reference (substrate) refractive index, and $x_0(z)$ is the axis bending profile. In writing the previous equation, we considered for the sake of simplicity a one, transverse spatial coordinate x, however, a similar analysis may be extended to two-transverse spatial dimensions. After a change of reference frame and a gauge transformation, defined by the relations

$$x' = x - x_0(z), \quad z' = z, \quad \phi(x', z') = \psi(x', z') \exp\left[-i(n_s/\hbar)\dot{x}_0(z')x' - i\varphi(z')\right] \tag{13.2}$$

where the dot indicates the derivative with respect to z' and

$$\varphi(z') = (n_s/2\hbar) \int_0^{z'} d\xi \dot{x}_0^2(\xi), \tag{13.3}$$

Equation 13.1 is transformed into the following Schrödinger-like equation for a particle of mass n_s in the potential $V(x') \simeq n_s - n(x')$ driven by an external force $F(z')$ (see, e.g., Longhi (2005) [31]):

$$i\hbar \frac{\partial \phi}{\partial z'} = -\frac{\hbar^2}{2n_s} \frac{\partial^2 \phi}{\partial x'^2} + V(x')\phi - F(z')x'\phi. \tag{13.4}$$

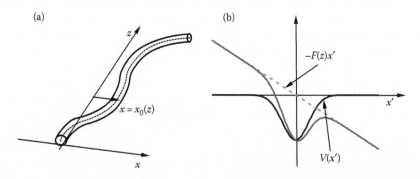

(a)   (b)

FIGURE 13.1 (a) Schematic of a weakly guiding optical waveguide with a curved axis. The function $x_0(z)$ defines the bending profile of waveguide axis. (b) Behavior of the effective potential $V(x') - F(z')x'$ entering in the Schrödinger equation 13.4 in the waveguide reference frame.

where $F(z') = -n_s\ddot{x}_0(z')$. The transformation (Equation 13.2) corresponds to a change of the reference frame from the (inertial) laboratory reference frame to the (noninertial) waveguide reference frame, and it was originally introduced in atomic physics by Kramers and Henneberger to study electron dynamics in noninertial reference frames [47–49]. Looking at Equation 13.4, one realizes that in the optical analog the Planck constant h is played by the wavelength λ of photons, whereas the temporal variable of the quantum problem is mapped into the spatial propagation coordinate z'. The external force $F(z')$ entering in Equation 13.4 is a fictitious force for light waves arising from the the noninertiality of reference frame (x', z')—where the waveguide appears to be straight—and it is related to the axis bending profile by the Newtonian equation of motion

$$F(z') = -n_s\ddot{x}_0(z'). \tag{13.5}$$

The force $F(z')$ may be also viewed as describing the interaction of an electron in the binding potential $V(x')$ (such as an electron trapped in a quantum well, double well or in a lattice potential) with the electric field $E(z') = F(z')/e$ of an electromagnetic (EM) wave in the electric dipole approximation, with a resulting effective potential $V(x') - F(z')x'$ [see Figure 13.1b]. In particular, a sinusoidal axis bending $x_0(z') = A\cos(2\pi z'/\Lambda)$ mimics the electronic interaction with a monochromatic EM wave with frequency $\omega = 2\pi/\Lambda$ and electric field amplitude $E_0 = n_s A\omega^2/e$. A multimode curved dielectric waveguide provides therefore, the optical analog of an artificial atom driven by an external laser field, and phenomena such as Rabi oscillations, atomic ionization, coherent population transfer, and adiabatic passage, can be therefore studied, at the classical level, using light waves rather than matter waves. Similarly, light propagation in a chain of two or more curved optical waveguides is suited to mimic driven quantum tunneling phenomena in coupled quantum wells.

13.3   COHERENT DESTRUCTION OF OPTICAL TUNNELING

Driven quantum tunneling, that is, the control of quantum tunneling by external driving fields, has received a continuous and increasing interest in different areas of physics [23]. The simplest physical system to study driven quantum tunneling is perhaps a quantum particle in a double-well potential driven by an external sinusoidal force of frequency ω, which has provided a paradigmatic model to study tunneling control. Depending on the strength and frequency of the driving field, a periodic tilting of the double-well potential may lead to suppression or enhancement of tunneling. In particular, for certain parameter ratios between the amplitude and frequency of the driving, tunneling can be brought to a standstill [24], a phenomenon which is refereed to as "coherent destruction of tunneling" (CDT). The first experimental demonstration of quantum tunneling control in a double-well potential has been reported solely very recently using cold atoms trapped in optical lattices [50]. The simplest optical analog of quantum tunneling in a double-well potential is provided by light propagation in an optical-directional coupler. A synchronous-optical coupler is basically composed by a couple of two straight evanescently coupled optical waveguides [31], in which light waves initially launched into one of the two waveguides periodically tunnel back and forth between the two waveguides owing to the propagation constant shift $(E_2 - E_1)/\hbar$ of symmetric and antisymmetric supermodes of the coupler. According to Equation 13.4, an external ac force for light waves is introduced by sinusoidal bending of the optical coupler with spatial period $\Lambda = 2\pi/\omega$ and amplitude A, as shown in Figure 13.2a. The experimental demonstration of CDT for light waves in sinusoidally curved optical directional couplers was reported in Della Valle et al. (2007) [18]. A typical energy level diagram of a double-well potential, in absence of the external driving field and corresponding to an optical coupler, is shown in Figure 13.2b. The two lowest-energy levels E_1 and E_2 corresponds to the symmetric and antisymmetric supermodes of the optical coupler, whereas the other energy levels correspond to higher order modes. When the external driving is switched on, that is the coupler is sinusoidally bent, the energy levels of the time-independent Hamiltonian are replaced by the quasienergy levels of the time-periodic Hamiltonian (for the definition of quasienergies we refer

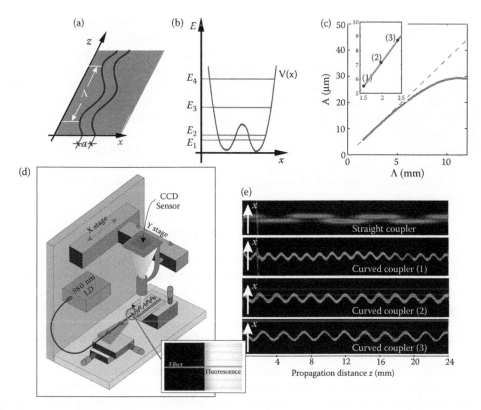

FIGURE 13.2 Coherent destruction of photonic tunneling in a sinusoidally curved synchronous directional coupler. (a) Schematic of the optical coupler. The axis bending profile is sinsuoidal with amplitude A and period Λ. (b) Typical energy level diagram of a double-well potential. (c) Manifold of quasienergy crossing in the (A, Λ) plane. (d) Experimental set-up for fluorescence light imaging: the green fluorescence emitted by Er ions is collected by a microscope objective and imaged onto a CCD sensor. (e) Measured fluorescence patterns in a straight directional coupler (upper figure) and in three sinusoidally curved couplers (bottom figures) for geometric parameters corresponding to points (1), (2), and (3) of Figure 13.2c.

to, e.g., Grifoni and Hänggi (1998) [23]). In this case, it can be shown that CDT occurs approximately for a modulation frequency ω in the range $(E_2 - E_1) < \hbar\omega < (E_3 - E_2)$ and for a modulation amplitude that corresponds to exact crossing between the quasienergies ε_1 and ε_2 associated to the lowest tunnel doublet, that is, for $\varepsilon_1 = \varepsilon_2$ [23]. The manifold of quasienergy level crossing $\varepsilon_2 = \varepsilon_1$ in the (A, Λ) plane is shown in Figure 13.2c by the solid line [18]. An approximate expression of the quasienergy splitting $\Delta\varepsilon = \varepsilon_2 - \varepsilon_1$, which is valid in the high-frequency limit $\hbar\omega \gg (E_2 - E_1)$, can be calculated by coupled-mode theory and reads [31]

$$\Delta\varepsilon = (E_2 - E_1)J_0\left(\frac{4\pi^2 n_s aA}{\lambda\Lambda}\right), \tag{13.6}$$

where a is the distance between the two waveguides. Note that, under such an approximation, the first manifold of quasienergy crossing corresponds to first zero of J_0 Bessel function and is represented by the straight dashed line in Figure 13.2c. The experimental demonstration of CDT for light waves, reported in Della Valle et al. (2007) [18], was simply based on fluorescence imaging of light propagation in a series of sinusoidally curved optical directional couplers manufactured by the ion-exchange technique in an active Er-doped phosphate glass. Single waveguide excitation at 980 nm is accomplished by fiber butt coupling, and spatial mapping of light propagation along the couplers is achieved from the top of the sample using a CCD camera connected to a microscope, as shown in Figure 13.2d. Light waves at 980 nm are partially absorbed by the Er^{3+} ions, yielding a green upconversion fluorescence which is monitored by the CCD. Since the fluorescence is proportional to the local photon density, the recorded fluorescence patterns map the profile of $|\psi|^2$ in the waveguide reference frame. The onset of CDT is clearly shown in Figure 13.2e after comparison of the light intensity patterns in the straight and curved waveguide couplers: as in the straight coupler light waves tunnel back and forth between the two waveguides, in the sinusoidally curved couplers optical tunneling is suppressed provided that the period Λ and amplitude A are chosen on the quasi-energy crossing manifold. In fact, the fluorescence images taken for the sinusoidally curved couplers clearly show that light remains trapped in the initially-excited (upper) waveguide, following the sinusoidally bent path of its optical axis without tunneling into the adjacent (bottom) waveguide.

13.4   OPTICAL ANALOG OF FIELD-INDUCED BARRIER TRANSPARENCY

In quantum mechanical tunneling, field-induced barrier transparency refers to an extremely large (almost close to 100%) tunneling probability of a free particle to cross a periodically driven potential barrier, in spite of the fact that the probability of tunneling through the static-potential barrier is almost zero [25,51,52]. A simple physical explanation of the field-induced barrier transparency effect can be obtained in the high-frequency limit. In this regime, particle scattering from the rapidly oscillating potential barrier $V(x,t)$ can be described, at leading order, by the scattering from the static potential barrier $V_{av}(x)$ defined by the time-average (over one oscillation cycle) of $V(x,t)$. If the amplitude of the oscillation is large enough, $V_{av}(x)$ shows a double-barrier profile, and barrier transparency then corresponds to resonant tunneling across the equivalent two-barrier system [25]. The optical analog of such a dynamical effect can be realized for photonic tunneling in FTIR through a periodically modulated glass–air–glass interface [32]. Let us consider a double interface which connects a region with a low-index dielectric medium (e.g., air) with two regions of a high-index medium with refractive index $n_s = \sqrt{\varepsilon_s}$ (e.g., glass). The two interfaces are assumed to be periodically curved along the z direction, with period Λ, as shown in Figure 13.3a. Assuming invariance along the y direction, the relative dielectric permittivity can be written as $\varepsilon(x,z) = \varepsilon_s - \Delta\varepsilon(x,z)$ with $\Delta\varepsilon(x,z) = f(x - x_0(z))$, where $x_0(z)$ is a periodic function describing the curvature profile of the interfaces and $f(x) > 0$ describes the profile of refractive index lowering in the dip; for example, a sharp discontinuity of refractive index due to a thin layer of air of width $2w$, one has $f(x) = 0$ for $|x| > w$ and $f(x) = \varepsilon_s - 1$ for $|x| < w$. We assume, for the sake of definiteness, that a TE plane wave ($E_x = E_z = H_y = 0$) at frequency ω is incident upon the double interface at an angle θ_i, producing diffracted fields that propagate in both the forward and the backward directions (see Figure 13.3a). From Maxwell's equations, the electric and magnetic field components along the y and z axes satisfy the coupled equations:

$$\frac{\partial E_y}{\partial x} = i\omega\mu_0 H_z \qquad \frac{\partial H_z}{\partial x} = \frac{i}{\omega\mu_0}\frac{\partial^2 E_y}{\partial z^2} + i\omega\varepsilon_0\varepsilon(x,z)E_y. \tag{13.7}$$

Owing to the periodicity of ε, the determination of the reflected and transmitted fields can be done using an expansion technique commonly used, as an example, in the theory of deep gratings [53]. The solution to Equation 13.7 can be expanded as

$$E_y(x,z) = \exp(ik_z z) \sum_{n=-\infty}^{\infty} [a_n^+(x)\exp(iQ_n x) + a_n^-(x)\exp(-iQ_n x)]\exp(2\pi i n z/\Lambda) \tag{13.8}$$

$$H_z(x,z) = \exp(ik_z z) \sum_{n=-\infty}^{\infty} [Q_n/(\mu_0\omega)][a_n^+(x)\exp(iQ_n x) - a_n^-(x)\exp(-iQ_n x)]\exp(2\pi i n z/\Lambda),$$

$$\tag{13.9}$$

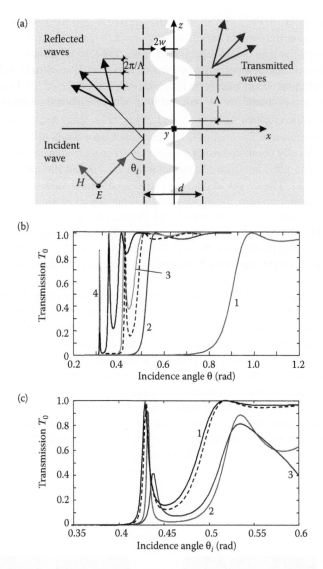

FIGURE 13.3 The optical analog of field-induced barrier transparency in photonic tunneling across a modulated glass–air–glass interface. (a) Scattering geometry for resonant FTIR. (b) Zero-order power transmission versus incidence angle in a sinusoidally curved air–glass interface ($n_s = 1.5$) for a few values of modulation depth A. Curve 1: $A = 0$ (flat interfaces); curve 2: $A = 0.8\lambda$; curve 3: $A = 1.2\lambda$; curve 4: $A = 2\lambda$. The other parameters are: $w = \lambda/2$ and $\Lambda = 0.3\lambda$. (c) Power transmission versus incidence angle for a few values of modulation period Λ. Curve 1: $\Lambda = 0.2\lambda$; curve 2: $\Lambda = 0.5\lambda$; curve 3: $\Lambda = \lambda$. Other parameters are: $w = 0.5\lambda$ and $A = 1.2\lambda$. The dashed curves in the figures are the power transmission curves predicted assuming a cycle-average dielectric permittivity $\varepsilon_{av}(x)$ for $A = 1.2\lambda$ in (b) and $\Lambda = 0.2\lambda$ in (c).

where $k_z \equiv (\omega/c_0)n_s\cos\theta_i$ is the wave vector component of the incident field along the z direction, c_0 is the speed of light in vacuum, and $Q_n \equiv [(\omega/c_0)^2\varepsilon_s - (k_z + 2\pi n/\Lambda)^2]^{1/2}$ are the transverse wave numbers of scattered waves ($\mathrm{Im}(Q_n) > 0$ for evanescent waves). $a_n^+(x)$, $a_n^-(x)$ are the amplitudes of forward and backward waves, which satisfy the coupled-mode equations:

$$\frac{da_n^\pm}{dx} = \mp\frac{i}{2Q_n}\left(\frac{\omega}{c_0}\right)^2\sum_m \Delta\varepsilon_{n-m}(x) \times \left\{a_m^\pm \exp[i\pm(Q_m - Q_n)x] + a_m^\mp\exp[\mp i(Q_m + Q_n)x]\right\} \quad (13.10)$$

In Equation 13.10, $\Delta\varepsilon_n(x) = (1/\Lambda)\int_0^\Lambda \Delta\varepsilon(x,z)\exp(-2\pi inz/\Lambda)$ are the Fourier components of $\Delta\varepsilon$, and propagative waves correspond to indices n for which $|k_z + 2\pi n/\Lambda| < (\omega/c_0)\varepsilon_s$. When the modulation period Λ is smaller than the wavelength $\lambda = 2\pi c_0/\omega$ of the incident wave, or in the paraxial regime discussed below, higher order diffracted waves are of small amplitude and, at leading order, transmission and reflection occurs mainly on the zero-order terms $a_0^\pm(\pm d/2)$. In such cases, at leading order the scattering problem can be reduced to a one-dimensional problem ruled out by the cycled-average dielectric constant $\varepsilon_{av}(x) = \varepsilon_s - (1/\Lambda)\int_0^\Lambda \Delta\varepsilon(x,z)\,dz$. As an example, Figure 13.3b shows the power transmission $T_0 = |a_0^+(d/2)|^2$ versus incidence angle θ_i for the zero-order transmission wave, numerically computed using a scattering matrix technique to solve Equation 13.10, for the case of a thin air–glass interface of constant width $2w = \lambda$ and sinusoidally curved according to $x_0(z) = A\sin(2\pi z/\Lambda)$. The curves in the figure refer to different values of modulation depth A. Note that, as A increases from zero, the incident angle (limit angle) at which total internal reflection (TIR) occurs decreases and, when A becomes larger than w, one or more transmission peaks appear in the TIR region. The physical origin of these resonant peaks may be understood by considering the one-dimensional scattering problem for the cycled-averaged dielectric constant $\varepsilon_{av}(x)$: for A smaller than w, $\varepsilon_{av}(x)$ shows a single dip which broadens as A increases but decreases in amplitude, thus explaining the lowering of incident angle for TIR; however, when A becomes larger than w, $\varepsilon_{av}(x)$ shows two dips, separated by $\simeq 2A$. This corresponds to the existence of an effective double air-gap structure, which shows Fabry–Perot-like transmission resonances at incident angles θ_i satisfying the constructive interference condition $2Ak_x \simeq l\pi$, where $k_x = (\omega/c_0)n_s \sin\theta_i$ and l is an integer number. This phenomenon is fully equivalent to the resonant tunneling in quantum mechanics. We checked our explanation of transmission peaks as a resonant tunneling effect by comparing the exact transmission curve calculated from the two-dimensional scattering problem Equation 13.10 with the curve calculated for the cycle-averaged $\varepsilon_{av}(x)$ problem (compare curve 3 and the dashed curve in Figure 13.3b). It should be noted that, as the period Λ increases, higher order transmission terms $T_n = |a_n^+(d/2)|^2$ may become nonnegligible and the zero-order transmission curve $T_0(\theta_i)$ may largely deviate from the one predicted by the cycled-averaged model, however, resonances are still visible (see Figure 13.3c). Resonant FTIR tunneling can be observed as well using light beams with finite transverse size. As an example, resonant transmission of a Gaussian beam along a sinusoidally curved interface is shown in Figure 13.4 for grazing incidence and for an interface with

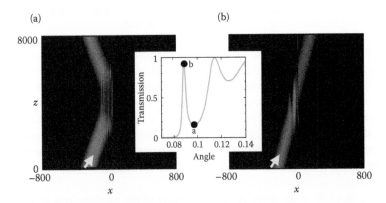

FIGURE 13.4 Photonic tunneling enhancement for a monochromatic Gaussian beam (full width at half maximum FWHM $\simeq 235\lambda$) on a sinusoidally curved FTIR interface ($A = 4\lambda$, $n_s = 1.5$, $w = \lambda$, $\Delta\varepsilon = 0.0891$ and $\Lambda = 4\lambda$) for two different incidence angles. In (a) $\theta_i = 0.0970$ rad, and in (b) $\theta_i = 0.0890$ rad. The inset is the power transmission curve showing a resonant peak at $\theta_i \simeq 0.0890$. The spatial variables x and z are in units of wavelength λ.

low contrast index. Note that, when resonant tunneling occurs (Figure 13.4b), a high intensity and narrow light stripe is visible at the interface, corresponding to Fabry–Perot light trapping associated to high transmission. The case of low contrast index ($\Delta\varepsilon \ll \varepsilon_s$) and of grazing incidence ($\theta_i \to 0$), such as that shown in Figure 13.4, is particularly relevant in order to clarify the quantum–optical analogy of FTIR at periodically curved interfaces with the above-mentioned field-induced barrier transparency. For a period Λ sufficiently larger than the wavelength λ, a paraxial model for light propagation along the interface (z axis) can be used. After setting $E_y = \psi(x,z)\exp[i(\omega/c_0)n_s z]$, the paraxial wave equation for the field amplitude ψ reads

$$i\hbar\frac{\partial\psi}{\partial z} = -\frac{\hbar^2}{2n_s}\frac{\partial^2\psi}{\partial x^2} + V(x - x_0(z))\psi, \tag{13.11}$$

where $\hbar = \lambda/(2\pi)$ and $V(x) = \Delta\varepsilon(x)/(2n_s)$. Note that, in this limit, the low-index medium embedded between the high-index medium acts as a photonic barrier whose axis is periodically curved. Equation 13.11 is formally analogous to the Schrödinger equation describing tunneling of a charged particle across a potential barrier $V(x)$, subjected to an external AC field, written in the Kramers–Henneberger reference frame [25]. The resonant transmission observed in FTIR with periodic modulation of the interfaces is thus fully equivalent to field-induced barrier transparency in quantum tunneling [25]—though the probability of tunneling in a static barrier may be almost zero, an external high-frequency AC field may render the barrier transparent at some energies owing to the appearance of quasibound (long-lived) states for the cycle-averaged barrier $V_{av}(x)$. Experimental realization of photonic tunneling enhancement due to field-induced barrier transparency has been not reported yet, however demonstration of this phenomenon with light may be easier than for matter waves [52] in truly quantum systems.

13.5   COHERENT PHOTONIC TUNNELING BY ADIABATIC PASSAGE

The possibility of coherently transporting in space a quantum particle among a chain of quantum wells is an interesting example of quantum tunneling control [33,34,54–57]. The transport scheme, which is referred to as "coherent tunneling adiabatic passage" (CTAP), is based on dynamical tuning of the tunneling rates between adjacent quantum units by changing either the distance or the height of the neighboring potential wells following a counterintuitive scheme which is reminiscent of the celebrated stimulated Raman adiabatic passage (STIRAP) technique [58,59], originally developed for transferring population between two long-lived atomic or molecular energy levels optically connected to a third auxiliary state. The application of CTAP for controlling photonic tunneling and for light transfer among evanescently coupled optical waveguides has been theoretically proposed and experimentally demonstrated in a series of recent works [38–42]. To understand the physical principle underlying photonic CTAP, it is worth recalling the problem of coherent population transfer in a three-level atomic systems by means of two delayed optical pulses [58,59]. Let us consider three energy levels $|1\rangle$, $|2\rangle$, and $|3\rangle$ of an atomic or molecular system, where $|1\rangle$ and $|2\rangle$ are two metastable states optically connected to an auxiliary excited state $|3\rangle$. The optical transition $|1\rangle \leftrightarrow |2\rangle$ is assumed to be electric-dipole forbidden. Assume that initially the population occupies level $|1\rangle$, and we like to completely transfer the population to level $|2\rangle$ by use of two laser pulses, the pump (P) and Stokes (S) pulses, which are in resonance with the two electric-dipole allowed transitions $|1\rangle \leftrightarrow |3\rangle$ and $|3\rangle \leftrightarrow |2\rangle$, respectively, as shown in Figure 13.5. The condition of resonance can be partially relaxed, however, it is fundamental that the two-photon resonance condition is met [58]. If $\Omega_P(t)$ and $\Omega_S(t)$ denote the Rabi frequencies of pump and Stokes pulses, which are assumed to be not chirped, in the rotating-wave approximation the evolution equations for the

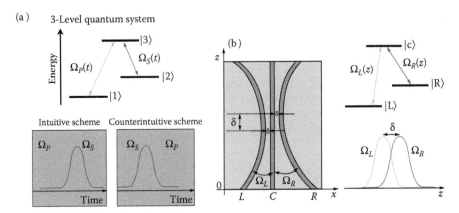

FIGURE 13.5 (a) Coherent population transfer in a three-level atomic system via coherent interaction with two resonant optical pulses (the pump and Stokes pulses) with Rabi frequencies $\Omega_P(t)$ and $\Omega_S(t)$. STIRAP corresponds to the counterintuitive pulse sequence order shown in the right panel. (b) Coherent tunneling by adiabatic passage in a triplet optical waveguide structure that mimics atomic STIRAP. Ω_L and Ω_R are the coupling rates between adjacent waveguides, which play the same role as the pump and Stokes laser pulses of atomic STIRAP. Time is replaced by the spatial propagation distance z.

amplitude probabilities c_1, c_2, and c_3 of the three levels read [58]

$$i\frac{dc_1}{dt} = -\frac{\Omega_P(t)}{2}c_3 \tag{13.12}$$

$$i\frac{dc_2}{dt} = -\frac{\Omega_S(t)}{2}c_3 \tag{13.13}$$

$$i\frac{dc_3}{dt} = -\frac{\Omega_P(t)}{2}c_1 - \frac{\Omega_S(t)}{2}c_2. \tag{13.14}$$

In the above equations, we assumed coherent interaction between the atom and the laser fields, neglecting dephasing effects and decay of level $|3\rangle$. The latter effect may be simply included, if needed, by adding a phenomenological term $-(\gamma/2)c_3$ on the right-hand side of Equation 13.14, where γ is the decay rate of the level $|3\rangle$ to other levels of the atom. Initial conditions, corresponding to a population in level $|1\rangle$, are $c_1(0) = 1$ and $c_2(0) = c_3(0) = 0$. A possible scheme to transfer population from level $|1\rangle$ to level $|2\rangle$ consists of sending first a pump pulse $\Omega_P(t)$ of area π (i.e., $\int dt\Omega_P(t) = \pi$), which transfers population from level $|1\rangle$ to the intermediate level $|3\rangle$, and then a Stokes pulse $\Omega_S(t)$ of area π which transfers population from the intermediate state $|3\rangle$ to level $|2\rangle$ [left panel in Figure 13.5a]. Such a pulse scheme, which is referred to as the *intuitive* pulse sequence scheme, suffers from several drawbacks, such as the need to carefully control the area of the pulses and the detrimental effect of possible decay of intermediate level $|3\rangle$, which is populated during the transfer process [58,59]. A powerful technique to overcome such limitations is represented by STIRAP, in which the temporal sequence of pump and Stokes pulses is reversed, as indicated in the right panel of Figure 13.5a. In such a pulse sequence scheme, which is referred to as the *counter-intuitive* pulse scheme, population transfer is robust against changes in the shape or area of the pulses and even against decay of the intermediate level $|3\rangle$, because such a level is not populated during the transfer process. STIRAP requires a certain temporal overlapping between pump and Stokes pulses and, very important, the evolution of system Equations 13.12 through 13.14 must be adiabatic. Population transfer via STIRAP is based on the existence of an adiabatic dark state (also called trapped state) for Equations 13.12 through 13.14, given by

$$c_1 = \frac{\Omega_S}{\sqrt{\Omega_P^2 + \Omega_S^2}}, \quad c_2 = \frac{-\Omega_P}{\sqrt{\Omega_P^2 + \Omega_S^2}}, \quad c_3 = 0. \tag{13.15}$$

For the counterintuitive pulse scheme, at $t \to -\infty$ one has $\Omega_P(t)/\Omega_S(t) \to 0$, and therefore, the dark state Equation 13.15 corresponds to $c_1 = 1$, $c_2 = c_3 = 0$, that is, to the population on level $|1\rangle$. At $t \to \infty$, one has conversely $\Omega_S(t)/\Omega_P(t) \to 0$, and therefore the dark state Equation 13.15 corresponds to $c_1 = c_3 = 0$, $c_2 = -1$, to the population on level $|2\rangle$. If the conditions for adiabatic evolution are satisfied, during the sequence of Stokes and pump pulses the atomic system evolves remaining in its dark state, and therefore population is fully transferred from level $|1\rangle$ to level $|2\rangle$ without never exciting the intermediate level $|3\rangle$. The condition for adiabatic evolution may be roughly expressed in saying that the temporal overlap interval $\Delta\tau$ of pump and Stokes pulses should be much larger that the maximum value of $\Omega_{\text{eff}} = \sqrt{\Omega_P^2 + \Omega_S^2}$ [58]. This simplest STIRAP protocol of atomic physics can be applied to photonic tunneling by adiabatic passage in a triplet waveguide system, as shown in Figure 13.5b. The optical system here simply consists of three evanescently coupled optical waveguides L, C and R in the geometry depicted on the left panel of Figure 13.5b. A central straight waveguide C, which plays the role of the intermediate atomic state $|3\rangle$, is side coupled to two circularly and oppositely curved left L and right R waveguides, which are displaced along the longitudinal z axis by a distance δ. Therefore, the minimum distance ρ between waveguides L and C, and between waveguides R and C, are reached at propagation distances z which are displaced by δ, as shown on the right panel of Figure 13.5b. Our objective is to completely transfer light between the outer waveguides L and R, a problem fully analogous to the previously discussed atomic population transfer. Indeed, coupled mode equations describing light transfer among the three waveguides are analogous to the STIRAP equations 13.12 through 13.14, where z plays the role of time and the Rabi frequencies of pump and Stokes pulses are played by the coupling amplitudes $\Omega_L(z)$, $\Omega_R(z)$ between adjacent waveguides [see Figure 13.5b, right panel]. Note that, as the distances between waveguides L-C and C-R change along the propagation distance z, the coupling amplitudes $\Omega_L(z)$ and $\Omega_R(z)$ correspondingly describe two bell-shaped and equal waveforms, shifted each other by δ, which mimic pump and Stokes laser pulses. If waveguide R is excited in its fundamental mode at the input plane $z = 0$, as Ω_L precedes Ω_R the equivalent sequence of pump and Stokes pulses follows a counterintuitive scheme, and light is fully transferred to waveguide L with negligible excitation of the central waveguide C, as shown on the left panel of Figure 13.6. Conversely, if waveguide L is excited in its fundamental mode at the input plane $z = 0$, the intuitive sequence of pump and Stokes pulses is now mimicked, and transfer among the three waveguides follows a rather involved path which is shown in the right panel of Figure 13.6. The light transfer mechanism shown in the left panel of Figure 13.6, in which light transfer between two outer waveguides is possible via an intermediate (but never excited) waveguide, is the optical analog of atomic STIRAP, and it is therefore relatively tolerant against moderate changes in the waveguide geometry.

Extensions of photonic tunneling by adiabatic passage in coupled optical waveguides, which mimics coherent population transfer and related phenomena of atomic physics, include the optical analogs of multilevel population adiabatic passage [39], straddle STIRAP [41], Raman chirped adiabatic passage [60], and population transfer via a continuum [43,61]. Here we limit to briefly discuss photonic tunneling between two channel waveguides via a continuum. This is a rather intriguing and somehow unexpected possibility of coherent photonic transport which is closely related to adiabatic population transfer between two bound states of an atomic or molecular system via a *continuum* of states. The extension of the STIRAP technique, in which the discrete intermediate state is replaced by a continuum of states, was initially proposed by Carroll and Hioe [62] and was followed by a lively debate on its practical feasibility, until the very recent experimental demonstration of STIRAP via continuum reported by Peters et al. [63]. Similarly to STIRAP in the continuum, photonic

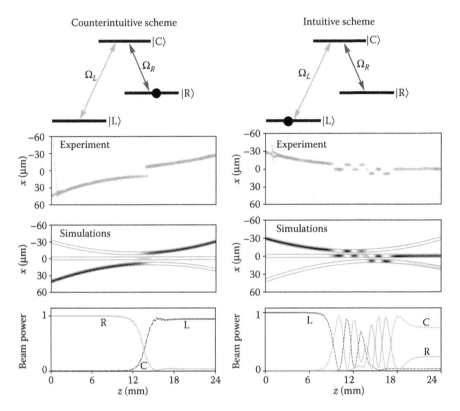

FIGURE 13.6 Experimental demonstration of coherent photonic tunneling by adiabatic passage in a triplet waveguide system. Beam propagation simulations and measured fluorescence images are depicted that show light transfer under beam excitation mimicking a counterintuitive (STIRAP) pulse sequence (left panel) and an intuitive pulse sequence (right panel).

tunneling via a continuum of states is based on the existence of a trapped (or dark) state embedded in the continuum, arising from a destructive Fano-like interference, and in the adiabatic evolution of the system in this state. A paradigmatic optical system to realize photonic tunneling via a continuum is shown in Figure 13.7a and consists of a couple of two equal channel waveguides W_1 and W_2 which are evanescently-coupled to a slab waveguide S supporting a continuous set of modes. The fundamental waveguide modes of channel waveguides represent, in the quantum–optical analogy, two discrete energy levels $|1\rangle$ and $|2\rangle$ (with the same energy) coupled to the continuous set of modes $|k\rangle$ supported by the slab waveguide. Indicating by β_0 the propagation constant shift of the fundamental mode of the channel waveguides from the reference value $2\pi n_s/\lambda$ (n_s is the bulk refractive index of the medium) and expanding the electric field envelope as a superposition of eigenmodes of channel $[u_{1,2}(x,y)]$ and slab $[u_k(x,y) = (2\pi)^{-1/2}q(y)\exp(ikx)]$ waveguides with z–dependent coefficients according to

$$\psi(x,y,z) \simeq \left[a_1(z)u_1(x,y) + a_2(z)u_2(x,y) + \int dk\, b(k,z)u_k(x,y) \right] \exp(-i\beta_0 z), \qquad (13.16)$$

the following coupled-mode equations can be derived for the modal amplitudes [43]

$$i\dot{a}_1 = \int dk\, g_1(k)b_k \qquad (13.17)$$

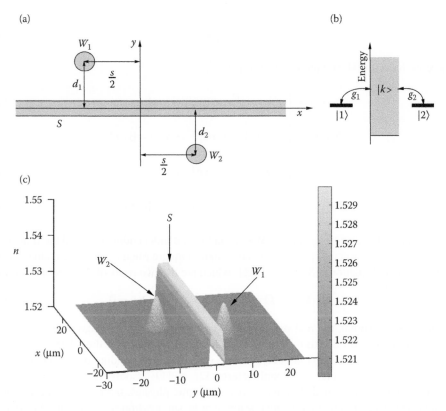

FIGURE 13.7 Photonic tunneling via a continuum in a slab-channel waveguide structure. (a) Schematic of the optical system; (b) quantum mechanical analog; (c) typical behavior of refractive index profile $n(x,y)$ of the structure.

$$i\dot{a}_2 = \int dk g_2(k) b_k \tag{13.18}$$

$$i\dot{b}_k = [\beta(k) - \beta_0]b_k + g_1^*(k)a_1 + g_2^*(k)a_2. \tag{13.19}$$

In the previous equations, k is the wave number in the transverse x direction of slab mode ($|k| \ll 2\pi n_s/\lambda$ in the paraxial approximation) , $\beta(k) = \beta_S + \hbar k^2/(2n_s)$, and $q(y)$, β_S are the bound slab eigenmode and corresponding eigenvalue of the one-dimensional equation $-[\hbar^2/(2n_s)]d^2q/dy^2 + V_{SLAB}(y)q = \hbar\beta_S q$ with the normalization $\int dy|q(y)|^2 = 1$. The coefficients $g_1(k)$ and $g_2(k)$ provide the coupling between channel and slab waveguide modes, and can be expressed in terms of spatial overlapping integrals (for details see Longhi (2008) [43]). Note that the spectrum $\hbar\beta(k)$ is nearly unstructured and bounded from below. To discuss photonic tunneling via the continuum, we assume that the two degenerate bound states $|1\rangle$ and $|2\rangle$ of energy $\hbar\beta_0$ are embedded in the continuum, as schematically shown in Figure 13.7b. This case corresponds to a slab waveguide S with an index change Δn_S smaller that that of the channel waveguides Δn_g, for which $\beta_S < \beta_0$. Starting from coupled-mode equations 13.17 through 13.19, one can eliminate the continuum degrees of freedom by making the Markovian approximation according to a standard procedure well-known in the atomic physics context (see, e.g., Knight et al. (1990) [64]). This yields the following coupled-mode equations for the modal amplitudes a_1 and a_2 of light waves trapped in channel waveguides W_1 and W_2:

$$\dot{a}_1 = -\Omega_{11}a_1 - \Omega_{12}a_2 \tag{13.20}$$

$$\dot{a}_2 = -\Omega_{21}a_1 - \Omega_{22}a_2, \tag{13.21}$$

where we have set

$$\Omega_{ln} = \int_0^\infty d\tau \Phi_{ln}(\tau) \tag{13.22}$$

and where we have introduced the memory functions

$$\Phi_{11}(\tau) = \int dk |g_1(k)|^2 \exp\{-i[\beta(k) - \beta_0]\tau\}, \tag{13.23}$$

$$\Phi_{12}(\tau) = \int dk g_1(k) g_2^*(k) \exp\{-i[\beta(k) - \beta_0]\tau\}, \tag{13.24}$$

$$\Phi_{21}(\tau) = \int dk g_1^*(k) g_2(k) \exp\{-i[\beta(k) - \beta_0]\tau\} \tag{13.25}$$

$$\Phi_{22}(\tau) = \int dk |g_2(k)|^2 \exp\{-i[\beta(k) - \beta_0]\tau\}. \tag{13.26}$$

For straight channel waveguides W_1 and W_2, d_1 and d_2 are independent of z and the transfer properties of light waves between W_1 and W_2 via the continuum are ruled by the eigenvalues σ_1 and σ_2 of the linear system Equations 13.20 and 13.21, which are the two roots of the algebraic equation

$$\sigma^2 + (\Omega_{11} + \Omega_{22})\sigma + \Omega_{11}\Omega_{22} - \Omega_{12}\Omega_{21} = 0. \tag{13.27}$$

From a physical viewpoint, the real parts of Ω_{11} and Ω_{22} represent the decay rates of light waves in waveguides W_1 and W_2 due to their coupling with the continuum S if the other waveguide were absent, whereas the imaginary parts of Ω_{11} and Ω_{22} are the corresponding decay-induced shifts of the mode propagation constants. The coupling terms Ω_{12} and Ω_{21} modify the decay rates and, under special circumstances, a trapped state may exist. In atomic physics, the existence of a trapped (or dark) state corresponds to ionization suppression, that is, suppression of decay into the continuum, via a Fano-like destructive interference effect (see, e.g., Knight et al. (1990) [64]). A trapped state obviously exists whenever the real part of one of the two eigenvalues σ_1 and σ_2 vanishes, a condition which is met under special circumstances. For the optical system of Figure 13.7a, one can show [43] that a trapped state in the continuum, corresponding to $\sigma_1 = 0$ and given by

$$a_1 = \frac{\Omega_{12}}{\sqrt{|\Omega_{11}|^2 + |\Omega_{12}|^2}}, \tag{13.28}$$

$$a_2 = -\frac{\Omega_{11}}{\sqrt{|\Omega_{11}|^2 + |\Omega_{12}|^2}}, \tag{13.29}$$

does rigorously exist whenever the horizontal spacing s of waveguides W_1 and W_2 [defined in Figure 13.7a] vanishes, or approximately for $k_0 s = \pi, 2\pi, 3\pi, \ldots$, where the wave number k_0 satisfies the phase-matching condition $\beta(k_0) = \beta_0$—that is,

$$k_0 = \sqrt{\frac{2n_s(\beta_0 - \beta_s)}{\hbar}}. \tag{13.30}$$

Suppose now that the distances d_1 and d_2 of waveguides W_1 and W_2 from the slab waveguide are slowly varied along the propagation distance z following a path that mimics the counterintuitive sequence of pump and Stokes pulses of atomic STIRAP. In practice this can be simply achieved by slightly curving the channel waveguides W_1 and W_2 in the (y, z) plane, as previously described for the three-level CTAP. Under adiabatic evolution conditions, according to Equations 13.28 and 13.29 light injected into channel waveguide W_1 is fully transferred into waveguide W_2 at the output plane. Taking into account nonadiabatic effects or in case of a nearly trapped state, some light may be irreversibly lost into the continuum and the transfer is, therefore, not complete. Numerical simulations based on beam propagation of the scalar wave equation for the channel-slab waveguide photonic structure confirms the possibility of coherent tunneling by adiabatic passage via a continuum. Examples of efficient adiabatic light transfer, based on the existence of a dark state embedded in the continuum, are shown in Figure 13.8. Similar results are predicted when the two channel waveguides W_1 and W_2 are indirectly coupled via a different kind of continuum, such as via a waveguide array which mimics a tight-binding continuum [61]. An experimental demonstration of efficient light transfer via a tight-binding continuum, which enables to visualize the coherent adiabatic transfer process, has been recently reported in Dreisow et al. (2009) [65] using femtosecond laser written waveguides.

13.6   PHOTONIC TUNNELING IN OPTICAL LATTICES: OPTICAL ANALOG OF THE QUANTUM ZENO EFFECT

Photonic tunneling in optical lattices has been recently proposed [44,45] and experimentally demonstrated [19,22] to visualize with light waves some basic phenomena of quantum mechanical decay, such as nonexponential decay features due to non-Markovian effects, and to provide an optical analog of macroscopic quantum tunneling and of decay control by frequent observations of the system. Control of quantum mechanical decay of a discrete state $|\chi\rangle$ coupled to a continuum of states $|\omega\rangle$ (reservoir) represents a rather general issue in different areas of physics. A typical example of such a decay process is represented by spontaneous emission of excited-state atomic hydrogen in vacuum. Contrary to a common belief, the survival probability $P(t)$ of a decaying unstable state $|\chi\rangle$ always deviates from a pure exponential decay law both at short time scale, where the decay is always parabolic, and at long time scale, where the decay is algebraic [66]. Such deviations are due to the spectral boundness from below of the continuum and are a signature of a memory effect, that is, of non-Markovian dynamics. Short-time deviations of the decay law led to the prediction that frequent observations of the system can either decelerate or accelerate its decay (quantum Zeno and anti-Zeno effects [67–70]). In particular, there exists a characteristic time τ_Z (the Zeno time) such that frequent observations of the system at successive time intervals $\tau < \tau_Z$ decelerate the decay process, whereas observations at time intervals $\tau > \tau_Z$ leads to deceleration of the decay (anti-Zeno effect) [70]. The relevant Hamiltonian H that describes the decay of the discrete state $|\chi\rangle$ into a continuum of states $|\omega\rangle$ can be written as $H = H_0 + H_I$, where [69,70]

$$H_0 = \hbar\omega_\chi|\chi\rangle + \hbar\int d\omega\,\omega|\omega\rangle\langle\omega| \tag{13.31}$$

is the free Hamiltonian, $\hbar$ is the Planck's constant,

$$H_I = \hbar\int d\omega\,[g(\omega)|\omega\rangle\langle\chi| + H.c.] \tag{13.32}$$

is the interaction Hamiltonian, $g(\omega)$ is the spectral coupling function, $\hbar\omega_\chi$ and $\hbar\omega$ are the energies of the discrete and continuous states $|\chi\rangle$ and $|\omega\rangle$ (see Figure 13.9a). If the system is initially prepared in the discrete state $|\chi\rangle$, the survival probability $P(t)$ can be calculated as $P(t) = |c_\chi(t)|^2$, where the amplitude c_χ satisfies the integro-differential equation

$$\frac{dc_\chi}{dt} = -\int_0^t dt'\,c_\chi(t')\Phi(t-t')\exp[i\omega_\chi(t-t')], \tag{13.33}$$

and $\Phi(\tau) = \int d\omega|g(\omega)|^2\exp(i\omega\tau)$ is the reservoir (memory) response function. The usual exponential decay rule $P(t) = \exp(-Rt)$, with a decay rate R given by $R = 2\pi|g(\omega_\chi)|^2$ according to the golden rule result, is attained in the Markovian limit of Equation 13.33, where memory effects are neglected. The Zeno time τ_Z can be estimated by a simple geometric construction from the intersection of the exact decay law with the approximate exponential one [70], as shown in Figure 13.9b.

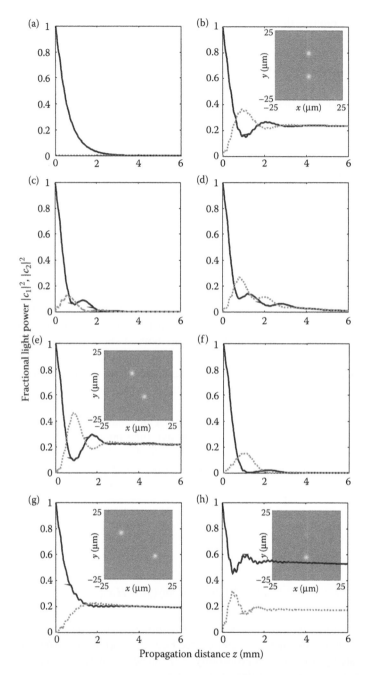

FIGURE 13.8 Photonic tunneling via a continuum in a slab-channel waveguide system with refractive index change $\Delta n_S = 0.005$ and $\Delta n_g = 0.009$ (see Figure 13.7c for the precise refractive index profile). The figures show the behavior of fractional light powers trapped in the channel waveguides W_1 (solid curves) and W_2 (dotted curves) versus propagation distance z. In (a) $d_1 = 8\,\mu m$ and waveguide W_2 is very far from the slab waveguide; note that the decay law $|a_1(z)|^2$ is almost exponential according to the Markovian approximation. In (b)–(g) $d_1 = d_2 = 8\,\mu m$ and the behavior of $|c_1(z)|^2$ and $|c_2(z)|^2$ is shown at increasing values of horizontal displacement s: $s = 0$ in (b); $s = 6\,\mu m$ in (c); $s = 9\,\mu m$ in (d); $s = 12.2\,\mu m$ in (e); $s = 18\,\mu m$ in (f); and $s = 24.4\,\mu m$ in (g). Note the incomplete decay in (b), (e) and (g), which is related to the existence of a trapped state. Panel (h) shows the existence of an unbalanced trapped state for $s = 0$, $d_1 = 8\,\mu m$ and $d_2 = 7\,\mu m$. The insets in figures (b), (e), (g), and (h) depict the transverse profile of the field intensity $|\psi|^2$ at the $z = 6\,cm$ output plane.

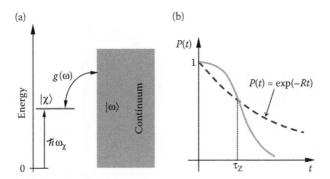

FIGURE 13.9 Quantum mechanical decay of an unstable discrete state coupled to a continuum of state (reservoir). (a) Energy level diagram. (b) Behavior of the survival probability $P(t)$ (solid curve), showing deviations from the exponential decay (dashed curve) predicted in the Markovian approximation. The intersection τ_Z of the two curves defines the Zeno time.

Frequent observations of the system, on a time scale τ shorter than τ_Z, can decelerate the decay process. Unfortunately, for most of quantum mechanical decay processes in microscopic physics the Zeno region is too short (for instance, for spontaneous emission of atomic hydrogen in vacuum the atom should be observed in the sub-femtosecond time scale [71]), and observation-induced decay acceleration (anti-Zeno effect) is much more ubiquitous, as shown by Kofman and Kurizki in a rather general way [69]. To date, there are few experimental evidences of nonexponential decay features and measured-induced acceleration or deceleration of the decay in truly unstable quantum mechanical systems, all based on quantum tunneling of cold atoms in tilted or accelerated optical lattices [35,72]. Owing to the similarity between macroscopic quantum tunneling of matter waves in optical lattices and photonic tunneling in waveguide lattices, the latter systems may provide an easily accessible and engineerable laboratory to visualize with light waves the dynamical aspects embodied in quantum tunneling phenomena and related Zeno or anti-Zeno dynamics. Nonexponential decay features and Zeno effect for photonic tunneling has been recently proposed in Longhi (2006) [44], followed by an experimental demonstration [22]. The optical system used to demonstrate the optical analog of the quantum Zeno effect consists of a straight optical waveguide W, side-coupled to a semi-infinite array of waveguides S in the geometry depicted in Figure 13.10a. The waveguide W is the optical analog of the discrete state $|\chi\rangle$, whereas the semiarray S plays the role of a tight-binding continuum $|\omega\rangle$, into which $|\chi\rangle$ decays owing to evanescent wave coupling. If waveguide W is excited in its fundamental mode at the input plane $z = 0$, the fractional beam power $P(z)$ trapped in waveguide W thus decays along the propagation distance owing to light leakage into the semiarray S. As the continuum is usually spectrally narrow and bounded from both below and above, non-Markovian effects and nonexponential features if decay may be strong. In order to reproduce the optical analog of the quantum Zeno effect, one can adopt the array configuration shown in Figure 13.10b, in which the waveguide W is alternately side-coupled to semi-infinite arrays S_1, S_2, S_3,... of finite length τ. Each interruption basically mimics a collapse of the wave function, that is, an ideal quantum measurement according to von Neumann's postulate of quantum mechanics. In fact, after each cut of a semiarray—for instance, of semiarray S_1, light trapped in S_1 is transferred into the bulk substrate over a short propagation distance few hundreds microns) and thus after each cut the memory in the continuum is erased. Quantitative measurements of light transport along the photonic systems, either in the configurations of Figure 13.10a or b, have been performed by scanning tunneling optical microscopy [22], and typical results are shown in Figure 13.10c and d. From the maps of light intensity evolution, one can determine the survival probability of photons in waveguide W for the two structures, which is shown in Figure 13.11. Note that, in presence of

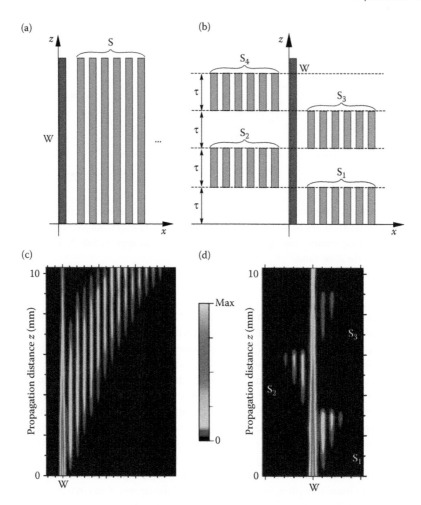

FIGURE 13.10 The optical analog of the quantum Zeno effect based on photonic tunneling in waveguide lattices. (a) and (b) Schematic of the photonic structures. In (a) light waves in a waveguide W decay owing to evanescent coupling to a semi-infinite array S. In (b) the decay process is repeatedly interrupted by alternations of the semiarray, which mimic an ideal quantum measurement with the collapse of the wave function. For $\tau < \tau_Z$ deceleration of the decay in waveguide W is expected. (c) and (d) Experimental results showing light decay maps in the photonic structures as measured by scanning tunneling optical microscopy.

interruptions mimicking system measurements, the decay is slowed down—that is, an optical analog of the quantum Zeno effect is realized.

A related issue is that of dynamical control of quantum mechanical decay and decoherence based on modulation of the coupling g to the continuum [36], which generalizes the concept of "frequent observation" of a system. The decay rate, in this case, deviates from its "olden rule" value R owing to an interplay between coupling modulation and non-Markovian character of the decay. Such a dynamical control scheme can be demonstrated for optical tunneling in the waveguide configuration of Figure 13.12, where the boundary waveguide W is periodically curved in such a way that the distance between W and S (and hence the coupling strength g) varies with propagation distance z [45]. Light transfer among the waveguides in the structure of Figure 13.12 can be described by the following coupled-mode equations for the mode amplitudes c_n of the various

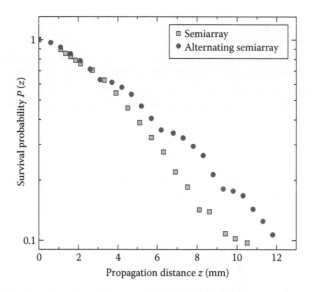

FIGURE 13.11 Survival probability to find a photon in waveguide W versus propagation distance z as obtained from the experimental maps of Figures 13.10c and d. Deceleration of the photonic decay is clearly observed in the alternating semiarray configuration.

waveguides [45]

$$i\dot{c}_0 = -\Delta_0(z)c_1 + \sigma c_0, \quad i\dot{c}_1 = -\Delta_0(z)c_0 - \Delta c_2,$$
$$i\dot{c}_n = -\Delta(c_{n+1} + c_{n-1}) \quad (n \geq 2) \tag{13.34}$$

where Δ is the hopping amplitude between adjacent waveguides in the array, $n = 0$ corresponds to the boundary waveguide, σ accounts for different propagation constants between the boundary waveguide and the array waveguides, $\Delta_0(z)$ is the z-varying coupling rate of the boundary waveguide with its adjacent waveguide, and the dot stands for the derivative with respect to z. Similarly to the previously discussed structure, Equation 13.34 describes the coupling of an unstable discrete state with a tight-binding continuum possessing an energy band of width $4\hbar\Delta$, $\hbar\sigma$ being the detuning of the energy of the discrete state from the center of the continuous band and z playing the role of time in the quantum mechanical analog. However, the coupling $\Delta_0(z)$ of the discrete state with the continuum is now modulated according to the bending profile of the boundary waveguide. For a

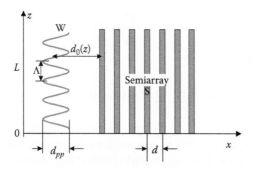

FIGURE 13.12 Schematic of the waveguide lattice to control photonic tunneling of boundary waveguide W to the semiarray S via discrete-to-continuum modulation coupling.

straight waveguide ($\Delta_0 = $ const), a light beam injected into the boundary waveguide tunnels into the array during propagation (discrete diffraction [73]), and complete decay is attained provided that $|\sigma/\Delta| < 2 - (\Delta_0/\Delta)^2$ (see Longhi (2006) [74]). The mechanism of tunneling decay control by modulation of the coupling Δ_0 is at best captured by rewriting Equations 13.34 in the canonical form of Kofman and Kurizki (2001) [36]. To this aim, let us indicate by $\omega = -2\Delta \cos \kappa$ the dispersion relation of the lattice band ($0 \leq \kappa \leq \pi$) and let us introduce, in place of $c_n(z)$ for $n \geq 1$, the continuous amplitude

$$\varphi(\omega, z) = \frac{1}{\alpha(\omega)} \sum_{n=1}^{\infty} c_n(z) \sin(n\kappa), \tag{13.35}$$

where $\alpha(\omega) = (\pi\Delta)^{1/2}[1 - (\omega/2\Delta)^2]^{1/4}$ is a normalization factor. Then it can be easily shown that $c_0(z)$ and $\varphi(\omega, z)$ satisfy the coupled equations

$$i\dot{c}_0 = \sigma c_0 + \varepsilon(z) \int_{-2\Delta}^{2\Delta} d\omega \, g(\omega)\varphi(\omega, z) \tag{13.36}$$

$$i\dot{\varphi} = \omega\varphi + \varepsilon(z)g(\omega)c_0 \tag{13.37}$$

where $\Delta_0(z) = \varepsilon(z)\Delta_{\text{peak}}$, Δ_{peak} is the maximum value of $\Delta_0(z)$, $\varepsilon(z)$ is the normalized modulation profile, and

$$g(\omega) = -\frac{\Delta_{\text{peak}}}{\sqrt{\pi\Delta}} \left[1 - \left(\frac{\omega}{2\Delta}\right)^2\right]^{1/4} \tag{13.38}$$

is the spectral coupling amplitude [$g(\omega) = 0$ for $|\omega| > 2\Delta$]. Equations 13.36 and 13.37 in their present form can be derived from the Hamiltonian $H = H_0 + H_I(z)$, with $H_0 = \hbar\sigma|a\rangle\langle a| + \int d\omega \, \hbar\omega|\omega\rangle\langle\omega|$ and $H_I(z) = \varepsilon(z)\hbar \int d\omega g(\omega)[|\omega\rangle\langle a| + |a\rangle\langle\omega|]$, describing the interaction of the discrete state $|a\rangle$, of energy $\hbar\sigma$, with a band of continuous states $|\omega\rangle$, of energy $-2\hbar\Delta \leq \hbar\omega \leq 2\hbar\Delta$. As compared to Equation 13.32, the interaction $H_I(z)$ is now modulated, and the coupling spectrum is given by $G(\omega) = |g(\omega)|^2$. If the boundary waveguide is excited at the input plane $z = 0$, after setting $c_0(z) = A(z)\exp(-i\sigma z)$ the evolution of the mode amplitude A along the waveguide is ruled by the exact integro-differential equation [36]

$$\frac{dA}{dz} = -\varepsilon(z) \int_0^z dz' \varepsilon(z') A(z') \exp[i\sigma(z - z')] \Phi(z - z'), \tag{13.39}$$

where $\Phi(\tau) = \int d\omega G(\omega)\exp(-i\omega\tau)$ is the reservoir response function. In the weak coupling limit, following the analysis of Kofman and Kurizki (2001) [36] one can derive for the decay law the universal form

$$|c_0(z)|^2 \simeq \exp[-R(z)Q(z)] \tag{13.40}$$

where

$$Q(z) = \int_0^z dz' \varepsilon^2(z') \tag{13.41}$$

is the effective interaction length,

$$R(z) = \int d\omega G(\omega) F_z(\omega - \sigma) \tag{13.42}$$

is the effective decay rate, and $F_z(\omega) = |\int_0^z dz' \varepsilon(z')\exp(-i\omega z')|^2/Q(z)$. For a periodic modulation of $\Delta_0(z)$ with spatial frequency $\Omega = 2\pi/\Lambda$, at propagation lengths z much larger than both Λ and an effective correlation time (see Kofman and Kurizki (2001) [36]), one has $Q(z) \simeq (\sum_n |\varepsilon_n|^2) z$ for the interaction length and

$$R = \frac{2\pi \sum_n |\varepsilon_n|^2 G(\sigma + n\Omega)}{\sum_n |\varepsilon_n|^2} \tag{13.43}$$

for the decay rate, where ε_n are the Fourier coefficients of $\varepsilon(z)$—that is, $\varepsilon(z) = \sum_n \varepsilon_n \exp(in\Omega z)$. Note that, for a nonmodulated waveguide [$\varepsilon(z) = 1$], Equation 13.43 yields for the decay rate the golden-rule result $R = 2\pi G(\sigma)$. In this case one has $R = 0$ for $|\sigma| > 2\Delta$, which corresponds to a regime of fractional decay—analogous to the one previously studied for spontaneous emission of atoms in photonic crystals [75]—and related to the existence of bound surface states [74]. In our optical context, fractional decay means that light initially injected into waveguide W is not fully transferred via evanescently coupling to the semi-array S, rather a fraction of the initial optical power remains trapped in waveguide W and because of the existence of a surface state. When the coupling of the waveguide with the continuum is periodically modulated one expects: (1) a deceleration of the decay rate when $\sigma = 0$; (2) an acceleration of the decay rate when σ is close to the boundary of the continuum, that is, for $\sigma \simeq \pm 2\Delta$, provided that a strong spectral sideband of $\varepsilon(z)$ falls near the band center; (3) suppression of fractional decay when $|\sigma| \gtrsim \sim 2\Delta$. The experimental demonstration of dynamical tunneling control via discrete-to-continuum coupling modulation has been reported in Dreisow et al. (2008) [19]. In the experiment, the waveguides were fabricated by the femtosecond laser writing technique [76] in fused silica glass. While the waveguides in S are processed with a fixed writing velocity, the boundary waveguides W may be written slower. This yields a slightly increased effective refractive index and, thus, a positive shift σ of propagation constant as compared to the waveguides in the semiarray. The excitation of the boundary waveguides was accomplished with fiber butt coupling of a He–Ne laser at $\lambda = 633$ nm. To visualize the flow of light, the waveguides were written in OH-rich fused silica (Suprasil 311) which emits fluorescence light at 650 nm wavelength from color centers created inside the modifications [77]. From the fluorescence pattern, which is imaged onto a CCD camera mounted above the sample, the decay properties for the light power trapped in W can be obtained after removing absorption and internal losses, subtracting background noise and normalizing the amplitudes [19]. As an example, Figure 13.13 shows the suppression of fractional decay of waveguide W into the semi-infinite array S when the distance between W and S is periodically modulated. Conditions leading to either acceleration or deceleration of photonic tunneling have been also demonstrated (see Dreisow et al. (2008) [19] for more details).

13.7 PHOTONIC LANDAU–ZENER TUNNELING

Originally introduced by Landau in the context of atomic scattering processes [78] and by Zener in the study of the electronic properties of a biatomic molecule [79], the Landau–Zener (LZ) transition represents a fundamental nonadiabatic dynamical process which occurs at the intersection of two energy levels that repel each other. Owing to its general character, the LZ model has been encountered in a countless physical fields and systems; we just mention here LZ transitions for Rydberg atoms [80], molecular nanomagnets [81,82], current-driven Josephson junctions [83], cold atoms and Bose–Einstein condensates in accelerated optical lattices [84–86], and Cooper-pair box qubits [87]. Classical analogs of LZ transitions have been also investigated, including polarization rotation in an optical cavity [88,89] and LZ tunneling of light waves in coupled waveguides [90,91]. Extensions of the original LZ model have been studied in the past few years to account for the effects of nonlinearities [92], finite-coupling duration effects [93], multistate dynamics [94,95], or decoherence, noise and dissipation [96–98].

In spite of the vast literature on LZ models, direct observations of time-resolved evolution of level occupancy during LZ transition are rare [88,89]. An optical directional coupler made of two closely spaced waveguides (waveguide spacing d) with a cubically bent axis (see the inset in Figure 13.14a) provides a simple photonic system that enables to visualize LZ dynamics of a two-level system with linear energy level crossing and with constant coupling of finite duration [20,91]. To mimic LZ tunneling with linear crossing of energy levels, let us consider a cubically bent profile

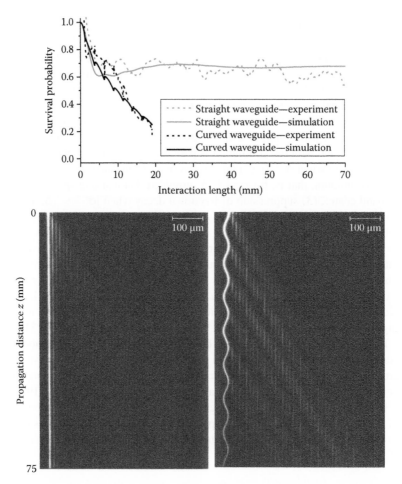

FIGURE 13.13 Experimental demonstration of suppression of fractional decay in the waveguide-semiarray structure of Figure 13.12. The upper figure shows the measured (dotted curves) and simulated (solid curves) survival probabilities of photons in waveguide W versus propagation distance when the distance W-S is constant (W is straight) or modulated (the axis of W is periodically curved). The measured fluorescence light distributions in the two cases is also shown at the bottom.

$x_0(z)$ for the waveguide axis given by

$$x_0(z) = \frac{A}{L_1^3}(z - L_1)^3 - A \tag{13.44}$$

$(0 < z < L)$, where $2A$ $(A \ll L)$ is the full lateral shift of the waveguides between input $(z = 0)$ and output $(z = L)$ planes and L_1 is the position at which the axis curvature $\ddot{x}_0$ vanishes. In the tight-binding approximation, the amplitudes $c_1(z')$ and $c_2(z')$ of light waves trapped in the two waveguides read [91]

$$i\frac{d}{dz'}\begin{pmatrix} c_1 \\ c_2 \end{pmatrix} = \begin{pmatrix} \eta^2 z' & \Omega_0 \\ \Omega_0 & -\eta^2 z' \end{pmatrix}\begin{pmatrix} c_1 \\ c_2 \end{pmatrix}, \tag{13.45}$$

where $z' = z - L_1$ $(-L_1 < z' < L - L_1)$, Ω_0 is the coupling strength between the two waveguides of the coupler, and the parameters η^2 governs the adiabaticity of level crossing and reads explicitly

$$\eta^2 = \frac{3dAn_s}{\hbar L_1^3}. \tag{13.46}$$

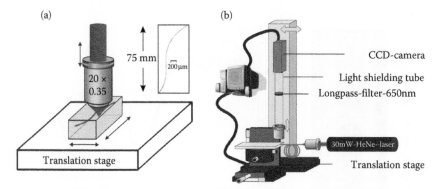

FIGURE 13.14 (a) Schematic of the femtosecond laser writing set-up for the fabrication of the curved waveguide coupler (shown in the inset). (b) Experimental set-up for light fluorescence imaging.

In its present form, Equation 13.45 describes LZ tunneling with linear crossing of energy level, at a rate η^2, and with constant coupling Ω_0 of finite duration L. The solution to Equation 13.45 can be expressed in terms of parabolic cylinder function $D_v(z')$ according to (see Vitanov and Garraway (1996) [93] for details)

$$c_1(z') = aD_v(\sqrt{2}\eta z' e^{-i\pi/4}) + bD_v(\sqrt{2}\eta z' e^{i3\pi/4}) \tag{13.47}$$

$$c_2(z') = \frac{\Omega_0}{\eta\sqrt{2}} e^{-i\pi/4} \left[-aD_{-1+v}(\sqrt{2}\eta z' e^{-i\pi/4}) + bD_{-1+v}(-\sqrt{2}\eta z' e^{-i\pi/4}), \tag{13.48}\right.$$

where we have set $v = i\Omega_0^2/2\eta^2$. The constants a and b entering in Equations 13.47 and 13.48 have to be determined once the initial conditions $c_1(-L_1)$ and $c_2(-L_1)$ are assigned. For excitation of one of the two waveguides at the input plane, the initial conditions read $c_1(-L_1) = 1$ and $c_2(-L_1) = 0$. Moreover, the nearly symmetric LZ tunneling $L_1 \sim L/2$ is typically considered.

The experimental visualization of LZ tunneling in curved directional couplers has been reported in Dreisow et al. (2009) [20]. In the experiment, a set of cubically curved waveguide couplers were manufactured by fs laser microstructuring [76], in which an ultrashort laser pulse is focused with a $20 \times microscope$ objective (0.35 numerical aperture) inside fused silica. By moving the sample transversally to the laser beam (Figure 13.14a) micro-modifications along almost arbitrary curves can be formed. The use of fused silica glass with a high content of silanol leads to massive formation of nonbridging oxygen-hole color centers. When excited with a He–Ne laser at 633 nm wavelength, these color centers emit fluorescence light around 650 nm, which can be conveniently detected from the top of the sample (Figure 13.14b). The recorded fluorescence images were digitally straightened and, from them, the evolution of light power trapped in the two waveguides were retrieved. Typical results are shown in Figure 13.15a and b and compared to theoretical predictions of Equations 13.47 and 13.48. The figures clearly indicate how the detailed LZ dynamics strongly depends on the adiabatic parameter η^2.

13.8 CONCLUSIONS

Optical (or photonic) tunnelling usually refers to light propagation across forbidden photonic barriers, such as in photonic band gap structures or undersized waveguides, to light passage via evanescent coupling in FTIR, or to evanescent field coupling among waveguides. In this chapter, we have reviewed some recent theoretical and experimental advances on photonic tunneling in coupled optical waveguides, that provide an experimentally accessible laboratory tool to visualize classical analogs of several tunneling phenomena encountered in quantum mechanical systems. The

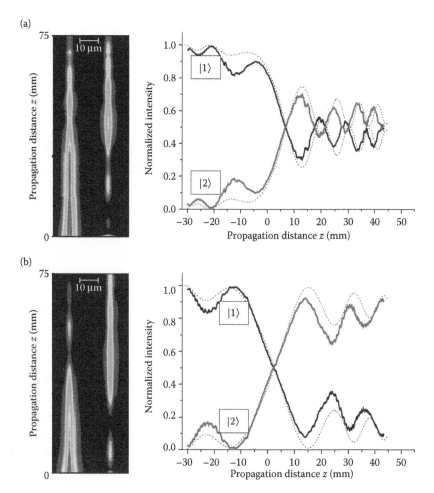

FIGURE 13.15 Measured fluorescence images of light tunneling in cubically bent waveguide couplers (left) and corresponding behavior of normalized light power versus propagation distance in the two waveguides $|1\rangle$ and $|2\rangle$ of the coupler (right). In (a), $d = 17\,\mu m$, $A = 300\,\mu m$ and $L_1 = 31.25\,mm$, whereas in (b) $d = 17\,\mu m$, $A = 500\,\mu m$ and $L_1 = 31.25\,mm$, resulting in a slower level crossing.

topics addressed in this chapter include specifically coherent destruction of tunneling in periodically curved directional couplers, the optical analog of field-induced barrier transparency, coherent tunneling of light by adiabatic passage in coupled waveguides, control of photonic tunneling in waveguide arrays and the optical analog of the quantum Zeno effect, and the photonic analog of Landau–Zener tunneling in the two-level crossing problem.

REFERENCES

1. A.M. Steinberg and R.Y. Chiao, Tunneling delay times in one and two dimensions. *Phys. Rev. A* 49, 3283, 1994.
2. Ch. Spielmann, R. Szipocs, A. Stingl, and F. Krausz, Tunneling of optical pulses through photonic band gaps. *Phys. Rev. Lett.* 73, 2308, 1994.
3. Ph. Balcou and L. Dutriaux, Dual optical tunneling times in frustrated total internal reflection. *Phys. Rev. Lett.* 78, 851, 1997.

4. R.Y. Chiao and A.M. Steinberg, Tunneling times and superluminality. In: E. Wolf (Ed.), *Progress in Optics*, vol. XXXVII, Elsevier, Amsterdam, 1997, p. 345.

5. A.A. Stahlhofen, Photonic tunneling time in frustrated total internal reflection. *Phys. Rev. A* 62, 012112, 2000.

6. J.J. Carey, J. Zawadzka, D.A. Jaroszynski, and K. Wynne, Noncausal time response in frustrated total internal reflection? *Phys. Rev. Lett.* 84, 1431, 2000.

7. S. Longhi, M. Marano, M. Belmonte, and P. Laporta, Superluminal pulse propagation in linear and non-linear photonic grating structures. *IEEE J. Selected Top. Quantum Electron.* 9, 4, 2003.

8. M.T. Reiten, K. McClatchey, D. Grischkowsky, and R.A. Cheville, Incidence-angle selection and spatial reshaping of terahertz pulses in optical tunneling. *Opt. Lett.* 26, 1900, 2001.

9. Ch. Hirlimann, B. Thomas, and D. Boose, Induced optical tunneling. *Europhys. Lett.* 69, 48, 2005.

10. B.R. Johnson, Theory of morphology-dependent resonances: Shape resonances and width formulas. *J. Opt. Soc. Am. A* 10, 343, 1993.

11. J.U. Nockel and A.D. Stone, Ray and wave chaos in asymmetric resonant optical cavities. *Nature* 385, 45, 1997.

12. V.A. Podolskiy and E.E. Narimanov, Semiclassical description of chaos-assisted tunneling. *Phys. Rev. Lett.* 91, 263601, 2003.

13. V.A. Podolskiy and E.E. Narimanov, Chaos-assisted tunneling in dielectric microcavities. *Opt. Lett.* 30, 474, 2005.

14. H.E. Tureci, H.G.L. Schwefel, P. Jacquod, and A.D. Stone, Modes of wave-chaotic dielectric resonators. *Progr. Opt.* 47, 75, 2005.

15. A. Yariv, *Optical Electronics* (4th edn). Saunders College Publishing, New York, 1991, pp. 519–529.

16. J.R. Black and A. Ankiewicz, Fiber-optic analogies with mechanics. *Am. J. Phys.* 53, 554, 1985.

17. D. Gloge and D. Marcuse, Formal quantum theory of light rays. *J. Opt. Soc. Am.* 59, 1629, 1969.

18. G. Della Valle, M. Ornigotti, E. Cianci, V. Foglietti, P. Laporta, and S. Longhi, Visualization of coherent destruction of tunneling in an optical double well system. *Phys. Rev. Lett.* 98, 263601, 2007.

19. F. Dreisow, A. Szameit, M. Heinrich, T. Pertsch, S. Nolte, A. Tünnermann, and S. Longhi, Decay control via discrete-to-continuum coupling modulation in an optical waveguide system. *Phys. Rev. Lett.* 101, 143602, 2008.

20. F. Dreisow, A. Szameit, M. Heinrich, S. Nolte, A. Tünnermann, M. Ornigotti, and S. Longhi, Direct observation of Landau-Zener tunneling in a curved optical waveguide coupler. *Phys. Rev. A* 79, 055802, 2009.

21. G. Della Valle, S. Longhi, P. Laporta, P. Biagioni, L. Duo, and M. Finazzi, Discrete diffraction in wave-guide arrays: A quantitative analysis by tunneling optical microscopy. *Appl. Phys. Lett.* 90, 261118, 2007.

22. P. Biagioni, G. Della Valle, M. Ornigotti, M. Finazzi, L. Duó, P. Laporta, and S. Longhi, Experimental demonstration of the optical Zeno effect by scanning tunneling optical microscopy. *Opt. Express* 16, 3762, 2008.

23. M. Grifoni and P. Hänggi, Driven quantum tunneling. *Phys. Rep.* 304, 229, 1998.

24. F. Grossmann, T. Dittrich, P. Jung, and P. Hänggi, Coherent destruction of tunneling. *Phys. Rev. Lett.* 67, 516, 1991.

25. I. Vorobeichik, R. Lefebvre, and N. Moiseyev, Field induced barrier transparency. *Europhys. Lett.* 41, 111, 1998.

26. S. Longhi, D. Janner, M. Marano, and P. Laporta, Quantum-mechanical analogy of beam propagation in waveguides with a bent axis: Dynamic-mode stabilization and radiation-loss suppression. *Phys. Rev. E* 67, 036601, 2003.

27. S. Longhi, M. Marangoni, D. Janner, R. Ramponi, P. Laporta, E. Cianci, and V. Foglietti, Observation of wave packet dichotomy and adiabatic stabilization in an optical waveguide. *Phys. Rev. Lett.* 94, 073002, 2005.

28. S. Longhi, Bloch dynamics of light waves in helical optical waveguide arrays. *Phys. Rev. B* 76, 195119, 2007.

29. S. Longhi, Light transfer control and diffraction management in circular fibre waveguide arrays. *J. Phys. B: At. Mol. Opt. Phys.* 40, 4477, 2007.

30. S. Longhi, Quantum optical analogies using photonic structures. *Laser & Photon. Rev.* 3, 243, 2009.

31. S. Longhi, Coherent destruction of tunneling in waveguide directional couplers. *Phys. Rev. A* 71, 065801, 2005.

32. S. Longhi, Resonant tunneling in frustrated total internal reflection. *Opt. Lett.* 30, 2781, 2005.
33. K. Eckert, M. Lewenstein, R. Corbalan, G. Birkl, W. Ertmer, and J. Mompart, Three-level atom optics via the tunneling interaction. *Phys. Rev. A* 70, 023606, 2004.
34. A.D. Greentree, J.H. Cole, A.R. Hamilton, and L.C.L. Hollenberg, Coherent electronic transfer in quantum dot systems using adiabatic passage. *Phys. Rev. B* 70, 235317, 2004.
35. M.C. Fischer, B. Gutierrez-Medina, and M.G. Raizen, Observation of the quantum Zeno and anti-Zeno effects in an unstable system. *Phys. Rev. Lett.* 87, 040402, 2001.
36. A.G. Kofman and G. Kurizki, Universal dynamical control of quantum mechanical decay: Modulation of the coupling to the continuum. *Phys. Rev. Lett.* 87, 270405, 2001.
37. A. Barone, G. Kurizki, and A. G. Kofman, Dynamical control of macroscopic quantum tunneling. *Phys. Rev. Lett.* 92, 200403, 2004.
38. E. Paspalakis, Adiabatic three-waveguide directional coupler. *Opt. Commun.* 258, 31, 2006.
39. S. Longhi, Optical realization of multilevel adiabatic population transfer in curved waveguide arrays. *Phys. Lett. A* 359, 166, 2006.
40. S. Longhi, G. Della Valle, M. Ornigotti, and P. Laporta, Coherent tunneling by adiabatic passage in an optical waveguide system. *Phys. Rev. B* 76, 201101(R), 2007.
41. G. Della Valle, M. Ornigotti, T. Toney Fernandez, P. Laporta, S. Longhi, A. Coppa, and V. Foglietti, Adiabatic light transfer via dressed states in optical waveguide arrays. *Appl. Phys. Lett.* 92, 011106, 2008.
42. Y. Lahini, F. Pozzi, M. Sorel, R. Morandotti, D.N. Christodoulides, and Y. Silberberg, Effect of nonlinearity on adiabatic evolution of light. *Phys. Rev. Lett.* 101, 193901, 2008.
43. S. Longhi, Transfer of light waves in optical waveguides via a continuum. *Phys. Rev. A* 78, 013815, 2008.
44. S. Longhi, Nonexponential decay via tunneling in tight-binding lattices and the optical Zeno effect. *Phys. Rev. Lett.* 97, 110402, 2006.
45. S. Longhi, Control of photon tunneling in optical waveguides. *Opt. Lett.* 32, 557, 2007.
46. S. Longhi, Landau-Zener dynamics in a curved optical directional coupler. *J. Opt. B: Quantum Semiclassical Opt.* 7, L9, 2005.
47. H.A. Kramers, *Collected Scientific Papers*, North-Holland, Amsterdam, 1956, p. 866.
48. W.C. Henneberger, Perturbation method for atoms in intense light beams. *Phys. Rev. Lett.* 21, 838, 1968.
49. M. Pont, N.R. Walet, M. Gavrila, and C.W. McCurdy, Dichotomy of the hydrogen atom in superintense, high-frequency laser fields. *Phys. Rev. Lett.* 61, 939, 1988.
50. E. Kierig, U. Schnorrberger, A. Schietinger, J. Tomkovic, and M.K. Oberthaler, Single-particle tunneling in strongly driven double-well potentials. *Phys. Rev. Lett.* 100, 190405, 2008.
51. I. Vorobeichik and N. Moiseyev, Revealing broad overlapping resonances by strong laser fields. *Phys. Rev. A* 59, 1699, 1999.
52. M.L. Chiofalo, M. Artoni, and G.C. La Rocca, Atom resonant tunneling through a moving barrier. *New Phys.* 5, 78, 2003.
53. O. Forslund and S. He, Electromagnetic scattering from an inhomogeneous grating using a wave-splitting approach. *Progr. Electrom. Res. PIER* 19, 147, 1998.
54. J. Siewert and T. Brandes, Applications of adiabatic passage in solid-state devices. *Adv. Solid State Phys.* 44, 181, 2004.
55. K. Eckert, J. Mompart, R. Corbalan, M. Lewenstein, and G. Birkl, Three level atom optics in dipole traps and waveguides. *Opt. Commun.* 264, 264, 2006.
56. E.M. Graefe, H.J. Korsch, and D. Witthaut, Mean-field dynamics of a Bose-Einstein condensate in a time-dependent triple-well trap: Nonlinear eigenstates, Landau-Zener models, and stimulated Raman adiabatic passage. *Phys. Rev. A* 73, 013617, 2006.
57. L.C.L. Hollenberg, A.D. Greentree, A.G. Fowler, and C.J. Wellard, Two-dimensional architectures for donor-based quantum computing. *Phys. Rev. B* 74, 045311, 2006.
58. K. Bergmann, H. Theuer, and B.W. Shore, Coherent population transfer among quantum states of atoms and molecules. *Rev. Mod. Phys.* 70, 1003, 1998.
59. N.Y. Vitanov, T. Halfmann, B.W. Shore, and K. Bergmann, Laser-induced population transfer by adiabatic passage techniques. *Annu. Rev. Phys. Chem.* 52, 753, 2001.
60. S. Longhi, Photonic transport via chirped adiabatic passage in optical waveguides. *J. Phys. B: At. Mol. Opt. Phys.* 40, F189, 2007.

61. S. Longhi, Optical analogue of coherent population trapping via a continuum in optical waveguide arrays. *J. Mod. Opt.* 56, 729, 2009.
62. C.E. Carroll and F.T. Hioe, Coherent population transfer via the continuum. *Phys. Rev. Lett.* 68, 3523, 1992; Selective excitation via the continuum and suppression of ionization. *Phys. Rev. A* 47, 571, 1993.
63. T. Peters, L.P. Yatsenko, and T. Halfmann, Experimental demonstration of selective coherent population transfer via a continuum. *Phys. Rev. Lett.* 95, 103601, 2005.
    See also: T. Peters and T. Halfmann, Stimulated Raman adiabatic passage via the ionization continuum in helium: Experiment and theory. *Opt. Commun.* 271, 475, 2007.
64. P.L. Knight, M.A. Lander, and B.J. Dalton, Laser-induced continuum structures. *Phys. Rep.* 190, 1, 1990.
65. F. Dreisow, A. Szameit, M. Heinrich, R. Keil, S. Nolte, and A. Tünnermann, Adiabatic transfer of light via a continuum in optical waveguides. *Opt. Lett.* 34, 2405, 2009.
66. L. Fonda, G.C. Ghirardi, and A. Rimini, Decay theory of unstable quantum systems. *Rep. Progr. Phys.* 41, 587, 1978.
67. B. Misra and E.C.G. Sudarshan, The Zeno's paradox in quantum theory. *J. Math. Phys.* 18, 756, 1977.
68. P. Knight, Watching a laser hot-pot. *Nature* (London) 344, 493, 1990.
69. A.G. Kofman and G. Kurizki, Acceleration of quantum decay processes by frequent observations. *Nature* (London) 405, 546, 2000.
70. P. Facchi, H. Nakazato, and S. Pascazio, From the quantum Zeno to the inverse quantum Zeno effect. *Phys. Rev. Lett.* 86, 2699, 2001.
71. P. Facchi and S. Pascazio, Temporal behavior and quantum Zeno time of an excited state of the hydrogen atom. *Phys. Lett. A* 241, 139, 1998.
72. S.R. Wilkinson, C.F. Bharucha, M.C. Fischer, K.W. Madison, P.R. Morrow, Q. Niu, B. Sundaram, and M.G. Raizen, Experimental evidence for non-exponential decay in quantum tunnelling. *Nature* (London) 387, 575, 1997.
73. D.N. Christodoulides, F. Lederer, and Y. Silberberg, Discretizing light behaviour in linear and nonlinear waveguide lattices. *Nature* (London) 424, 817, 2003.
74. S. Longhi, Tunneling escape in optical waveguide arrays with a boundary defect. *Phys. Rev. E* 74, 026602, 2006.
75. A.G. Kofman, G. Kurizki, and B. Sherman, Spontaneous and induced atomic decay in photonic band structures. *J. Mod. Opt.* 41, 353, 1994.
76. K. Itoh, W. Watanabe, S. Nolte, and C.B. Schaffer, Ultrafast processes for bulk modification of transparent materials. *MRS Bull.* 31, 620, 2006.
77. A. Szameit, F. Dreisow, H. Hartung, S. Nolte, and A. Tünnermann, Quasi-incoherent propagation in waveguide arrays. *Appl. Phys. Lett.* 90, 241113, 2007.
78. L.D. Landau, On the theory of transfer of energy at collisions I. *Phys. Z. Sowjetunion* 1, 89, 1932.
79. C. Zener, Non-adiabatic crossing of energy levels. *Proc. R. Soc. A* 137, 696, 1932.
80. J.R. Rubbmark, M.M. Kash, M.G. Littman, and D. Kleppner, Dynamical effects at avoided level crossings: A study of the Landau-Zener effect using Rydberg atoms. *Phys. Rev. A* 23, 3107, 1981.
81. W. Wernsdorfer and R. Sessoli, Quantum phase interference and parity effects in magnetic molecular clusters. *Science* 284, 133, 1999.
82. P. Földi, M.G. Benedict, J. Milton Pereira, and F.M. Peeters, Dynamics of molecular nanomagnets in time-dependent external magnetic fields: Beyond the Landau-Zener-Stueckelberg model. *Phys. Rev. B* 75, 104430, 2007.
83. K. Mullen, E. Ben-Jacob, and Z. Schuss, Combined effect of Zener and quasiparticle transitions on the dynamics of mesoscopic Josephson junctions. *Phys. Rev. Lett.* 60, 1097, 1988.
84. M. Jona-Lasinio, O. Morsch, M. Cristiani, N. Malossi, J.H. Müller, E. Courtade, M. Anderlini, and E. Arimondo, Asymmetric Landau-Zener tunneling in a periodic potential. *Phys. Rev. Lett.* 91, 230406, 2003.
85. S.J. Woo, S. Choi, and N.P. Bigelow, Controlling quasiparticle excitations in a trapped Bose-Einstein condensate. *Phys. Rev. A* 72, 021605(R), 2005.
86. A. Zenesini, H. Lignier, G. Tayebirad, J. Radogostowicz, D. Ciampini, R. Mannella, S. Wimberger, O. Morsch, and E. Arimondo, Time-resolved measurement of Landau-Zener tunneling in periodic potentials. *Phys. Rev. Lett.* 103, 090403, 2009.
87. W.D. Oliver, Y. Yu, J.C. Lee, K.K. Berggren, L.S. Levitov, and T.P. Orlando, Mach-Zehnder interferometry in a strongly driven superconducting qubit. *Science* 310, 1653, 2005.

88. R.J.C. Spreeuw, N.J. van Druten, M.W. Beijersbergen, E.R. Eliel, and J.P. Woerdman, Classical realization of a strongly driven two-level system. *Phys. Rev. Lett.* 65, 2642, 1990.

89. D. Bouwmeester, N.H. Dekker, F.E.v. Dorsselaer, C.A. Schrama, P.M. Visser, and J.P. Woerdman, Observation of Landau-Zener dynamics in classical optical systems. *Phys. Rev. A* 51, 646, 1995.

90. R. Khomeriki and S. Ruffo, Nonadiabatic Landau-Zener tunneling in waveguide arrays with a step in the refractive index. *Phys. Rev. Lett.* 94, 113904, 2005.

91. S. Longhi, Landau-Zener dynamics in a curved optical directional coupler. *J. Opt. B: Quantum Semiclass. Opt.* 7, L9, 2005.

92. J. Liu, Libin Fu, Bi-Yiao Ou, Shin-Gang Chen, Dae-Il Choi, Biao Wu, and Qian Niu, Theory of nonlinear Landau-Zener tunneling. *Phys. Rev. A* 66, 023404, 2002.

93. N.V. Vitanov and B. M. Garraway, Landau-Zener model: Effects of finite coupling duration. *Phys. Rev. A* 53, 4288, 1996.

94. V.L. Pokrovsky and N.A. Sinitsyn, Landau-Zener transitions in a linear chain. *Phys. Rev. B* 65, 153105, 2002.

95. M.V. Volkov and V.N. Ostrovsky, Exact results for survival probability in the multistate Landau-Zener model. *J. Phys. B: At. Mol. Opt. Phys.* 37, 4069, 2004.

96. E. Shimshoni and A. Stern, Dephasing of interference in Landau-Zener transitions. *Phys. Rev. B* 47, 9523, 1993.

97. K. Saito, M. Wubs, S. Kohler, Y. Kayanuma, and P. Hänggi, Dissipative Landau-Zener transitions of a qubit: Bath-specific and universal behavior. *Phys. Rev. B* 75, 214308, 2007.

98. V.L. Pokrovsky and D. Sun, Fast quantum noise in the Landau-Zener transition. *Phys. Rev. B* 76, 024310, 2007.

14 Quantum Discrete Breathers

Ricardo A. Pinto and Sergej Flach

CONTENTS

14.1 INTRODUCTION

In solid-state physics, the phenomenon of localization is usually perceived as arising from extrinsic disorder that breaks the discrete translational invariance of the perfect crystal lattice. Familiar examples include the localized vibrational modes around impurities or defects in crystals and Anderson localization of waves in disordered media [1]. The usual perception among solid-state researchers is that, in perfect lattices excitations must be extended objects as well, essentially plane wave like. Such firmly entrenched perceptions were severely jolted since the discovery of discrete breathers (DBs), also known as intrinsic localized modes (ILM). These states are typical excitations in perfectly periodic but strongly nonlinear systems, and are characterized by being spatially localized, at variance to plane wave states [2–4].

DB-like excitations, being generic objects, have been observed in a large variety of lattice systems that include bond excitations in molecules, lattice vibrations and spin excitations in solids, charge flow in coupled Josephson junctions, light propagation in interacting optical waveguides, cantilever vibrations in michromechanical arrays, cold atom dynamics in Bose–Einstein condensates loaded on optical lattices, among others. They have been extensively studied, and a high level of understanding about their properties has been reached.

Two decades of intensive research have polished our theoretical understanding of DBs in classical nonlinear lattices. Less is known about their quantum counterparts—quantum breathers (QBs). This chapter is devoted to a review of the more recent studies in this field. The concept of QBs is closely related with the theme of dynamical tunneling in phase space.

14.1.1 A FEW FACTS ABOUT CLASSICAL DB

Let us study the combined effect of nonlinearity and discreteness on the spatial localization of a DB on a basic level. For that we look into the dynamics of a one-dimensional chain of interacting (scalar) oscillators with the Hamiltonian

$$H = \sum_n \left[\frac{1}{2} p_n^2 + V(x_n) + W(x_n - x_{n-1}) \right]. \tag{14.1}$$

The integer n marks the lattice site number of a possibly infinite chain, and x_n and p_n are the canonically conjugated coordinate and momentum of a degree of freedom associated with site number n. The on-site potential V and the interaction potential W satisfy $V'(0) = W'(0) = 0$, $V''(0), W''(0) \geq 0$. This choice ensures that the classical ground state $x_n = p_n = 0$ is a minimum of the energy H. The equations of motion read

$$\dot{x}_n = p_n, \quad \dot{p}_n = -V'(x_n) - W'(x_n - x_{n-1}) + W'(x_{n+1} - x_n). \tag{14.2}$$

Let us linearize the equations of motion around the classical ground state. We obtain a set of linear coupled differential equations with solutions being small-amplitude plane waves:

$$x_n(t) \sim e^{i(\omega_q t - qn)}, \quad \omega_q^2 = V''(0) + 4W''(0)\sin^2\left(\frac{q}{2}\right). \tag{14.3}$$

These waves are characterized by a wave number q and a corresponding frequency ω_q. All allowed plane-wave frequencies fill a part of the real axis which is coined linear spectrum. Owing to the underlying lattice the frequency ω_q depends periodically on q and its absolute value has always a *finite upper bound*. The maximum (Debye) frequency of small-amplitude waves $\omega_\pi = \sqrt{V''(0) + 4W''(0)}$. Depending on the choice of the potential $V(x)$, ω_q can be either acoustic- or optic-like, $V(0) = 0$ and $V(0) \neq 0$, respectively. In the first case, the linear spectrum covers the interval $-\omega_\pi \leq \omega_q \leq \omega_\pi$ which includes $\omega_{q=0} = 0$. In the latter case, an additional (finite) gap opens for $|\omega_q|$ below the value $\omega_0 = \sqrt{V''(0)}$.

For large-amplitude excitations the linearization of the equations of motion is not correct anymore. Similar to the case of a single anharmonic oscillator, the frequency of possible time-periodic excitations will depend on the amplitude of the excitation, and thus may be located outside the linear spectrum. Let us assume that a time-periodic and spatially localized state—that is, a *DB*, $\hat{x}_n(t + T_b) = \hat{x}_n(t)$ exists as an exact solution of Equation 14.2 with the period $T_b = 2\pi/\Omega_b$. Due to its time periodicity, we can expand $\hat{x}_n(t)$ into a Fourier series

$$\hat{x}_n(t) = \sum_k A_{kn} e^{ik\Omega_b t}. \tag{14.4}$$

The Fourier coefficients are by assumption also localized in space

$$A_{k,|n|\to\infty} \to 0. \tag{14.5}$$

Inserting this ansatz into the equations of motion Equation 14.2 and linearizing the resulting algebraic equations for Fourier coefficients in the spatial breather tails (where the amplitudes are by assumption small), we arrive at the following linear algebraic equations:

$$k^2 \Omega_b^2 A_{kn} = V''(0) A_{kn} + W''(0)(2A_{kn} - A_{k,n-1} - A_{k,n+1}). \tag{14.6}$$

If $k\Omega_b = \omega_q$, the solution to (Equation 14.6) is $A_{k,n} = c_1 e^{iqn} + c_2 e^{-iqn}$. Any nonzero (whatever small) amplitude $A_{k,n}$ will thus oscillate without further spatial decay, contradicting the initial assumption. If however,

$$k\Omega_b \neq \omega_q \tag{14.7}$$

for any integer k and any q, then the general solution to Equation 14.6 is given by $A_{kn} = c_1 \kappa^n + c_2 \kappa^{-n}$ where κ is a real number depending on ω_q, Ω_b and k. It always admits a (actually exponential) spatial decay by choosing either c_1 or c_2 to be nonzero. In order to fulfill Equation 14.7 for at least one real value of Ω_b and any integer k, we have to request $|\omega_q|$ to be bounded from above. That is precisely the reason why the spatial lattice is needed. In contrast most spatially continuous field equations will have linear spectra which are unbounded. That makes resonances of higher order harmonics of a localized excitation with the linear spectrum unavoidable. The nonresonance condition Equation 14.7 is thus an (almost) necessary condition for obtaining a time-periodic localized state on a Hamiltonian lattice [2].

The performed analysis can be extended to more general classes of discrete lattices, including for example, long-range interactions between sites, more degrees of freedom per each site, higher dimensional lattices and so on. But the resulting nonresonance condition (Equation 14.7) keeps its generality, illustrating the key role of discreteness and nonlinearity for the existence of DBs.

Let us show DB solutions for various lattices. We start with a chain (Equation 14.1) with the functions

$$V(x) = x^2 + x^3 + \frac{1}{4}x^4, \quad W(x) = 0.1x^2. \tag{14.8}$$

The spectrum ω_q is optic-like and shown in Figure 14.1. DB solutions can have frequencies Ω_b which are located both below and above the linear spectrum. The time-reversal symmetry of Equation 14.2 allows to search for DB displacements $x_n(t = 0)$ when all velocities $\dot{x}_n(t = 0) = 0$. These initial displacements are computed with high accuracy (see following sections) and plotted in the insets in Figure 14.1 [3]. We show solutions to two DB frequencies located above and below ω_q— their actual values are marked with the arrows. To each DB frequency we show two different spatial DB patterns—among an infinite number of other possibilities, as we will see below. The high-frequency DBs ($\Omega_b \approx 1.66$) occur for large-amplitude, high-energy motion with adjacent particles moving out of phase. Low-frequency DBs ($\Omega_b \approx 1.26$) occur for small-amplitude motion with adjacent particles moving in phase.

In Figure 14.2 we show two DB solutions for a Fermi–Pasta–Ulam chain of particles coupled via anharmonic springs $V(x) = 0, W(x) = \frac{1}{2}x^2 + \frac{1}{4}x^4$ (cf. Figure 14.1) which has an acoustic-type spectrum [5]. The DB frequency is in both cases $\Omega_b = 4.5$. Again the displacements x_n are shown for an initial time when all velocities vanish. In the inset, we plot the strain $u_n = x_n - x_{n-1}$ on a log-normal scale. The DB solutions are exponentially localized in space.

Finally, we show DB solutions for a *two-dimensional* square lattice of anharmonic oscillators with nearest-neighbor coupling. The equations of motion read

$$\ddot{x}_{i,j} = k(x_{i+1,j} + x_{i-1,j} - 2x_{i,j}) + k(x_{i,j+1} + x_{i,j-1} - 2x_{i,j}) - x_{i,j} - x_{i,j}^3 \tag{14.9}$$

with oscillator potentials $V(x) = \frac{1}{2}x^2 + \frac{1}{4}x^4$. In Figure 14.3, we plot the oscillator displacements with all velocities equal to zero for three different DB frequencies and $k = 0.05$ [6]. For all cases adjacent oscillators move out of phase.

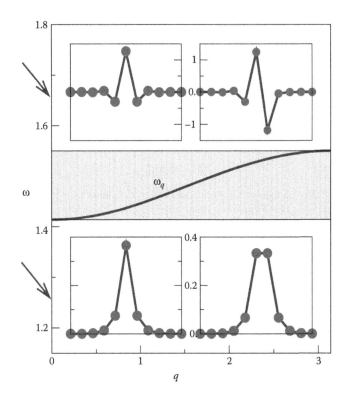

FIGURE 14.1 The frequency versus wavenumber dependence of the linear spectrum for a one-dimensional chain of anharmonic oscillators with potentials (Equation 14.8). Chosen DB frequencies are marked with arrows and lie outside the linear spectrum ω_q. Circles indicate the oscillator displacements for a given DB solution, with all velocities equal to zero. Lines connecting circles are guides for the eye. (Adapted from D. K. Campbell, S. Flach, and Y. S. Kivshar, *Phys. Today* 57 (1), 43–49, 2004.)

We conclude this section emphasizing that DB solutions can be typically localized on a few lattice sites, regardless of the lattice dimension. Thus, little overall coherence is needed to excite a state nearby—just a few sites have to oscillate coherently, the rest of the lattice does not participate strongly in the excitation.

14.1.2 FROM CLASSICAL TO QUANTUM

A natural question is what remains of DBs if the corresponding quantum problem is considered [7–9]. The many-body Schrödinger equation is linear and translationally invariant, therefore, all eigenstates must obey the Bloch theorem. Thus we cannot expect eigenstates of the Hamiltonian to be spatially localized (on the lattice). What is the correspondence between the quantum eigenvalue problem and the classical dynamical evolution?

The concept of tunneling is a possible answer to this puzzle. Naively speaking we quantize the family of periodic orbits associated with a DB located somewhere on the lattice. Notice that there are as many such families as there are lattice sites. The quantization (e.g., Bohr–Sommerfeld) yields some eigenvalues. Since we can perform the same procedure with any family of DB periodic orbits which differ only in their location on the lattice, we obtain N-fold degeneracy for every thus obtained eigenvalue, where N stands for the number of lattice sites. Unless we consider the trivial case of uncoupled lattice sites, these degeneracies will be lifted. Consequently, we will instead obtain bands of states with finite bandwidth. These bands will be called QB bands. The inverse

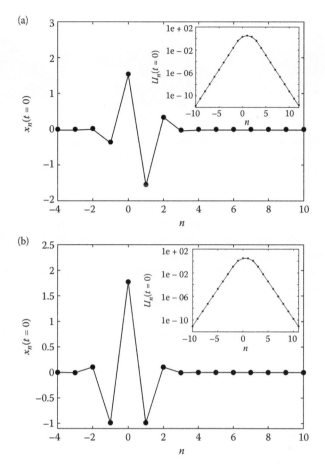

FIGURE 14.2 DB solutions for a Fermi–Pasta–Ulam chain. These states are frequently referred to as the Page mode (a) and the Sievers–Takeno mode (b). (Adapted from S. Flach and A. V. Gorbach, *Chaos* 15, 015112, 2005.)

tunneling time of a semiclassical breather from one site to a neighboring one is a measure of the bandwidth.

We can then formulate the following expectation: if a classical nonlinear Hamiltonian lattice possesses DBs, its quantum counterpart should show up with nearly degenerate bands of eigenstates, if the classical limit is considered. The number of states in such a band is N, and the eigenfunctions are given by Bloch-like superpositions of the semiclassical eigenfunctions obtained using the mentioned Bohr–Sommerfeld quantization of the classical periodic orbits. By nearly degenerate we mean that the bandwidth of a QB band is much smaller than the spacing between different breather bands and the average level spacing in the given energy domain, and the classical limit implies large eigenvalues.

Another property of a QB state is that such a state shows up with exponential localization in appropriate correlation functions [10]. This approach selects all many-particle bound states, no matter how deep one is in the quantum regime. In this sense QB states belong to the class of many-particle bound states.

For large energies and N the density of states becomes large too. What will happen to the expected QB bands then? Will the hybridization with other nonbreather states destroy the particle-like nature of the QB, or not? What is the impact of the nonintegrability of most systems allowing

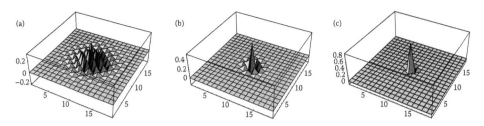

FIGURE 14.3 Displacements of DBs on a two-dimensional lattice Equation 14.9 with $k = 0.05$, all velocities equal to zero. (a) $\Omega_b = 1.188$; (b) $\Omega_b = 1.207$; (c) $\Omega_b = 1.319$. (Adapted from M. Eleftheriou and S. Flach, *Physica D* 202, 142, 2005.)

for classical breather solutions? Since the quantum case corresponds to a quantization of the classical phase space, we could expect that chaotic trajectories lying nearby classical breather solutions might affect the corresponding quantum eigenstates.

From a computational point of view we are very much restricted in our abilities to study QBs. Ideally, we would like to study quantum properties of a lattice problem in the high-energy domain (to make contact with classical states) and for large lattices. This is typically impossible, since solving the quantum problem amounts to diagonalizing the Hamiltonian matrix with rank b^N where b is the number of states per site, which should be large to make contact with classical dynamics. Thus, typically QB states have been so far obtained numerically for small one-dimensional systems [10–12].

14.2 QB MODELS

14.2.1 Bose–Hubbard Chain

Let us discuss QBs within the widely used quantum discrete nonlinear Schrödinger model (also called Bose–Hubbard model) with the Hamiltonian [13]

$$H = -\sum_{l=1}^{N} \left[\frac{1}{2} a_l^\dagger a_l^\dagger a_l a_l + C(a_l^\dagger a_{l+1} + h.c.) \right] \tag{14.10}$$

and the commutation relations

$$a_l a_m^\dagger - a_m^\dagger a_l = \delta_{lm}, \tag{14.11}$$

with δ_{lm} being the standard Kronecker symbol. This Hamiltonian conserves the total number of particles

$$B = \sum_l n_l, \quad n_l = a_l^\dagger a_l. \tag{14.12}$$

For b particles and N sites the number of basis states is

$$\frac{(b+N-1)!}{b!(N-1)!}. \tag{14.13}$$

For $b = 0$ there is just one trivial state of an empty lattice. For $b = 1$ there are N states which correspond to one-boson excitations. These states are similar to classical extended wave states. For $b = 2$ the problem is still exactly solvable, because it corresponds to a two-body problem on a lattice. A corresponding numerical solution is sketched in Figure 14.4. Note the wide two-particle continuum, and a single band located below. This single band corresponds to quasiparticle states

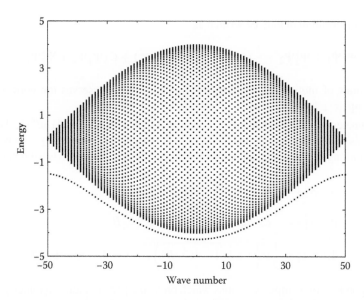

FIGURE 14.4 Spectrum of the quantum DNLS with $b = 2$ and $N = 101$. The energy eigenvalues are plotted versus the wavenumber of the eigenstate. (Adapted from A. C. Scott, J. C. Eilbeck, and H. Gilhøj, *Physica D* 78, 194, 1994.)

characterized by one single quantum number (related to the wavenumber q). These states are two-particle bound states. The dispersion of this band is given by [13]

$$E = -\sqrt{1 + 16C^2 \cos^2\left(\frac{q}{2}\right)}. \tag{14.14}$$

Any eigenstate from this two-particle bound state band is characterized by exponential localization of correlations, that is, when represented in some set of basis states, the amplitude or overlap with a basis state where the two particles are separated by some number of sites is exponentially decreasing with increasing separation distance. Note that a compact bound state is obtained for $q = \pm\pi$, that is, for these wave numbers basis states with nonzero separation distance do not contribute to the eigenstate at all [14].

14.2.2 DIMER

A series of papers was devoted to the properties of the quantum dimer [15–18]. This system describes the dynamics of bosons fluctuating between two sites. The number of bosons is conserved, and together with the conservation of energy the system appears to be integrable. Of course, one cannot consider spatial localization in such a model. However, a reduced form of the discrete translational symmetry—namely the permutational symmetry of the two sites—can be imposed. Together with the addition of nonlinear terms in the classical equations of motion the dimer allows for classical trajectories which are not invariant under permutation. The phase space can be completely analyzed, all isolated periodic orbits (IPOs) can be found. There appears exactly one bifurcation on one family of IPOs, which leads to the appearance of a separatrix in phase space. The separatrix separates three regions—one invariant and two noninvariant under permutations. The subsequent analysis of the quantum dimer demonstrated the existence of pairs of eigenstates with nearly equal eigenenergies [15]. The separatrix and the bifurcation in the classical phase space can be traced in the spectrum of the quantum dimer [17].

The classical Hamiltonian may be written as

$$H = \Psi_1^* \Psi_1 + \Psi_2^* \Psi_2 + \frac{1}{2} \left((\Psi_1^* \Psi_1)^2 + (\Psi_2^* \Psi_2)^2 \right) + C \left(\Psi_1^* \Psi_2 + \Psi_2^* \Psi_1 \right) \tag{14.15}$$

with the equations of motion $\dot{\Psi}_{1,2} = i \partial H / \partial \Psi_{1,2}^*$. The model conserves the norm (or number of particles) $B = |\Psi_1|^2 + |\Psi_2|^2$.

Let us parameterize the phase space of Equation 14.15 with $\Psi_{1,2} = A_{1,2} e^{i\phi_{1,2}}$, $A_{1,2} \geq 0$. It follows that $A_{1,2}$ is time independent and $\phi_1 = \phi_2 + \Delta$ with $\Delta = 0, \pi$ and $\dot{\phi}_{1,2} = \omega$ being also time independent. Solving the algebraic equations for the amplitudes of the IPOs we obtain

$$\text{I}: A_{1,2}^2 = \frac{1}{2}B, \quad \Delta = 0, \quad \omega = 1 + C + \frac{1}{2}B, \tag{14.16}$$

$$\text{II}: A_{1,2}^2 = \frac{1}{2}B, \quad \Delta = \pi, \quad \omega = 1 - C + \frac{1}{2}B, \tag{14.17}$$

$$\text{III}: A_1^2 = \frac{1}{2}B \left(1 \pm \sqrt{1 - 4C^2/B^2} \right), \quad \Delta = 0, \quad \omega = 1 + B. \tag{14.18}$$

IPO III corresponds to two elliptic solutions which break the permutational symmetry. IPO III exists for $B \geq B_b$ with $B_b = 2C$ and occurs through a bifurcation from IPO I. The corresponding separatrix manifold is uniquely defined by the energy of IPO I at a given value of $B \geq B_b$. This manifold separates three regions in phase space—two with symmetry broken solutions, each one containing one of the IPOs III, and one with symmetry conserving solutions containing the elliptic IPO II. The separatrix manifold itself contains the hyperbolic IPO I. For $B \leq B_b$ only two IPOs exist—IPO I and II, with both of them being of elliptic character. Remarkably there exists no other IPOs, and the mentioned bifurcation and separatrix manifolds are the only ones present in the classical phase space of Equation 14.15.

To conclude the analysis of the classical part, we list the energy properties of the different phase space parts separated by the separatrix manifold. First it is straightforward to show that the IPOs (Equations 14.16 through 14.18) correspond to maxima, minima or saddle points of the energy in the allowed energy interval for a given value of B, with no other extrema or saddle points present. It follows

$$E_1 = H(\text{IPO I}) = B + \frac{1}{4}B^2 + CB, \tag{14.19}$$

$$E_2 = H(\text{IPO II}) = B + \frac{1}{4}B^2 - CB, \tag{14.20}$$

$$E_3 = H(\text{IPO III}) = B + \frac{1}{2}B^2 + C^2. \tag{14.21}$$

For $B < B_b$ we have $E_1 > E_2$ (IPO I—maximum, IPO II—minimum). For $B \geq B_b$ it follows $E_3 > E_1 > E_2$ (IPO III—maxima, IPO I—saddle, IPO II—minimum). If $B < B_b$, then all trajectories are symmetry conserving. If $B \geq B_b$, then trajectories with energies $E_1 < E \leq E_3$ are symmetry breaking, and trajectories with $E_2 \leq E \leq E_1$ are symmetry conserving.

The quantum eigenvalue problem amounts to replacing the complex functions Ψ, Ψ^* in Equation 14.15 by the boson annihilation and creation operators $a, a^\dagger$ with the standard commutation relations (to enforce the invariance under the exchange $\Psi \leftrightarrow \Psi^*$ the substitution has to be done on rewriting $\Psi\Psi^* = 1/2(\Psi\Psi^* + \Psi^*\Psi)$):

$$H = \frac{5}{4} + \frac{3}{2} \left(a_1^\dagger a_1 + a_2^\dagger a_2 \right) + \frac{1}{2} \left((a_1^\dagger a_1)^2 + (a_2^\dagger a_2)^2 \right) + C \left(a_1^\dagger a_2 + a_2^\dagger a_1 \right). \tag{14.22}$$

Note that $\hbar = 1$ here, and the eigenvalues b of $B = a_1^\dagger a_1 + a_2^\dagger a_2$ are integers. Since B commutes with H we can diagonalize the Hamiltonian on the basis of eigenfunctions of B. Each value of b

corresponds to a subspace of the dimension $(b+1)$ in the space of eigenfunctions of B. These eigenfunctions are products of the number states $|n\rangle$ of each degree of freedom and can be characterized by a symbol $|n,m\rangle$ with n bosons in the site 1 and m bosons in the site 2. For a given value of b it follows $m = b - n$. So we can actually label each state by just one number n: $|n,(b-n)\rangle \equiv |n\rangle$. Consequently, the eigenvalue problem at fixed b amounts to diagonalizing the matrix

$$
H_{nm} = \begin{cases}
\dfrac{5}{4} + \dfrac{3}{2}b + \dfrac{1}{2}\left(n^2 + (b-n)^2\right) & n = m \\[2mm]
C\sqrt{n(b+1-n)} & n = m+1 \\[2mm]
C\sqrt{(n+1)(b-n)} & n = m-1 \\[2mm]
0 & \text{else}
\end{cases} \tag{14.23}
$$

where $n,m = 0,1,2,\ldots,b$. Notice that the matrix H_{nm} is a symmetric band matrix. The additional symmetry $H_{nm} = H_{(b-n),(b-m)}$ is a consequence of the permutational symmetry of H. For $C = 0$ the matrix H_{nm} is diagonal, with the property that each eigenvalue is doubly degenerate (except for the state $|b/2\rangle$ for even values of b). The classical phase space contains only symmetry broken trajectories, with the exception of IPO II and the separatrix with IPO I (in fact in this limit the separatrix manifold is nothing but a resonant torus containing both IPOs I and II). With the exception of the separatrix manifold, all tori break permutational symmetry and come in two groups separated by the separatrix. Then quantizing each group will lead to pairs of degenerate eigenvalues—one from each group. There is a clear correspondence to the spectrum of the diagonal $(C = 0)$ matrix H_{nm}. The eigenvalues $H_{00} = H_{bb}$ correspond to the quantized IPOs III. With increasing n the eigenvalues $H_{nn} = H_{(b-n),(b-n)}$ correspond to quantized tori further away from the IPO III. Finally, the states with $n = b/2$ for even b or $n = (b-1)/2$ for odd b are tori most close to the separatrix. Switching the side diagonals on by increasing C will lead to a splitting of all pairs of eigenvalues. In the case of small values of b these splittings have no correspondence to classical system properties. However, in the limit of large b we enter the semiclassical regime, and due to the integrability of the system, eigenfunctions should correspond to tori in the classical phase space which satisfy the Einstein–Brillouin–Keller quantization rules. Increasing C from zero will lead to a splitting ΔE_n of the eigenvalue doublets of $C = 0$. In other words, we find pairs of eigenvalues, which are related to each other through the symmetry of their eigenvectors and (for small enough C) through the small value of the splitting. These splittings have been calculated numerically and using perturbation theory [15,17]. In the limit of large b the splittings are exponentially small for energies above the classical separatrix energy (i.e., for classical trajectories which are not invariant under permutation). If the eigenenergies are lowered below the classical separatrix energy, the splittings grow rapidly up to the mean level spacing.

In Figure 14.5 the results of a diagonalization of a system with 600 particles ($b = 600$) is shown [17]. The inset shows the density of states versus energy, which nicely confirms the predicted singularity at the energy of the separatrix of the classical counterpart. In order to compute the exponentially small splittings, we may use, for example, a Mathematica routine which allows to choose arbitrary values for the precision of computations. Here we chose precision 512. In Figure 14.6 the numerically computed splittings are compared to perturbation theory results. As expected, the splittings become extremely small above the separatrix. Consequently these states will follow for long time the dynamics of a classical broken symmetry state.

14.2.3 TRIMER

The integrability of the dimer does not allow a study of the influence of chaos (i.e., nonintegrability) on the tunneling properties of the mentioned pairs of eigenstates. A natural extension of the dimer to a trimer adds a third degree of freedom without adding a new integral of motion. Consequently,

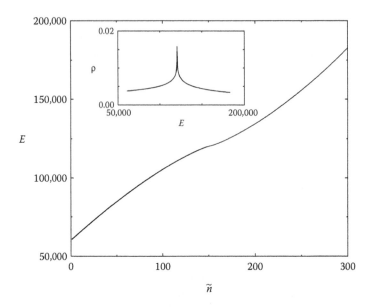

FIGURE 14.5 Eigenvalues versus ordered state number $\tilde{n}$ for symmetric and antisymmetric states ($0 < \tilde{n} < b/2$ for both types of states). Parameters: $b = 600$ and $C = 50$. Inset: Density of states versus energy. (Adapted from S. Aubry et al., *Phys. Rev. Lett.* 76, 1607, 1996.)

the trimer is nonintegrable. A still comparatively simple numerical quantization of the trimer allows to study the behavior of many tunneling states in the large-energy domain of the eigenvalue spectrum [19].

Similarly to the dimer, the quantum trimer Hamiltonian is represented in the form

$$H = \frac{15}{8} + \frac{3}{2}(a_1^\dagger a_1 + a_2^\dagger a_2 + a_3^\dagger a_3) + \frac{1}{2}\left[(a_1^\dagger a_1)^2 + (a_2^\dagger a_2)^2\right]$$
$$+ C(a_1^\dagger a_2 + a_2^\dagger a_1) + \delta(a_1^\dagger a_3 + a_3^\dagger a_1 + a_2^\dagger a_3 + a_3^\dagger a_2). \tag{14.24}$$

Again $B = a_1^\dagger a_1 + a_2^\dagger a_2 + a_3^\dagger a_3$ commutes with the Hamiltonian, thus we can diagonalize Equation 14.24 in the basis of eigenfunctions of B. For any finite eigenvalue b of B the number of states is finite, namely $(b + 1)(b + 2)/2$. Thus, the infinite dimensional Hilbert space separates into an infinite set of finite dimensional subspaces, each subspace containing only vectors with a given eigenvalue b. These eigenfunctions are products of the number states $|n\rangle$ of each degree of freedom and can be characterized by a symbol $|n, m, l\rangle$ where we have n bosons on site 1, m bosons on site 2, and l bosons on site 3. For a given value b it follows that $l = b - m - n$. So we can actually label each state by just two numbers (n, m): $|n, m, (b - n - m)\rangle \equiv |n, m\rangle$. Note that the third site added to the dimer is different from the first two sites. There is no boson–boson interaction on this site. Thus site 3 serves simply as a boson reservoir for the dimer. Dimer bosons may now fluctuate from the dimer to the reservoir. The trimer has the same permutational symmetry as the dimer.

The matrix elements of Equation 14.24 between states from different b subspaces vanish. Thus for any given b the task amounts to diagonalizing a finite dimensional matrix. The matrix has a tridiagonal block structure, with each diagonal block being a dimer matrix Equation 14.23. The nonzero off-diagonal blocks contain interaction terms proportional to δ. We consider symmetric $|\Psi\rangle_s$ and antisymmetric $|\Psi\rangle_a$ states. The structure of the corresponding symmetric and antisymmetric decompositions of H is similar to H itself. In the following we will present results for $b = 40$. We will also drop the first two terms of the RHS in Equation 14.24, because these only lead to a shift of the energy spectrum. Since we evaluate the matrix elements explicitly, we need only a few seconds

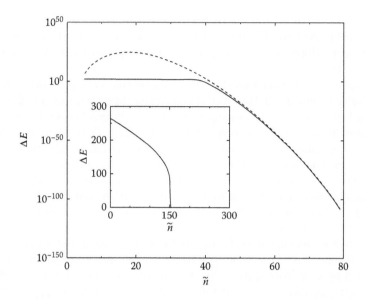

FIGURE 14.6 Eigenvalue splittings versus $\tilde{n}$ for $b = 150$ and $C = 10$. Solid line—numerical result; dashed line—perturbation theory. Inset: Same for $b = 600$ and $C = 50$. Only numerical results are shown. (Adapted from S. Aubry et al., *Phys. Rev. Lett.* 76, 1607, 1996.)

to obtain all eigenvalues and eigenvectors with the help of standard Fortran routines. In Figure 14.7 we plot a part of the energy spectrum as a function of δ for $C = 2$ [19]. As discussed above, the Hamiltonian decomposes into noninteracting blocks for $\delta = 0$, each block corresponding to a dimer with a boson number between 0 and b. For $\delta \neq 0$ the block–block interaction leads to typical features in the spectrum, like, for example, avoided crossings. The full quantum energy spectrum extends

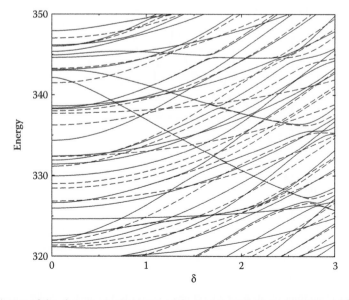

FIGURE 14.7 A part of the eigenenergy spectrum of the quantum trimer as a function of δ with $b = 40$ and $C = 2$. Lines connect data points for a given state. Solid lines—symmetric eigenstates; thick dashed lines—antisymmetric eigenstates. (Adapted from S. Flach and V. Fleurov, *J. Phys.: Cond. Matt.* 9, 7039, 1997.)

roughly over 10^3, leading to an averaged spacing of order 10^0. Also the upper third of the spectrum is diluted compared to the lower two thirds. The correspondence to the classical model is obtained with the use of the transformation $E_{cl} = E_{qm}/b^2 + 1$ and for parameters C/b and δ/b (the classical value for B is $B = 1$).

The main result of this computation so far is that tunneling pairs of eigenstates of the dimer persist in the nonintegrable regime $\delta \neq 0$. However, at certain pair-dependent values of δ a pair breaks up. From the plot in Figure 14.7 we cannot judge how the pair splittings behave. In Figure 14.8 we plot the pair splitting of the pair which has energy ≈ 342 at $\delta = 0$ [20]. Denote with x, y, z the eigenvalues of the site number operators n_1, n_2, n_3. We may consider the quantum states of the trimer at $\delta = 0$ when z is a good quantum number and then follow the evolution of these states with increasing δ. The state for $\delta = 0$ can be traced back to $C = 0$ and be thus characterized in addition by x and y. The chosen pair states are then characterized by $x = 26(0)$, $y = 0(26)$ and $z = 14$ for $C = \delta = 0$. Note that this pair survives approximately 30 avoided crossings before it is finally destroyed at coupling strength $\delta \approx 2.67$ as seen in Figure 14.7.

From Figure 14.8 we find that the splitting rapidly increases gaining about eight orders of magnitude when δ changes from 0 to slightly above 0.5. Then this rapid but nevertheless smooth rise is interrupted by very sharp spikes when the splitting ΔE rises by several orders of magnitude with δ changing by mere percents and then abruptly changes in the opposite direction sometimes even overshooting its prespike value. Such spikes, some larger, some smaller, repeat with increasing δ until the splitting value approaches the mean level spacing of order one. Only then one may say that the pair is destroyed since it can be hardly distinguished among the other trimer levels.

Another observation is presented in Figure 14.9 [20]. We plot the intensity distribution of the logarithm of the squared symmetric wave function of our chosen pair for five different values of $\delta = 0, 0.3, 0.636, 1.0, 1.8$ (their locations are indicated by filled circles in Figure 14.8). We use the eigenstates of B as basis states. They can be represented as $|x, y, z>$ where x, y, z are the particle numbers on sites 1, 2, 3, respectively. Due to the commutation of B with H two site occupation

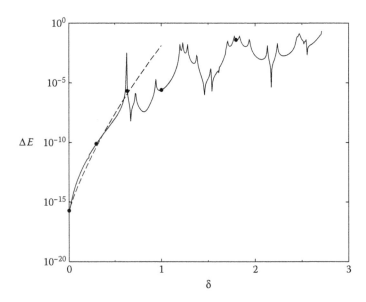

FIGURE 14.8 Level splitting versus δ for a level pair as described in the text. Solid line—numerical result; dashed line—semiclassical approximation; filled circles—location of wave function analysis in Figure 14.9. (Adapted from S. Flach, V. Fleurov, and A. A. Ovchinnikov, *Phys. Rev. B* 63, 094304, 2001.)

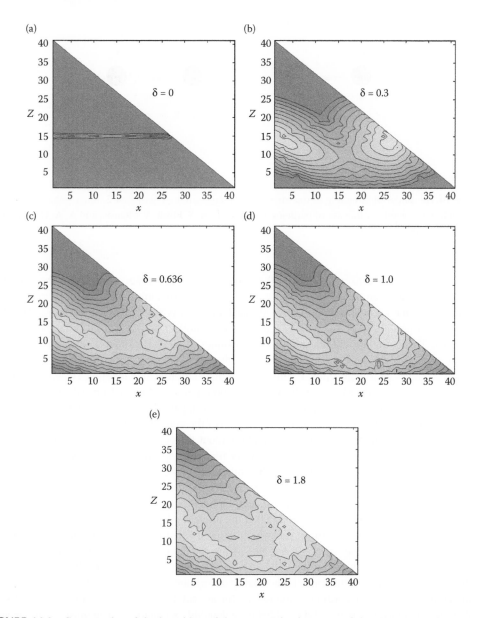

FIGURE 14.9 Contour plot of the logarithm of the symmetric eigenstate of the chosen tunneling pair (cf. Figure 14.7) for five different values of $\delta = 0, 0.3, 0.636, 1.0, 1.8$ (their location is indicated by filled circles in Figure 14.8). (a): Three equidistant grid lines are used; (b–e): 10 grid lines are used. Minimum value of squared wave function is 10^{-30}, maximum value is about 1. (Adapted from S. Flach, V. Fleurov, and A. A. Ovchinnikov, *Phys. Rev. B* 63, 094304, 2001.)

numbers are enough if the total particle number is fixed. Thus the final encoding of states (for a given value of b) can be chosen as $|x, z\rangle$. The abscissa in Figure 14.9 is x and the ordinate is z. Thus the intensity plots provide us with information about the order of particle flow in the course the tunneling process. For $\delta = 0$ (Figure 14.9a) the only possibility for the 26 particles on site 1 is to directly tunnel to site 2. Site 3 is decoupled with its 14 particles not participating in the process. The squared wave function takes the form of a compact rim in the (x, z) plane which is

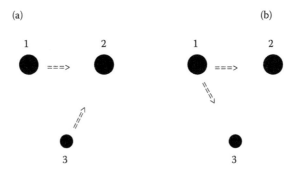

FIGURE 14.10 Order of tunneling in the trimer. Filled large circles-sites 1 and 2, filled small circle-site 3. Arrows indicate direction of transfer of particles. (Adapted from S. Flach, V. Fleurov, and A. A. Ovchinnikov, *Phys. Rev. B* 63, 094304, 2001.)

parallel to the x axis. Nonzero values of the wave function are observed only on the rim. This direct tunneling has been described in Section 14.2.2. When switching on some nonzero coupling to the third site, the particle number on the dimer (sites 1,2) is not conserved anymore. The third site serves as a particle reservoir which is able either to collect particles from or supply particles to the dimer. This coupling will allow for nonzero values of the wave function away from the rim. But most importantly, it will change the shape of the rim. We observe that the rim is bent down to smaller z values with increasing δ. This implies that the order of tunneling (when, e.g., going from large to small x values) is as follows: First, some particles tunnel from site 1 to site 2 and simultaneously from sites 3 to 2 (Figure 14.10a). Afterward, particles flow from site 1 to both sites 2 and 3 (Figure 14.10b). With increasing δ the structure of the wave-function intensity becomes more and more complex, possibly revealing information about the classical phase-space flow structure. Thus we observe three intriguing features. First, the tunneling splitting increases by eight orders of magnitude when δ increases from zero to 0.5. This seems to be unexpected, since at those values perturbation theory in δ should be applicable (at least Figure 14.7 indicates that this should be true for the levels themselves). The semiclassical explanation of this result was obtained in Flach et al. (2001) [20]. The second observation is that the tunneling begins with a flow of particles from the bath (site 3) directly to the empty site which is to be filled (with simultaneous flow from the filled dimer site to the empty one). At the end of the tunneling process the initially filled dimer site is giving particles back to the bath site. Again this is an unexpected result, since it implies that the particle number on the dimer is increasing during the tunneling, which seems to decrease the tunneling probability, according to the results for an isolated dimer. These first two results are closely connected (see Flach et al. (2001) [20] for a detailed explanation). The third result concerns the resonant structure on top of the smooth variation in Figure 14.8. The resonant enhancements and suppressions of tunneling are related to avoided crossings. Their presence implies that a fine tuning of the system parameters may strongly suppress or enhance tunneling which may be useful for spectroscopic devices. In Figure 14.11 we show the four various possibilities of avoided crossings between a pair and a single level and between two pairs, and the schematic outcome for the tunneling splitting [20]. If the interaction to further more distant states in the spectrum is added, the tunneling splitting can become exactly zero [21] for some specific value of the control parameter. In such a rare situation the tunneling is suppressed for all times.

14.2.4 QUANTUM ROTO-BREATHERS

When discussing classical breather solutions we have been touching some aspects of roto-breathers, including their property of being not invariant under time-reversal symmetry. In a recent study,

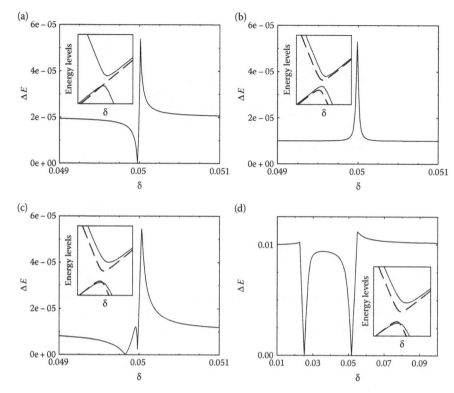

FIGURE 14.11 Level splitting variation at avoided crossings. Inset: Variation of individual eigenvalues participating in the avoided crossing. Solid lines—symmetric eigenstates; dashed lines—antisymmetric eigenstates. (Adapted from S. Flach, V. Fleurov, and A. A. Ovchinnikov, *Phys. Rev. B* 63, 094304, 2001.)

Dorignac et al. have performed [22] an analysis of the corresponding quantum roto-breather properties in a dimer with the Hamiltonian

$$H = \sum_{i=1}^{2} \left\{ \frac{p_i^2}{2} + \alpha(1 - \cos x_i) \right\} + \varepsilon(1 - \cos(x_1 - x_2)). \qquad (14.25)$$

The classical roto-breather solution consists of one pendulum rotating and the other oscillating with a given period T_b. Since the model has two symmetries—permutation of the indices and time-reversal symmetry—which may be both broken by classical trajectories, the irreducible representations of quantum eigenstates contain four symmetry sectors (with possible combinations of symmetric or antisymmetric states with respect to the two symmetry operations). Consequently, a quantum rotobreather state is belonging to a quadruplet of weakly split states rather than to a pair as discussed above. The schematic representation of the appearance of such a quadruplet is shown in Figure 14.12 [22]. The obtained quadruplet has an additional fine structure as compared to the tunneling pair of the above considered dimer and trimer. The four levels in the quadruplet define three characteristic tunneling processes. Two of them are energy or momentum transfer from one pendulum to the other one, while the third one corresponds to total momentum reversal (which restores time reversal symmetry). The dependence of the corresponding tunneling rates on the coupling ε is shown for a specific quadruplet from Dorignac and Flach (2002) [22] in Figure 14.13. For very weak coupling $\varepsilon \ll 1$ the fastest tunneling process will be momentum reversal, since tunneling between the pendula is blocked. However, as soon as the coupling is increased, the momentum reversal turns into the slowest process, with breather tunneling from one pendulum to the other one being orders

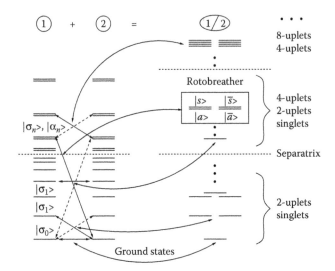

FIGURE 14.12 Schematic representation of the sum of two pendula spectra. Straight solid arrows indicate the levels to be added and dashed arrows the symmetric (permutation) operation. The result is indicated in the global spectrum by a curved arrow. The construction of the quantum rotobreather state is explicitly represented. (Adapted from J. Dorignac and S. Flach, *Phys. Rev. B* 65, 214305, 2002.)

of magnitude faster. Note that again resonant features on these splitting curves are observed, which are related to avoided crossings.

14.2.5 LARGE LATTICES WITH FLUCTUATING NUMBERS OF QUANTA

A number of publications are devoted to the properties of QBs in chains and two-dimensional lattices of coupled anharmonic oscillators. For the respective one-dimensional case, the Hamiltonian

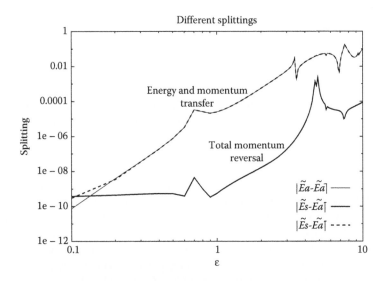

FIGURE 14.13 Dependence of different splittings of a quadruplet on ε. Only three of them have been displayed, each being associated with a given tunneling process. (Adapted from J. Dorignac and S. Flach, *Phys. Rev. B* 65, 214305, 2002.)

is given by

$$H = \sum_n \left[\frac{1}{2} p_n^2 + V(x_n) + W(x_n - x_{n-1}) \right].$$ (14.26)

Here $V(x) = \frac{1}{2} x^2 + \frac{1}{4} v_4 x^4$ (or similar) and the nearest-neighbor coupling $W(x) = \frac{1}{2} C x^2$. The classical version of such models conserves only the energy, but not any equivalent of a norm. Therefore, no matter whether one uses creation and annihilation operators of the harmonic oscillator [23], or similar operators which diagonalize the single anharmonic oscillator problem [24], the resulting Hamiltonian matrix will not commute with the corresponding number operator. Calculations will typically be restricted to 4–6 quanta, and lattice sizes of the order of 30 for $d = 1$, 13×13 for $d = 2$ [23]. With these parameters one can calculate properties of QB states, which correspond to typically two quanta which are bound (with unavoidable states with different number of quanta, contributing as well). For large enough v_4 a complete gap opens between the two-quanta continuum and QB states [10,23] (Figures 14.14 and 14.15). When decreasing the anharmonic constant v_4,

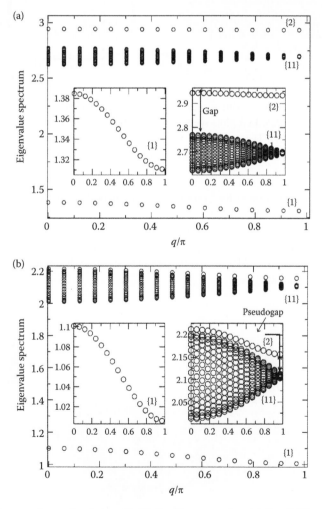

FIGURE 14.14 Eigenspectrum of a chain with 33 sites for parameters (a) $C = 0.05$, $v_4 = 0.2$, and (b) $C = 0.05$, $v_4 = 0.02$. The insets show magnifications of the fundamental branch (left) and overtone region (right). The QB branch is marked by (2), and the two-phonon band by (11). (Adapted from L. Proville, *Phys. Rev. B* 71, 104306, 2005.)

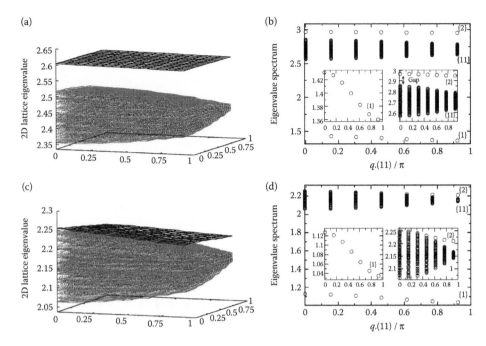

FIGURE 14.15 Eigenspectrum of a lattice with 13×13 sites for parameters (top) $C = 0.025$, $v_4 = 0.1$, and (bottom) $C = 0.025$, $v_4 = 0.025$. Left plots-spectra over the whole Brilloin zone. Right plots-profiles of the spectra along the direction [11]. The insets show the magnifications of the phonon branch (left) and the QB energy region (right). (Adapted from L. Proville, *Phys. Rev. B* 71, 104306, 2005.)

Proville found, that the gap closes for certain wave numbers, but persists for others, becoming a pseudogap [23,25] (Figure 14.14).

Involved calculations of the dynamical structure factor (e.g., available by neutron scattering in crystals) have shown, that signatures of QBs are imprinted in these integral characteristics of the underlying lattice dynamics [10,24], yet the working out of these differences may become a subtle task (see Figure 14.16 for example).

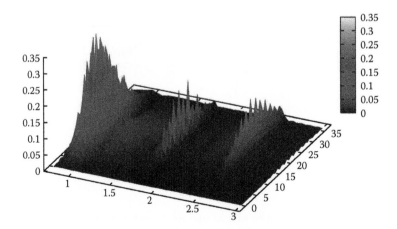

FIGURE 14.16 A 3D plot of the inelastic structure factor $S(q, \omega)$ as a function of the dimensionless energy transfer $0 \leq \omega \leq 3$ and the scalar product of the transfer momentum **q** and the polarization **u**. (Adapted from L. Proville, *Phys. Rev. B* 72, 184301, 2005.)

Finally, Fleurov et al. (1998) [26] estimated the influence of the tunnel splitting of a dimer, when embedded in an infinite chain. This situation is close to the tunneling of a very localized DB, so that the nonlinearity (interaction between bosons) can be taken into account only on the two sites, which participate in the tunneling, while the nonlinearity can be neglected on all other sites. Using path integral techniques, the computed tunneling splitting has been shown to become smaller as compared to the case of an isolated dimer. This is due to the fact, that a DB in an infinite chain has a core and a localized tail. That tail has to be carried through the tunneling process as well, and in analogy with a single particle tunneling in a double well, the tail increases the effective mass of such a particle. Consequently the exponential tail of a DB in an infinite chain tends to decrease its ability to perform quantum tunneling motion, yet it never leads to a full suppression of tunneling [26].

14.3 QB PROPERTIES

14.3.1 EVOLUTION OF QUANTUM LOCALIZED STATES

Suppose that we initially excite only one site in the trimer from above. If this initial state has strong enough overlap with tunneling pair eigenstates, its evolution in time should show distinct properties as compared to the case when the overlap is vanishing, or when there are simply no tunneling pair states available. Several results have been reported. First, a quantum echo was observed in Flach and Fleurov (1997) [19] by calculating the survival probability of the initial state as a function of time. That quantity measures the probability to find the system in the initial state at later times. If the initial state has strong overlap with many eigenstates, it is expected to quickly decohere into these different states. Yet, if a substantial overlap with QBs takes place, the survival probability first rapidly decays to zero, but echoes up after regular time intervals (Figure 14.17, left plot). If one simply measures the dependence of the number of quanta, then a similar situation will show up with a very slow beating of the occupation numbers in time, if the overlap of the initial state and a tunneling pair is strong [21,27] (see Figure 14.17, right plot).

Suppose we have a large lattice, and put initially many quanta on one site. Then any tunneling of this packet as a whole will occur on very long time scales. On time scales much shorter, we may describe the excitation as a classical DB state plus a small perturbation. Treating the perturbation quantum mechanically, one could expect that the time-periodic DB acts as a constant source of quantum radiation for the quantized phonon field. It turns out to be impossible, for the same reasons as in the purely classical treatment (see Flach et al. (2005) [30]). This result implies, that there is almost no other source of decay for a localized initial state in a quantum lattice, but to slowly tunnel as a whole along the lattice, if nonlinearities allow for the formation of exact classical DB states [31]. Numerical calculations for such a case, but with few quanta, were performed by Proville [32], and, similar to the above trimer discussion, showed that if QB states exist in the system, then localized excitations stay localized for times which are much longer than the typical phonon diffusion times in the absence of anharmonicity.

14.3.2 SPLITTING AND CORRELATIONS

QBs are nearly degenerate eigenstates. For the dimer and the trimer, they come in symmetric–antisymmetric pairs. So one may compute the nearest-neighbor energy spacing (tunneling splitting) between pairs of symmetric–antisymmetric eigenstates in order to identify QBs. Since QBs correspond to classical orbits that are characterized by energy localization, they may be identified by defining correlation functions. For large lattices it has been shown that QBs have exponentially localized correlation functions, in full analogy to their classical counterparts.

For the dimer and the trimer, the correlation functions may be defined as follows:

$$f_\mu(1,2) = \langle \hat{n}_1 \hat{n}_2 \rangle_\mu, \quad f_\mu(1,1) = \langle \hat{n}_1^2 \rangle_\mu, \tag{14.27}$$

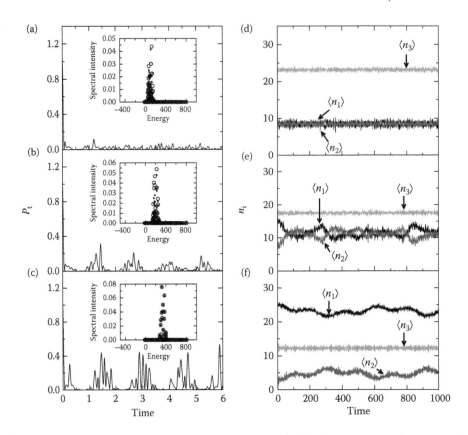

FIGURE 14.17 Left plot: Survival probability of the initial state $|\Psi_0\rangle = |20 + v, 0, 20 - v\rangle$. $v =$ (a) -6, (b) 0, and (c) 6. Insets: spectral intensity of the initial state $|\Psi_0\rangle$. Filled circles-symmetric eigenstates; open circles-antisymmetric eigenstates. Right plot: Time evolution of expectation values of the number of bosons at each site of the trimer for different initial states $|\Psi_0\rangle = |20 + v, 0, 20 - v\rangle$. $v =$ (d) -6, (e) 0, and (f) 6. (Adapted from R. A. Pinto and S. Flach, *Phys. Rev. A* 73, 022717, 2006.)

where $\hat{n}_i = \hat{a}_i^\dagger \hat{a}_i$, and $\langle \hat{A} \rangle_\mu = \langle \chi_\mu | \hat{A} | \chi_\mu \rangle$, $\{|\chi_\mu\rangle\}$ being the set of eigenstates of the system. The ratio $0 \leq f_\mu(1,2)/f_\mu(1,1) \leq 1$ measures the site correlation of quanta: it is small when quanta are site-correlated (i.e., when many quanta are located on one site there are almost none on the other one) and close to unity otherwise.

For the dimer case, the relation $b = n_1 + n_2$, leads to

$$f_\mu(1,2) = b\langle \hat{n}_1 \rangle_\mu - \langle \hat{n}_1^2 \rangle_\mu. \tag{14.28}$$

In Figure 14.18 (left) we show the energy splitting and the correlation function of the eigenstates. We see that beyond a threshold (dashed line), the splitting drops exponentially fast with energy. The corresponding pairs of eigenstates, which are tunneling pairs, are site correlated. Thus they are QBs. Their correlation functions show a fast decrease for energies above the threshold. In these states many quanta are localized on one site of the dimer and the tunneling time of such an excitation from one site to the other (given by the inverse energy splitting between the eigenstates of the pair) is exponentially large. As shown in Aubry et al. (1996) [17], this energy threshold is close to the threshold for the existence of DBs in the corresponding classical model.

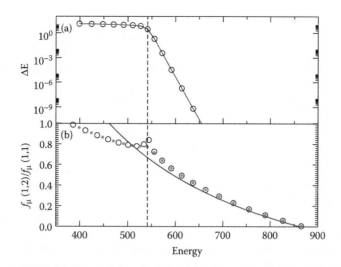

FIGURE 14.18 (a) Energy splitting and (b) correlation function vs. energy of the eigenstates of the dimer (open circles, symmetric eigenstates; solid circles, antisymmetric eigenstates). The vertical dashed line marks the energy threshold for appearance of QB states. The thin solid line in (a) is a guide for the eye, whereas in (b) it is the estimation using Equation 14.30. Here $b = 40$ and $C = 2$.

QBs are close to symmetric (S) and antisymmetric (A) eigenstates of the $C = 0$ case given by

$$|n_1, n_2\rangle_{S,A} = \frac{1}{\sqrt{2}}(|n_1, n_2\rangle \pm |n_2, n_1\rangle), \tag{14.29}$$

with $n_{1,2} \gg n_{2,1}$. So we may estimate the dependence of the correlation functions on n_1 using the eigenstates Equation 14.29 and $b = n_1 + n_2$. The result is

$$\frac{f(1,2)_{n_1}}{f(1,1)_{n_1}} = \frac{2n_1(b - n_1)}{n_1^2 + (b - n_1)^2}, \tag{14.30}$$

where we note that it is equal to unity when $n_1 = b/2$, and vanishes when $n_1 = 0, b$.

Using the relation between the eigenenergy ε of the $C = 0$ case and the number $n_1 \ (= 1, 2, \ldots, b)$

$$\varepsilon_{n_1} = \frac{5}{4} + \frac{3}{2}b + \frac{1}{2}\left[n_1^2 + (b - n_1)^2\right], \tag{14.31}$$

one may obtain the energy dependence of the correlation function (Equation 14.30), which is plotted in Figure 14.18b (thin solid line). We can see that beyond the energy threshold for appearance of QBs, the numerical results are close to the estimation (Equation 14.30).

14.3.3 ENTANGLEMENT

QBs may also be differentiated from other quantum states when measuring the degree of entanglement [33,34]. For the dimer and the trimer the degree of entanglement in the eigenstates may be measured by minimizing the distance of a given state to the space of product states of the dimer part (expanded by the product basis $\{|n_1\rangle \otimes |n_2\rangle\}$), which depends on the largest eigenvalue of the corresponding reduced density matrix [29,35–37]:

$$\Delta = \sum_{n_1, n_2}^{N} (\chi_{n_1, n_2} - f_{n_1} g_{n_2})^2, \tag{14.32}$$

where for the case of the dimer $\chi_{n_1,n_2} = \langle n_1,n_2 | \chi \rangle$, and for the trimer $\chi_{n_1,n_2} = \langle n_1, n_2, (n_3 = b - n_1 - n_2) | \chi \rangle$. The functions f_{n_1} and g_{n_2} are such that Δ is minimum [29]. Δ measures how far a given eigenstate of the system is from being a product of single-site states, and has values $0 < \Delta < 1$. This measure has a direct relation to the distance of a given eigenstate from a possible one obtained after performing a Hartree approximation [35].

For the dimer, since QB states are close to eigenstates of the $C = 0$ case

$$|\chi\rangle_{QB} \simeq \frac{1}{\sqrt{2}}(|n,0\rangle \pm |0,n\rangle), \tag{14.33}$$

with $n \lesssim b$, one expects that the degree of entanglement in QB states is similar to the degree of entanglement in such states. Since only two basis states are involved, it cannot be a state of maximum entanglement. For $C = 0$ the eigenstates of the system are the basis states given by (Equation 14.29), where for $n_1 = n_2$ it follows that $\Delta = 0$, and for $n_1 \neq n_2$ (which includes the state in Equation 14.33) $\Delta = 0.5$.

In previous works in a similar quantum dimer model [33,34], it was shown that at the energy threshold for appearance of QB states the entanglement (in this case measured in a different way) becomes maximum and then decreases with energy. From this, and the above reasoning, we expect that QB states show decreasing entanglement Δ with energy, tending to 0.5. Results in left panel a of Figure 14.19 agree with this expectation.

For $C = 0$, the entanglement has the values 0 and 0.5 corresponding to the basis states $|b/2, b/2\rangle$ and $|n, b - n\rangle$ ($n \neq b/2$) with equal and distinct number of quanta at each site respectively. When $C > 0$, the eigenstates become linear superpositions of the basis states and the entanglement rises, being larger as long as more basis states are involved in building up an eigenstate. This can be seen in the right panel of Figure 14.19, where we plot the density $\rho(n_1, n_2 = b - n_1) = |\langle n_1, b - n_1 | \chi \rangle|^2 \equiv \rho(n_1)$ of four symmetric eigenstates marked by labeled arrows in the left panel: the low-energy eigenstate marked by the arrow 1 consists mainly of one basis state, $|b/2, b/2\rangle$, as seen in the right panel a, hence the entanglement is relatively small. When going up in energy the entanglement in the eigenstates quickly increases, becoming maximum at the energy threshold, and then decreases. An eigenstate just before the threshold like the one marked by the arrow 2 in the left panel involves many basis states fulfilling $n_1 + n_2 = b = 40$ (right panel b), hence the entanglement is large. However, for a QB state lying in the energy region beyond the threshold, like the one marked by the arrow 3 in the left panel, the number of involved basis states, and thus the entanglement starts to

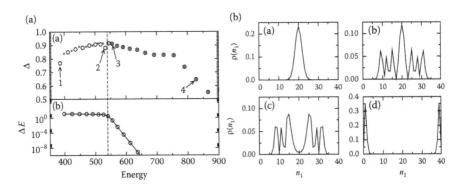

FIGURE 14.19 Left panel (a) Entanglement of the eigenstates and (b) energy splitting as a function of energy in the dimer (open circles, symmetric eigenstates; solid circles, antisymmetric eigenstates). The vertical dashed line marks the energy threshold for appearance of QB states. Here $b = 40$, and $C = 2$. Right panel: The density of the symmetric eigenstates marked by labeled arrows in left panel-a: (a) S-0 (arrow 1), (b) S-7 (arrow 2), (c) S-9 (arrow 3), (d) S-19 (arrow 4).

decrease (right panel c). Finally, in high-energy eigenstates like the one marked by the arrow 4 in the left panel, which has the form shown in Equation 14.33 (right panel d), the entanglement is even smaller and gets close to 0.5 as expected.

From the above results we see that by measuring entanglement one may gain information not only about the energy threshold for existence of QBs (also visible when measuring the energy splitting and correlation function), but also about how many basis states overlap strongly with the eigenstate under consideration. We also computed the *von Neumann entropy* [38], which is another standard measure of entanglement, and the results were consistent with those discussed above.

14.4 QUANTUM EDGE-LOCALIZED STATES

Most of the studies about QBs in large lattices (with few bosons) were done considering a system with periodic boundary conditions, and thus, translational invariant. However, usually real systems have to be modeled with open boundary conditions. Hence, it is natural to wonder what happens with QBs when the lattice has finite size, and therefore, no translational invariance.

In the classical case, it has been shown that in finite nonlinear lattices, the breaking of the translational symmetry may lead to the formation of so called *nonlinear edge states*. These are excitations which are localized at the edges of the lattice and they have been studied, in particular, in nonlinear optics experiments employing optical waveguides, being coined with the name of *discrete surface solitons* [39]. In particular, it is well known that the discrete nonlinear Schrödinger equation (DNLS) has time-periodic solutions localized at the edge of the lattice [40,41]. It is therefore expected that the large-boson limit of the open-boundary Bose–Hubbard model will show eigenstates in which the bosons are localized at the edge.

Numerical studies by Pouthier were done to answer the edge-localization question in a lattice with a few bosons [42], where the mean-field approximation (DNLS equation) cannot *a priori* be expected to provide the correct intuition. The answer turns out to be subtle—this phenomenon is not present for the case of two particles, but appears when the particle number is three or more [42]. Further studies focused on the energy spectrum and eigenstates (Figure 14.20) were done by Pinto et al. (2009) [43], where degenerate perturbation theory allowed to explain why edge-localized states exist only if the number of particles is three or more [43].

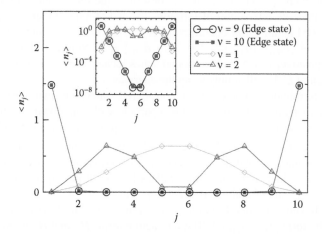

FIGURE 14.20 Spatial profile of site occupancies for several eigenstates of the three-boson Bose–Hubbard chain. The index ν counts the eigenstates from the lowest-energy one ($\nu = 1$). Inset shows the same plot in semilog scale, the linear behavior indicating exponential localization in the edge states. (Adapted from R. A. Pinto, M. Haque, and S. Flach, *Phys. Rev. A* 79, 052118, 2009.)

14.5   QBs WITH FERMIONS

Advances in experimental techniques of manipulation of ultracold atoms in optical lattices make it feasible to explore the physics of few-body interactions. Systems with few quantum particles on lattices have new unexpected features as compared to the condensed matter case of many-body interactions, where excitation energies are typically small compared to the Fermi energy. Therefore, it is of interest to study binding properties of fermionic pairs with total spin zero, as recently presented in Nguenang and Flach (2009) [44]. We use the extended Hubbard model, which contains two interaction scales—the on-site interaction U and the nearest-neighbor intersite interaction V. The nonlocal interaction V is added in condensed matter physics to emulate remnants of the Coulomb interaction due to nonperfect screening of electronic charges. For fermionic ultracold atoms or molecules with magnetic or electric dipole–dipole interactions, it can be tuned with respect to the local interaction U by modifying the trap geometry of a condensate, additional external DC electric fields, combinations with fast rotating external fields, and so on (for a review and relevant references see Baranov (2008) [45]).

Consider a one-dimensional lattice with f sites and periodic boundary conditions described by the extended Hubbard model with the following Hamiltonian:

$$\hat{H} = \hat{H}_0 + \hat{H}_U + \hat{H}_V, \tag{14.34}$$

where

$$\hat{H}_0 = -\sum_{j,\sigma} \hat{a}_{j,\sigma}^+ (\hat{a}_{j-1,\sigma} + \hat{a}_{j+1,\sigma}), \tag{14.35}$$

$$\hat{H}_U = -U \sum_j \hat{n}_{j,\uparrow} \hat{n}_{j,\downarrow}, \quad \hat{n}_{j,\sigma} = \hat{a}_{j,\sigma}^+ \hat{a}_{j,\sigma}, \tag{14.36}$$

$$\hat{H}_V = -V \sum_j \hat{n}_j \hat{n}_{j+1}, \quad \hat{n}_j = \hat{n}_{j,\uparrow} + \hat{n}_{j,\downarrow}. \tag{14.37}$$

$\hat{H}_0$ describes the nearest-neighbor hopping of fermions along the lattice. Here the symbols $\sigma = \uparrow, \downarrow$ stand for a fermion with spin up or down. $\hat{H}_U$ describes the onsite interaction between the particles, and $\hat{H}_V$ the intersite interaction of fermions located at adjacent sites. $\hat{a}_{j,\sigma}^+$ and $\hat{a}_{j,\sigma}$ are the fermionic creation and annihilation operators satisfying the corresponding anticommutation relations: $\{\hat{a}_{j,\sigma}^+, \hat{a}_{l,\sigma'}\} = \delta_{j,l} \delta_{\sigma,\sigma'}$, $\{\hat{a}_{j,\sigma}^+, \hat{a}_{l,\sigma'}^+\} = \{\hat{a}_{j,\sigma}, \hat{a}_{l,\sigma'}\} = 0$. Note that throughout this work we consider U and V positive, which leads to bound states located below the two-particle continuum. A change of the sign of U, V will simply swap the energies.

To observe the fermionic character of the considered states, any two-particle number state is generated from the vacuum $|O\rangle$ by first creating a particle with spin down, and then a particle with spin up: for example, $\hat{a}_{2,\uparrow}^+ \hat{a}_{1,\downarrow}^+ |O\rangle$ creates a particle with spin down on site 1 and one with spin up on site 2, while $\hat{a}_{2,\uparrow}^+ \hat{a}_{2,\downarrow}^+ |O\rangle$ creates both particles with spin down and up on site 2.

Due to periodic boundary conditions the Hamiltonian equation 14.34 commutes also with the translation operator $\hat{T}$, which shifts all lattice indices by one. It has eigenvalues $\tau = \exp(ik)$, with Bloch wave number $k = 2\pi\nu/f$ and $\nu = 0, 1, 2, ..., f-1$.

For the case of having only one fermion (either spin up or spin down) in the lattice ($n = 1$), a number state has the form $|j\rangle = \hat{a}_{j,\sigma}^+ |O\rangle$. The interaction terms $\hat{H}_U$ and $\hat{H}_V$ do not contribute. For a given wave number k, the eigenstate to Equation 14.34 is, therefore, given by

$$|\Psi_1\rangle = \frac{1}{\sqrt{f}} \sum_{s=1}^f \left(\frac{\hat{T}}{\tau}\right)^{s-1} |1\rangle. \tag{14.38}$$

The corresponding eigenenergy

$$\varepsilon_k = -2\cos(k). \tag{14.39}$$

For two particles, the number state method involves $N_s = f^2$ basis states, which is the number of ways one can distribute two fermions with opposite spins over the f sites including possible double occupancy of a site. Below we consider only cases of odd f for simplicity. Extension to even values of f is straightforward. The details of the calculations can be found in Nguenang and Flach (2009) [44].

In Figure 14.21 we show the energy spectrum of the Hamiltonian matrix obtained by numerical diagonalization for the interaction parameters $U = 2$, $V = 2$, and $f = 101$. At $U = 0$ and $V = 0$, the spectrum is given by the two fermion continuum, whose eigenstates are characterized by the two fermions independently moving along the lattice. In this case the eigenenergies are the sum of the two single-particle energies:

$$E^0_{k_1,k_2} = -2[\cos(k_1) + \cos(k_2)], \tag{14.40}$$

with $k_{1,2} = \pi v_{1,2}/(f+1)$, $v_{1,2} = 1, \ldots, f$. The Bloch wave number $k = k_1 + k_2 \mod 2\pi$. Therefore, if $k = \pm\pi$, the continuum degenerates into points. The continuum is bounded by the hull curves $h_\pm(k) = \pm 4\cos k/2$. The same two-particle continuum is still observed in Figure 14.21 for nonzero interaction. However, in addition to the continuum, we observe one, two or three bound states dropping out of the continuum, which depends on the wave number. For any nonzero U and V, all three bound states drop out of the continuum at $k = \pm\pi$. One of them stays bounded for all values of k. The two other ones merge with the continuum at some critical value of $|k|$ upon approaching $k = 0$ as observed in Figure 14.21. Note that for $k = \pm\pi$ and $U = V$, all three bound states are degenerate.

Upon increasing U and V, we observe that a second bound state band separates from the continuum for all k (Figure 14.22). At the same time, when $U \neq V$, the degeneracy at $k = \pm\pi$ is reduced to two.

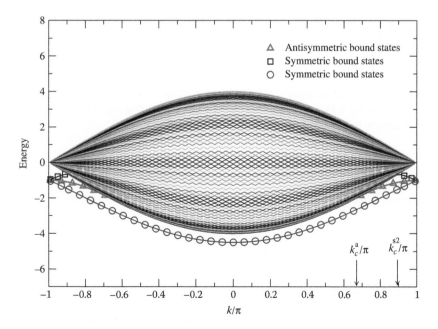

FIGURE 14.21 Energy spectrum of the two fermion states. The eigenvalues are plotted as a function of the wave number k. Here $U = 1, V = 1, f = 101$. Symbols are from analytical derivation, lines are the result of numerical diagonalization. The arrows indicate the location of the critical wave numbers. (Adapted from J.-P. Nguenang and S. Flach, *Phys. Rev. A* 80, 015601, 2009.)

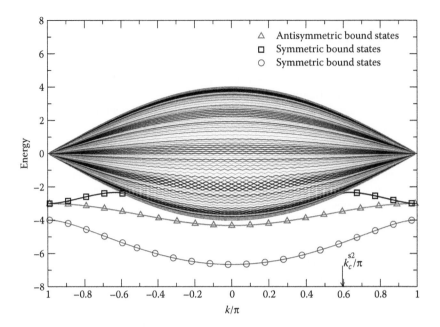

FIGURE 14.22 Energy spectrum of the two fermion states. The eigenvalues are plotted as a function of the wave number k. Here $U = 4, V = 3, f = 101$. The symbols are from analytical derivation, lines are the results of numerical diagonalization. The arrow indicates the location of the critical wave number. (Adapted from J.-P. Nguenang and S. Flach, *Phys. Rev. A* 80, 015601, 2009.)

Finally, for even larger values of U and V, all three bound state bands completely separate from the continuum (Figure 14.23).

14.5.1 SYMMETRIC AND ANTISYMMETRIC STATE REPRESENTATION

In order to obtain analytical estimates on the properties of the observed bound states, we use the fact that the Hamiltonian for a two fermion state is invariant under flipping the spins of both particles. We define symmetric basis states

$$|\Phi_{j,s}\rangle = \frac{1}{\sqrt{2}}(|\Phi_{j,+}\rangle + |\Phi_{j,-}\rangle) \tag{14.41}$$

and antisymmetric states

$$|\Phi_{j,a}\rangle = \frac{1}{\sqrt{2}}(|\Phi_{j,+}\rangle - |\Phi_{j,-}\rangle). \tag{14.42}$$

Note that $|\Phi_1\rangle$ is a symmetric state as well.

14.5.2 ANTISYMMETRIC BOUND STATES

The antisymmetric states exclude double occupation. Therefore, the spectrum is identical with the one of two spinless fermions [13]. Following the derivations in Scott et al. (1994) [13] we find that the antisymmetric bound state, if it exists, has an energy

$$E_2^a(k) = -\left(V + \frac{4}{V}\cos^2\left(\frac{k}{2}\right)\right). \tag{14.43}$$

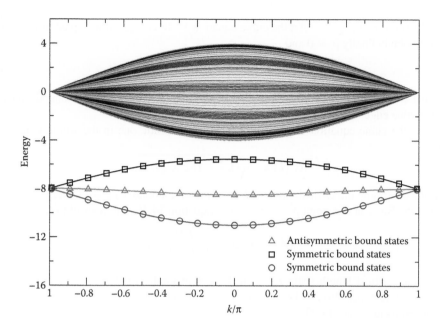

FIGURE 14.23 Energy spectrum for the two fermion states. The eigenvalues are plotted as a function of the wave number k. Here $U = 8, V = 8, f = 101$. The symbols are from analytical derivation, lines are the results of numerical diagonalization. (Adapted from J.-P. Nguenang and S. Flach, *Phys. Rev. A* 80, 015601, 2009.)

This result is valid as long as the bound state energy stays outside of the continuum. The critical value of k at which validity is lost, is obtained by requesting $|E_2^a(k)| = |h_\pm(k)|$. It follows $V = 2\cos(k/2)$. Therefore, the antisymmetric bound state merges with the continuum at a critical wave number

$$k_c^a = 2\arccos\left(\frac{V}{2}\right), \tag{14.44}$$

setting a critical length scale $\lambda_c^a = 2\pi/k_c^a$. For $V = 1$ it follows $k_c^a/\pi \approx 0.667$ (see Figure 14.21).

Equation 14.43 is in excellent agreement with the numerical data in Figures 14.21 through 14.23 (cf. open triangles). We also note, that the antisymmetric bound state is located between the two symmetric bound states, which we discuss next.

14.5.3 SYMMETRIC BOUND STATES

A bound state can be searched for by assuming an unnormalized eigenvector of the form $|c, 1, \mu, \mu^2, \mu^3, \ldots\rangle$ with $|\mu| \equiv \rho \leq 1$. We obtain [44]

$$E_2^s(k) = -2\left(\rho + \frac{1}{\rho}\right)\cos k/2. \tag{14.45}$$

The parameter ρ satisfies a cubic equation

$$a\rho^3 + b\rho^2 + c\rho + d = 0 \tag{14.46}$$

with the real coefficients $a, b, c,$ and d given by $a = 2V\cos(k/2), b = 4\cos^2(k/2) - UV, c = 2(U + V)\cos(k/2), d = -4\cos^2(k/2)$. We plot the results in Figures 14.21 through 14.23 (cf. open circles and squares). We obtain excellent agreement.

At the Brilloin zone edge $k = \pm\pi$ the cubic equation 14.46 is reduced to a quadratic one, and can be solved to obtain finally $\rho \to 0$ and

$$E_2^{s1}(k \to \pm\pi) = -U, \quad E_2^{s2}(k \to \pm\pi) = -V. \tag{14.47}$$

In particular we find for $k = \pm\pi$ that $E_2^{s2} = E_2^a$. In addition, if $U = V$, all three bound states degenerate at the zone edge.

If $V = 0$, the cubic equation 14.46 is reduced to a quadratic one in the whole range of k and yields [13]

$$E_2^{s1}(k) = -\sqrt{U^2 + 16\cos^2(k/2)}. \tag{14.48}$$

Next, we determine the critical value of k for which the bound state with energy E_2^{s2} is joining the continuum. Since at this point $\rho = 1$, we solve Equation 14.46 with respect to k_c and find

$$k_c^{s2} = 2\arccos\left(\frac{UV}{2(U + 2V)}\right) \tag{14.49}$$

setting another critical length scale $\lambda_c^s = 2\pi/k_c^s$. For example, for $U = V = 1$ $k_c^{s2}/\pi \approx 0.89$, in excellent agreement with Figure 14.21. For $U = 4$ and $V = 3$ we find $k_c^{s2}/\pi \approx 0.59$ confirming numerical results in Figure 14.22.

14.6 SMALL JOSEPHSON JUNCTION NETWORKS

Recent studies of Pinto et al. [28,29] deal with QB excitations in two capacitively coupled Josephson junctions. Such systems are currently under experimental investigation, being candidates for quantum information processing, and show remarkably long coherence times up to 100 ns for few quanta excitations. The system does not conserve the number of excited quanta, and can be best compared with the above Bose–Hubbard trimer. QB signatures are found simultaneously in the spectra (tunneling splittings), correlation functions, entanglement, and quanta number fluctuations.

We address the excitation of QBs in a system of two coupled Josephson junctions in the quantum regime [29]. Josephson junctions are nonlinear devices that show macroscopic quantum behavior, and nowadays they can be manipulated with high precision, in such a way that the energy flow between coupled junctions can be resolved in time.

Josephson junctions behave like anharmonic oscillators, and by lowering the temperature one can bring them into the quantum regime, where their quantization leads to energy levels which are nonequidistant because of the anharmonicity. These levels can be separately excited by using microwaves pulses, and the energy distribution between the junctions can be measured in time using subsequent pulses [46]. So far these techniques have been used for experiments on quantum information processing with Josephson junctions [47–49], but we think that arrays of Josephson junctions in the quantum regime also might be used as playgrounds for experiments on quantum dynamics of excitations in nonlinear lattices.

The system is sketched in Figure 14.24a—two Josephson Junctions (JJs) are coupled by a capacitance C_c, and they are biased by the same current I_b. The strength of the coupling due to the capacitor is $\zeta = C_c/(C_c + C_J)$. The dynamics of a biased JJ is analogous to the dynamics of a particle with a mass proportional to the junction capacitance C_J, moving on a tilted washboard potential

$$U(\varphi) = -I_c \frac{\Phi_0}{2\pi}\cos\varphi - I_b\varphi\frac{\Phi_0}{2\pi}, \tag{14.50}$$

which is sketched in Figure 14.24b. Here φ is the phase difference between the macroscopic wave functions in both superconducting electrodes of the junction, I_b is the bias current, I_c is the critical current of the junction, and $\Phi_0 = h/2e$ is the flux quantum. When the energy of the particle is large enough to overcome the barrier ΔU (that depends on the bias current I_b) it escapes and moves down

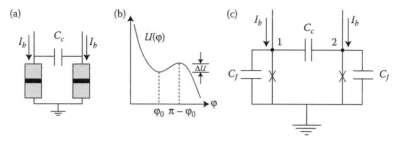

FIGURE 14.24 (a) Sketch of the two capacitively coupled JJs. (b) Sketch of the washboard potential for a single current-biased JJ. (c) Circuit diagram for two ideal capacitively coupled JJs.

the potential, switching the junction into a resistive state with a nonzero voltage proportional to $\dot{\varphi}$ across it. Quantization of the system leads to discrete energy levels inside the potential wells, which are nonequidistant because of the anharmonicity. Note that even if there is not enough energy to classically overcome the barrier, the particle may perform a quantum escape and tunnel outside the well, thus switching the junction into the resistive state. Thus each state inside the well is characterized by a bias and state-dependent lifetime, or its inverse—the escape rate.

The Hamiltonian of the system is

$$H = \frac{P_1^2}{2m} + \frac{P_2^2}{2m} + U(\varphi_1) + U(\varphi_2) + \frac{\zeta}{m} P_1 P_2, \qquad (14.51)$$

where

$$m = C_J(1 + \zeta)\left(\frac{\Phi_0}{2\pi}\right)^2, \qquad (14.52)$$

$$P_{1,2} = (C_c + C_J)\left(\frac{\Phi_0}{2\pi}\right)^2 (\dot{\varphi}_{1,2} - \zeta\dot{\varphi}_{2,1}). \qquad (14.53)$$

Note that the conjugate momenta $P_{1,2}$ are proportional to the charge at the nodes of the circuit (which are labeled in Figure 14.24c).

In the quantum case the energy eigenvalues and the eigenstates of the system were computed and analyzed in Pinto and Flach (2008) [29]. In Figure 14.25 we show the nearest-neighbor energy spacing (tunneling splitting) and the correlation function of the eigenstates. For this, and all the rest, we used $I_c = 13.3\,\mu\text{A}$, $C_J = 4.3$ pF, and $\zeta = 0.1$, which are typical values in experiments. We see that in the central part of the spectrum the energy splitting becomes small in comparison to the average. The corresponding pairs of eigenstates, which are tunneling pairs, are site correlated, and thus QBs. In these states many quanta are localized on one junction and the tunneling time of such an excitation from one junction to the other (given by the inverse energy splitting between the eigenstates of the pair) can be exponentially large and depend sensitively on the number of quanta excited.

Note that the tunneling of quanta between the JJs occurs without an obvious potential energy barrier being present (the interaction between the junctions is only through their momenta). This process has been coined *dynamical tunneling* [50–52], to distinguish from the usual tunneling through a potential barrier. In dynamical tunneling, the barrier—a so-called invariant separatrix manifold—is only visible in phase space, where it separates two regions of regular classical motion between which the tunneling process takes place. Therefore, when referring to the tunneling between the JJs, we implicitly mean that it is dynamical.

The fact that the strongest site correlated eigenstates occur in the central part of the energy spectrum may be easily explained as follows: let N be the highest excited state in a single junction, with

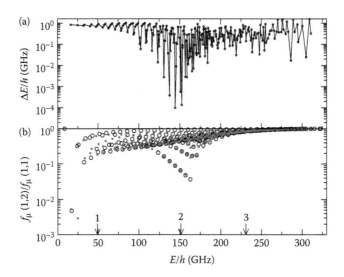

FIGURE 14.25 (a) Energy splitting and (b) correlation function versus energy of the eigenstates of the two-junctions system (open circles, symmetric eigenstates; filled circles, antisymmetric eigenstates). The labeled arrows mark the energy corresponding to the peak of the spectral intensity in Figure 14.26b,d, and f. The parameters are $\gamma = I_b/I_c = 0.945$ and $\zeta = 0.1$ (22 levels per junction).

a corresponding maximum energy ΔU (Figure 14.24). For two junctions the energy of the system with both junctions in the Nth state is $2\Delta U$, which roughly is the width of the full spectrum. Thus states of the form $|N,0\rangle$ and $|0,N\rangle$ that have energy ΔU are located approximately in the middle.

Having the eigenvalues and eigenstates, we compute the time evolution of different initially localized excitations, and the expectation value of the number of quanta at each junction $\langle \hat{n}_i \rangle(t) = \langle \Psi(t)|\hat{n}_i|\Psi(t)\rangle$. Results are shown in Figure 14.26a,c, and e. We also compute the spectral intensity $I_\mu^0 = |\langle \chi_\mu | \Psi_0 \rangle|^2$, which measures the strength of overlap of the initial state $|\Psi_0\rangle$ with the eigenstates. Results are shown in Figure 14.26b,d, and f, where we can see a peak in each case, which corresponds to the arrows in Figure 14.25b. We can see that the initial state $|\Psi_0\rangle = |0,5\rangle$ overlaps with eigenstates with an energy splitting between them being relatively large and hence the tunneling time of the initially localized excitation is short. For the case $|\Psi_0\rangle = |0,19\rangle$ QBs are excited: The excitation overlaps strongly with tunneling pairs of eigenstates in the central part of the spectrum, which are site correlated and nearly degenerate. The tunneling time of such an excitation is very long, and thus keeps the quanta localized on their initial excitation site for corresponding times. Finally, the initial state $|\Psi_0\rangle = |9,19\rangle$ overlaps with weakly site-correlated eigenstates with large energy splitting. Hence, the tunneling time is short again.

We computed also the time evolution of the expectation values of the number of quanta for initial conditions which are coherent or incoherent (mixtures) superpositions of product basis states with equal weights. This is relevant for experiments, since it may be hard to excite one junction to a determined state but easier to excite several states of the junction at the same time. We used coherent superpositions (characterized by well-defined states $|\Psi_0\rangle$), and mixtures (characterized by their corresponding density operators $\hat{\rho}_0$), of four basis states around the already used initial states: For the state $|0,5\rangle$ we superposed the basis states $|0,5\rangle$, $|0,6\rangle$, $|0,7\rangle$, and $|0,8\rangle$, for $|0,19\rangle$ the basis states $|0,20\rangle$, $|0,19\rangle$, $|0,18\rangle$, and $|0,17\rangle$, and for $|9,19\rangle$ the basis states $|9,20\rangle$, $|9,19\rangle$, $|9,18\rangle$, and $|9,17\rangle$. Both for superposition and mixture of basis states, the results are qualitatively similar to those shown in Figure 14.26. Therefore, we expect that some imprecision in exciting an initial state in the junctions would not affect in a relevant way the results. We may also conclude that the excitation of QB states does not rely on the phase coherence.

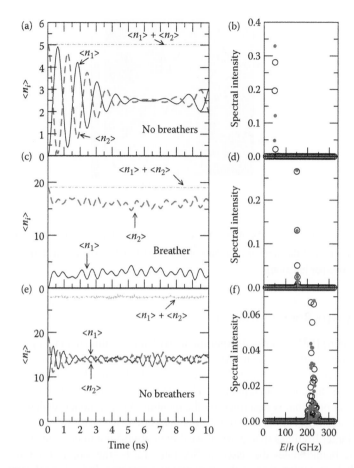

FIGURE 14.26 Time evolution of expectation values of the number of quanta at each junction (left panels) for different initial excitations with corresponding spectral intensities (right panels). (a) and (b): $|\Psi_0\rangle = |0,5\rangle$; (c) and (d): $|\Psi_0\rangle = |0,19\rangle$; (e) and (f): $|\Psi_0\rangle = |9,19\rangle$. Open circles, symmetric eigenstates; filled circles, antisymmetric eigenstates. The energies of the peaks in the spectral intensity are marked by labeled arrows in Figure 14.25b. The parameters are $\gamma = Ib/Ic = 0.945$ and $\zeta = 0.1$ (22 levels per junction).

Let us estimate how many quanta should be excited in the junctions in order to obtain QBs (tunneling pairs). We compute the density $\rho(n_1, n_2) = |\langle n_1, n_2 | \chi \rangle|^2$ of the asymmetric state $|\chi\rangle = (|\chi_b^{(S)}\rangle + |\chi_b^{(A)}\rangle)/\sqrt{2}$, where $|\chi_b^{(S,A)}\rangle$ are the eigenstates belonging to a tunneling pair [29]. In Figure 14.27 we show a contour plot of the logarithm of the density for the tunneling pair with energy marked by the arrow labeled by number two in Figure 14.25b. We see that the density has its maximum around $n_1 = 19$ and $n_2 = 0$ which is consistent with the result shown in Figure 14.26c and d where QBs were excited by using this combination of quanta in the junctions.

14.7 EXPERIMENTAL REALIZATIONS

There is a fast growing amount of experimental and related theoretical work on applying the quantum DB concept to many different branches in physics, like Bose–Einstein condensates, crystals and molecules, surfaces, and others. We will discuss some of these at length, while others will be reviewed briefly.

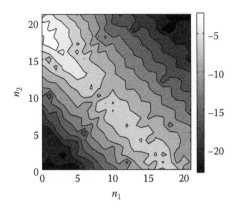

FIGURE 14.27 Contour plot of the logarithm of the density of the asymmetric state $|\chi\rangle = (|\chi_b^{(S)}\rangle +$ $|\chi_b^{(A)}\rangle)/\sqrt{2}$ as a function of the number of quanta at junctions 1 and 2. The parameters are $\gamma = I_b/I_c = 0.945$ and $\zeta = 0.1$ (22 levels per junction).

14.7.1 REPULSIVELY BOUND ATOM PAIRS

Winkler et al. (2006) [53] performed experiments with a three-dimensional optical lattice with initially each site being either not occupied, or being occupied by two Rb atoms bound in a pair due to attractive interaction. A magnetic field sweep across the Feshbach resonance changes the sign of interaction, turning attraction into repulsion. The dynamics of ultracold atoms loaded into the lowest band of the optical potential is described by the quantum DNLS model, which is equivalent to the Bose–Hubbard model (Equation 14.10). Lifetime measurements have shown, that repulsive pairs of Rb atoms have larger lifetimes than pairs of weakly or almost not interacting atoms (Figure 14.28). The two-particle bound states discussed in Section 14.2.1—the simplest versions of a quantum DB—are the obvious explanation of the experimental findings. Indeed, neglecting Landau–Zener transitions to higher-lying bands in the optical potential, the Bose–Hubbard model is justified. The sign of the interaction does not play any role, since it only changes quantum DBs from being low lying to being excited states, not affecting their localization properties. The most simple argument of why two quanta (or atoms) placed initially close to each other, do not separate despite they repel each other, is based on the fact that if they would do so, the (large) interaction energy should be converted into kinetic energy, which is restricted to be less than two times the width of the single particle band. In other words, repulsively bound atom pairs are a straightforward consequence of quantum DB states with two quanta.

Another sophisticated experimental investigation aimed at measuring the quasimomentum distribution of atom pairs in various regimes by mapping it onto a spatial distribution, which was finally measured using standard absorption imaging (Figure 14.29). Therefore, predictions of such states, which were made more than 30 years ago by Ovchinnikov (1969/1970) [54], were confirmed experimentally with ultracold repulsive atoms.

14.7.2 MOLECULES

Intramolecular vibrational energy redistribution (IVR) has been a central issue in the field of chemical physics for many decades. In particular, pathways and rates are of importance there, since understanding them allows to describe for example, the dynamics of various chemical reactions, and dissociation processes [55]. Spectroscopical studies, where single vibrational quanta are excited, allow to measure the frequencies of normal vibrational modes—that is, to characterize the dynamics of a molecule for small amplitude vibrations. These normal modes consist of coherent combinations

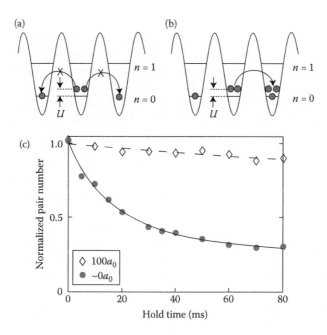

FIGURE 14.28 (a) Repulsive interaction between two atoms sharing a lattice site gives rise to an interaction energy U. Breaking up of the pair is suppressed owing to the lattice band structure and energy conservation. This is the simplest version of a quantum DB. (b) The DB can tunnel along the lattice. (c) Long lifetimes of strongly repulsive atoms. The plot shows the remaining fraction of pairs for strong interaction (open diamonds) and for almost vanishing interaction (filled circles). (Adapted from K. Winkler et al., *Nature* 441, 853, 2006.)

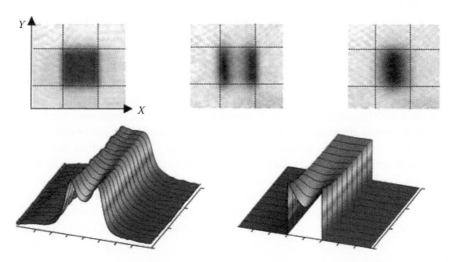

FIGURE 14.29 (Upper row) Absorption images of the atomic distribution after release from the 3D lattice and a subsequent 15 ms time of flight. The horizontal and vertical black lines enclose the first Brillouin zone. (upper left) Lattice sites are occupied by single atoms; (upper middle) repulsively bound atom pairs; (upper right) attractively bound atom pairs; (bottom row) quasimomentum distribution for pairs in one direction as a function of the lattice depth after integrating out the other direction. (Bottom left) Experiment; (bottom right) numerical calculations. (Adapted from K. Winkler et al., *Nature* 441, 853, 2006.)

of vibrational excitations of several bonds (or rotational groups) in a molecule. However, in order, for example, to dissociate a molecule, a many quanta excitation is needed, and nonlinearities will certainly become important. It was realized then, that strong vibrational excitations of molecules are much better described by so-called *local modes*—that is, basically one or few bond vibrational excitations. That transition from normal to local modes remained a puzzle for a long time. A practically complete modern theoretical account on these issues can be found in a recent monography by Ovchinnikov et al. [56]. On its most abstract level, the transition from normal to local modes is identical with the bifurcation in the dimer model. Thus, local modes are essentially DBs or slight perturbations of them. Note, that the connection between local modes, breathers and periodic orbits has been recently studied by Farantos in the context of large biological molecules [57]. DBs (ILMs) have been theoretically predicted to exist in ionic crystals [58], ways of optical excitation of DBs (ILMs) have been proposed [59], and their possible presence in hydrocarbon structures has been discussed [60].

Exciting local modes in molecules with discrete symmetries leads to small tunneling splittings of excitation levels [56], and goes back to the work of Child and Lawton [61], see also a recent comprehensive review by Keshavamurthy [62] and references therein. On its most abstract level, this effect is identical with the tunneling splitting in the permutationally symmetric dimer model discussed in Section 14.2.2.

An early example of experimental evidence of DB excitations in molecules comes from spectroscopical studies of visible red absorption spectra of benzene, naphtalene, and anthracene by Swofford et al. [64]. The C–H stretching vibrations have been excited to the sixth quantum level, and red shifts of the lines show that instead of a delocalized excitation of six bonds to the first level (yielding six quanta), the excitation resides on just one of the six available bonds. While it can tunnel (as a quantum DB) to the other bonds, this tunneling time is a new large time scale in the problem, strongly affecting, for example, dissociation rates.

A recent study of femtosecond infrared pump-probe spectroscopy of the N–H mode of a stable α-helix (poly-γ-benzyl-L-glutamate [PBLG]) revealed two excited-state absorption bands, which disappear upon unfolding of the helix [63]. [PBLG] forms extremely stable, long α-helices in both helicogenic solvents and films grown from these solvents. The monomeric unit of PBLG is a nonnatural amino acid with a long side chain that stabilizes the helix. PBLG has served as the standard model helix since the very early days of structural investigations of proteins. Figure 14.30a (gray line) shows the absorption spectrum of the helix at 293 K, which is dominated by the strong N–H stretching band at 3290 cm^{-1}. Figure 14.30b (gray line) shows the pump–probe response 600 fs after excitation with an ultrashort broadband pulse. One negative (3280 cm^{-1}) and two positive bands (3160 and 3005 cm^{-1}) are observed. If the N–H stretching modes were isolated, a negative band associated with bleach and stimulated emission, and a positive band associated with excited-state absorption, would be expected. This is indeed observed here for the unfolded molecule. In contrast, the observation of *two* positive bands for the intact helix rather than just one, is exceptional. Edler et al. [63] argue that these features cannot be explained due to intensity dependencies or Fermi resonances. A consistent explanation is reached by assuming that two vibron states are excited, and these vibrons may form two different types of bound states, self trapped either on the same site, or on neighbouring ones. The latter states originate from the acoustic phonons of the helix, which correlate adjacent sites (see Figure 14.31 and also Hamm and Edler (2004) [65]).

14.7.3 Crystal Surfaces

Depositing atoms or molecules on crystal surfaces can be controlled experimentally, and as a result a planar regular two-dimensional lattice structure of the deposited material can be obtained. Guyot–Sionnest [66] used hydrogen to be deposited on Si(111) surfaces. The Si–H bonds can be excited using pump-probe techniques with infrared dye lasers. There is substantial interaction between the Si–H bonds on the Si(111) surface. The pump excites one phonon (quantum), while the tunable

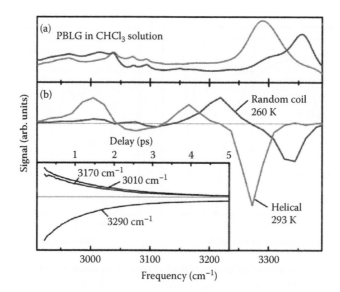

FIGURE 14.30 (a) Absorption spectra of PBLG in chloroform at 293 K (gray line, helical conformation) and at 260 K (dark line, random coil). (b) Pump-probe spectra 600 fs after excitation under the same conditions. Inset: Decay of negative and both positive bands at 293 K. (Adapted from J. Edler et al., *Phys. Rev. Lett.* 93, 106405, 2004.)

probe frequency finds a substantial red shift of the two-phonon excitation, and allows to conclude, that two-phonon bound states are observed.

Another set of experiments was performed by Jakob [67–69]. Carbon monoxide (CO) was deposited on a Ru(001) single crystal surface. The C–O stretching modes constitute a two-dimensional array of weakly interacting anharmonic oscillators with 4.7 Å intermolecular distance. Intermolecular coupling is provided by means of the electric field of the oscillating dipoles. Experimental spectra at 30 K are shown in Figure 14.32. The one phonon mode frequency is at 2031 cm^{-1}. This has to be compared to the naturally abundant $^{13}C^{16}O$ frequency at 1941 cm^{-1}. The corresponding blue shift for the adsorbate is thus due to additional stiffness provided by the Ru surface

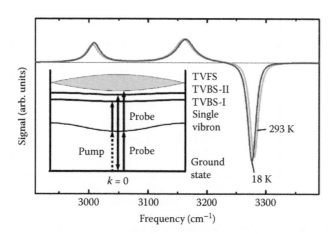

FIGURE 14.31 Simulated pump-probe spectrum for 293 K (gray line) and 18 K (dark line). Inset: schematic of the energy levels. (Adapted from J. Edler et al., *Phys. Rev. Lett.* 93, 106405, 2004.)

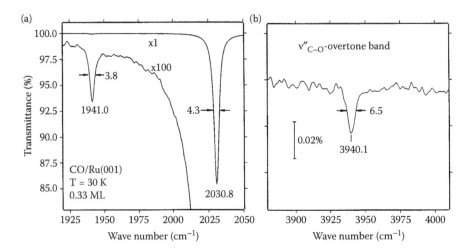

FIGURE 14.32 (a) Infrared absorption spectra of the C–O stretching mode at 30 K. The corresponding mode of naturally abundant $^{13}C^{16}O$ is displayed in an enlarged vertical scale; (b) the overtone band observed at less than twice the frequency of the fundamental mode. (Adapted from P. Jakob, *Phys. Rev. Lett.* 77, 4229, 1996.)

coupling. Excitation of two uncorrelated phonons would yield a two phonon continuum at about 4062 cm^{-1}. The narrow line observed at 3940 cm^{-1} can be thus attributed to a two-phonon bound state, or a quantum DB excitation.

The temperature dependence of the line positions also clearly shows, that the two-phonon bound state line softens much slower than the line of the one-phonon delocalized state (Figure 14.33). This is, among other facts, a strong indication that the observed red shift of the overtone line is due to the formation of a localized two-phonon bound state, or a (quantum) DB.

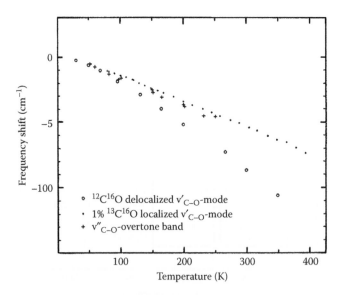

FIGURE 14.33 Frequency shifts of the vibrational bands with temperature: crosses—overtone band; dots—fundamental of naturally abundant $^{13}C^{16}O$; open circles—delocalized fundamental of $^{12}C^{16}O$. (Adapted from P. Jakob, *Phys. Rev. Lett.* 77, 4229, 1996.)

14.7.4 IN THE BULK OF SOLIDS

Vibrational spectra in the overtone or combination region of molecular crystals have been studied intensively in the 1970s and 1980s. A pioneering theoretical proposal was due to Agranovich, who predicted the existence of two-exciton bound states in various molecular crystal materials [70]. Experimental studies of infrared absorption spectra for CO_2 crystals were conducted by Dows et al. (1973) [71] and gave evidence of two-phonon bound states. Dressler et al. studied the slow vibrational relaxation of N_2, which also indicates the presence of many-phonon bound states [72]. In a remarkable theoretical paper, Bogani calculated the spectrum of two phonon excitations in molecular crystals [73], to some extent one of the first accurate calculations of quantum DBs. More recently, Bini et al. reconsidered the theory of three-phonon bound states in crystal CO_2 [74]. While there certainly are many other results worth to be mentioned, we recommend reading related chapters in Ovchinnikov et al. (2001) [56] and Agranovich and Hochstrasser (1983) [75].

The pioneering studies of Swanson et al. (1999) [76] have shown that up to seven phonons can bind and form a localized state. The system of choice was a PtCl-based crystal—a halide-bridged mixed-valence transition metal complex, which is a model low-dimensional electronic material where the ground states can be systematically tuned (with chemistry, doping, pressure, and temperature). It is a very strong charge-density wave (CDW) example. The material is a well-formed crystal with a homogeneous lattice consisting of quasi1D chains (see Figure 14.34). The CDW ground state consists of alternating Pt^{2+} and Pt^{4+} sites with a corresponding distortion of the chloride ions toward the Pt^{4+} site. Resonance Raman spectra were used to probe both ground and photoexcited states. They probe the fundamental Cl–Pt–Cl stretch and the progression of many overtones. At low temperatures, the fundamental exhibits a fine structure with up to six discrete, well-resolved modes. The analysis of the evolution of the spectral structure in the overtones was performed for isotopically pure samples, in order to avoid exciting localized states due to isotopic disorder. The fundamental and overtone spectra for the pure $Pt^{35}Cl$ sample are shown in Figure 14.35. The data

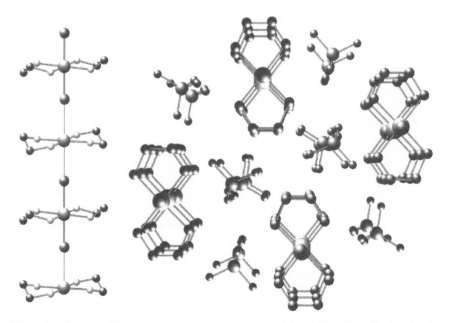

FIGURE 14.34 Structure of the PtCl crystal. One PtCl chain is shown on the left. Each Pt atom is coordinated by two ethylenediamine units in a near square planar geometry, while Cl ions connect pt sites along the chain. The packing arrangement of the 1D chains and their ClO_4^- counterions is shown on the right. (Adapted from B. I. Swanson et al., *Phys. Rev. Lett.* 82, 3288, 1999.)

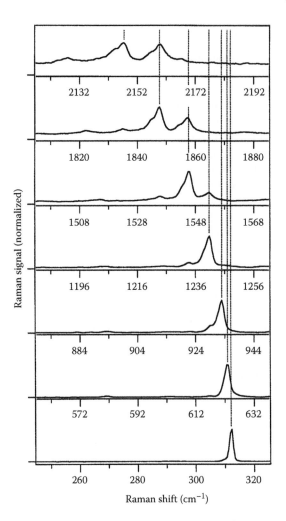

FIGURE 14.35 Fundamental and overtone spectra of isotopically pure Pt^{35}Cl. Moving upward in each panel, each x axis is offset by the appropriate integral multiple of the 312 cm^{-1} fundamental frequency. All spectra have been scaled vertically to equal peak intensities. (Adapted from B. I. Swanson et al., *Phys. Rev. Lett.* 82, 3288, 1999.)

are presented in a stack plot in which each successive trace is offset along the horizontal axis by increasing multiples of the fundamental frequency 312 cm^{-1}. Such plots clearly expose the relation of features in the overtone spectrum to multiples of the fundamental peak. The lowest energy dominant feature in each trace (marked by vertical lines) demonstrates a strongly increasing anharmonic redshift. Further, at higher overtones, each of these dominant peaks recurs, offset by the fundamental frequency, in the next trace above. A simple interpretation is that the lowest-energy dominant peaks in the overtone spectra correspond to all quanta of vibrational energy localized in approximately one PtCl$_2$ unit, while the higher energy peaks correspond to having all quanta but one in a localized PtCl$_2$ unit combined with one quanta in the more extended fundamental. The schematic process of the energy transfer is shown in Figure 14.36 and has been analyzed theoretically in Kladko et al. (1999) [77].

An incoming photon at frequency ν is exciting an electron from a Pt^{2+} to a Pt^{4+} site. The Cl ion between them starts oscillating. Finally, the electron relaxes back to its original position, and releases a photon with frequency ν'. The energy difference is remaining in a localized vibration.

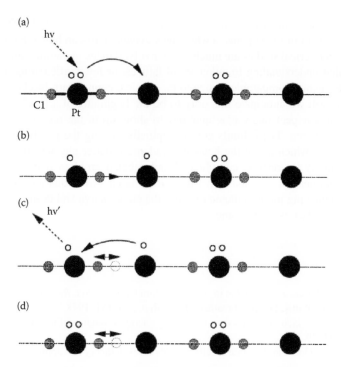

FIGURE 14.36 A simple picture of a resonant Raman scattering event in the localized atomic limit. Large filled circles mark Pt ions, small gray circles mark Cl ions. Open circles mark the positions of electrons. (Adapted from K. Kladko, J. Malek, and A. R. Bishop, *J. Phys.: Condens. Matter* 11, L415, 1999.)

The effect of isotope disorder was treated by Kalosakas et al. [78]. The experimentally observed redshifts were also theoretically described by Fehske et al. [79] using Peierls–Hubbard models, and by Wellein et al. using the Holstein model [80].

Inelastic x-ray and neutron scattering was used by Manley et al. [81–83] to probe phonon dispersion in α-uranium single crystals. Variation of temperature showed softening, and the abrupt appearance of a new dynamical mode, without a typically observed phase transition. The authors argue that this mode is a DB, and forms due to strong electron–phonon interaction.

Russell and Eilbeck reported evidence for moving breathers in the layered crystal muscovite at 300 K [84]. Breathers were created by bombardment of the crystal surface with heavy ions. Ejection of atoms at the opposite (shielded) crystal surface was attributed to breathers, which were able to carry the vibrational energy without dispersing over more than 10^7 unit cells of the crystal.

Finally Abrasonis et al. [85] reported on anomalous bulk diffusion of interstitial nitrogen in steel microcrystals. N ions were deposited in a micron-thick layer, and are trapped by local structures, with a characteristic binding energy. Ar ion bombardment increases the N mobility at depths far beyond the ion penetration depth. The authors see evidence for coherent transfer of vibrational energy deep into the bulk of the material.

14.8 CONCLUSIONS AND OUTLOOK

Progress in the understanding of classical DBs evolved for two decades, and has significantly improved our understanding of the complexity in the dynamics of anharmonic lattice dynamics. The little input it needs to form DBs—spatial discreteness and nonlinearity—were demonstrated to generate an impressive list of experimental observations in a large variety of different physical systems, with length and time scales ranging over many orders of magnitude. The quantization of the classical equations of motion opened a whole new avenue of research on QBs. For obvious computational reasons numerical studies are much more involved in the quantum domain, restricting our potential of detailed understanding to either small lattices, or few participating quanta. Nevertheless, a set of experiments, comparable in size to its classical counterpart, has been reviewed where QBs are detected. Still we anticipate this only to be the beginning of a much more sophisticated undertaking. We may expect many new qualities to show up in the regime of large lattices and many participating quanta. These limits are conceptually closing the gap to quantum interacting many body problems, which are at the heart of condensed matter physics, and quantum computing. Yet another highly interesting, complicated and debated problem is the combined influence of discreteness, nonlinearity, and disorder. Linear disordered wave equations allow for Anderson localization. It is a thrilling modern theme to study the constructive and destructive interplay of all three ingredients. The best is yet to come.

REFERENCES

1. P. W. Anderson, Absence of diffusion in certain random lattices. *Phys. Rev.* 109, 1492, 1958.
2. S. Flach and C. R. Willis, Discrete breathers. *Phys. Rep.* 295, 181, 1998.
3. D. K. Campbell, S. Flach, Y. S. Kivshar, Localizing energy through nonlinearity and discreteness. *Phys Today* 57 (1), 43–49, 2004.
4. S. Flach and A. V. Gorbach, Discrete breathers—advances in theory and applications. *Phys. Rep.* 467, 1, 2008.
5. S. Flach and A. V. Gorbach, Discrete breathers in Fermi–Pasta–Ulam lattices. *Chaos* 15, 015112, 2005.
6. M. Eleftheriou and S. Flach, Interaction of discrete breathers with thermal fluctuations. *Physica D* 202, 142, 2005.
7. S. Aubry, Breathers in nonlinear lattices: Existence, linear stability and quantization. *Physica D* 103, 201, 1997.
8. R. S. MacKay, Discrete breathers: Classical and quantum. *Physica A* 288, 174, 2000.
9. V. Fleurov, Discrete quantum breathers—What do we know about them? *Chaos* 13, 676, 2003.
10. W. Z. Wang, J. Tinka Gammel, A. R. Bishop, and M. I. Salkola, Quantum breathers in a nonlinear lattice. *Phys. Rev. Lett.* 76, 3598, 1996.
11. S. A. Schöefield, R. E. Wyatt, and P. G. Wolynes, Computational study of many-dimensional quantum vibrational energy redistribution. I. Statistics of the survival probability. *J. Chem. Phys.* 105, 940, 1996.
12. P. D. Miller, A. C. Scott, J. Carr, and J. C. Eilbeck, Binding energies for discrete nonlinear Schrödinger equations. *Phys. Scr.* 44, 509, 1991.
13. A. C. Scott, J. C. Eilbeck, and H. Gilhøj, Quantum lattice solitons. *Physica D* 78, 194, 1994.
14. J. C. Eilbeck, in *Localization and Energy Transfer in Nonlinear Systems*, L. Vazquez, R. S. MacKay, and M. P. Zorzano (Eds.), World Scientific, Singapore, 2003, p. 177.
15. L. Bernstein, J. C. Eilbeck, and A. C. Scott, The quantum theory of local modes in a coupled system of nonlinear oscillators. *Nonlinearity* 3, 293, 1990.
16. L. J. Bernstein, Quantizing a self-trapping transition. *Physica D* 68, 174, 1993.
17. S. Aubry, S. Flach, K. Kladko, and E. Olbrich, Manifestation of classical bifurcation in the spectrum of the integrable quantum dimmer. *Phys. Rev. Lett.* 76, 1607, 1996.
18. G. Kalosakas and A. R. Bishop, Small-tunneling-amplitude boson-Hubbard dimer: Stationary states. *Phys. Rev. A* 65, 043616, 2002.
19. S. Flach and V. Fleurov, Tunneling in the nonintegrable trimer—a step towards quantum breathers. *J. Phys.: Cond. Matt.* 9, 7039, 1997.
20. S. Flach, V. Fleurov, and A. A. Ovchinnikov, Tunneling of localized excitations: Giant enhancement due to fluctuations. *Phys. Rev. B* 63, 094304, 2001.
21. R. A. Pinto and S. Flach, Quantum dynamics of localized excitations in a symmetric trimer molecule. *Phys. Rev. A* 73, 022717, 2006.
22. J. Dorignac and S. Flach, Tunneling of quantum rotobreathers. *Phys. Rev. B* 65, 214305, 2002.
23. L. Proville, Biphonons in the Klein–Gordon lattice. *Phys. Rev. B* 71, 104306, 2005.

24. L. Proville, Dynamical structure factor of a nonlinear Klein–Gordon lattice. *Phys. Rev. B* 72, 184301, 2005.
25. L. Proville, Two-phonon pseudogap in the Klein–Gordon lattice. *Europhys. Lett.* 69, 763, 2005.
26. V. Fleurov, R. Schilling, and S. Flach, Tunneling of a quantum breather in a one-dimensional chain. *Phys. Rev. E* 58, 339, 1998.
27. G. Kalosakas, A. R. Bishop, V. M. Kenkre, Multiple time-scale quantum dynamics of many interacting bosons in a dimer. *J. Phys. B* 36, 3233, 2003.
28. R. A. Pinto and S. Flach, Quantum localized modes in capacitively coupled Josephson junctions. *EPL* 79, 66002, 2007.
29. R. A. Pinto and S. Flach, Quantum breathers in capacitively coupled Josephson junctions: Correlations, number conservation, and entanglement. *Phys. Rev. B* 77, 024308, 2008.
30. S. Flach, V. Fleurov, and A. V. Gorbach, Classical and quantum radiation of perturbed discrete breathers. *Phys. Rev. B* 71, 064302, 2005.
31. L. S. Schulman, D. Tolkunov, and E. Mihokova, Stability of quantum breathers. *Phys. Rev. Lett.* 96, 065501, 2006; Structure and time-dependence of quantum breathers. *Chem. Phys.* 322, 55, 2006; L.S. Schulman, D. Tolkunov, and E. Mihókova, *Chem. Phys.* 322, 55, 2006.
32. L. Proville, Quantum breather in a nonlinear Klein–Gordon lattice. *Physica D* 216, 191, 2006.
33. A. P. Tonel, J. Links, and A. Foerster, Quantum dynamics of a model for two Josephson-coupled Bose–Einstein condensates. *J. Phys. A: Math. Gen.* 38, 1235, 2005.
34. L. B. Fu and J. Liu, Quantum entanglement manifestation of transition to nonlinear self-trapping for Bose–Einstein condensates in a symmetric double well. *Phys. Rev. A* 74, 063614, 2006.
35. T.-C. Wei and P. M. Goldbart, Geometric measure of entanglement and applications to bipartite and multipartite quantum states. *Phys. Rev. A* 68, 042307, 2003.
36. A. Shimony, Degree of entanglement. *Ann. NY. Acad. Sci.* 755, 675, 1995.
37. H. Barnum and N. Linden, Monotones and invariants for multi-particle quantum states. *J. Phys. A.: Math. Gen.* 34, 6787, 2001.
38. A. P. Hines, R. H. McKenzie, and G. J. Milburn, Entanglement of two-mode Bose–Einstein condensates. *Phys. Rev. A* 67, 013609, 2003.
39. K. G. Makris, S. Suntsov, D. N. Christodoulides, G. I. Stegeman, and A. Hache, Discrete surface solitons. *Opt. Lett.* 30, 2466, 2005; M. Molina, I. L. Garanovich, A. A. Sukhorukov, and Yu. Kivshar, Discrete surface solitons in semi-infinite binary waveguide arrays. *Opt. Lett.* 31, 2332, 2006.
40. M. I. Molina, R. A. Vicencio, and Y. S. Kivshar, Discrete solitons and nonlinear surface modes in semi-infinite waveguide arrays. *Opt. Lett.* 31, 1693, 2006.
41. S. Suntsov, K. G. Makris, D. N. Christodoulides, G. I. Stegeman, A. Haché, R. Morandotti, H. Yang, G. Salamo, and M. Sorel, Observation of discrete surface solitons. *Phys. Rev. Lett.* 96, 063901, 2006; E. Smirnov, M. Stepic, C. E. Rüter, D. Kip, and V. Shandarov, Observation of staggered surface solitary waves in one-dimensional waveguide arrays. *Opt. Lett.* 31, 2338, 2006; C. R. Rosberg, D. N. Neshev, W. Krolikowski, A. Mitchell, R. A. Vicencio, M. I. Molina, and Y. S. Kivshar, Observation of surface gap solitons in semi-infinite waveguide arrays. *Phys. Rev. Lett.* 97, 083901, 2006.
42. V. Pouthier, Boundary-induced energy localization in a nonlinear quantum lattice. *Phys. Rev. B* 76, 224302, 2007.
43. R. A. Pinto, M. Haque, and S. Flach, Edge-localized states in quantum one-dimensional lattices. *Phys. Rev. A* 79, 052118, 2009.
44. J.-P. Nguenang and S. Flach, Fermionic bound state on a one-dimensional lattice. *Phys. Rev. A* 80, 015601, 2009.
45. M. A. Baranov, Theoretical progress in many-body physics with ultracold dipolar gases. *Phys. Rep.* 464, 71, 2008.
46. R. McDermott, R. W. Simmonds, M. Steffen, K. B. Cooper, K. Cicak, K. D. Osborn, Seongshik Oh, D. P. Pappas, and J. M. Martinis, Simultaneous state measurement of coupled Josephson phase qubits. *Science* 307, 1299, 2005; M. Steffen, M. Ansmann, R. C. Bialczak, N. Katz, E. Lucero, R. McDermott, M. Neeley, E. M. Weig, A. N. Cleland, and J. M. Martinis, Measurement of the entanglement of two superconducting qubits via state tomography. *Science* 313, 1423, 2006.
47. A. Leggett, B. Ruggiero, and P. Silvestrini (Eds.), *Quantum Computing and Quantum Bits in Mesoscopic Systems*, Kluwer Academic/Plenum Publishers, New York, 2004.

48. D. Estève, J. M. Raimond, and J. Dalibard (Eds.), *Quantum Entanglement and Information Processing, Les Houches 2003*, Elsevier, Amsterdam, 2004.

49. J. Q. You and F. Nori, Superconducting circuits and quantum information. *Physics Today*, 58, 42–47, 2005.

50. M. Davis and E. Heller, Quantum dynamical tunneling in bound states. *J. Chem. Phys.* 75, 246, 1981.

51. S. Keshavamurthy, Dynamical tunneling in molecules: Quantum routes to energy flow. *Int. Rev. Phys. Chem.* 26, 521, 2007.

52. S. Keshavamurthy, On dynamical tunneling and classical resonances. *J. Chem. Phys.* 122, 114109, 2005.

53. K. Winkler, G. Thalhammer, F. Lang, R. Grimm, J. Ecker Denshlag, A. J. Daley, A. Kantian, H. P. Büchler, and P. Zoller, Repulsively bound atom pairs in an optical lattice. *Nature* 441, 853, 2006.

54. A. A. Ovchinnikov, Localized long-lived vibrational states in molecular crystals. *Zh. Eksp. Teor. Fiz./Soviet Phys. JETP* 57/30:263, 147, 1969/1970.

55. M. Joyeux, S. Yu. Grebenshchikov, J. Bredenbeck, R. Schinke, and S. C. Farantos, Intramolecular dynamics along isomerization and dissociation pathways. *Adv. Chem. Phys.* 130, 267, 2005.

56. A. A. Ovchinnikov, N. S. Erikhman, and K. A. Pronin, *Vibrational-Rotational Excitations in Nonlinear Molecular Systems*, Kluwer Academic/Plenum Publishers, 2001.

57. S. C. Farantos, Periodic orbits in biological molecules: Phase space structures and selectivity in alanine dipeptide. *J. Chem. Phys.* 126, 175101, 2007.

58. S. A. Kisilev, A. J. Sievers and G. V. Chester, Eigenvectors of strongly anharmonic intrinsic gap modes in three-dimensional ionic crystals. *Physica D* 123, 393, 1998.

59. T. Rossler and J. B. Page, Driven intrinsic localized modes and their stability in anharmonic lattices with realistic potentials. *Physica B* 220, 387, 1996; Creation of intrinsic localized modes via optical control of anharmonic lattices. *Phys. Rev. Lett.* 78, 1287, 1997; Optical creation of vibrational intrinsic localized modes in anharmonic lattices with realistic interatomic potentials. *Phys. Rev. B* 62, 11460, 2000.

60. G. Kopidakis and S. Aubry, Discrete breathers in realistic models: Hydrocarbon structures. *Physica B* 296, 237, 2001.

61. M. S. Child and R. T. Lawton, Local mode degeneracies in the vibrational spectrum of H_2O. *Chem. Phys. Lett.* 87, 217, 1982.

62. S. Keshavamurthy, Dynamical tunneling in molecules: Quantum routes to energy flow. *Int. Rev. Phys. Chem.* 26, 521, 2007.

63. J. Edler, R. Pfister, V. Pouthier, C. Falvo, and P. Hamm, Direct observation of self-trapped vibrational states in α-helices. *Phys. Rev. Lett.* 93, 106405, 2004.

64. R. L. Swofford, M. E. Long, and A. C. Albrecht, Benzene, naphthalene, and anthracene in the visible region by thermal lensing spectroscopy and the local mode model. *J. Chem. Phys.* 65, 179, 1976.

65. P. Hamm and J. Edler, in: T. Dauxois, A. Litvak-Hinenzon, R. MacKay, A. Spanoudaki (Eds.), *Energy Localization and Transfer*, World Scientific, Singapore, 2004.

66. P. Guyot-Sionnest, Two-phonon bound state for the hydrogen vibration on the H/Si(111) surface. *Phys. Rev. Lett.* 67, 2323, 1991.

67. P. Jakob, Dynamics of the C–O stretch overtone vibration of CO/Ru(001). *Phys. Rev. Lett.* 77, 4229, 1996.

68. P. Jakob, Localization and relaxation effects of weakly interacting anharmonic oscillators. *Physica D* 119, 109, 1998.

69. P. Jakob, Lateral interactions in (NO+O) coadsorbate layers on Ru(001): Fundamental and overtone modes. *Surf. Sci.* 428, 309, 1999.

70. V. M. Agranovich, Theory of biexcitons in molecular crystals for infrared region. *Sov. Phys.-Solid St.* 12, 438, 1970.

71. D. A. Dows and V. Schettino, 2-Phonon infrared-absorption spectra in crystalline carbon-dioxide. *J. Chem. Phys.* 58, 5009, 1973.

72. K. Dressler, O. Oehler, and D. A. Smith, Measurement of slow vibrational relaxation and fast vibrational energy transfer in solid N_2. *Phys. Rev. Lett.* 34, 1364, 1975.

73. F. Bogani, 2-Phonon resonance and bound-states in molecular crystals. 1. General theory. *J. Phys. C.: Solid State Phys.* 11, 1283, 1978.

74. R. Bini, P. R. Salvi, V. Schettino, and H. J. Jodl, The spectroscopy and relaxation dynamics of 3-phonon bound-state in crystal CO_2. *J. Chem. Phys.* 98, 164, 1993.

75. V. M. Agranovich and R. M. Hochstrasser, *Spectroscopy and Excitation Dynamics of Condensed Molecular Crystals*, North-Holland, Amsterdam, 1983.

76. B. I. Swanson, J. A. Brozik, S. P. Love, G. F. Strouse, A. P. Shreve, A. R. Bishop, W.-Z. Wang, and M. I. Salkola, Observation of intrinsically localized modes in a discrete low-dimensional material. *Phys. Rev. Lett.* 82, 3288, 1999.

77. K. Kladko, J. Malek, and A. R. Bishop, Intrinsic localized modes in the charge-transfer solid PtCl. *J. Phys.: Condens. Matter* 11, L415, 1999.

78. G. Kalosakas, A. R. Bishop, and A. P. Shreve, Nonlinear disorder model for Raman profiles in naturally abundant PtCl. *Phys. Rev. B* 66, 894303, 2002.

79. H. Fehske, G. Wellein, H. Buettner, A. R. Bishop, and M. I. Salkola, Local mode behavior in quasi-1D CDW systems. *Physica B* 281, 673, 2000.

80. G. Wellein and H. Fehske, Self-trapping problem of electrons or excitons in one dimension. *Phys. Rev. B* 58, 6208, 1998.

81. M. E. Manley, M. Yethiraj, H. Sinn, H. M. Volz, A. Alatas, J. C. Lashley, W. L. Hults, G. H. Lander, and J. L. Smith, Formation of a new dynamical mode in α-uranium observed by inelastic x-ray and neutron scattering. *Phys. Rev. Lett.* 96, 125501, 2006.

82. M. E. Manley, M. Yethiraj, H. Sinn, H. M. Volz, A. Alatas, J. C. Lashley, W. L. Hults, G. H. Lander, D. J. Thoma, and J. L. Smith, Intrinsically localized vibrations and the mechanical properties of alpha-uranium. *J. Alloys Comp.* 444–445, 129, 2007.

83. M. E. Manley, J. W. Lynn, Y. Chen, and G. H. Lander, Intrinsically localized mode in α-U as a precursor to a solid-state phase transition. *Phys. Rev. B* 77, 852301, 2008.

84. F. M. Russell and J. C. Eilbeck, Evidence for moving breathers in a layered crystal insulator at 300K. *EPL* 78, 10004, 2007.

85. G. Abrasonis, W. Moeller, and X. X. Ma, Anomalous ion accelerated bulk diffusion of interstitial nitrogen. *Phys. Rev. Lett.* 96, 065901, 2006.

15 Tunneling in Open Quantum Systems

Alvise Verso and Joachim Ankerhold

CONTENTS

15.1 INTRODUCTION

Processes involving quantum tunneling can be found in a huge variety of physical and chemical systems, ranging in length from the mesoscopic scale of a few microns to the subatomic scale of a few fermi. It has been realized already three decades ago that the interaction of the tunneling degree of freedom with environmental degrees of freedom plays a crucial role. Namely, any realistic description has to account for the fact that a purely isolated system is always an idealization, an issue which tends to become more and more relevant with the growing complexity of the system and particularly with increasing system size. A quantitative understanding of experimental observations must include the presence of, for example, electromagnetic modes in electrical circuits, vibrational modes in molecular aggregates, or phonon backgrounds in solid-state systems. Moreover, as a genuine quantum effect tunneling processes serve as paradigm to analyze the boundary between the microscopic and the macroscopic world or at least to elucidate how the latter one emerges from the former one.

Experimental studies started in the 1980s mainly in the new field of mesoscopic physics with the detection of the switching out of the zero voltage state in Josephson junctions (JJs). In these systems the phase difference between the two superconducting reservoirs is a collective degree of freedom the physics of which is equivalent to that of a fictitious particle in a tilted periodic potential. The tunneling of this phase out of one of the metastable wells, then coined macroscopic quantum tunneling, attracted a substantial amount of research and triggered theoretical developments to describe tunneling in open systems. Presently, advanced fabrication techniques have led to the design and tailoring of quantum systems in atomic, molecular, and solid-state physics which allow for the observation of quantum effects in general and tunneling in particular with unprecedented accuracy. In fact, tunneling processes have even been exploited as sensitive detection mechanisms close to the quantum limit [1]. Theory is again challenged to provide a deeper understanding of the interaction of these systems with their surrounding.

Here, we give a brief overview of basic ideas and theoretical concepts for tunneling in presence of dissipation, established ones and recent developments, and illustrate them through specific examples. A formally exact description of the dynamics of dissipative quantum systems is given in terms of path integrals. The explicit evaluation of this expression, however, is even numerically possible only in few cases, which serve as benchmark results for approximate approaches. In the context of tunneling rates there are basically two situations where simplifications can be achieved. In one case, the coupling to the environment is sufficiently strong so that on the reactant side of the tunneling system a local thermodynamic state is preserved over long periods of time. Then, thermodynamical methodologies apply and heavily rely on semiclassical techniques to treat imaginary time path integrals. Another case is the domain of weak friction and elevated temperatures, where thermodynamic methods fail, but approximate equations of motion can be derived from the path integral expression. Again semiclassical techniques are of importance to provide explicit expressions for decay rates. This type of approach can also be extended to systems driven by time-periodic forces where escape over and tunneling through dynamical barriers occurs.

15.2 DYNAMICS OF DISSIPATIVE QUANTUM SYSTEMS

Dissipative quantum systems are described with system + reservoir models [2,3], where the position q of a system with potential $V(q)$ is bilinearly coupled to positions x_α of environmental degrees of freedom. Typically, the number of these reservoir degrees of freedom is macroscopically large and they are assumed to reside in thermodynamic equilibrium at inverse temperature $\beta = 1/k_B T$. In this situation, the relevant system degree of freedom is effectively subject to fluctuating forces which obey Gaussian statistics. It is thus possible to mimic the reservoir by a quasicontinuum of harmonic oscillators, independent of its actual microscopic realization. In turn, all properties of the bath are captured by the first and the second cumulants of the bath-force interacting with the system.

Accordingly, one considers a Hamiltonian of the form $H = H_S + H_R + H_I$ with a system part H_S, a reservoir H_R and an interaction H_I—that is,

$$H_S = \frac{p^2}{2m} + V(q),$$

$$H_R = \sum_\alpha \frac{p_\alpha^2}{2m_\alpha} + \frac{m_\alpha \omega_\alpha^2}{2} x_\alpha^2,$$ (15.1)

$$H_I = -q \sum_\alpha c_\alpha x_\alpha + q^2 \sum_\alpha \frac{c_\alpha^2}{2m_\alpha \omega_\alpha^2},$$

where the translational invariant form of the coupling between system and reservoir avoids coupling-induced potential renormalizations. The dynamics of the relevant system is described by the reduced density operator with the environmental modes traced out

$$\rho(t) = \mathrm{Tr}_R \left\{ e^{-\frac{i}{\hbar}Ht} W(0) e^{+\frac{i}{\hbar}Ht} \right\}$$ (15.2)

and with $W(0)$ being the initial state of the total compound. The only nonperturbative way to deal with the elimination of the bath degrees of freedom is to apply the path integral approach. The coordinate representation of Equation 15.2 with $\rho(q_f, q_f', t) \equiv \langle q_f | \rho(t) | q_f' \rangle$ follows as

$$\rho(q_f, q_f', t) = \int dq_i \, dq_i' \, d\vec{x}_f \, d\vec{x}_i \, d\vec{x}_i' \, G(q_f, \vec{x}_f, t, q_i, \vec{x}_i) \langle q_i, \vec{x}_i | W(0) | q_i', \vec{x}_i' \rangle \, G(q_f', \vec{x}_f, t, q_i', \vec{x}_i')^*, \quad (15.3)$$

where the $*$ means complex conjugation and $\vec{x}$ collects the x_α. The time-dependent transition amplitudes (propagators) on the RHS are expressed as

$$G(q_f,\vec{x}_f,t,q_i,\vec{x}_i) = \int \mathcal{D}[q]\,\mathcal{D}[\vec{x}]\,e^{iS[q,\vec{x}]/\hbar}, \tag{15.4}$$

with the total action $S = S_S + S_R + S_I$ according to the three parts of the Hamiltonian (Equation 15.1). The sum goes over all paths running in time t from $q(0) = q_i, \vec{x}(0) = \vec{x}_i$ to $q(t) = q_f, \vec{x}(t) = \vec{x}_f$. To carry out all integrations over the bath degrees of freedom in Equation 15.4 explicitly, the initial state must be specified.

In the ordinary Feynman–Vernon theory [4] this state is assumed to be a factorizing state, $W(0) = \rho_S(0)\exp(-\beta H_R)/Z_R$ (Z_R is the bath partition function), so that each one, system and equilibrated bath, lives in splendid isolation at $t = 0$. While this assumption may be justified in the weak damping/high-temperature limit or in certain experimental situations, in general, it fails particularly for condensed phase systems and for moderate to strong friction and/or lower temperatures. A more general approach is to work with correlated initial states [5], that is, $W(0) = \sum_j O_j \exp(-\beta H)O'_j/Z$ with preparation operators O_j and O'_j acting onto the system degree of freedom only and the total partition function Z. In fact, the corresponding classical model can be shown to reproduce the well-known generalized Langevin equation.

To keep the formulation transparent we focus in the sequel on the case where the preparation operators depend exclusively on coordinate and refer to [3,5] for generalizations. As an example think about a position measurement with a Gaussian slit, in which case the preparation operators are Gaussian weighted projection operators onto position. Hence, one has

$$\rho(q_i,q'_i,t = 0) = \rho_\beta(q_i,q'_i)\lambda(q_i,q'_i), \tag{15.5}$$

with the preparation function $\lambda(q,q') = \sum_j \langle q|O_j|q\rangle\langle q'|O'_j|q'\rangle$ and the reduced thermal equilibrium density $\rho_\beta = \mathrm{Tr}_R\{\exp(-\beta H)\}/Z$. Accordingly, in Equation 15.3 the initial state is represented as a path integral in imaginary time (Euclidian path integral), which sums over system and bath paths connecting the respective endpoints in the time interval $\hbar\beta$.

Having fixed the initial state, the integrations over the bath degrees of freedom in Equation 15.3 can now be performed exactly. One finds

$$\rho(q_f,q'_f,t) = \int dq_i dq'_i\, J(q_f,q'_f,t,q_i,q'_i)\,\lambda(q_i,q'_i), \tag{15.6}$$

where the propagating function $J(\cdot)$ is a threefold path integral over the system degree of freedom only

$$J(q_f,q'_f,t,q_i,q'_i) = Z^{-1}\int \mathcal{D}[q]\,\mathcal{D}[q']\,\mathcal{D}[\bar{q}]\,e^{i\Sigma_0[q,q',\bar{q}]/\hbar}\,\langle e^{i\Sigma_I[q,q',\bar{q},\vec{x},\vec{x}',\bar{\vec{x}}]/\hbar}\rangle_R. \tag{15.7}$$

The two real-time paths $q(s)$ and $q'(s)$ connect in time t the initial points q_i and q'_i with end points q_f and q'_f, while the imaginary time path $\bar{q}(\sigma)$ runs from q_i to q'_i in the interval $\hbar\beta$ (cf. Figure 15.1). The contribution of each path is weighted with the total bare action $\Sigma_0 = S_S[q] - S_S[q'] + i\bar{S}_S[\bar{q}]$ ($\bar{S}$ denotes the Euclidian action) and with the expectation value $\langle e^{i\Sigma_I/\hbar}\rangle_R$ of the coupling $\Sigma_I = S_I[q,\vec{x}] - S_I[q',\vec{x}'] + i\bar{S}_I[\bar{q},\bar{\vec{x}}]$ with respect to the equilibrium distribution of the reservoir. According to Equation 15.1 one has, for example, $S_I[q,\vec{x}] = \int_0^t ds\,q(s)\xi(s) - \mu\int_0^t ds\,q(s)^2$ with the bath force $\xi = \sum_\alpha c_\alpha x_\alpha$ and $\mu = \sum_\alpha c_\alpha^2/(m_\alpha\omega_\alpha^2)$. Hence, $\langle e^{i\Sigma_I/\hbar}\rangle_R$ is the generating functional of the thermal distribution of the bath. For the harmonic model considered here, the first cumulant vanishes $\langle\xi(t)\rangle_R = 0$ and one obtains the so-called influence functional $\langle e^{i\Sigma_I/\hbar}\rangle_R = e^{-\Phi/\hbar}$ with

$$\Phi[\bar{q}] = \int dz\int_{z>z'} dz'\,\tilde{q}(z)\,K(z-z')\,\tilde{q}(z') + \frac{i}{2}\mu\int dz\tilde{q}(z)^2. \tag{15.8}$$

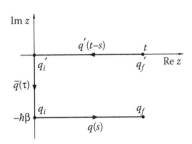

FIGURE 15.1 Real and imaginary time paths in the complex time plane $Z = s - i\tau$ contributing to the propagating function (Equation 15.7).

The ordered time integration is understood along the contour (Figure 15.1): $z = s$ for s from $t \to 0$, $z = -i\tau$ for τ from $0 \to \hbar\beta$, $z = -i\hbar\beta + s$ for s from $0 \to t$ with

$$\tilde{q}(z) = \begin{cases} q'(s) & \text{for } z = s & 0 \leq s \leq t \\ \bar{q}(\tau) & \text{for } z = -i\tau & 0 \leq \tau \leq \hbar\beta \\ q(s) & \text{for } z = -i\hbar\beta + s & 0 \leq s \leq t. \end{cases} \tag{15.9}$$

The effective impact of the bath is completely controlled by the second cumulant $K(z) = \langle\xi(z)\xi(0)\rangle_R/\hbar$, that is,

$$K(z) = \int_0^\infty \frac{d\omega}{\pi} I(\omega) \frac{\cosh[\omega(\hbar\beta - iz)]}{\sinh(\omega\hbar\beta/2)}, \tag{15.10}$$

where $I(\omega) = (\pi/2)\sum_\alpha c_\alpha^2/(m_\alpha\omega_\alpha)\,\delta(\omega - \omega_\alpha)$ denotes the spectral density of the environment. In particular, for real times the kernel $K(s) = K'(s) + iK''(s)$ is related to the macroscopic damping kernel entering the classical generalized Langevin equation

$$\gamma(s) = \frac{2}{m} \int_0^\infty \frac{d\omega}{\pi} \frac{I(\omega)}{\omega} \cos(\omega s) \tag{15.11}$$

via $K''(s) = (M/2)\,d\gamma(s)/ds$ and $K'(s) \to M\gamma(s)/\hbar\beta$ in the classical limit.

Equation 15.6 is a formally exact result for the time-dependent reduced dynamics. Its explicit evaluation, however, is very challenging particularly for processes involving quantum tunneling. Namely, due to the nonlocal time interactions mediated by the reservoir and captured by the influence functional, in general, a time-local evolution equation for the reduced density matrix does not exist. In this situation, progress has been made in basically three directions: (1) application of numerical approaches as for example, quantum Monte Carlo techniques, which in principle provide numerically exact results but are plagued by the so-called dynamical sign problem, (2) derivation of approximate time evolution equations from the exact path integral expression in certain ranges of parameter space, (3) development of semiclassical techniques according to the Wentzel–Kramers–Brillouin (WKB)-machinery.

While approach (1) has been used very successfully to describe, for example, the dynamics of tight-binding systems [3], for tunneling in continuous systems it faces two interconnected problems. Namely, one needs to consider the dynamics on sufficiently long time scales *and* with high accuracy to extract exponentially small tunneling rates. To meet these criteria simultaneously, is very demanding and computationally time-consuming. Practically, only very few systems have thus been analyzed. Approach (2) leads to various types of master equations in the complementary domains of weak [6] and strong friction [7]. In both cases tunneling rates have been derived. Approach (3) sounds very appealing since path integrals offer a natural starting point for an $\hbar$-expansion. Unfortunately, a direct semiclassical evaluation of the time-dependent density (Equation 15.6) is not feasible. While the minimal action paths can be determined at least numerically, the standard

strategies to relate the Gaussian fluctuations around them to the corresponding minimal action or to an equation of motion (Gelfand–Yaglom) do not apply due to the bath induced time retardation and time irreversibility. Moreover, a systematic $\hbar$-expansion is not straightforward since the minimal action paths do *not* reproduce in the classical limit the classical Langevin equation [3]. If one keeps in mind that though in condensed phase systems tunneling often occurs from a thermal state on the reactant side, a very powerful semiclassical technique has been developed based purely on the imaginary time-path integral representation of the partition function. This formalism, known as the imaginary part of the free energy method ("Im F") [3,8,9], requires sufficiently high-energy barriers and sufficiently strong friction or low temperatures.

In the sequel, we thus proceed as follows: we first consider a case, where the complete tunneling dynamics can be obtained, namely, a situation where tunneling is restricted to the parabolic top of a barrier potential. This result is already nontrivial and confirms corresponding predictions obtained in the 1980s within the Im F technique. The latter approach will be introduced in Section 15.3 including its recent generalization to non-Gaussian baths. Section 15.4 deals with semiclassical type of time evolution equations applicable in the regime of low friction and moderate temperatures, where the Im F approach fails. This formulation can be extended to extract rates for dynamical tunneling in driven systems close to a bistability, which have been implemented lately as highly sensitive detectors.

15.3 BARRIER DYNAMICS IN REAL-TIME

We consider a quantum particle in a metastable potential subject to a thermal environment. Initially, the system is prepared in a local thermal equilibrium in the well. Then, for a sufficiently high barrier potential $V_b \gg k_B T, \hbar\omega_0$ the dynamics of the reduced density will approach a steady state within a plateau range of time from which the escape rate can be obtained. For sufficiently high temperatures and sufficiently strong dissipation this steady state distribution coincides in the well region with the local thermodynamic state and deviates from it only in a boundary layer around the barrier top. Accordingly, the tunneling dynamics is obtained from the time evolution in an inverted harmonic potential and from the thermodynamic state in the harmonic well, both situations, for which the threefold path integral (Equation 15.6) can be solved exactly [10].

To be specific, let us consider an archetypical metastable well potential

$$V(q) = \frac{m\omega_0^2}{2} q^2 \left(1 - \frac{2q}{3q_b}\right) \tag{15.12}$$

with a harmonic well around $q = 0$ with frequency ω_0 and a barrier top with energy $V_b = V(q_b)$ located at $q = q_b$ with frequency $-\omega_b^2 \equiv -\omega_0^2$. The initial state is a thermal state restricted to the left of the barrier top

$$\rho(x_i, r_i) = \rho_\beta(x_i, r_i)\,\theta(q_b - r_i), \tag{15.13}$$

where for convenience we introduced sum and difference coordinates $r = (q + q')/2$ and $x = q - q'$, respectively, and θ denotes the step function. From the imaginary time path integral one first finds the density distribution around the parabolic barrier top as

$$\rho_\beta^{(b)}(x, r) = \frac{1}{Z\sqrt{\omega_0^2 m\beta\Lambda_b}}\sqrt{\frac{m}{2\pi\hbar^2\beta}}\left(\prod_{n=1}^{\infty} v_n^2 u_n^{(b)}\right)\exp\left[-\beta V_b - \frac{(r - q_b)^2}{2\Lambda_b} - \frac{\Omega_b x^2}{2\hbar}\right] \tag{15.14}$$

with the "variances"

$$\Lambda_b = \frac{1}{m\beta}\sum_{n=-\infty}^{\infty} u_n^{(b)}, \quad \Omega_b = \frac{m}{\beta}\sum_{n=-\infty}^{\infty}\left(|v_n|\hat{\gamma}(|v_n|) - \omega_b^2\right)u_n^{(b)}, \tag{15.15}$$

which contain the Matsubara frequencies $\nu_n = 2\pi n/\hbar\beta$ and $u_n^{(b)} = 1/[\nu_n^2 - \omega_b^2 + |\nu_n|\hat{\gamma}(|\nu_n|)]$. Friction enters through the Laplace transform $\hat{\gamma}(z)$ of the classical damping kernel $\gamma(t)$ Equation 15.11. As a function of temperature Λ_b displays a nontrivial behavior, which can already be read off from its nondissipative limit $\Lambda_b^{(0)} = -(\hbar/m)\cot(\omega_b\hbar\beta/2)$. Apparently, with lowering temperature $\Lambda_b^{(0)}$ vanishes for the first time at a critical temperature $T_c^{(0)} = \hbar\omega_b/(\pi k_B)$ which is *twice* the non-dissipative crossover temperature T_0 introduced below. The effect of friction is to push the critical temperature toward lower temperatures $T_c < T_c^{(0)}$. At T_c the local harmonic approximation for the equilibrium distribution (Equation 15.13) breaks down even in the close vicinity of the barrier top and the global shape of the potential (Equation 15.12) must be taken into account.

Now, starting with this initial state the density matrix $\rho(x_f, r_f, t)$ reaches for longer times a quasi-stationary state, the so-called flux state, which can be cast into the form

$$\rho_{\mathrm{fl}}(x_f, r_f) = \rho_\beta(x_f, r_f) g_{\mathrm{fl}}(x_f, r_f), \tag{15.16}$$

where

$$g_{\mathrm{fl}}(x, r) = \frac{1}{2}\mathrm{erfc}\left[(-r + q_b + ix\hbar|\Lambda_b|\omega_R/m)/\sqrt{2|\Lambda_b|}\right] \tag{15.17}$$

describes deviations from the equilibrium state around $q = q_b$ on the length scale $l_{\mathrm{fl}} = \sqrt{2|\Lambda_b|}$ (see Figure 15.2). This state is approached on large time scales compared to the inverse of the Grote–Hynes frequency ω_R, which is the largest positive root of $\omega_R^2 + \omega_R\hat{\gamma}(\omega_R) - \omega_b^2 = 0$ and captures the local real-time dynamics around the barrier top. Accordingly, the steady state reduces to the thermal state to the left of the barrier top but within the parabolic range of the potential if l_{fl} is sufficiently smaller than the anharmonicity length scale q_b. This in turn provides the range of validity of calculations based on local harmonic properties (see the Im F method discussed in the next section). Roughly speaking, for a given temperature above the critical temperature T_c friction must be sufficiently strong. Then, deep inside the well around $q = 0$, one reaches the thermal state of a harmonic oscillator $\rho_\beta^{(0)}(x, r)$ obtained from $\rho_\beta^{(b)}(x, r)$ upon replacing $\omega_b \to i\omega_0$ and putting $q_b = V_b = 0$. To the right of the barrier $g \to 0$, thus describing an exponentially decreasing population on the product side. The escape rate out of the well is given as the probability flux

$$J_{\mathrm{fl}} = \frac{1}{2m}\langle p\delta(q - q_b) + \delta(q - q_b)p\rangle_{\mathrm{fl}} \tag{15.18}$$

with respect to the flux state and normalized to the population in the well. The latter one follows from identifying the normalization Z in Equation 15.14 with the harmonic partition function

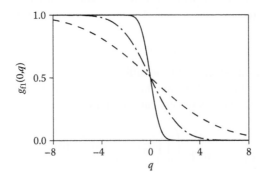

FIGURE 15.2 Distribution $g(x, r)$ describing deviations from thermal equilibrium in Equation 15.16. Shown is the diagonal part $g_{\mathrm{fl}}(0, q)$ for various temperatures $\omega_b\hbar\beta = 0.1$ (dashed), 1 (dotted-dashed), and 2 (solid) for ohmic friction with $\gamma/\omega_b = 1$. Position is scaled with $\sqrt{\hbar/m\omega_b}$.

in the well. This way, the escape rate is found to read $\Gamma = \Gamma_{cl} f_q$ with the classical rate $\Gamma_{cl} = (\omega_0 \omega_R / 2\pi \omega_b) \exp(-\beta V_b)$ and the quantum factor

$$f_q = \prod_{n=1}^{\infty} \frac{u_n^{(b)}}{u_n^{(0)}} = \prod_{n=1}^{\infty} \frac{v_n^2 + \omega_0^2 + v_n \hat{\gamma}(v_n)}{v_n^2 - \omega_b^2 + v_n \hat{\gamma}(v_n)} \tag{15.19}$$

describing the impact of quantum fluctuations. For high temperatures one has $f_q \to 1$ and the barrier escape is purely due to thermal activation over the barrier. For lower temperatures $f_q > 1$ so that barrier escape is enhanced partially due to zero point fluctuations in the well, partially due to tunneling through the top of the barrier. At the so-called crossover temperature T_0, however, where $v_1^2 - \omega_b^2 + v_1 \hat{\gamma}(v_1) = 0$, the rate expression Equation 15.19 breaks down. For instance, for ohmic friction $\hat{\gamma}(z) = \gamma$ one derives from the positive solution to $\lambda_0^2 - \omega_b^2 + \lambda_0 \gamma = 0$ that $T_0 = \hbar \lambda_0 / 2\pi k_B$. For vanishing friction this leads to $T_0^{(0)} = \hbar \omega_b / 2\pi k_B = T_c^{(0)}/2$. Interestingly, while for temperatures below the critical temperature T_c the steady-state distribution is determined by global properties of the potential, the escape process is dominated by tunneling only below the crossover temperature $T_0 < T_c$ [11,12]. Both temperatures are lowered with increasing friction corresponding to the fact that dissipation tends to drive a quantum system back toward the classical regime even though the reservoir is quantum mechanical in nature as well.

To calculate tunneling rates within a dynamical calculation for temperatures below the crossover temperature is a challenging and yet unsolved problem. The crucial point is that even for nondissipative systems a semiclassical expression for the time-evolution operator (Equation 15.4) in the long time limit where deep tunneling occurs, is not known. Some progress has been made by analyzing the phase space dynamics of minimal action paths in the complex plane [7,11,12] or by using initial value representations for the quantum propagator [13].

15.4 THERMODYNAMICAL APPROACH

Physically, a full dynamical treatment is not always necessary. It seems intuitively clear that this is the case if the state from which tunneling occurs remains close to a local thermal equilibrium. The most prominent approach is the imaginary part of the free energy method ("Im F") [3,8,9], which applies over the whole temperature range and is based upon a semiclassical treatment of imaginary time path integrals. In the range where it applies, it provides an extremely elegant and powerful formulation.

15.4.1 GENERAL FORMULATION AND GAUSSIAN HEAT BATHS

The underlying idea of the Im F method is this: inside a metastable well quasienergy levels $\varepsilon_n = E_n - i\hbar\Gamma_n/2$ exist, the finite lifetime of which are related to imaginary parts Γ_n with $E_n \gg \hbar\Gamma_n$. Equivalently, in a scattering experiment such states appear as resonances with finite widths. Hence, the partition function of the unstable system is (formally) calculated as

$$Z = \sum_{n=0}^{\infty} e^{-\beta \varepsilon_n} \approx \sum_{n=0}^{\infty} e^{-\beta E_n} - i \frac{\hbar \beta}{2} \sum_{n=0}^{\infty} \Gamma_n e^{-\beta E_n}.$$

Here, for energies near and above the barrier, the sum is taken as an integral. Obviously, the imaginary part in Z is proportional to the thermally averaged decay rate. In the nondissipative case, a careful WKB treatment proves that [8]

$$\Gamma = -\frac{2}{\hbar} \mu(T) \, \text{Im} \, F \tag{15.20}$$

with the free energy $F = -\ln Z / \beta$ and a temperature-dependent prefactor $\mu(T)$ with $\mu(T \geq T_0^{(0)}) = T_0^{(0)}/T$ above and $\mu(T < T_0^{(0)}) = 1$ below the crossover temperature $T_0^{(0)} = \hbar \omega_b / 2\pi k_B$. It is not a priori clear that the relation (Equation 15.20) also applies to finite dissipation (with $T_0^{(0)}$ replaced by the crossover temperature T_0 for finite friction) and, in fact, a rigorous proof has not been given yet. Qualitatively, one can at least formulate a necessary condition: a thermodynamic approach is supposed to be valid when a thermal state is preserved in the well region over periods of time long compared to intrawell relaxations. For temperatures above T_c the $\mathrm{Im}\, F$ produces results identical to those derived in the previous section within the full dynamical calculation. It is thus applicable if friction is sufficiently strong, but certainly fails for weak dissipation where the steady state broadens in position space to cover the entire well domain. For temperatures far below T_0 the population is confined to the ground state in the potential well so that even for weaker dissipation the method applies. All experimental results obtained so far are in complete agreement with its theoretical predictions which indicates that the $\mathrm{Im}\, F$ approach provides at least an extremely accurate approximation to a full dynamical theory.

The starting point is the representation of the partition function of the composite system in terms of Euclidian path integrals. Upon eliminating the reservoir degrees of freedom along the lines described in Section 15.2, one has

$$Z = \oint \mathcal{D}[\bar{q}] \; e^{-\bar{S}_S[\bar{q}]/\hbar} \, \langle e^{-S_I[\bar{q},\vec{\bar{x}}]/\hbar}\rangle_R \,, \tag{15.21}$$

where the integral sums over all periodic paths in the interval $\hbar\beta$. As above, the expectation value $\langle e^{-S_I[\bar{q},\vec{\bar{x}}]/\hbar}\rangle_R$ can explicitly be evaluated for a Gaussian bath distribution, that is, for reservoirs consisting of harmonic modes. In this case, one obtains the influence functional (Equation 15.8) for $z = i\tau, \tau \in [0,\hbar\beta]$ and $\langle e^{-S_I[\bar{q},\vec{\bar{x}}]/\hbar}\rangle_R = \exp(-\bar{\Phi}[\bar{q}]/\hbar)$. The remaining integral over the system paths with the effective action $\bar{\Sigma} = \bar{S}_S + \bar{\Phi}$ is then evaluated in a semiclassical fashion. In contrast to the conventional recipe, however, exponentially small imaginary parts must be retained against dominating real parts since the imaginary parts determine the escape rate. The fact that the partition function carries an imaginary contribution is a consequence of the instability of the system and follows from a proper steepest descent evaluation of the fluctuation path integral around the minimal action paths. For $T > T_0$ the system supports only the trivial minimal action paths $q_0(\tau) = 0$ with $\bar{\Sigma}[0] = 0$ and $q_b(\tau) = q_b$ with $\bar{\Sigma}[q_b] = \hbar\beta V_b$. Static fluctuations around the latter path are unstable and must be treated via an analytical continuation procedure [9]. Eventually, this reproduces the result (Equation 15.19) that diverges at T_0. The fact that the full dynamical approach discussed in the previous section is strongly influenced by anharmonic quantum fluctuations already at the higher temperature T_c, while this temperature does not explicitly appear in the final rate expression (Equation 15.19), indicates that the $\mathrm{Im}\, F$ calculation and the dynamical calculation deviate in the temperature range between T_c and T_0. These discrepancies are small and vanish for temperatures below T_0 [12], where the path integral (Equation 15.21) is dominated by the contribution of a newly emerging periodic orbit in the inverted barrier potential, the so-called bounce $\bar{q}_B(\tau)$ [9]. Its explicit trajectory must be calculated numerically as well as its corresponding minimal action $S_B \equiv \bar{\Sigma}[\bar{q}_B] \ll \hbar\beta V_b$. It can be shown that the bounce is again an unstable orbit giving rise to an imaginary contribution to the partition function. Hence, one arrives at

$$\Gamma = \sqrt{\frac{S_B}{2\pi\hbar}} \sqrt{\frac{\det[L_0]}{\det'[L_B]}}\, e^{-S_B/\hbar}\,. \tag{15.22}$$

Here, the first factor accounts for a zero-mode of the bounce orbit related to its invariance against phase shifts $\bar{q}_B(\tau) \to \bar{q}_B(\tau + \tau_0)$. The operators L_0 and L_B correspond to the second-order variational operator of the effective action $\bar{\Sigma}[q]$ around the trivial periodic orbit at the well bottom and the bounce orbit, respectively. The prime indicates that the zero mode has to be omitted. While explicit results can only be obtained numerically (see, for instance, Grabert et al. (1987) [9]), approximate ones can be given for weak dissipation and for vanishing temperature. In this case, one has $\Gamma = \Gamma_0 Y_2$

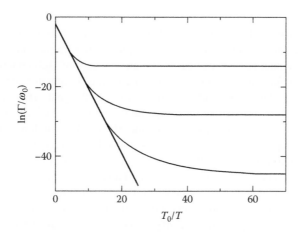

FIGURE 15.3 Quantum tunneling rate versus inverse temperature for the potential (Equation 15.12) and various ohmic friction strengths $\gamma/\omega_0 = 0$ (top), 1 (middle), 2 (bottom). Barrier height is $V_b/\hbar\omega_0 = 5$ and the thick straight line depicts the classical result.

with the WKB tunneling rate for vanishing dissipation Γ_0 and a dissipative factor stemming from the influence functional evaluated along the zero-friction bounce path $\bar{q}_B^{(0)}$, that is,

$$Y_2 = \exp\left[-\frac{1}{2}\int_{-\infty}^{\infty} d\tau \int_{-\infty}^{\infty} d\sigma \bar{q}_B^{(0)}(\tau)k(\tau - \sigma)\bar{q}_B^{(0)}(\sigma)\right]. \tag{15.23}$$

The imaginary time kernel is related to the second cumulant $K(i\tau)$ of the bath distribution (Equation 15.10) via $k(\tau) = \mu : \delta(\tau) : -K(i\tau)$ with the static contribution $\mu = 2\lim_{\hbar\beta \to 0} K(0)$ and the periodically continued δ function $: \delta(\tau) := \sum_n \delta(\tau - n\hbar\beta)$. This dissipative factor thus adds to the exponential of the bare tunneling rate which contains the bare bounce action $\bar{S}_S[\bar{q}_B^{(0)}]$. It can be shown that the exponential in Equation 15.23 is always positive so that dissipation due to position–position coupling always suppresses tunneling. In case of the metastable potential (Equation 15.12) and for ohmic friction the explicit evaluation gives

$$\Gamma_0 \propto e^{-36V_b/(5\hbar\omega)}, \quad Y_2 = e^{-162\zeta(3)V_b\gamma/(\pi^3\omega_0^2)}. \tag{15.24}$$

A typical Arrhenius plot $[\ln(\Gamma)$ vs. inverse temperature] is shown in Figure 15.3. The changeover from classical thermal escape to quantum tunneling appears for stronger friction at lower temperatures and is also smeared out. At the same time, friction suppresses tunneling substantially according to the explicit dependence of the bounce action on dissipation [cf. Equation 15.24].

These and further predictions have been verified experimentally in JJ devices [14]. We also note that alternative formulations, for example, multidimensional transition, state theory, are completely equivalent to the Im F method and provide identical results.

15.4.2 Extension to Reservoirs with Non-Gaussian Fluctuations

The results discussed so far assume a heat bath environment which obeys Gaussian statistics. As already pointed out, for a large majority of experimentally relevant situations this is indeed an accurate description. In the last few years, however, reservoirs with non-Gaussian characteristics have attracted substantial interest in the context of current noise generated by mesoscopic conductors (typically voltage biased) [15] such as tunnel contacts, atomic point contacts, and ballistic wires, to name but a few. In fact, it has been shown that higher than second-order current cumulants carry information about the transport process that cannot be gained from usual current–voltage measurements (I–V curves). The granularity of the elementary charge carriers gives rise to non-Gaussian deviations from the mean current, for example, in the simplest case of a tunnel junction corresponding to Poissonian noise statistics. While this concept of "full counting statistics" is fascinating and has also interesting relations to photon counting in quantum optics, the detection of higher-order current cumulants is very challenging due to small signals and strict filtering demands. As we have seen above for Gaussian noise, escape rates depend very sensitively on noise produced by the surrounding, particularly in the quantum regime. Hence, the switching out of the zero-voltage state in JJs has been proposed as detection process. Indeed, very recently the third cumulant produced by a tunnel junction has been retrieved from the asymmetry of the switching rate with the JJ being operated in the classical regime of over-the-barrier-activation. The extension to the tunneling regime, however, is not easy since the second-order cumulant of current noise gives rise to additional heating. One proposal to overcome this difficulty has been discussed in Ankerhold and Grabert (2005) [16] (the corresponding circuitry is sketched in Figure 15.4): the noise generating element is placed in parallel to the JJ so that only equilibrium current fluctuations reach the junction. No net current flows through the conductor and due to time reversibility all odd order cumulants vanish so that direct access to higher order even cumulants is obtained.

The tunneling rate at zero temperature is determined by the $\mathrm{Im}\,F$-recipe with the representation of the partition function as specified in Equation 15.21. The reservoir distribution is given by the current distribution in equilibrium and the coupling between the detector degree of freedom and the current noise is assumed to be weak. This way, the tunneling rate is formally given by

$$\Gamma = \Gamma_0 \, \langle e^{-S_I[\theta_B^{(0)}/2, I/e]/\hbar} \rangle_G \tag{15.25}$$

with the nondissipative bounce orbit $\theta_B^{(0)}$ and the current operator I of the conductor G. Note that the factor 1/2 appears in S_I since, according to the second Josephson relation $V_J = (\hbar/e)\dot{\theta}/2$ (phase across the JJ is θ), the voltage across the junction equals that across the conductor $V_J = V_G \equiv (\hbar/e)\dot{\varphi}$. (phase across the conductor is φ). The correction to the bare rate is again given by the generating functional of the reservoir, that is, the conductor for which in some cases even analytical expressions are available. For instance, in case of a tunnel junction with dimensionless conductance $g_T = \hbar/(2e^2 R_T) = \pi \sum_i T_i$ (tunneling resistance R_T and transmissions T_i) one derives for the corresponding influence functional

$$\Phi_T[\varphi] = -4g_T \int_{\mathcal{C}} dz \int_{\mathcal{C}} dz' \alpha(z - z') \sin^2\left[\frac{\varphi(z) - \varphi(z')}{2}\right] \tag{15.26}$$

with the kernel $\alpha(z) = \pi/[2(\hbar\beta)^2 \sinh^2(\pi z/\hbar\beta)]$. $\mathcal{C}$ denotes the integration contour introduced in Equation 15.9 consisting of two segments in real and one segment in imaginary time. For the $\mathrm{Im}\,F$ calculation only the contribution along this latter branch $z = -i\tau$ is relevant. By expanding the sine-function in a power series of the phase difference, the impact of higher-order cumulants

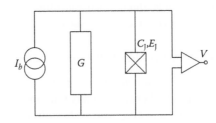

FIGURE 15.4 Electrical circuit with a mesoscopic conductor G in parallel to a JJ with capacitance C_J and coupling energy E_J biased by an external current I_b.

on the tunneling process can be studied. Thereby the kth power corresponds to the kth cumulant. Approximating a single well-barrier segment of the tilted washboard potential of a JJ by Equation 15.12 and switching from $q \to \theta$, one finds for the bounce orbit $\theta_B^{(0)}(\tau) = 3\theta_b/[2\cosh^2(\omega_0\tau/2)]$. This way, in lowest order (second order) we have $\Gamma^{(2)} = \Gamma_0 Y_2$ with $Y_2 = \exp[-g_T 54\zeta(3)\theta_b^2/(4\pi^3)]$ so that the known suppression of tunneling due to the second cumulant of the reservoir is regained. In next order (fourth order) one finds $\Gamma^{(4)} = \Gamma^{(2)} Y_4$ with

$$Y_4 = \exp\left[\frac{g_T A_4}{(4\pi)^3}(9\theta_b)^4\right] \tag{15.27}$$

and a numerical constant $A_4 = 66.3547\ldots$. Apparently, the fourth-order contribution *enhances* the tunneling rate. Moreover, it displays a characteristic dependence on the tilt (the applied bias current) of the washboard potential different from that of the Gaussian contribution Y_2. This in turn allows for the experimental detection of the fourth order cumulant of the current noise distribution.

Theoretically, the above scheme extends the standard Im F approach to reservoirs with non-Gaussian statistics still, however, residing in equilibrium. The calculation of tunneling rates in situations when a finite bias voltage is applied across the conductor so that its current distribution is stationary but out of equilibrium, has not been solved yet. It is related to the open problem of how to obtain deep tunneling rates within a real-time approach.

15.5 TUNNELING RATES FROM EQUATIONS OF MOTION

In the regime of weak friction and for temperatures above the crossover temperature, the Im F approach is not applicable since then the steady state covers in position space even the well region and does not reduce to a thermal equilibrium. Accordingly, the flux state distribution from which tunneling occurs is not known a priori. In the classical limit this situation refers to the energy-diffusive regime of Kramers' theory [17]. To analyze the impact of quantum fluctuations in this domain, one starts by deriving from the exact path integral expression (Equation 15.3) an approximate time evolution equation for the reduced density. Namely, in the domain $\gamma\hbar\beta \ll 1$ the time scale on which relaxation occurs, that is, $1/\gamma$, by far exceeds the retardation time scale $\hbar\beta$ of the reservoir so that on a coarse-grained time scale a time local equation of motion for the density does exist. This leads to master type of equations corresponding to second-order perturbation theory in the system-bath coupling (Born–Markov approximation). Explicit expressions are most conveniently obtained in the eigenstate representation of the bare system Hamiltonian H_S.

15.5.1 ESCAPE OVER A POTENTIAL BARRIER

Let us first illustrate the general concept for a standard situation, namely, a particle initially confined in a metastable potential (Equation 15.12). We first assume a high barrier $V_b \gg k_B T, \hbar\omega_0$ so that for weak damping the well supports a ladder of quasistationary discrete energy levels which reach the continuum for energies above the barrier top. Second, we consider higher temperatures $k_B T$ which sufficiently exceed $\hbar\omega_0$ so that the energy ladder is smeared out by thermal fluctuations. This is a typical semiclassical situation not only in the sense of barrier tunneling but also with respect to the dynamics inside the well. The crucial question, then, is: What are the dominating quantum effects that influence the escape rate?

For this purpose, it is convenient to introduce the occupation probability of a well state with energy E via

$$P(E,t) = \sum_{n=0}^{N} \delta(E - E_n) p_n(t), \tag{15.28}$$

where N is the number of states in the well. Here p_n is the occupation probability of a well eigenstate $|E_n\rangle$ with quasienergy E_n. It is thus identical to the diagonal part of the reduced density

Equation 15.2 matrix in the energy representation $p_n(t) = \langle E_n|\rho(t)|E_n\rangle$. The explicit construction of these eigenstates follows from a type of WKB-recipe as shown below. Starting with a set of master equations for the populations p_n, the corresponding time evolution equation for the occupation probability is found [18] to read

$$\dot{P}(E,t) = \int dE' \left[W_{E,E'} \frac{R(E')P(E',t)}{n(E')} - W_{E',E} \frac{R(E)P(E,t)}{n(E)} \right] - T(E) \frac{\omega(E)}{2\pi} P(E,t), \qquad (15.29)$$

with $\omega(E)$ being the frequency of a classical oscillation at energy E. This diffusion equation captures the incoming probability flux to and outgoing probability flux from the state with energy E according to intrawell transition rates [19]

$$W_{E,E'} = \frac{1}{\hbar^2} \int_{-\infty}^{\infty} dt \, \mathrm{Tr}_R \{ \langle E|H_I(t)|E'\rangle \langle E'|H_I|E\rangle \rho_{\beta,R} \}, \qquad (15.30)$$

reflection probabilities $R(E)$ from the barrier, and transmission probabilities $T(E) = 1 - R(E)$ through the barrier. In the transition rates the system–reservoir coupling appears in the interaction picture $H_I(t) = e^{i(H_S+H_R)t/\hbar} H_I e^{-i(H_S+H_R)t/\hbar}$ with $\rho_{\beta,R} = e^{-\beta H_R}/Z_R$ being the equilibrium bath density matrix. Further, $n(E)$ in Equation 15.29 denotes the density of states.

The transition rate (Equation 15.30) can be evaluated explicitly in case of the bilinear system–bath coupling as in Equation 15.1. One arrives at the golden rule type of formula,

$$W_{E,E'} = \frac{1}{\hbar^2} |Q_{\mathrm{qm}}(E',E)|^2 D(E-E'), \qquad (15.31)$$

with $Q_{\mathrm{qm}}(E',E) \equiv \langle E'|q|E\rangle$ and the bath absorption/emission rate captured by

$$D(E) = \hbar \int_{-\infty}^{\infty} dt \, K(t) e^{itE/\hbar} = \hbar I(E/\hbar)\bar{n}(E), \qquad (15.32)$$

with the Bose–Einstein distribution $\bar{n}(E) = 1/[\exp(\beta E) - 1]$. In accordance with an effectively Markovian dynamics we consider a purely ohmic environment with $I(\omega) = m\gamma\omega$. In order to calculate Equation 15.31 semiclassically, we have to construct the energy-dependent wave functions inside the well. In the energy range close to the barrier top, however, classical turning points to the left and to the right of the barrier are not sufficiently separated so that the standard WKB approximation is not applicable. In this situation, one exploits (as done already above) that any sufficiently smooth barrier potential can be approximated by a parabolic barrier with barrier frequency ω_b for which the Schrödinger equation can be solved exactly. The proper eigenfunctions are then matched asymptotically (sufficiently away from the barrier top in the well region) onto WKB wave functions to determine phases and amplitudes of the latter ones. This leads us to

$$\langle E|q\rangle = \frac{1}{2} \left[\langle E|q\rangle^- + r(E)\langle E|q\rangle^+ \right] \qquad (15.33)$$

with matrix elements

$$\langle E|q\rangle^{\pm} = \frac{N(E)}{\sqrt{\partial H_S(q,p)/\partial p}} e^{\pm \frac{i}{\hbar} S_0(E,q) \mp \frac{i\pi}{4}}, \qquad (15.34)$$

where $S_0(E,q) = \int_{q_1}^{q} p(E,q')dq'$ is the action of an orbit starting at a turning point q_1 and running in time t toward q with momentum $p(E,q')$ at energy E. The complex valued reflection amplitudes $r(E)$ of a parabolic barrier [20] are related to the reflection probabilities $R(E) = |r(E)|^2$ and the normalization is determined from $\langle E|E'\rangle = \delta(E-E')$ to read

$$N(E) = 2\sqrt{\frac{1}{\hbar\pi[R(E)+1]}}. \qquad (15.35)$$

These results match for lower energies onto the standard semiclassical ones. In particular, transmission and reflection coefficients over the full energy range are given by the uniform semiclassical expressions

$$T(E) = |t(E)|^2 = \frac{1}{1 + \exp[-\bar{S}(E)/\hbar]}$$

$$R(E) = |r(E)|^2 = \frac{\exp[-\bar{S}(E)/\hbar]}{1 + \exp[-\bar{S}(E)/\hbar]}, \tag{15.36}$$

where $\bar{S}(E)$ denotes the Euclidian action of a periodic orbit with energy E oscillating in the inverted barrier potential $-V(q)$. With the semiclassical wave function at hand and using the restricted interference approximation [21] one can now calculate the transition matrix elements which enter the transition rates in Equation 15.31.

Upon inserting these transition rates into the time evolution Equation 15.29, an expansion in powers of $\hbar$ can be performed, where one has to keep in mind, however, that for energies near the barrier top reflection and transmission probabilities are of order 1, particularly with $T(E = V_b) = R(E = V_b) = 1/2$. This way, we arrive at the semiclassical expression for the energy diffusion Equation 15.29 in the metastable well

$$\dot{P}(E,t) = \left\{ \frac{\partial}{\partial E} C(E) \gamma S_0(E) \left[1 + \frac{1}{\beta} \frac{\partial}{\partial E} \right] R(E) - T(E) \right\} \frac{\omega(E)}{2\pi} P(E,t), \tag{15.37}$$

with

$$C(E) = 2 \frac{1 + R(E)^2}{[1 + R(E)]^2}, \tag{15.38}$$

and $S_0(E)$ the action of a periodic orbit in the well with energy E. Note that for vanishing transmission ($R = 1, T = 0$) one recovers from the above expression the classical Kramers equation [17]. Corrections to the diffusion Equation 15.37 are at most of order $\hbar^2$.

The escape rate follows again from a quasistationary nonequilibrium state, this time from the quasistationary energy distribution $P_{st}(E)$ corresponding to Equation 15.37—that is,

$$\Gamma_{scl} = \int_0^\infty dE \, n(E) T(E) P_{st}(E) \tag{15.39}$$

with the semiclassical density of states $n(E) = 1/[\hbar\omega(E)]$. Taking into account Equation 15.37, one finds

$$\Gamma_{scl} = \frac{\sinh(\omega_0 \hbar \beta/2)}{(\omega_0 \hbar \beta/2)} |B| \, \Gamma_{cl}, \tag{15.40}$$

where the classical escape rate is

$$\Gamma_{cl} = \frac{\omega_0 \gamma S_0(V_b) \beta}{2\pi} e^{-\beta V_b} \tag{15.41}$$

and the coefficients read

$$B = -\frac{1}{4^\theta} \frac{{}_2F_1\left[\frac{1}{2} - \frac{\theta}{2} - a, \frac{1}{2} - \frac{\theta}{2} + a, 1 - \theta, -\frac{4}{9}\right]}{{}_2F_1\left[\frac{1}{2} + \frac{\theta}{2} - a, \frac{1}{2} + \frac{\theta}{2} + a, 1 + \theta, -\frac{4}{9}\right]}, \quad a = \sqrt{\frac{\beta \gamma S_0(V_b)(1-\theta)^2 + 360^2}{4\beta \gamma S_0(V_b)}} \tag{15.42}$$

with the abbreviation $\theta = \omega_b \hbar \beta$. In Equation 15.40, quantum fluctuations are captured by two factors: the first one describes zero-point fluctuations in the well, while the second one describes the impact of finite barrier transmission close to the top. Interestingly, for weak friction the latter one can actually prevail and lead to a reduction of the escape rate compared to the classical situation due to a finite reflection from the barrier also for energies $E \geq V_b$ (Figure 15.5).

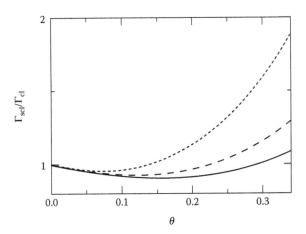

FIGURE 15.5 Escape rate from Equation 15.40 is normalized to the classical rate as a function of the inverse temperature θ for various values of the dimensionless friction strength: $\gamma/\omega_0 = 0.01$ (solid), $\gamma/\omega_0 = 0.005$ (dashed) and $\gamma/\omega_0 = 0.0015$ (dotted) and barrier hight $V_b\beta = 10$.

15.5.2 QUANTUM FLUCTUATIONS IN ESCAPE PROCESSES OVER DYNAMICAL BARRIERS

The situation discussed in the previous section where escape happens to occur over a static energy barrier is now generalized to a situation, where two stable basins in phase space are separated by a dynamical barrier. Specifically, we consider a system with a weakly anharmonic potential driven by an external time-periodic force (Duffing oscillator), namely,

$$H_S = \frac{p^2}{2m} + \frac{m}{2}\omega_0^2 q^2 - \frac{\Gamma}{4}q^4 + Fq\cos(\omega_F t).$$ (15.43)

Accordingly, for the anharmonic coefficient we assume $\Gamma/m \ll \omega_0^2/\langle q^2\rangle$ so that driving is almost resonant for $\delta\omega = \omega_F - \omega_0 \ll \omega_F$. Classically, when damping is taken into account two stable oscillations with different amplitudes and phases appear beyond a bifurcation threshold. The latter one depends on external parameters such as driving amplitude F and frequency mismatch $\delta\omega$. In phase space, these two stable states correspond to stable basins of attraction which are separated by an unstable domain. Thermal fluctuations, however, can induce switchings between the stable basins leading again to a rate process [22]. In fact, the sensitivity of this rate with respect to the curvature of the potential surface in Equation 15.43 has recently been exploited as working principle in a very sensitive detection device. In the so-called Josephson Bifurcation Amplifier (JBA) [23,24] a superconducting tunnel contact (JJ) is placed in parallel to a Cooper-pair box implementing a two-level system (qubit) and is driven by microwave fields. In the operational regime of the junction, the device can be described by the Hamiltonian (Equation 15.43), where the two qubit states lead to slightly different curvatures ω_0. Measurements of the switching of the JBA give thus direct access to the state of the qubit.

Theoretically, the difficulty for a rate description in this kind of system is twofold: first, the Hamiltonian of the isolated system H_S is time-dependent and therefore, energy is not conserved, and second, there is no static energy barrier. Due to the coupling to the environment, however, the system approaches, as already discussed for the classical case above, a steady-state situation such that the reduced density matrix takes the form $\rho(t) \sim \bar{\rho}(t)\cos(\omega_F t)$ with an only weakly time-dependent density $\bar{\rho}$. Hence, it is convenient to move to a rotating frame described by the unitary operator [25]

$$U(t) \equiv U_S(t)U_B(t) = e^{-i\hat{a}^\dagger\hat{a}\omega_F t - i\sum_n^N \hat{b}_n^\dagger\hat{b}_n\omega_F t},$$ (15.44)

where $\hat{a}$ and $\hat{b}_n$ are annihilation operators for *harmonic* oscillators in the system and in the bath, respectively. In the rotating frame, the total Hamiltonian reads

$$\tilde{H} = U^\dagger \left[H - i\hbar \frac{\partial}{\partial t} \right] U = \tilde{H}_S + \tilde{H}_R + \tilde{H}_I \qquad (15.45)$$

with H as specified in Equation 15.1 and H_S as in Equation 15.43. The system part follows upon discarding fast oscillating terms $\exp(\pm ik\omega_F t)$ with $|k| \geq 1$ as a *time-independent* Hamiltonian [26] of the form

$$\tilde{H}_S(P,Q) = m\omega_F \delta\omega C^2 \left[-\frac{1}{4} \left(\frac{Q^2}{C^2} + \frac{P^2}{(C\omega_F m)^2} - 1 \right)^2 + \frac{\sqrt{\alpha}}{C} Q \right] \qquad (15.46)$$

with $C = \sqrt{\frac{8\omega_F \delta\omega m}{3\Gamma}}$ and $\alpha = \frac{3F^2\Gamma}{32(\omega_0\delta\omega m)^3}$. This latter quantity plays the role of a bifurcation parameter: for $0 < \alpha < 4/27$ the Hamiltonian (Equation 15.46) has three extrema, where the two stable ones correspond in the laboratory frame to oscillations with low and high amplitude, respectively. They are separated by a phase-space barrier associated with an unstable extremum (see Figure 15.6). The remaining parts of the composite system (Equation 15.45) are written as [27]

$$\tilde{H}_R = \sum_{n=1}^{N} \left(\frac{p_n^2}{2\tilde{m}_n} + \frac{\tilde{m}_n}{2} \tilde{\omega}_n^2 x_n^2 \right),$$

$$\tilde{H}_I = -\sum_{n=1}^{N} \tilde{c}_n \left(x_n Q + \frac{p_n}{\tilde{\omega}_n \tilde{m}_n} \frac{P}{\omega_F m} \right) + \left[Q^2 + \frac{P^2}{(\omega_F m)^2} \right] \sum_{n=1}^{N} \frac{c_n^2}{4m_n\omega_n^2}, \qquad (15.47)$$

with new bath parameters

$$\tilde{m}_n = \frac{m_n}{1 - \omega_F/\omega_n}, \quad \tilde{\omega}_n = \omega_n - \omega_F, \quad \tilde{c}_n = \frac{c_n}{2}. \qquad (15.48)$$

These parameters determine an effective spectral density and an effective damping kernel (see Equation 15.10) in the rotating frame [27].

We now calculate the probability for a system prepared initially in one of the stable states, say the low-amplitude state, to switch to the other one, say the high-amplitude state. Since in the rotating frame the Hamiltonian takes a time-independent form, we can apply the approach presented in

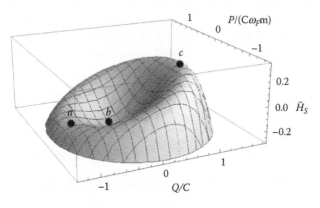

FIGURE 15.6 The Hamiltonian function (Equation 15.46) in the rotating frame, for $\alpha = 1/27$. The energy is scaled with $C^2\omega_F m\delta\omega$. The minimum (b) and the maximum (c) in $\tilde{H}_S$ correspond in the laboratory frame to the stable states with low and high amplitude, respectively, separated by a marginal state (a).

the first part of this section to determine this rate from an equation of motion. The difference here is that the system Hamiltonian is not of standard form and that the system–reservoir coupling carries an additional momentum–momentum interaction. Hence, the transition rates (Equation 15.30) appearing in the master Equation 15.29 read

$$W_{E,E'} = \frac{D_{QQ}(E - E')}{\hbar^2} \left[|Q_{\rm qm}(E',E)|^2 + \frac{1}{\omega_F^2 m^2} |P_{\rm qm}(E',E)|^2 \right]$$
$$+ \frac{D_{QP}(E - E')}{\hbar^2 \omega_F m} 2i \,{\rm Im}\{Q_{\rm qm}(E',E)^* P_{\rm qm}(E',E)\}, \tag{15.49}$$

with $P_{\rm qm}(E',E) \equiv \langle E'|P|E \rangle$. Here, the bath functions D_{XY} are defined according to Equation 15.32 with correlations $K_{XY}(t) = \langle \xi_X(t)\xi_Y(0) \rangle / \hbar$ containing those bath forces ξ_X, ξ_Y that couple to system operators X, Y in Equation 15.47. For the escape process near the bifurcation threshold the energy level spacings of the well states in Equation 15.46 are small compared to ω_F and one arrives at [27]

$$D_{QQ}(E) = \tilde{\gamma}\{n_\beta(E_F + E)(E_F + E) + [n_\beta(E_F - E) + 1](E_F - E)\},$$
$$D_{QP}(E) = i\tilde{\gamma}\{-n_\beta(E_F + E)(E_F + E) + [n_\beta(E_F - E) + 1](E_F - E)\}, \tag{15.50}$$

where $E_F = \hbar\omega_F$ and $\tilde{\gamma} = \gamma/4$ is the effective friction constant in the rotating frame. Notably, the above expressions display that physically two channels of bath modes are accessible for emission or absorption of quanta, namely, one with energy $E_F + E$ and one with energy $E_F - E$ [27].

Now, semiclassical wave functions in the well are constructed as discussed above in Equation 15.34. With their help, transition rates are evaluated and one arrives at the semiclassical energy diffusion equation

$$\dot{P}(E,t) = \left[\frac{\partial}{\partial E} \tilde{\gamma} \left(\bar{\Delta}(E) + \bar{\kappa}(E)\frac{\partial}{\partial E} \right) R(E) - T(E) \right] \frac{\omega(E)}{2\pi} P(E,t) \tag{15.51}$$

with

$$\bar{\Delta}(E) = \Delta(E)C(E) - 2\hbar v \Delta^{(1)}(E), \tag{15.52}$$

$$\bar{\kappa}(E) = \Delta(E)C(E)\frac{\kappa}{2} + \hbar\omega_F \Delta^{(1)}(E). \tag{15.53}$$

Here, the coefficient $\kappa = \hbar\omega_F \coth(\hbar\omega_F \beta/2)$ results from the expansion of D_{QQ} in Equation 15.50 and the coefficient $\hbar v = [\hbar\beta\omega_F - \sinh(\hbar\omega_F \beta)]/[\cosh(\hbar\omega_F \beta) - 1]$ from D_{QP} in Equation 15.50. The factor $C(E)$ is again due to the finite barrier transmission. Further, Δ can be interpreted as a generalized action and originates from the first term in Equation 15.49, while $\Delta^{(1)}$ is related to the "unconventional" term ${\rm Im}\{Q_{\rm qm}(E',E)^* P_{\rm qm}(E',E)\}$ in the transition rate Equation 15.49. These two functions are given by

$$\Delta(E) = m \oint dQ \dot{Q} + \frac{1}{m\omega_F^2} \oint dP \dot{P} \tag{15.54}$$

$$\Delta^{(1)}(E) = \frac{\omega}{\pi\omega_F} \lim_{\varepsilon \to 0^+} \int_0^{2\pi/\omega} dt\, dt'\, \dot{Q}(t)\dot{P}(t') \frac{\sin[\omega(t-t')]}{1 + e^{-2\varepsilon} - 2e^{-\varepsilon}\cos[\omega(t-t')]}. \tag{15.55}$$

Corrections to Equation 15.51 are of the same order as in Equation 15.37, namely, of order $\hbar^2$ or smaller.

This way, the escape rate gains leading order quantum corrections in the form

$$\Gamma_{\rm scl} = \Gamma_{\rm cl} \left[1 + \frac{\hbar\omega_F}{\kappa\pi} \left(-b_1 \frac{\omega_b}{\omega_F} + b_2 \right) \right]. \tag{15.56}$$

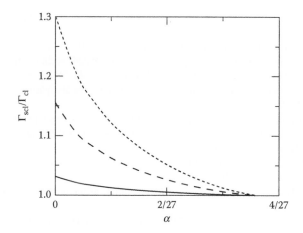

FIGURE 15.7 Escape rate from Equation 15.56 is normalized to the classical rate as a function of the bifurcation parameter α for $\hbar\omega_F\beta/(2\pi) = 0.01$ (solid), $\hbar\omega_F\beta/(2\pi) = 0.05$ (dashed), and $\hbar\omega_F\beta/(2\pi) = 0.1$ (dotted). For all the three lines is $\beta(m\omega_F\delta\omega)^2/\Gamma = 20$, $\delta\omega/\omega_0 = 0.1$ and the dimensionless friction constant $\beta\tilde{\gamma}mD^2\delta\omega = 0.001$.

The classical rate Γ_{cl} can directly be inferred from Equation 15.41 by replacing

$$S_0 \to \Delta, \quad \gamma \to \tilde{\gamma}, \quad \beta \to 2/\kappa, \tag{15.57}$$

with the barrier height $V_b = E_a - E_b$, where the energies E_a and E_b refer to the unstable point (a) and the stable point (b) in Figure 15.6, respectively, and the well frequency ω_0 is determined by Equation 15.46. The quantum factor contains the coefficient $b_1 \simeq 1.04$ originating from a finite barrier transmission/reflection and the bath-induced coefficient

$$b_2 = \frac{4\pi}{\omega_F\kappa}(\nu\kappa + \omega_F) \int_{E_a}^{E_b} dE \frac{\Delta^{(1)}(E)}{\Delta(E)}, \tag{15.58}$$

originating from the position–momentum contribution in the transition rates (Equation 15.49). The range of validity of this rate expression is given by those values of α which are sufficiently larger than 0 (where the motion near the barrier top is overdamped) and sufficiently smaller than $\alpha = 4/27$ (where the barrier height tends to zero). Explicit results for the rate (Equation 15.56) are depicted in Figure 15.7 for various values of temperature. Interestingly, the two types of quantum fluctuations have opposite effects on the rate expression: while a finite reflection for energies above the barrier top leads to a suppression of the escape probability, bath induced fluctuations produce an increase. In contrast to the situation of a static energy barrier discussed in the previous section (see Figure 15.5), these latter corrections always prevail in the relevant range of parameters. The corresponding rate enhancement is substantial and grows with decreasing temperature.

REFERENCES

1. D. Vion, A. Aassime, A. Cottet, P. Joyez, H. Pothier, C. Urbina, D. Esteve, and M. H. Devoret, Manipulating the quantum state of an electrical circuit. *Science* 296, 886, 2002.
2. A. O. Caldeira and A. J. Leggett, Path integral approach to quantum Brownian motion. *Physica A* 121, 587, 1983.
3. U. Weiss, *Quantum Dissipative Systems*, World Scientific, Singapore, 2007.
4. R. P. Feynman and F. L. Vernon, The theory of a general quantum system interacting with a linear dissipative system. *Ann. Phys.* 24, 118, 1963.

5. H. Grabert, P. Schramm, and G.-L. Ingold, Quantum Brownian motion: The functional integral approach. *Phys. Rep.* 168, 115, 1988.

6. C. W. Gardiner and P. Zoller, *Quantum Noise*, Springer, Berlin, 1991.

7. J. Ankerhold, *Quantum Tunneling of Complex Systems*, STMP 224, Springer, Berlin, 2007.

8. I. K: Affleck, Quantum-statistical metastability. *Phys. Rev. Lett.* 46, 388, 1981.

9. H. Grabert, P. Olschowski, and U. Weiss, Quantum decay rates for dissipative systems at finite temperatures. *Phys. Rev. B* 36, 1931, 1987.

10. J. Ankerhold, H. Grabert, and G.-L. Ingold, Dissipative quantum systems with a potential barrier: General theory and the parabolic barrier. *Phys. Rev. E* 51, 4267, 1995; J. Ankerhold, and H. Grabert, Quantum effects in barrier dynamics. *Chem. Phys.* 204, 27, 1996.

11. J. Ankerhold and H. Grabert, Quantum tunneling and the semiclassical real time evolution of the density matrix. *Europhys. Lett.* 47, 285, 1999.

12. J. Ankerhold and H. Grabert, Semiclassical time evolution of the density matrix and tunneling. *Phys. Rev. E* 61, 3450, 2000.

13. S. Zhang and E. Pollak, Monte Carlo method for evaluating the quantum real time propagator. *Phys. Rev. Lett.* 91, 190201, 2003.

14. M. H. Devoret, D. Esteve, C. Urbina, J. Martinis, A. Cleland, and J. Clarke, in *Quantum Tunneling in Solids*, Yu. Kagan and A. J. Leggett (eds.), Elsevier, Amsterdam, 1992.

15. Y. V. Nazarov (ed.), *Quantum Noise in Mesoscopic Physics*, NATO Science Series in Mathematics, Physics and Chemistry, Kluwer, Dordrecht, 2003.

16. J. Ankerhold, and H. Grabert, Comment on "Quantum tunneling at zero temperature in the strong friction limit." *Phys. Rev. Lett.* 95, 186601, 2005.

17. H. A. Kramers, Brownian motion in a field of force and the diffusion model of chemical reactions. *Physica* 7, 284, 1940.

18. A. Verso and J. Ankerhold, Semiclassical theory of energy diffusive escape. *Phys. Rev. A* 79, 22115, 2009.

19. H. P. Breuer and F. Petruccione, *The Theory of Open Quantum Systems*, Claredon Press, Oxford 2002.

20. O. Atabek, R. Lefebvre, M. Garcia Sucre, J. Gomez-Llorente, and H. Taylor, Quantum localization over a potential barrier. *Int. J. Quant. Chem.* 40, 211, 1991.

21. R. M. More and K. H. Warren, Semiclassical calculation of matrix elements. *Ann. Phys.* 207, 282, 1991.

22. M. I. Dykman and M. A. Krivoglaz, Fluctuations in nonlinear systems near bifurcations corresponding to the appearance of new stable states. *Physica A* 104, 480, 1980; M. I. Dykman, C. M. Maloney, V. N. Smelyanskiy, and M. Silverstein, Fluctuational phase-flip transitions in parametrically driven oscillators. *Phys. Rev. E* 57, 5202, 1998.

23. I. Siddiqi, R. Vijay, F. Pierre, C. M. Wilson, M. Metcalfe, C. Rigetti, L. Frunzio, and M. H. Devoret, RF-Driven Josephson bifurcation amplifier for quantum measurement. *Phys. Rev. Lett.* 93, 207002, 2004.

24. M. B. Metcalfe, E. Boaknin, V. Manucharyan, R. Vijay, I. Siddiqi, C. Rigetti, L. Frunzio, R. J. Schoelkopf, and M. H. Devoret, Measuring the decoherence of a quantronium qubit with the cavity bifurcation amplifier. *Phys. Rev. B* 76, 174516, 2007.

25. I. Serban and F. K. Wilhelm, Dynamical tunneling in macroscopic systems. *Phys. Rev. Lett.* 99, 137001, 2007.

26. M. I. Dykman, Critical exponents in metastable decay via quantum activation. *Phys. Rev. E* 75, 011101, 2007.

27. A. Verso and J. Ankerhold, Dissipation in a rotating frame: Master equation, effective temperature, and Lamb shift. *Phys. Rev. A* 81, 022110, 2010.

Index

A